Spon's
Architects' and Builders' Price Book

Spon's Architects' and Builders' Price Book

EDITED BY

DAVIS, BELFIELD and EVEREST
Chartered Quantity Surveyors

1981

One hundred and sixth Edition

LONDON

E. & F. N. SPON LTD

First published 1873
One hundred and sixth edition 1980
E. & F. N. Spon Ltd
11 New Fetter Lane, London, EC4P 4EE
© *1980 E. & F. N. Spon Ltd*
Printed in Great Britain by
Richard Clay (*The Chaucer Press*), *Ltd,*
Bungay, Suffolk

ISBN 0 419 12090 4
ISSN 0306 3046

British Library Cataloguing in Publication Data

Spon's architects' and builders' price book.
  1981: 106th ed.
  1 Building – Estimates – Great Britain –
  Periodicals
  I. Davis, Belfield and Everest (*Firm*)
  692′ .5   TH435   80–41263

ISBN 0–419–12090–4
ISSN 0306–3046

Preface

Since the last edition of Spon's, a year ago, the building industry in common with most other industries has suffered from very high inflation. Over the year the tender price of building – that is the price paid by the Employer for his building – which many had thought would have begun to level off some months ago has continued to rise to a year-on-year increase of around 28%.

A combination of increases in the basic price of labour and in the cost of materials together with an unexpectedly high level of activity has kept prices high. The full effect of the current depression has yet to hit the building industry. Even when the country at large is going through the doldrums it takes some time for such a pattern to be fully reflected in the fortunes of the industry, such is the time span involved between the decision to build and the finished product. Moreover, rarely are contracts halted once they have started on site. This results in the cycle of changed demand in the building industry lagging behind that of other industries.

However, the reduced work flow recently experienced by architects and quantity surveyors will now be manifesting itself to contractors, and with the industry's annual wage awards behind us and a reduction in the rate of inflation in raw material prices becoming evident, a levelling off of tender prices from now on is inevitable. It may be some time before confidence is seen to return to the country as a whole by which time decisions to invest in new buildings will come too late; by the time sites are then assembled and approvals obtained and early design and documentation work completed prices in the building industry may have taken off again. The availability of skilled men has been progressively reducing over the years with the result that increases in price levels, which follow increases in demand, occur more rapidly and noticeably than reductions during a period of depression.

Clients who have long term plans to invest in buildings would be well advised to place their orders now and to exploit this inherent time lag in the pricing cycle. Recognition of this principle is likely to do much to ensure a more even supply of work to the building industry.

It is against this background that we have begun a programme of total reappraisal of the content of the Price Book. In this 106th edition, in addition to a total revision of all the price data contained in it, we have introduced some major changes and expansion in some important sections. First the section on 'Building Prices Per Square Metre', used when feasibility studies are carried out prior to the decision to build, has been considerably increased in scope. Second the 'Preliminaries' section of the 'Prices For Measured Work' has been completely revised in detail and its content extended; it is now in line with the 1980 Edition of Standard Form of Building Contract. Third, conscious of the current preoccupation of the industry with renovation and maintenance, the section on 'Alterations and Additions' has been extended.

In the general revision of prices we have taken into account the revised wage agreement in the building and allied industries which came into force on 30 June 1980 and material

v

prices which were generally current in April/May 1980. The revised wage agreements are valid for 12 months, that is until mid 1981, but material prices remain subject to change and readers are advised to check their validity during the currency of this edition.

The primary purpose of the book is to provide average prices for building work given certain assumptions together with ancillary information. It is not intended that the prices should be used in the preparation of actual tenders without consideration of the particular conditions such as the location and size of the job, the availability of labour and the degree of competition. It is essential for a full understanding of the book that readers should study the notes which precede each section.

For the benefit of new readers a brief guide to the book follows:

Part 1 General
This section contains Rates of Wages, Daywork and Prime Cost, Fees for Professional Services and Building Costs and Tender Prices Index.

Part II Prices
This section contains Market Prices of Materials, Constants of Labour and Materials, Prices for Measured Work and Alterations and Additions.

Part III Approximate Estimating
This section gives information on prices per square metre for various types of building, cost limits and allowances, approximate estimates, elemental cost plans for two typical buildings and comparative prices.

Part IV European Section
This section contains a summary of tendering procedures and cost information for ten European countries.

Part V Property Insurance
This section contains a procedure for valuing property for insurance purposes on a total reinstatement basis.

While every effort is made to ensure the accuracy of the information given in this publication neither the Editors nor Publishers in any way accept liability for loss of any kind resulting from the use made by any person of such information.

In conclusion the Editors wish to record their appreciation of the assistance given by many individuals and organizations.

DAVIS, BELFIELD and EVEREST
Chartered Quantity Surveyors
5 Golden Square
London W1R 3AE

STOP PRESS

Rates of Wages

There have been no changes in the rates of wages in the Building, Civil Engineering and Plumbing Mechanical Engineering Services Industries.

Daywork and Prime Cost

The Schedule of Basic Plant Charges, page 19, is being revised and is expected to be published towards the end of 1980 and take effect from the beginning of 1981.

Building Costs and Tender Price Index

Building Costs, page 155
  Provisional index for third quarter 1980, 436

Market Prices of Materials

Materials prices will inevitably have changed and the readers attention is particularly drawn to the preamble to this section on page 158.

Cost Limits and Allowances

Local Authority Buildings, page 424, an increase of $12\frac{1}{2}\%$ has been announced by the DHSS to take effect from 15 May 1980.

Contents

PART I
GENERAL

PART II
PRICES

ix

PART III
APPROXIMATE ESTIMATING

PART IV
EUROPEAN SECTION

PART V
PROPERTY INSURANCE

Index

PART I

General

This part of the book contains general information relating to the building industry including the following:

Rates of Wages

BUILDING INDUSTRY

Authorized rates of wages etc. in the building industry in England, Wales and Scotland agreed by the National Joint Council for the Building Industry

AND EFFECTIVE FROM 30th JUNE 1980

Subject to the conditions prescribed in the Working Rule Agreement guaranteed minimum weekly earnings shall be as follows:

	Craft operatives £	Labourers £
LONDON AND LIVERPOOL DISTRICT .	80·60	68·80
GRADE A AND SCOTLAND	80·40	68·60

These guaranteed minimum weekly earnings shall be made up as follows:

	Craft operatives £	Labourers £
LONDON AND LIVERPOOL DISTRICT		
Standard basic rates of wages	69·20	59·00
Guaranteed minimum bonus payment	11·40	9·80
Guaranteed minimum weekly earnings	80·60	68·80

NOTES

1. The guaranteed minimum bonus payment shall be set off against existing bonus payments of all kinds or against existing extra payments other than those prescribed in the Working Rule Agreement. The entitlement to the guaranteed minimum bonus shall be *pro rata* to normal working hours for which the operative is available for work.

3

BUILDING INDUSTRY

	Craft Operatives £	Labourers £
GRADE A AND SCOTLAND		
Standard basic rates of wages	69·00	58·80
Guaranteed minimum bonus payment	11·40	9·80
Guaranteed minimum weekly earnings	80·40	68·60

Young Labourers

The rates of wages for young labourers shall be the following proportions of the labourers' rates:

At 16 years of age　50%
At 17 years of age　70%
At 18 years of age　100%

Watchmen

The weekly pay for watchmen shall be equivalent to the labourers' guaranteed minimum weekly earnings provided that not less than five shifts (day or night) are worked.

Apprentices/Trainees

The rates of wages for apprentices/trainees are calculated as a proportion of the craftsmen's rates depending on age and/or period of service and mode of training.

ROAD HAULAGE WORKERS EMPLOYED IN THE BUILDING INDUSTRY

Authorized rates of pay for road haulage workers in the building industry agreed between the National Federation of Building Trades Employers and the Transport and General Workers Union

AND EFFECTIVE FROM 30th JUNE 1980

Employers	*Operatives*
The National Federation of Building Trades Employers, 82 New Cavendish Street, London, W1M 8AD *Telephone:* 01-580 5588	The Transport and General Workers Union, Transport House, Smith Square, Westminster, London, SW1P 3JB *Telephone:* 01-828 7788

DEFINITIONS OF GRADES

London

London Region as defined by the National Joint Council for the Building Industry.

Grade 1

The Grade A districts as defined by the National Joint Council for the Building Industry, including Liverpool and District (i.e. the area covered by the Liverpool Regional Federation of Building Trades Employers).

	LONDON Per week £	GRADE 1 Per week £
(*i*) Drivers of vehicles of gross vehicle weight		
Up to 3·5 tonnes	69·20	69·00
Over 3·5 tonnes and up to and including 7·5 tonnes	69·60	69·40
Over 7·5 tonnes and up to and including 10 tonnes	70·00	69·80
Over 10 tonnes and up to and including 16 tonnes	70·40	70·20
Over 16 tonnes and up to and including 24 tonnes	70·80	70·60
Over 24 tonnes carrying capacity . . .	71·20	71·00
(*ii*) Mates and Statutory Attendants (18 years of age and over)	68·80	68·60

Subject to certain conditions workers covered by this agreement are entitled to additional payments corresponding to the NJCBI's Guaranteed Minimum Bonus.

The gross vehicle weight in every case is to include, where applicable, the unladen weight of a trailer designed to carry goods.

BUILDING INDUSTRY – ISLE OF MAN

Authorized rates of wages in the building industry in the Isle of Man agreed by the Isle of Man Joint Industrial Council for the Building Industry

AND EFFECTIVE FROM 30th JUNE 1980

SECRETARIES

Employers	*Operatives*
A. Hill, F.C.A., F.C.I.S.,	Mrs. Kissack,
Kensington House,	Transport House,
Rosemount,	Prospect Hill,
Douglas,	Douglas,
Isle of Man	Isle of Man
Telephone: 0624 24331	*Telephone:* 0624 21156

	Standard basic wage rate per 40-hour week £	*Guaranteed minimum bonus payment* £	*Guaranteed minimum weekly earnings* £
Craftsmen	69·00	11·40	80·40
Labourers	58·80	9·80	68·60

BUILDING AND CIVIL ENGINEERING INDUSTRY – NORTHERN IRELAND

Authorized rates of wages in the building and civil engineering industry in Northern Ireland as agreed by the Joint Council for the Building and Civil Engineering Industry (Northern Ireland)

AND EFFECTIVE FROM 30th JUNE 1980

Joint Council for the Building and Civil Engineering Industry (Northern Ireland)
Industrial Relations Division,
Netherleigh,
Massey Avenue,
Belfast, BT4 2JP
Telephone: 0232 63244

	Standard basic wage rate per 40-hour week £	*Guaranteed minimum bonus payment* £	*Guaranteed minimum weekly earnings* £
Craftsmen	69·00	11·40	80·40
Labourers	58·80	9·80	68·60

CIVIL ENGINEERING INDUSTRY

Authorized rates of wages in the civil engineering industry as agreed by Civil Engineering Construction Conciliation Board

SECRETARIES

Employers	*Operatives*
B. J. Weller, B.A.,	L. C. Kemp,
Cowdray House,	Transport House,
6, Portugal Street,	Smith Square,
London, WC2A 2HH	Westminster,
Telephone: 01-404 4020	London, SW1P 3JB
	Telephone: 01-828 7788

Rates of pay effective from 30th June 1980

Per hour
£

General Operatives:
  London Super Grade . . . . 1·47½
  Class I . . . . 1·47

Craft Operatives
  London Super Grade . . . . 1·73
  Liverpool Grade . . . . 1·73
  Class I . . . . 1·72½

Guaranteed bonus from 30th June 1980

Per week
£

General Operatives . . . . 9·80
Craft Operatives . . . . 11·40

Watchmen (based on five shifts)

Per week
£

London Super Grade . . . . 68·80
Class I . . . . 68·60

PLUMBING MECHANICAL ENGINEERING SERVICES

Authorized rates of wages agreed by the Joint Industry Board for the Plumbing Mechanical Engineering Services Industry in England and Wales on 23rd July 1979

The Joint Industry Board for Plumbing Mechanical
    Engineering Services in England and Wales,
    Brook House, Brook Street,
    St Neots,
    Huntingdon, Cambs, PE19 2HW
    Telephone: 0480 76925

	Hourly rate with effect from	
	4th Feb. 1980 £	*2nd Feb. 1981* £
Technical plumber	2·55	2·82
Advanced plumber	2·24	2·48
Trained plumber	2·04	2·24
Apprentices		
1st year of training	0·68	0·75
2nd ,, ,, ,, ...	1·02	1·12
3rd ,, ,, ,, ...	1·28	1·40
4th ,, ,, ,, ...	1·63	1·79
An Apprentice in his third or fourth year who gains his City & Guilds Advanced Craft Certificate in accordance with the requirements of the Board	1·84	2·02

NOTE

From 2nd February 1981 wage rates will be further increased by an adjustment calculated on the increase in the Retail Price Index between September 1979 and September 1980.

Daywork and Prime Cost

When work is carried out which cannot be valued in any other way it is customary to assess the value on a cost basis with an allowance to cover overheads and profit. The basis of costing is a matter for agreement between the parties concerned, but definitions of prime cost for the building industry have been prepared and published jointly by the Royal Institution of Chartered Surveyors and the National Federation of Building Trades Employers for the convenience of those who wish to use them. These documents are reproduced on the following pages by kind permission of the publishers.

Also reproduced in this section is the daywork schedule published by the Federation of Civil Engineering Contractors.

For larger Prime Cost contracts the reader is referred to the form of contract issued by the Royal Institute of British Architects.

BUILDING INDUSTRY

Definition of prime cost of daywork carried out under a building contract

This Definition of Prime Cost is published by the Royal Institution of Chartered Surveyors and the National Federation of Building Trades Employers, for convenience and for use by people who choose to use it. Members of the National Federation of Building Trades Employers are not in any way debarred from defining Prime Cost and rendering their accounts for work carried out on that basis in any way they choose. Building owners are advised to reach agreement with contractors on the Definition of Prime Cost to be used prior to issuing instructions.

SECTION 1 – APPLICATION

1.1. This definition provides a basis for the valuation of daywork executed under such building contracts as provide for its use (e.g. contracts embodying the Standard Forms issued by the Joint Contracts Tribunal).

1.2. It is not applicable in any other circumstances, such as jobbing or other work carried out as a separate or main contract nor in the case of daywork executed during the Defects Liability Period of contracts embodying the above mentioned Standard Forms.

SECTION 2 – COMPOSITION OF TOTAL CHARGES

2.1. The prime cost of daywork comprises the sum of the following costs:
    (*a*) Labour as defined in Section 3.
    (*b*) Materials and goods as defined in Section 4.
    (*c*) Plant as defined in Section 5.

2.2. Incidental costs, overheads and profit as defined in Section 6, as provided in the building contract and expressed therein as percentage adjustments are applicable to each of 2.1 (*a*)–(*c*).

11

Daywork and Prime Cost

BUILDING INDUSTRY

SECTION 3 – LABOUR

3.1. The standard wage rates, emoluments and expenses referred to below and the standard working hours referred to in 3.2 are those laid down for the time being in the rules or decisions of the National Joint Council for the Building Industry and the terms of the Building and Civil Engineering Annual and Public Holiday Agreements applicable to the works, or the rules or decisions or agreements of such body, other than the National Joint Council for the Building Industry, as may be applicable relating to the class of labour concerned at the time when and in the area where the daywork is executed.

3.2. Hourly base rates for labour are computed by dividing the annual prime cost of labour, based upon standard working hours and as defined in 3.4 (*a*)–(*i*), by the number of standard working hours per annum.

3.3. The hourly rates computed in accordance with 3.2 shall be applied in respect of the time spent by operatives directly engaged on daywork, including those operating mechanical plant and transport and erecting and dismantling other plant (unless otherwise expressly provided in the building contract).

3.4. The annual prime cost of labour comprises the following:

(*a*) Guaranteed minimum weekly earnings (e.g. Standard Basic Rate of Wages, Joint Board Supplement and Guaranteed Minimum Bonus Payment in the case of N.J.C.B.I. rules).

(*b*) All other guaranteed minimum payments (unless included in Section 6).

(*c*) Differentials or extra payments in respect of skill, responsibility, discomfort, inconvenience or risk (excluding those in respect of supervisory responsibility – see 3.5).

(*d*) Payments in respect of public holidays.

(*e*) Any amounts which may become payable by the Contractor to or in respect of operatives arising from the operation of the rules referred to in 3.1 which are not provided for in 3.4 (*a*)–(*d*) or in Section 6.

(*f*) Employer's National Insurance contributions applicable to 3.4 (*a*)–(*e*).

(*g*) Employer's contributions to annual holiday credits.

(*h*) Employer's contributions to death benefit scheme.

(*i*) Any contribution, levy or tax imposed by statute, payable by the Contractor in his capacity as an employer.

3.5. **Note:**

Differentials or extra payments in respect of supervisory responsibility are excluded from the annual prime cost (see Section 6). The time of principals, foremen, gangers, leading hands and similar categories, when working manually, is admissible under this Section at the appropriate rates for the trades concerned.

ECTION 4 – MATERIALS AND GOODS

4.1. The prime cost of materials and goods obtained from stockists or manufacturers is the invoice cost after deduction of all trade discounts but including cash discounts not exceeding 5 per cent and includes the cost of delivery to site.

4.2. The prime cost of materials and goods supplied from the Contractor's stock is based upon the current market prices plus any appropriate handling charges.

BUILDING INDUSTRY

4.3. Any Value Added Tax which is treated, or is capable of being treated, as input tax (as defined in the Finance Act, 1972) by the Contractor is excluded.

SECTION 5 – PLANT

5.1. The rates for plant shall be as provided in the building contract.

5.2. The costs included in this Section comprise the following:

(*a*) Use of mechanical plant and transport for the time employed on daywork.

(*b*) Use of non-mechanical plant (excluding non-mechanical hand tools) for the time employed on daywork.

5.3. **Note:**

The use of non-mechanical hand tools and of erected scaffolding, staging, trestles or the like is excluded (see Section 6).

SECTION 6 – INCIDENTAL COSTS, OVERHEADS AND PROFIT

6.1. The percentage adjustments provided in the building contract, which are applicable to each of the totals of Sections 3, 4 and 5, comprise the following:

(*a*) Head Office charges.

(*b*) Site staff, including site supervision.

(*c*) The additional cost of overtime (other than that referred to in 6.2).

(*d*) Time lost due to inclement weather.

(*e*) The additional cost of bonuses and all other incentive payments in excess of any guaranteed minimum included in 3.4 (*a*).

(*f*) Apprentices study time.

(*g*) Subsistence and periodic allowances.

(*h*) Fares and travelling allowances.

(*i*) Sick pay or insurance in respect thereof.

(*j*) Third-party and employers' liability insurance.

(*k*) Liability in respect of redundancy payments to employees.

(*l*) Employers' National Insurance contributions not included in Section 3.4.

(*m*) Tool allowances.

(*n*) Use, repair and sharpening of non-mechanical hand tools.

(*o*) Use of erected scaffolding, staging, trestles or the like.

(*p*) Use of tarpaulins, protective clothing, artificial lighting, safety and welfare facilities, storage and the like that may be available on the site.

(*q*) Any variation to basic rates required by the Contractor in cases where the building contract provides for the use of a specified schedule of basic plant charges (to the extent that no other provision is made for such variation).

(*r*) All other liabilities and obligations whatsoever not specifically referred to in this Section nor chargeable under any other Section.

(*s*) Profit.

6.2. **Note:**

The additional cost of overtime, where specifically ordered by the Architect/Supervising Officer shall only be chargeable in the terms of prior written agreement between the parties to the building contract.

BUILDING INDUSTRY

Example of calculation of typical standard hourly base rate (as defined in Section 3) for N.J.C.B.I. building craftsman and labourer in Grade A areas at 1st July, 1975.

		Rate £	Craftsman £	Rate £	Labourer £
Guaranteed minimum weekly earnings					
Standard Basic Rate . . .	49 wks	37·00	1,813·00	31·40	1,538·60
Joint Board Supplement . .	49 wks	5·00	245·00	4·20	205·80
Guaranteed Minimum Bonus .	49 wks	4·00	196·00	3·60	176·40
			2,254·00		1,920·80
Employer's National Insurance Contribution at 8·5%			191·59		163·27
			2,445·59		2,084·07
Employer's Contributions to:					
C.I.T.B. annual levy . . .			15·00		3·00
Annual holiday credits .	49 wks	2·80	137·20	2·80	137·20
Public holidays (included in guaranteed minimum weekly earnings above)			—		—
Death benefit scheme . .	49 wks	0·10	4·90	0·10	4·90
Annual labour cost as defined in Section 3			£2,602·69		£2,229·17

Hourly rate of labour as defined in Section 3, Clause 3.02 . . .

$$\frac{£2602·69}{1904} = £1·37 \qquad \frac{£2229·17}{1904} = £1·17$$

Note:

1. Standard working hours per annum calculated as follows:

52 weeks @ 40 hours	=		2080
Less			
3 weeks holiday @ 40 hours	=	120	
7 days public holidays @ 8 hours	=	56	176
			1904

2. It should be noted that all labour costs incurred by the Contractor in his capacity as an employer other than those contained in the hourly rate, are to be taken into account under Section 6.

3. The above example is for the convenience of users only and does not form part of the Definition; all the basic costs are subject to re-examination according to the time when and in the area where the daywork is executed.

BUILDING INDUSTRY

NOTE: *For the convenience of readers the example which appears on the previous page has been up-dated by the Editors as at 4th August 1980.*

	Rate	Rate £	Craft operative £	Rate £	Labourer £
Guaranteed minimum weekly earnings					
Standard Basic Rate . . .	47⅖ wks	69·00	3,298·30	58·80	2,810·64
Guaranteed Minimum Bonus .	47⅖ wks	11·40	544·92	9·80	468·44
			3,843·12		3,279·08
Employer's National Insurance Contribution at 13·7% . . .			526·51		449·23
			4,369·63		3,728·31
Employer's Contributions to:					
C.I.T.B. annual levy . . .			33·60		6·00
Annual holiday credits .	48 wks	8·35	400·80	8·35	400·80
Public holidays (including in guaranteed minimum weekly earnings above)			—		—
Death benefit scheme . .	48 wks	0·10	4·80	0·10	4·80
Annual labour cost as defined in Section 3			4,808·83		4,139·91

Hourly rate of labour as defined in Section 3, Clause 3.02 . . .

$$\frac{£4,808·83}{1848} = £2·60 \qquad \frac{£4,139·91}{1848} = £2·24$$

BUILDING INDUSTRY

A definition of Prime Cost of Building works of a jobbing or maintenance character

This Definition of Prime Cost is published by the Royal Institution of Chartered Surveyors and the National Federation of Building Trades Employers for convenience and for use by people who choose to use it, but members of the National Federation of Building Trades Employers are not in any way debarred from defining Prime Cost and rendering their accounts for work carried out on that basis in any way they choose.

This definition covers jobbing or maintenance work carried out as a main or separate contract. It does not cover daywork carried out under or incidental to a building contract.

SECTION A – LABOUR

Payments in respect of wages and salaries made to persons directly engaged upon the work either on the site or in the Contractor's workshops and yard, but excluding all off-site office staff.

The payments referred to above shall include:

(*i*) operatives' wages and all other payments in respect of operatives (not included under Section C) properly made in accordance with the Working Rule Agreement current at the time when, and at the places where, the work is in progress; and

(*ii*) wages paid to foremen (including visiting, head, walking, working and similar foremen), and wages of chargehands, leading men, timekeepers, watchmen and the like; and

(*iii*) time of working principals at the time rates for the trades practised when actually working with their hands;

(*iv*) payments above the standard rates by way of bonus, incentive or the like agreed between the parties or normally paid by the contractor; and

(*v*) payments in respect of overtime necessitated by the work or agreed between the parties or normally worked; and

(*vi*) payments (not covered by (*i*), (*ii*), (*iii*), (*iv*) and (*v*) above) made to operatives or foremen in order to comply with the provisions of any statute, regulation, order or direction.

(*vii*) The actual cost of employer's contributions paid by the contractor in respect of:

(*a*) Holidays with pay and Public Holidays;

(*b*) National Insurance;

(*c*) National Graduated Pension Scheme;

(*d*) Any compulsory contributions similar to those in sub-paragraphs (*a*) to (*c*) above.

Alternatively, the cost of the above-named employer's contributions may be covered by the addition of a percentage on the total cost of (*i*) to (*vi*) above.

The word 'contribution' in this Definition of Prime Cost is deemed to include Selective Employment Tax and any other similar tax or imposition.

BUILDING INDUSTRY

SECTION B – MATERIALS AND GOODS

(*i*) *Materials and goods not supplied from the contractor's stock*

(*a*) The invoice cost of materials and goods purchased specifically for the works.

(*b*) The cost of crates, packaging and carriage.

(*c*) The cost of returning crates and packaging less any credit given for the same.

(*ii*) *Materials supplied from the contractor's stock*

At prices current at the date of supply by the contractor.

(*iii*) *Materials worked on in the contractor's workshops*

The cost to the contractor of all materials shown to be directly and solely used for the job, plus justifiable allowance for waste.

(*iv*) *Materials and goods supplied by the building owner*

The value as agreed between the parties of materials and goods so supplied (see items 3 and 6). Materials and goods obtained from the works and re-used in the building are not deemed to be materials supplied by the building owner.

The cost of materials and goods referred to in paragraphs (*i*), (*ii*) and (*iii*) above is the cost less all trade discounts but including all discounts for cash not exceeding 5%.

SECTION C – PLANT, CONSUMABLE STORES AND SERVICES

The cost of hire at rates to be agreed between the parties or in the absence of prior agreement at rates not exceeding those normally applied in the locality at the time when the works are carried out less 5% where the contractor uses his own plant.

(*i*) Machinery in workshops.

(*ii*) Mechanical plant and power-operated tools.

(*iii*) Scaffolding and scaffold boards.

(*iv*) Non-mechanical plant excluding hand tools.

(*v*) Haulage.

The cost of hire at rates as described in the previous paragraph or on the basis of use and waste.

(*vi*) Tarpaulins and dust sheets.

(*vii*) Temporary roadways, shoring, planking and strutting, hoarding, centering, temporary fans, partitions or the like.

The cost of the following:

(*viii*) The use of brushes, sponges, leathers and the like.

(*ix*) All necessary resharpening of tools occasioned by the works.

(*x*) Water charges for the works, including temporary plumbing and storage.

(*xi*) Installation and running cost of temporary electric and other power and lighting.

(*xii*) Fuel and consumable stores for mechanical plant and power-operated tools.

(*xiii*) Fuel and equipment for drying out the works and fuel for testing mechanical services.

(*xiv*) Fees, charges and royalties properly incurred by reason of the work.

BUILDING INDUSTRY

(*xv*) Testing of materials.

(*xvi*) The use of temporary buildings including rates and telephones and including heating and lighting if not charged under paragraph (*xi*) above.

(*xvii*) The use of canteens, sanitary accommodation, protective clothing, safety equipment and other provision for welfare in accordance with the current Working Rule Agreement and any Act of Parliament, statutory instrument, rule, order, regulation or bye-law.

(*xviii*) Any necessary travelling, lodging and other similar expenses properly incurred, if not chargeable under Section A.

(*xix*) Premiums for any special insurances which the parties may agree to be necessary having regard to the nature of the work.

SECTION D – WORK SUB-CONTRACTED

The cost of work carried out by sub-contractors, whether nominated by the building owner or appointed by the contractor (including any cash discount offered by the sub-contractor to the contractor up to a maximum of 2½%, but excluding all trade discounts) shall be shown separately.

Except where estimates have been obtained and approved in advance, the cost of sub-contractors' work shall be shown in a similar manner to that of the contractor's work.

SPECIMEN ACCOUNT

If this Definition of Prime Cost is followed, overheads and profit may be dealt with by means of percentages on any or all of the totals of Prime Cost in Sections A, B, C and D.

The contractor's account would then be in the following form:

                                       £

1. **Labour** (as defined in Section A)
2. **Add** . . . % required for Overheads

                                 £

3. **Materials** (as defined in Section B)
4. **Plant, etc.** (as defined in Section C)

                                 £

5. **Add** . . . % required for profit

                                 £

6. **Deduct** value of any materials supplied by the Building Owner as defined in Section B (*iv*)

                                 £

7. **Sub-contractors' accounts** (as defined in Section D)
8. **Add** . . . % required for profit on sub-contractors' Accounts

                           Total £

BUILDING INDUSTRY

SCHEDULE OF BASIC PLANT CHARGES (JULY 1975 ISSUE)

This Schedule is published by the Royal Institution of Chartered Surveyors and is for use in connection with Dayworks under a Building Contract.

EXPLANATORY NOTES

1. The rates in the Schedule are intended to apply solely to daywork carried out under and incidental to a Building Contract. They are NOT intended to apply to:

   (*i*) jobbing or any other work carried out as a main or separate contract; or
   (*ii*) work carried out after the date of commencement of the Defects Liability Period;

2. The rates in the Schedule are basic and may be subject to an overall adjustment to be quoted by the Contractor prior to the placing of the Contract.

3. The rates apply to plant and machinery already on site, whether hired or owned by the Contractor. They do not apply to plant not on site and specifically hired for daywork.

4. The rates, unless otherwise stated, include the cost of fuel of every description, lubricating oils, grease, maintenance, sharpening of tools, replacement of spare parts, all consumable stores and for licences and insurances applicable to items of plant. They do not include the costs of drivers and attendants.

5. The hourly rates in the Schedule are not intended to establish minimum charges. The rates should be applied to the time during which the plant is actually engaged in daywork.

6. Whether or not plant is chargeable on daywork depends on the daywork agreement in use, and the inclusion of an item of plant in this schedule does not necessarily indicate that that item is chargeable.

7. Rates for other plant not included in the Schedule shall be settled at prices reasonably related to the rates so included.

NOTE. *All rates in the schedule are expressed per hour and were calculated during the second quarter of 1974.*

BUILDING INDUSTRY

MECHANICAL PLANT AND TOOLS

Item of plant	Description	Unit	Rate per hour £
BAR BENDING AND SHEARING MACHINES			
Power-driven machine . . .	Up to 2 in. (50 mm) dia. rods	Each	0·46
BRICK SAWS			
Brick saw (use of abrasive disc to be charged net and credited) . . .	Power driven (Bench type Clipper or similar)	,,	0·38
COMPRESSORS			
Portable compressor (machine only) .	*Nominal delivery of free air per min. at 100 lb. sq. in.(7 kg/sq. cm.) pressure. Up to and including.*		
	cu. ft. *(cu. m.)*		
	100 (2·83)	,,	0·50
	101–130 (2·86– 3·70)	,,	0·68
	131–180 (3·73– 5·10)	,,	0·80
	181–220 (5·13– 6·30)	,,	1·12
	221–260 (6·33– 7·40)	,,	1·45
	261–320 (7·43– 9·10)	,,	1·85
	321–380 (9·13–10·70)	,,	2·18
	500–599 (14·20–16·97)	,,	2·66
	600–630 (17·00–17·90)	,,	3·00
Mobile compressor	101–150 (2·86– 4·24)	,,	1·00
(machine plus lorry only) . .	151–250 (4·27– 5·66)	,,	1·40
Tractor-mounted compressor . .	101–120 (2·86– 3·46)	,,	1·10
COMPRESSED-AIR EQUIPMENT (with and including up to 50 ft. (15·3 m) of hose)			
Heavy breaker and six steels . .		,,	0·09
Light pneumatic pick and six steels .		,,	0·08
Pneumatic clay spade and one blade .		,,	0·09
Hand-held rock drill and rod . .	35 lb. (15·9 kg) Class	,,	0·10
	45 lb. (20·5 kg) ,,	,,	0·12
(Bits to be paid for at net cost)	55 lb. (25·0 kg) ,,	,,	0·15
Drill, rotary	Up to ¾ in. (19 mm)	,,	0·10
Reversible drill		,,	0·12
Sander/Grinder		,,	0·10
Chipping hammer or similar . .		,,	0·06
Additional hoses	Per 50 ft. (15·3 m) length	,,	0·02

BUILDING INDUSTRY

MECHANICAL PLANT AND TOOLS – *continued*

Item of plant	Description	Unit	Rate per hour £
CONCRETE OR MORTAR MIXERS			
Concrete mixer	Diesel, electric or petrol		
	cu. ft. (cu. m.)		
Open drum without hopper . .	3/2 (0·09/0·06)	Each	0·15
	4/3 (0·12/0·09)	,,	0·17
	5/3½ (0·15/0·10)	,,	0·20
Open drum with hopper . . .	7/5 (0·20/0·15)	,,	0·37
Closed drum with hopper . . .	10/7 (0·30/0·20)	,,	0·52
Closed or reversing drum with hopper, weigher and feed shovel . . .	10/7 (0·30/0·20)	,,	0·70
Closed drum with hopper . . .	14/10 (0·40/0·30)	,,	0·68
Closed or reversing drum with hopper, weigher and feed shovel . . .	14/10 (0·40/0·30)	,,	0·83
Closed drum with hopper . . .	18/12 (0·50/0·35)	,,	0·70
Closed or reversing drum with hopper, weigher and feed shovel . . .	18/12 (0·50/0·35)	,,	0·98
Closed drum with hopper . . .	21/14 (0·60/0·40)	,,	0·82
Closed or reversing drum with hopper, weigher and feed shovel . . .	21/14 (0·60/0·40)	,,	1·17
Portable batch swing weighing gear .	8 (0·24)	,,	0·18
	14 (0·40)	,,	0·18
Roller pan mortar mixer . . .	Power-driven 4½ ft. (1·37 m)	,,	0·30
Drag feed shovel scraper – Dalli type or similar		,,	0·18
CONCRETE EQUIPMENT	cu. ft. (cu. m.)		
Concrete placer including discharge box and air receiver (excluding air compressor)	8½ (0·24)	,,	0·50
	13 (0·37)	,,	0·70
Piping per 10 ft. (3·1 m) length . .	6 in. (150 mm) dia.	,,	0·03
Extra over piping for bend . .	6 in. (150 mm) dia.	,,	0·06
Vibrator poker type	Petrol or electric driven	,,	0·26
	Air, excluding compressor	,,	0·10
Vibrator, external	Air, excluding compressor	,,	0·04
Vibrator-tamper		,,	0·27
Power float		,,	0·40
	cu. yd. (cu. m.)		
Wet Storage hopper	Up to 1 (0·80)	,,	0·10
	Up to 3 (2·30)	,,	0·13
CONVEYOR BELTS	Power operated up to 25 ft. (7·60 m) long 16 in. (400 mm) wide	,,	0·50

BUILDING INDUSTRY

MECHANICAL PLANT AND TOOLS – *continued*

Item of Plant	Description	Unit	Rate per hour £
CRANES	*Maximum load*		
Mobile, rubber tyred . . .	Up to 15 cwt. (762 kg)	Each	0·65
	1 ton (tonne)	,,	0·80
	2 tons (tonnes)	,,	1·00
	3 tons (tonnes)	,,	1·25
	4 tons (tonnes)	,,	1·75
	6 tons (tonnes)	,,	2·00
Static tower	Capacity metre/tonnes		
	10	,,	1·27
	15	,,	1·40
	25	,,	1·90
	60	,,	3·75
	150	,,	9·15
Track mounted			Add to the above rates 16%
CRANE EQUIPMENT	*cu. yd.* *(cu. m.)*		
Skip, muck	Up to ½ (0·40)	,,	0·02
	½ to ¾ (0·40 to 0·60)	,,	0·03
	1 (0·80)	,,	0·04
	2 (1·60)	,,	0·06
Skip, concrete	Up to ½ (0·40)	,,	0·02
	½ to ¾ (0·40 to 0·60)	,,	0·03
	1 (0·80)	,,	0·04
Skip, concrete lay down or roll over .	½ to ¾ (0·40 to 0·60)	,,	0·04
	1 (0·80)	,,	0·07
	2 (1·60)	,,	0·09
DUMPERS AND POWER BARROWS	*Maker's Rated Capacity*		
Dumper (site use only excluding Tax, Insurance and extra cost of D.E.R.V., etc., when operating on highway) .	*cu. yd./cwt.* *(cu. m./kg)*		
	½ / 10 (0·40/508)	,,	0·33
	¾ / 15 (0·60/762)	,,	0·41
	1 / 20 (0·80/1016)	,,	0·46
	1½ / 30 (1·20/1529)	,,	0·72
	2 / 40 (1·60/2032)	,,	0·81
	2½ / 50 (2·00/2540)	,,	1·00
	3 / 60 (2·30/3048)	,,	1·40
Power-driven barrow . . .	Not exceeding 8 cu. ft. (0·25 cu. m.)	,,	0·20
ELECTRIC HAND TOOLS (excluding generator but including transformer)			
Drills – rotary	Up to ⅜ in. (10 mm)	,,	0·05
	Up to ¾ in. (19 mm)	,,	0·07
Hammer drill (light). . . .		,,	0·11
Hammer drill kit (Kango Type). .	Light	,,	0·16
	Medium	,,	0·25
	Heavy	,,	0·30
Pipe drilling tackle	Ordinary type	Set	0·17
	Under pressure type	,,	0·42
Saw, circular		Each	0·15
Sand/Grinder lightweight . . .		,,	0·08
Disc cutter or chaser heavy duty .		,,	0·19

BUILDING INDUSTRY

MECHANICAL PLANT AND TOOLS – *continued*

Item of Plant	*Description*	*Unit*	*Rate per hour* £
EXCAVATORS	*Maker's Rated Capacity*		
Universal rope operated with single equipment	*cu. yd.* (*cu. m.*)		
	¼ (0·20)	Each	1·21
	⅜ (0·30)	,,	1·30
	½ (0·40)	,,	1·65
	⅝ (0·50)	,,	1·80
	¾ (0·60)	,,	2·35
Hydraulic full-circle slew excavator			
Hymac 480 or similar . . .	½ (0·40)	,,	2·50
Hymac 580 ,, ,, .	⅝ (0·50)	,,	3·50
Wheeled-tractor type hydraulic excavator. JCB Type C3 or similar . .		,,	2·75
FLOOR POLISHERS	With attachments for polishing, scraping or scrubbing	,,	0·33
FLOOR GRINDERS (stones to be charged net)	Diesel	,,	0·45
GENERATING SETS. . .	1 kVA	,,	0·28
	3 ,,	,,	0·33
	5 ,,	,,	0·37
	10 ,,	,,	0·48
	15 ,,	,,	0·68
	30 ,,	,,	0·87
	60 ,,	,,	1·40
GRADER MOTORS			
12–15-ft. (3·75–4·57-m) blade width .	90–140 h.p.	,,	3·81
HAMMERS, CARTRIDGE	Excluding cartridges and studs	,,	0·10
HEATERS – SPACE	Paraffin/electric Btu/hr		
	50,000	,,	0·24
	100,000	,,	0·48
	200,000	,,	0·70
	300,000	,,	1·14

BUILDING INDUSTRY

MECHANICAL PLANT AND TOOLS – *continued*

Item of Plant	*Description*		*Unit*	*Rate per hour* £
HOISTS				
Scaffold	Up to 5 cwt. (250 kg)		Each	0·15
Cantilever or centre slung scaffold tower type				
Mast and guide rails up to: . .				
ft. (*m*).	*cwt.*	(*kg*)		
56 (17·069)	5–10	(254–508)	,,	0·32
60 (18·288)	10–15	(508–762)	,,	0·41
72 (21·946)	20–30	(1016–1524)	,,	0·50
72 (21·946)	40–50	(2032–2540)	,,	0·55
Extra for additional mast sections – up to 10 ft (3·048 m) long . . .			,,	0·02
Extra for safety gate . . .			,,	0·01
Passenger, mast and guide rails up to 150 ft. (45·72 m) single cage . .	22½ cwt. (1144 kg) 15 men		,,	2·25
Extra for additional 6 ft. (1·83 m) tower section			,,	0·02
Extra for landing gate . . .			,,	0·04
LORRIES	*Plated Gross Vehicle Weight*			
	ton/tonnes			
Fixed body	Up to 3·50		,,	1·39
	,, 4·20		,,	1·43
	,, 5·50		,,	1·52
	,, 6·50		,,	1·68
	,, 7·50		,,	1·85
Tipper	,, 3·50		,,	1·82
	,, 4·20		,,	1·86
	,, 5·50		,,	1·96
	,, 6·50		,,	2·13
	,, 7·50		,,	2·31
PAINT SPRAYING EQUIPMENT				
(excluding compressor) . . .	1-gun type with 25 ft. (7·6 m) hose and 2-gallon (9-litre) pressure tank		,,	0·11
	2-gun type with 50 ft. (15·3 m) hose and 5-gal. (22·7-litre) pressure tank		,,	0·20
PIPE WORK EQUIPMENT				
Pipe bender, power driven . .	50–150 mm dia.		,,	0·40
Pipe cutting machine . . .	Hydraulic		,,	0·34
Pipe defrosting equipment . .	Electrical		Set	0·81
Pipe testing equipment . . .	Compressed Air		,,	0·34
	Hydraulic		,,	0·17

BUILDING INDUSTRY

MECHANICAL PLANT AND TOOLS – *continued*

Item of Plant	*Description*		*Unit*	*Rate per hour* £
PUMPS (exclusive of all hoses)				
Single diaphragm	*in.* 2	*(mm)* (50)	Each	0·17
	3	(75)	,,	0·24
	4	(100)	,,	0·28
Double diaphragm	2	(50)	,,	0·30
	3	(75)	,,	0·32
	4	(100)	,,	0·39
Self-priming centrifugal . . .	3	(75)	,,	0·32
	4	(100)	,,	0·43
	6	(150)	,,	0·60
Sludge and sewage	2	(50)	,,	0·31
	3	(75)	,,	0·47
	4	(100)	,,	0·71
	6	(150)	,,	1·05
Sump	3	(75)	,,	0·20
Submersible electric	2	(50)	,,	0·29
	3	(75)	,,	0·44
	4	(100)	,,	0·53
PUMPING EQUIPMENT				
Pump hoses, per 25 ft. (7·62 m) flexible,	2	(50)	,,	0·04
suction or delivery, including coupling	3	(75)	,,	0·06
valve and strainer	4	(100)	,,	0·09
	6	(150)	,,	0·17
RAMMERS AND COMPACTORS				
Power rammer, Pegson or similar .			,,	0·33
Soil compactor, Wacker type Plate size				
11 in. × 13 in. (278 × 330 mm) .	Up to 5 cwt. (254 kg)		,,	0·49
16 in. × 16 in. (406 × 406 mm) .	Up to 15 cwt. (762 kg)		,,	0·64

BUILDING INDUSTRY

MECHANICAL PLANT AND TOOLS – *continued*

Item of Plant	Description		Unit	Rate per hour £

ROLLERS

		cwt.	(kg)		
Vibrating roller self-propelled	. .	6¾	(343)	Each	0·46
Vibrating roller twin roller	. .	12½	(635)	,,	0·66
		21¾	(1105)	,,	0·88
		ton/tonne			
Pavement rollers dead weight	. .	Up to 3		,,	1·08
		Up to 6		,,	1·43
		Up to 10		,,	1·65
Scarifier	Extra over roller including sharpening of tines			,,	1·10

SAWS, MECHANICAL

		in.	(m)		
Chain saw		18	(0·46)	,,	0·36
		30	(0·76)	,,	0·46
		Up to 60	(1·52)	,,	0·53
Bench saw		Up to 18 (0·46) blade		,,	0·28
		Up to 30 (0·76) blade		,,	0·42

SCRAPERS

		Maker's Rated Capacity			
		cu. yd.	(cu. m.)		
Scraper when hired with tractor	.	4	(3·10)	,,	0·85
		6	(4·70)	,,	1·05
		8/9	(6·20/ 6·90)	,,	1·10
		12/13	(9·00/10·80)	,,	1·55
Motorized scraper		7/9	(5·40/ 6·90)	,,	4·09
		14/18	(10·80/13·80)	,,	6·83

BUILDING INDUSTRY

MECHANICAL PLANT AND TOOLS – *continued*

Item of Plant	Description	Unit	Rate per hour £
SCREWING MACHINES	13– 50 mm dia.	Each	0·12
	25–100 mm dia.	„	0·20

Item of Plant	Description		Unit	Rate per hour £
TRACTORS	*cu. yd.*	*(cu. m.)*		
Shovel, tractor (crawler), any type of bucket	Up to ¾	(0·60)	„	2·70
	1	(0·80)	„	3·89
	1¼	(1·00)	„	4·04
	1½	(1·20)	„	4·04
	1¾	(1·40)	„	5·28
	2	(1·50)	„	5·84
Shovel, tractor (wheeled), two-wheel drive	Up to ¾	(0·60)	„	1·67
	1	(0·80)	„	1·99
Shovel, tractor (wheeled), four-wheel drive	Up to 1¼	(1·00)	„	3·65
	2¼	(1·80)	„	5·20
	2¾	(2·10)	„	5·75
	Maker's Rated drawbar horse-power			
Tractor (crawler) with dozer . .	Up to 25·9		„	1·73
	26 to 34·9		„	2·13
	35 to 44·9		„	2·70
	45 to 65·9		„	3·57
	66 to 89·9		„	4·63
	90 to 140		„	5·86
Hydraulic ripper				Add 15% to above tractor rates
Back hoe equipment . . .				Add 17½% to above tractor rates
Tractor-wheeled (rubber-tyred) . .	Light 35 h.p.		„	0·79
Agricultural type	Heavy 65 h.p.		„	1·38

WELDING AND CUTTING AND BURNING SETS

Item of Plant	Description	Unit	Rate per hour £
Welding and cutting set (including oxygen and acetylene, excluding underwater equipment and thermic boring).		„	1·50
Electric welding set (excluding electrodes)	Diesel 300 amp. Single operator	„	0·65
	Diesel 600 amp. Double operator	„	1·00
	Transformer electric 300 amp. Single operator	„	0·46

BUILDING INDUSTRY

NON-MECHANICAL PLANT

Item of Plant	Description	Unit	Rate per hour £
BAR-BENDING AND SHEARING MACHINES			
Bar Bending machine, hand operated .	Up to 1 in. (25 mm) dia. rods	Each	0·03
Shearing machine, hand operated .	Up to 1½ in. (38 mm) dia. rods	,,	0·04
BROTHER OR SLING CHAINS	Not exceeding 2 ton/tonne	Set	0·08
	Exceeding 2, not exceeding 5 ton/tonne	,,	0·13
	Exceeding 5, not exceeding 10 ton/tonne	,,	0·28
	Exceeding 20 ton/tonne	,,	0·85
DRAIN TESTING EQUIPMENT		,,	0·17
LIFTING AND JACKING GEAR	*ton/tonne*		
Pipe winch – including shear legs .	½	Set	0·12
	1	,,	0·14
Pipe winch – including gantry . .	2	,,	0·43
	3	,,	0·50
Chain blocks up to 20-ft. (6·10-m) lift	*ton/tonne*		
	1	Each	0·06
	2	,,	0·08
	3	,,	0·08
	4	,,	0·10
	5	,,	0·13
	6	,,	0·14
	7½	,,	0·17
	10	,,	0·21
Screw jack	Not exceeding 5 ton/tonne	,,	0·04
Hydraulic jack	Not exceeding 5 ton/tonne	,,	0·03
	,, ,, 10 ,, ,,	,,	0·05
	,, ,, 20 ,, ,,	,,	0·09
Pull lift (Tirfor type) . . .	¾ ton/tonne	,,	0·09
	1½ ,, ,,	,,	0·11
	3 ,, ,,	,,	0·20
PIPE BENDERS	13–75 mm. dia.	,,	0·17
	50–100 mm. dia.	,,	0·27

BUILDING INDUSTRY

NON-MECHANICAL PLANT – *continued*

Item of Plant	*Description*	*Unit*	*Rate per hour £*
PAINT BURNERS			
Paint-burning-off plant . . .	Calor gas or similar	Torch and hose	0·17
PLUMBER'S FURNACE	Calor gas or similar	Each	0·30
ROAD WORKS EQUIPMENT			
Barrier pole and trestles or similar .		10 ft.	0·01
Crossing plates (steel sheets) . .		m²	0·01
Danger lamp, including oil . .		Each	0·02
Traffic control lights (not pad operated)	Automatic equipment 150 yds (137·16 m) cable		
Mains supply		Set	0·44
With generator		,,	0·76
Warning sign		Each	0·01
Road cone		10	0·03
Flasher unit (including battery charging)		Each	0·04
Flashing bollard (including battery charging)		,,	0·06
ROLLERS			
Hand-roller	Up to 5 cwt. (254 kg)	,,	0·01
SCAFFOLDING			
Boards		100 ft. (30·48 m)	0·014
Castor wheels	Steel or rubber tyred	100	0·137
Fall ropes	Up to 200 ft. (61 m)	Each	0·02
Fittings (including couplers, base plates)	Steel or alloy	100	0·01
Ladders, pole	20 rung	Each	0·017
	30 ,,	,,	0·024
	40 ,,	,,	0·030
	Extended length		
	ft. (m)		
Ladders, extension . . .	20 (6·10)	Each	0·03
	26 (7·29)	,,	0·04
	35 (10·67)	,,	0·05
	40 (13·72)	,,	0·06
Putlogs	Steel or alloy	100	0·033
Tube	Steel	100 ft.	0·003
	Alloy	(30·48 m)	0·007
Staging, lightweight . . .		,,	0·115
Wheeled tower . . .	Working platform up to 25 ft. (7·62 m) high. Including castors and boards		
7 ft. × 7 ft. (2·13 m × 2·13 m) . .		Each	0·24
10 ft. × 10 ft. (3·05 m × 3·05 m)		,,	0·30
Split head		10	0·02
Small		,,	0·03
Medium		,,	0·04
Large			

BUILDING INDUSTRY

NON-MECHANICAL PLANT – *continued*

Item of Plant	Description	Unit	Rate per hour £
TARPAULINS		10 m²	0·01

TRENCH STRUTS AND SHEETS

Adjustable steel trench strut . .	All sizes from 1 ft. (305 mm) closed to 5 ft. 6 in. (1·68 m) extended	10	0·02
Steel trench sheet	5–14 ft. (1·52–4·27 m) lengths	m²	0·03

WINCHES

Crab winch (hand) and cable up to 100 ft. (30·48 m)	Up to 2 ton/tonne	Each	0·09
Special rope requirements to be charged extra	„ „ 5 „ / „	„	0·24

CIVIL ENGINEERING INDUSTRY

Daywork charges for the civil engineering industry in accordance with the schedule prepared by the Federation of Civil Engineering Contractors

WITH EFFECT FROM 21 APRIL 1980

SCHEDULES OF DAYWORKS CARRIED OUT INCIDENTAL TO CONTRACT WORK

(For operation throughout England, Scotland and Wales)

1. Labour   3. Plant
2. Materials   4. Supplementary charges

These schedules are the schedules referred to in Clause 52 (3) of the I.C.E. Conditions of Contract 5th Edition and have been prepared for use in connection with dayworks carried out incidental to contract work where no other rates have been agreed. They are not intended to be applicable for dayworks ordered to be carried out after the contract works have been substantially completed or to a contract to be carried out wholly on a daywork basis. The circumstances of such works vary so widely that the rates applicable call for special consideration and agreement between contractor and employing authority.

1. LABOUR

Add to the amount of wages paid to workmen 130%

NOTE.

(1) 'Amount of wages' means:

Wages, bonus, daily travelling allowances (fare and/or time), tool allowance and all prescribed payments including those in respect of time lost due to inclement weather paid to workmen at plain time rates and/or at overtime rates.

All payments shall be in accordance with the Working Rule Agreement of the Civil Engineering Construction Conciliation Board for Great Britain Rule Nos I to XXIII inclusive or that of other appropriate wage-fixing authorities current at the date of executing the work and where there are no prescribed payments or recognized wage-fixing authorities, the actual payments made to the workmen concerned.

(2) The percentage addition provides for all statutory charges at the date of publication and other charges including:

Standard Rate National Insurances.
Normal Contract Works, Third Party, and Employers' Liability Insurances.
Annual and Public Holidays with Pay.
Non-contributory Sick Pay Scheme.
Industrial Training Levy.
Redundancy Payments Contribution.
Contracts of Employment Act.
Site supervision and staff including foreman and walking gangers, but the time of the gangers or charge hands working with their gangs is to be paid for as for workmen.
Small tools – such as picks, shovels, barrows, trowels, hand saws, buckets, trestles, hammers, chisels and all items of a like nature.
Protective clothing.
Head Office charges and profit.

CIVIL ENGINEERING INDUSTRY

(3) All labour only subcontractors' accounts to be charged at full amount of invoice (without deduction of any cash discounts not exceeding $2\frac{1}{2}\%$) plus 64%.

(4) Subsistence or lodging allowances and periodic travel allowances (fare and/or time) paid to or incurred on behalf of workmen are chargeable at cost plus $12\frac{1}{2}\%$.

(5) The charges referred to above are as at date of publication and will be subject to periodic review.

2. MATERIALS

Add to the cost of materials delivered site . . . . $12\frac{1}{2}\%$

NOTES AND CONDITIONS

(1) The percentage addition provides for Head Office charges and profit.

(2) The cost of materials means the invoiced price of materials including delivery to site without deduction of any cash discounts not exceeding $2\frac{1}{2}\%$.

(3) Unloading of materials:
The percentage added to the cost of materials excludes the cost of handling.
An allowance for unloading into site stock or storage including wastage should be added where incurred.

(4) The charges referred to above are as at the date of publication and will be subject to periodic review.

3. PLANT

Add to the rates for plant

Mechanical plant, except lorries and vans	36%
Lorries and vans	48%
Plywood	27%
Bulk timber	22%
Softwood for planking, strutting	19%
Scaffold boards	5%
Other non-mechanical plant	31%

NOTES AND CONDITIONS

(1) These rates apply *only* to contractors' own plant (already on site) exclusive of driver and attendants, but inclusive of fuel and consumable stores, unless stated to be charged in addition, repairs and maintenance, insurance of plant, but excluding time spent on general servicing.

(2) Fuel distribution is not included in the rates quoted.

(3) Where metric capacities are indicated these are not necessarily exact conversions from their imperial equivalents, but cater for the variations arising from comparison of plant manufacturing firms' ratings.

(4) Minimum hire charge will be for the period quoted.

(5) Hire rates for mechanical or other special plant not normally classed as small tools and not included below shall be settled at prices reasonably related to the rates quoted.

(6) The rates for temporary track in Section 17 are for track already in position. Arrangements should be made on site for payment in cases where track has to be re-located or when used exclusively on daywork operations.

(7) The charges referred to above are as at the date of publication and will be subject to periodic review.

(8) To differentiate between mechanical and non-mechanical plant, consumables, etc., the latter is indicated by an asterisk (*).

CIVIL ENGINEERING INDUSTRY

Description	Unit	Hire rate £	Period

ASPHALT EQUIPMENT

Asphalt and/or coated macadam spreader

	Description	Unit	Hire rate £	Period
Crawler or wheeled	Up to 50 h.p.	Each	11·88	Hour
„ „	51 to 100 h.p.	„	15·70	„
„ „	101 to 150 h.p.	„	39·77	„
* Joint matcher(s). Separately or in conjunction with longitudinal beams . .	extra over machine	„	1·70	„
* Ditto with addition of transverse control devices or fully automatic equipment capable of control from reference guides .	„	„	4·03	„
Self-propelled, metered coated chipping applicator for asphalt work up to 3·7 m or 12 ft. width (including trailer) . .	15 h.p. incl. trailer	„	16·06	„
Heater planer		„	66·59	„
Cold planer mini		„	14·22	„
„ medium . . .		„	49·03	„
„ master . . .		„	121·11	„
* Asphalt road burner (portable) . .		„	2·18	„
* „ „ (towable) .		„	24·79	„
„ „ (self-propelled) .		„	45·96	„
* Asphalt kerb machine . . .		„	11·58	„

BAR BENDING AND BAR SHEARING MACHINES

For mild steel rods up to

	mm dia.		in. dia.		Hire rate	Period
* Bar bending machine, hand operated .	25	or	1	„	0·53	Day
Ditto, power driven .	38	„	1½	„	1·00	Hour
Ditto, ditto . .	51	„	2	„	1·27	„
* Bar shearing machine, hand operated .	25	„	1	„	0·76	Day
Ditto, power driven .	38	„	1½	„	1·03	Hour
Ditto, ditto . .	51	„	2	„	1·18	„

For high tensile steel rods up to

	mm dia.		in. dia.		Hire rate	Period
Bar bending machine, power driven . .	55	or	2⅛	„	3·23	„
Bar shearing machine, ditto . . .	55	„	2⅛	„	2·48	„

BOILERS

Evaporative capacity per hour
Natural draught 54·4 kg/120 lb
working pressure

					Hire rate	Period
Oil fired . . .	383 kg	or	845 lb.	„	5·79	„
„ . . .	513 kg	„	1,130 lb.	„	7·45	„
„ . . .	703 kg	„	1,550 lb.	„	9·81	„
„ . . .	771 kg	„	1,700 lb.	„	10·71	„
„ . . .	907 kg	„	2,000 lb.	„	12·51	„
„ . . .	1,361 kg	„	3,000 lb.	„	18·55	„

CIVIL ENGINEERING INDUSTRY

COMPRESSORS

NOTE: Cu. ft. or Cu. m.
piston displacement is
approx. 20% above the
nominal free air delivery.

Nominal delivery of free air per min.
at 45 kg or 100 lb. pressure
(for compressors with higher delivery safe
working pressure, rates to be negotiated)
Up to and including

Description			Unit	Hire rate £	Period
cu. m/min.		cu. ft./min.			
Portable compressor					
(machine only) . .	1·2	or	40	Each	0·97	Hour
„ „ · ·	1·5	„	53	„	1·04	„
„ „ · ·	2·0	„	70	„	1·85	„
„ „ · ·	2·3	„	80	„	2·01	„
„ „ · ·	2·6	„	90	„	2·22	„
„ „ · ·	3·2	„	110	„	2·49	„
„ „ · ·	3·7	„	130	„	2·88	„
„ „ · ·	4·0	„	140	„	3·01	„
„ „ · ·	4·6	„	160	„	3·19	„
„ „ · ·	5·1	„	180	„	3·57	„
„ „ · ·	6·0	„	210	„	4·57	„
„ „ · ·	7·4	„	260	„	4·80	„
„ „ · ·	9·1	„	320	„	6·11	„
„ „ · ·	10·7	„	380	„	7·14	„
„ „ · ·	11·4	„	400	„	7·49	„
„ „ · ·	12·2	„	430	„	8·35	„
„ „ · ·	13·6	„	480	„	9·56	„
„ „ · ·	17·0	„	600	„	10·68	„
(silenced) (machine only)	2·6	„	90	„	2·32	„
„ „ · ·	3·7	„	130	„	3·11	„
„ „ · ·	4·0	„	140	„	3·27	„
„ „ · ·	4·6	„	160	„	3·58	„
„ „ · ·	5·1	„	180	„	3·95	„
„ „ · ·	6·0	„	210	„	5·23	„
„ „ · ·	7·4	„	260	„	5·66	„
„ „ · ·	9·7	„	340	„	8·20	„
„ „ · ·	10·5	„	370	„	8·54	„
„ „ · ·	12·2	„	430	„	10·30	„
„ „ · ·	13·6	„	480	„	11·17	„
„ „ · ·	17·0	„	600	„	12·78	„
„ „ · ·	19·8	„	700	„	14·25	„
„ „ · ·	25·5	„	900	„	17·15	„

Lorry mounted com-
pressor (machine, plus
lorry only) (site and
public highway use –
including Tax, Insurance
and extra cost of
D.E R.V.)

	Range				
	cu. m/min.	cu. ft./min.			
Portable on lorry ·	1·44/2·83	or	50/100	„	4·99	„
„ „ ·	2·86/4·24	„	101/150	„	5·74	„
„ „ ·	4·27/5·66	„	151/200	„	6·13	„
„ silenced ·	1·44/2·83	„	50/100	„	5·08	„
„ „ ·	2·86/4·24	„	101/150	„	5·83	„
„ „ ·	4·27/5·66	„	151/200	„	6·24	„

CIVIL ENGINEERING INDUSTRY

	Description		Unit	Hire rate £	Period
COMPRESSORS – *continued*					

Tractor mounted compressor (machine plus rubber tyred tractor only) (site use only – *excluding* Tax, Insurance and extra cost of D.E.R.V.)

	Range		Unit	Hire rate £	Period
	cu. m/min.	cu. ft./min			
Portable on tractor .	1·44/2·83	or 50/100	Each	3·86	Hour
,, ,, . .	2·86/3·40	,, 101/120	,,	4·28	,,
,, ,, . .	3·43/4·24	,, 121/150	,,	4·71	,,
,, ,, . .	4·27/5·66	,, 151/200	,,	5·32	,,
,, ,, . .	5·69/7·12	,, 201/250	,,	6·37	,,
	cu. m	cu. ft.	,,		
* Air Receiver . .	0·28	,, 10	,,	0·80	Day
* ,, ,, . .	0·57	,, 20	,,	0·88	,,
* ,, ,, . .	0·85	,, 30	,,	1·04	,,
* ,, ,, . .	1·14	,, 40	,,	1·33	,,
* ,, ,, . .	1·42	,, 50	,,	1·62	,,
* ,, ,, . .	1·70	,, 60	,,	2·09	,,
* ,, ,, . .	1·98	,, 70	,,	2·21	,,
* ,, ,, . .	2·26	,, 80	,,	2·42	,,
* ,, ,, . .	2·83	,, 100	,,	2·88	,,
* ,, ,, . .	3·54	,, 125	,,	3·25	,,
* ,, ,, . .	4·25	,, 150	,,	3·40	,,
* ,, ,, . .	5·66	,, 200	,,	4·18	,,
* ,, ,, . .	8·49	,, 300	,,	6·04	,,

	Capacity		Unit	Hire rate £	Period
	dry/wet	dry/wet			
CONCRETE MIXERS	cu. m	cu. ft.			
Open drum, mixer only . .	0·09/0·06	or 3/2	,,	0·35	Hour
,, ,, ,, . .	0·12/0·09	,, 4/3	,,	0·43	,,
,, ,, ,, . .	0·15/0·10	,, 5/3½	,,	0·60	,,
Closed drum, ,, . .	0·20/0·15	,, 7/5	,,	0·93	,,
,, ,, ,, . .	0·25/0·18½	,, 8/6½	,,	1·10	,,
,, ,, ,, . .	9·30/0·20	,, 10/7	,,	1·65	,,
Ditto, with swing batch weighing gear	0·40/0·30	,, 14/10	,,	2·36	,,
,, ,, ,, ,, . .	0·50/0·35	,, 18/12	,,	2·83	,,
,, ,, ,, ,, . .	0·60/0·40	,, 21/14	,,	3·28	,,
Extra for power driven loading shovel	for 0·30	,, 10/7	,,	0·45	,,
,, ,, ,, ,, ,, . .		for other sizes	,,	0·52	,,
Mixer complete with integral batch weighing gear and power driven loading shovel . .	cu. m	cu. ft.			
	0·20/0·15	or 7/5	,,	1·80	,,
,, ,, ,, ,, ,, . .	0·30/0·20	,, 10/7	,,	2·20	,,
,, ,, ,, ,, ,, . .	0·40/0·30	,, 14/10	,,	3·00	,,
,, ,, ,, ,, ,, . .	0·50/0·35	,, 18/12	,,	3·30	,,
,, ,, ,, ,, ,, . .	0·60/0·40	,, 21/14	,,	3·80	,,
* Extra for aggregate feed apron .	single compartment		,,	0·09	,,
* ,, ,, ,, ,, .	2 ,,		,,	0·11	,,
* ,, ,, ,, ,, .	3 ,,		,,	0·12	,,

CIVIL ENGINEERING INDUSTRY

CONCRETE MIXERS – *continued*

Description			Unit	Hire rate £	Period
	unmixed capacity				
	cu. m	cu. ft.			
Pan mixer	0·06 or	2	Each	1·85	Hour
,, ...	0·12 ,,	4	,,	2·23	,,
,, ...	0·20 ,,	6·5	,,	2·97	,,
,, ...	0·40 ,,	14	,,	3·87	,,
,, ...	0·55 ,,	20	,,	5·08	,,
,, ...	0·85 ,,	30	,,	6·37	,,
,, ...	1·25 ,,	44	,,	9·10	,,
,, ...	1·50 ,,	54	,,	10·54	,,
* Extra for batch weighing gear .	0·20 to 0·55 ,,	6·5 to 20	,,	0·65	,,
* ,, ,, ,, ,, .	1·25 ,,	44	,,	0·94	,,
* ,, ,, ,, ,, .	1·50 ,,	54	,,	1·28	,,
	unmixed capacity				
Extra for power driven batch loader	0·20 ,,	6·5	,,	1·12	,,
,, ,, ,, ,, .	0·40 ,,	14	,,	1·14	,,
,, ,, ,, ,, .	0·55 ,,	20	,,	1·60	,,
,, ,, ,, ,, .	0·85 ,,	30	,,	2·03	,,
,, ,, ,, ,, .	1·25 ,,	44	,,	3·21	,,
,, ,, ,, ,, .	1·50 ,,	54	,,	3·90	,,
Roller pan mortar mixer .	0·22 ,,	8	,,	3·22	,,
Extra for power driven batch loader			,,	1·38	,,
	tonnes	tons			
Cement Silo (low level) . .	12·2 ,,	12	,,	0·91	,,
,, ,, ,, . .	25·4 ,,	25	,,	1·49	,,
,, ,, ,, . .	30·5 ,,	30	,,	1·66	,,
,, ,, ,, . .	50·8 ,,	50	,,	2·25	,,
* Extra for aeration equipment .	—		,,	0·08	,,
* Extra for capsule weighing equipment and swinging arm . .	—		,,	0·20	,,
* Extra for autofeed weighing equipment and screw conveyor .	—		,,	0·74	,,
* Cement silo (high level) with weighing equipment and swinging arm (manually operated) . .	25·4 or	25	,,	1·89	,,
* ,, ,, ,, ,, .	30·5 ,,	30	,,	2·03	,,
Cement silo (high level) with weighing equipment and screw conveyor (power operated) .	25·4 ,,	25	,,	2·63	,,
,, ,, ,, .	30·5 ,,	30	,,	2·79	,,

		maximum output cu. m/ cu. ft./ min. min.		hopper capacity cu. m cu. ft.		Unit	Hire rate £	Period
	hp							
Grout mixer, single drum .	. 15	0·06 or 2·12		0·12 or 4·25		,,	3·31	,,
,, ,, double drum .	. 15	0·11	4·0	0·12 ,, 4·25		,,	3·91	,,
,, ,, ,,	. 37	0·17	6·0	0·23 ,, 8·25		,,	7·07	,,
Loading hopper, batch weighing gear and power driven loading shovel	.	204 kg or		450 lb		,,	4·53	,,
Grout mixer, double drum .	. 74	0·22	7·20	0·48 16·00		,,	11·18	,,
,, ,, roller . .	. 3	0·01	0·38			,,	1·09	,,
,, ,, ,, . .	. 6	0·02	0·77			,,	1·84	,,
,, ,, ,, . .	. 12	0·04	1·54			,,	2·99	,,

CIVIL ENGINEERING INDUSTRY

Description					Unit	Hire rate £	Period

CONCRETE MIXERS – *continued*

	maximum output		*hopper capacity*				
	cu. m/ min.	cu. ft. min.	cu. m	cu. ft.			
Grouting machine including pump and hopper	1·5	or 0·04	1·5	or 0·04	Each	1·93	Hour
,, ,, ,, ,, .	2·5	,, 0·07	12·8	,, 0·363	,,	2·29	,,
,, ,, ,, ,, .	4·0	,, 0·11	12·8	,, 0·363	,,	3·04	,,

	maximum output						
	cu. m/min	cu. ft./min					
* Grout pump (excl. compressor) ram dia. 50 mm or 2 in. . .	0·031	or 1·1			,,	4·67	,,
* Ditto, ditto 63 mm or 2½ in.	0·048	,, 1·7			,,	5·12	,,
* Agitating tank for use with above	—				,,	0·72	,,

	mm	in.					
* Grout pump (hand operated) .	76	3	0·028 ,, 1·0		,,	0·41	,,
* ,, ,, ,, ,, .	38	1½	0·048 ,, 1·7		,,	0·37	,,

	mixed batch capacity						
	cu. m	cu. yds.					
Concrete transporter (lorry mounted) (lorry 4 × 2) . .	2·30	or 3·00			,,	7·58	,,
,, ,, ,, . .	3·00	,, 4·00			,,	8·26	,,
Extra for lorry 4 × 4	—				,,	1·80	,,
Concrete truck mixer (agitator) (incl. separate engine with mech. drive to drum) . . .	5·00	,, 6·50			,,	19·86	,,
,, ,, ,, ,, .	6·00	,, 8·00			,,	20·00	,,

CONCRETE EQUIPMENT

	output per hour maximum						
	cu. m.	cu. yds.					
Concrete pump (skid mounted) (exclusive of piping) . . .	20/26	or 26·2/34			,,	16·26	,,
,, ,, ,, ,, .	30/45	,, 39·3/58·9			:,	17·56	,,
,, ,, ,, ,, .	46/54	,, 60·2/70·6			,,	21·27	,,
,, ,, ,, ,, .	70/72	,, 91·6/94·2			,,	31·00	,,
,, ,, ,, ,, .	90/110	,, 117·7/143·9			,,	32·12	,,
Concrete pump (lorry mounted) (exclusive of piping) . . .	50	,, 65·4			,,	26·32	,,
,, ,, ,, ,, .	60	,, 78·5			,,	30·66	,,
,, ,, ,, ,, .	80	,, 104·6			,,	33·69	,,
,, ,, ,, ,, .	90/110	,, 117·7/143·9			,,	37·78	,,
Concrete pump (mobile trailer) (rubber tyred) (exclusive of piping)	15/26	,, 19·62/34			,,	17·74	,,
,, ,, ,, ,, .	30/34	,, 39·3/44·5			,,	18·07	,,
,, ,, ,, ,, .	38/45	,, 49·7/58·9			,,	21·57	,,
,, ,, ,, ,, .	50/75	,, 65·4/98·1			,,	28·03	,,
,, ,, ,, ,, .	90/110	,, 117·7/143·9			,,	30·61	,,

	mm dia.	dia. in.					
* Piping per 3·05 m or 10 ft. straight length	102	or 4			,,	0·22	Day
* ,, ,, ,, ,, .	127	,, 5			,,	0·25	,,
* ,, ,, ,, ,, .	152	,, 6			,,	0·30	,,
* Bends	102	,, 4			,,	0·40	,,
,, ,, ...	127	,, 5			,,	0·52	,,
* ,, ,, ...	152	,, 6			,,	0·94	,,

CIVIL ENGINEERING INDUSTRY

	Description		Unit	Hire rate £	Period
CONCRETE EQUIPMENT – *continued*					
* Flexible distributor hose per 3·96 m or 13 ft. length . . .	mm dia. 102	in. dia. 4	Each	2·89	Day
* ,, ,, ,, ,, .	127	5	,,	3·71	,,
* ,, ,, ,, ,, .	152	6	,,	4·40	,,
* Shut off pipe section . .	102	4	,,	0·98	,,
* ,, ,, . . .	127	5	,,	1·02	,,
* ,, ,, . . .	152	6	,,	1·10	,,
* Pipe cleaning equipment incl. cleaning bend rubber sponge ball and trap basket . . .	102	4	,,	1·62	,,
* ,, ,, ,, ,, .	127	5	,,	1·73	,,
* ,, ,, ,, ,, .	152	6	,,	1·83	,,
Vibrator poker petrol driven .	—		,,	0·85	Hour
,, ,, diesel driven .	—		,,	1·00	,,
,, ,, electric .	—		,,	0·66	,,
* ,, ,, air (excl. compressor)	—		,,	0·24	,,
Alternator/frequency convertor .	up to 2·5 kVA		,,	0·75	,,
,, ,, ,, .	,, 4·5 ,,		,,	0·91	,,
* Vibrator poker (H/F motor in head type)	—		,,	0·29	,,
Vibrator external type clamp on electric	small		,,	0·42	,,
,, ,, ,, ,, .	medium		,,	0·59	,,
,, ,, ,, ,, .	large		,,	0·97	,,
* ,, ,, ,, ,, air (excluding compressor) . .	—		,,	0·42	,,
Vibrator tamper type including single timber or metal screed board	screed length 3·10 m or 10 lin. ft.		,,	0·46	,,
* Extra for additional single screed length	additional 0·31 m or 1 lin. ft.		,,	0·03	,,
Vibrator tamper type including double screed board . . .	screed length 3·10 m or 10 lin. ft.		,,	0·63	,,
* Extra for additional double screed length	additional 0·31 m or 1 lin. ft.		,,	0·06	,,
Power float petrol driven . .	—		,,	0·86	,,
Ditto, electric driven . . .	—		,,	0·65	,,
Concrete saw (exclusive of blades and water supply to be charged in addition) manually propelled .	blade diameter mm or in. 356	14	,,	0·90	,,
,, ,, ,, ,, .	457	18	,,	1·20	,,
Self-propelled . . .	18 hp 457	18	,,	2·40	,,
Ditto, with hydraulic system for blade movement . . .	23 hp 457	18	,,	3·28	,,
,, ,, ,, ,, .	41 hp 457	18	,,	3·81	,,
Joint former hand propelled with vibratory blade . . .	blade width Up to 2·4 m or 8 ft.		,,	2·62	,,
,, ,, ,, ,, .	8·2 m or 27 ft.		,,	3·58	,,
Grinder pedestrian operated .	—		,,	1·24	,,
* Concrete pram dobbin rubber tyred tipping	Struck capacity cu. m cu. ft. 0·10 or 3		,,	0·62	Day
* ,, ,, ,, ,, .	0·14 ,, 5		,,	0·88	,,
* ,, ,, ,, ,, .	0·20 ,, 7		,,	1·01	,,
* ,, ,, ,, ,, .	0·28 ,, 10		,,	1·38	,,
Power driven barrow . . .	0·17 ,, 6		,,	0·64	Hour

CIVIL ENGINEERING INDUSTRY

	Description	Unit	Hire rate £	Period
CRANES	Maximum working load in accordance with			
Mobile rubber tyred (site use only) (*excluding* Tax, Insurance and extra cost of D.E.R.V.)	British Standard 1757 (1964) Clause II – Stability			
	tonnes tons			
Full circle slew	Up to 0·76 or ¾	Each	2·45	Hour
„ „	4·06 4	„	6·88	„
„ „	5·08 5	„	8·43	„
„ „	8·13 8	„	10·15	„
„ „	12·20 12	„	12·65	„
„ „	15·24 15	„	15·26	„
„ „	18·30 18	„	19·69	„
Mobile rubber tyred rough terrain type (site use only) (*excluding* Tax, Insurance and extra cost of D.E.R.V.)	8·13 or 8·0	„	11·58	„
„ „ „ „ .	12·8 12·7	„	14·22	„
„ „ „ „ .	16·4 16·1	„	16·59	„
„ „ „ „ .	18 17·7	„	20·38	„
„ „ „ „ .	20 19·7	„	22·33	„
„ „ „ „ .	23 22·6	„	25·14	„
Excavator crane tracked (site use only) (*excluding* Tax, Insurance and extra cost of D.E.R.V.) .	tonnes tons			
Full circle slew	Up to 6·10 or 6	„	7·88	„
„ „	15·3 15	„	9·54	„
„ „	21·3 21	„	10·03	„
„ „	35·6 35	„	12·72	„
„ „	54·9 54	„	19·78	„
„ „	60·1 60	„	30·06	„
„ „	81·3 80	„	37·29	„
„ „	101·6 100	„	40·44	„
„ „	111·8 110	„	43·46	„
Lorry mounted crane (site and public highway use) (*including* Tax, Insurance and extra cost of D.E.R.V.)	Maximum working load as B.S. 1757 (1964) Clause II Stability (outriggers down)			
	Up to tonnes tons			
Full circle slew	12·20 or 12	„	12·26	„
„ „	15·24 15	„	17·29	„
„ „	18·30 18	„	19·23	„
„ „	25·40 25	„	26·22	„
„ „	30·48 30	„	29·34	„
„ „	36·60 36	„	34·51	„
„ „	40·64 40	„	37·42	„
Loco crane – Steam standard gauge 1·44 m or 4 ft. 8½ in. .	tonnes tons 3·05 or 3	„	Rates	„
„ „ „ „ .	5·08 5	„	to be	„
„ „ „ „ .	7·11 7	„	negotiated	„
„ „ „ „ .	10·20 10	„		„
Loco crane – Diesel standard gauge 1·44 m or 4 ft. 8½ in. .	9·10 9	„		„
Tower crane – electric – (complete with ballast and/or kentledge). Standard trolley or luffing jib operating on straight/curved rail track. Capacity maximum lift in tons × maximum radius in metres at which it can be lifted	(height under hook above ground trolley jib cranes) (jib pivot height above ground for luffing jib cranes)			

CIVIL ENGINEERING INDUSTRY

CRANES – *continued*				Description				Unit	Hire rate £	Period
Metre/tonnes				m. or	ft.	to	m. or	ft.		
Up to 10	.	.	.	17·1	56	22·0	72	Each	4·10	Hour
„ 15	.	.	.	16·5	54	17·4	57	„	4·90	„
„ 20	.	.	.	18·3	60	20·1	66	„	6·76	„
„ 25	.	.	.	20·1	66	22·6	74	„	7·71	„
„ 31	.	.	.	22·0	72	24·7	81	„	8·70	„
„ 40	.	.	.	22·0	72	22·6	74	„	10·50	„
„ 50	.	.	.	22·0	72	22·6	74	„	12·66	„
„ 61	.	.	.	22·0	72	22·6	74	„	13·63	„
„ 70	.	.	.	22·0	72	22·6	74	„	15·14	„
„ 81	.	.	.	22·0	72	22·6	74	„	16·26	„
„ 110	.	.	.	22·0	72	22·6	74	„	20·00	„
„ 130	.	.	.	21·4	70	22·6	74	„	22·27	„
„ 150	.	.	.	21·4	70	22·6	74	„	24·24	„
„ 170	.	.	.	21·4	70	22·6	74	„	25·40	„
„ 180	.	.	.	21·4	70	23·8	78	„	25·90	„
„ 210	.	.	.	21·4	70	23·8	78	„	27·25	„
„ 220	.	.	.	21·4	70	23·8	78	„	27·52	„
„ 255	.	.	.	21·4	70	23·8	78	„	29·00	„

	Unit	Hire rate £	Period
Static tower crane to be charged at the following percentage of the above rates – Up to 20 metre/tonnes	—	„	80%	„
21 to 70 metre/tonnes . .	—	„	85%	„
71 metre/tonnes upwards .	—	„	90%	„
Climbing or extendable mounted base crane to be charged at the following percentage of the above rates:			
Up to 20 metre/tonnes . .	—	„	85%	„
21 metre/tonnes upwards .	—	„	80%	„

	Description	Unit	Hire rate £	Period
Extra for extended height of tower (incl. accessories) for trolley jib crane. Up to 60 m/t . . .	per 0·31 m. or lin. ft. of height	„	0·05	„
* 130 m/t . . .	„	„	0·06	„
* 170 m/t . . .	„	„	0·09	„
* 255 m/t . . .	„	„	0·11	„
Extra for extended jib length for trolley jib crane				
* Up to 60 m/t . . .	per 0·31 m. or 1 lin. ft.	„	0·04	„
* 130 m/t . . .	„	„	0·05	„
* 170 m/t . . .	„	„	0·06	„
* 255 m/t . . .	„	„	0·09	„
* Extra for extended height of tower (incl. accessories) for luffing crane.				
* Up to 30 metre/tonnes .	per 0·31 m. or lin. ft. of height	„	0·04	„
* 50 „ . .	„	„	0·06	„
* 70 „ . .	„	„	0·08	„
* Extra for extended jib length for luffing jib crane. Up to 15 m/t .				
* 20 m/t .	„	„	0·16	„
25 m/t .	„	„	0·20	„
* 30 m/t .	„	„	0·22	„
* 40 m/t .	„	„	0·25	„
* 50 m/t .	„	„	0·26	„
* 60 m/t .	„	„	0·28	„
	„	„	0·30	„

CIVIL ENGINEERING INDUSTRY

Description					Unit	Hire rate £	Period
CRANES – *continued*							
	Width of track between rail centres				per yd.		
* Tower crane rail track, timbers	m. or ft.		to	m. or ft.	or 0·91 m.		
and fastenings . . .	1·83	6	2·44	8	trk.	0·23	Day
* ,, ,, ,, ,, .	3·35	11	4·88	16	Each	0·30	,,
* ,, ,, ,, ,, .	5·18	17	6·71	22	,,	0·37	,,
* ,, ,, ,, ,, .	7·01	23	8·54	28	,,	0·40	·,,

CRANE EQUIPMENT

Description			Unit	Hire rate £	Period
	struck capacity up to and including				
* Crane Grab, all types, excavating	cu. m.	cu. ft.			
(normal weight) . . .	0·55 or	20	,,	1·41	Hour
* ,, ,, ,, ,, .	0·80	28	,,	1·73	,,
* ,, ,, ,, ,, .	1·25	45	,,	2·19	,,
* ,, (heavy duty) . .	0·55	20	,,	2·24	,,
* ,, ,, ,, ,, .	0·80	28	,,	2·74	,,
* ,, ,, ,, ,, .	1·25	45	,,	2·91	,,
* Crane grab, all types, rehandling					
(normal weight) . . .	0·55	20	,,	1·09	,,
* ,, ,, ,, ,, .	0·80	28	:,	1·12	,,
* ,, ,, ,, ,, .	1·25	45	,,	1·59	,,
* ,, (heavy duty) . .	0·55	20	,,	1·73	,,
* ,, ,, ,, . .	0·80	28	,,	2·13	,,
* ,, ,, ,, . .	1·25	45	,,	2·70	,,
	Up to and including				
	cu. m.	cu. yd.			
* Skip, muck tipping circular .	0·08 or	$\frac{3}{32}$	,,	0·05	:,
* ,, ,, ,, .	0·15	$\frac{3}{16}$	:,	0·05½	,,
* ,, ,, ,, .	0·20	$\frac{1}{4}$	,,	0·06	,,
* ,, ,, ,, .	0·30	$\frac{3}{8}$	9:	0·06½	,,
* ,, ,, ,, .	0·40	$\frac{1}{2}$	,,	0·07	,,
* ,, ,, ,, .	0·60	$\frac{3}{4}$	9:	0·10	,,
* ,, ,, ,, .	0·80	1	,,	0·11	,,
* ,, ,, ·, .	1·20	1½	,,	0·13	,,
* ,, ,, ,, .	1·60	2	,,	0·16	,,
* Skip, concrete tipping . .	0·15	$\frac{3}{16}$	,,	0·08½	,,
* ,, ,, ,, .	0·20	$\frac{1}{4}$	,,	0·09	,,
* ,, ,, ,, .	0·40	$\frac{1}{2}$	,,	0·11½	,,
* ,, ,, ,, .	0·60	$\frac{3}{4}$	,,	0·16½	,,
* ,, ,, ,, .	0·80	1	,,	0·17	,,
* ,, ,, ,, .	1·00	1½	,,	0·26	,,
	Up to and including				
* Skip concrete lay down or roll over	cu. m	cu. ft.			
standard front or bottom discharge	0·20 or	7	,,	0·17	,,
* ,, ,, ,, ,, .	0·30	10	,,	0·19	,,
* ,, ,, ,, ,, .	0·40	14	,,	0·21	,,
* ,, ,, ,, ,, .	0·60	21	,,	0·28	,,
* ,, ,, ,, ,, .	0·80	27	,,	0·32	,,
* ,, ,, ,, ,, .	1·20	41	,,	0·56	,,
* Skip concrete roll over geared or					
hydraulic hand operated clamshell	0·40	14	,,	0·29	,,
* ,, , ,, ,, .	0·60	21	,,	0·35	,,
* ,, ,, ,, ,, .	0·80	27	,,	0·37	,,
* ,, ,, ,, ,, .	1·20	41	,,	0·67	,,
* ,, ,, ,, ,, .	1·60	54	,,	0·85	,,

CIVIL ENGINEERING INDUSTRY

Description					Unit	Hire rate £	Period
CRANE EQUIPMENT – *continued*							
* Chain slings or brothers	chain diameter		safe working load				
1·83 metres or 6 ft. E.W.L.							
	mm. or in.		tonne or	ton			
* Single . . .	6	¼	1·27	1·25	Each	0·16	Day
* Double . . .	6	¼	1·78	1·75	,,	0·21	,,
* Single . . .	10	⅜	2·19	2·16	,,	0·19	,,
* Double . . .	10	⅜	3·24	3·19	,,	0·27	,,
* Single . . .	13	½	5·08	5·00	,,	0·25	,,
* Double . . .	13	½	7·36	7·24	,,	0·39	,,
* Single . . .	16	⅝	7·27	7·16	,,	0·39	,,
* Double . . .	16	⅝	11·18	11·00	,,	0·60	,,
* Single . . .	19	¾	11·68	11·50	,,	0·50	,,
* Double . . .	19	¾	15·42	15·18	,,	0·83	,,
* Single . . .	22	⅞	15·85	15·60	,,	0·65	,,
* Double . . .	22	⅞	21·46	21·12	,,	1·10	,,
	cu. m	or	cu. yd.				
* Grab tag lines . .	Up to 0·60		¾		,,	2·32	,,
* ,, ,, . .	1·20		1½		,,	3·26	,,
* ,, ,, . .	Over 1·20		1½		,,	5·10	,,
	Dead weight						
	kg	or	cwt.				
* Demolition ball . .	254		5		,,	0·82	,,
* ,, ,, . .	508		10		,,	1·03	,,
* ,, ,, . .	762		15		,,	1·30	,,
* ,, ,, . .	1,016		20		,,	1·57	,,
* ,, ,, . .	1,270		25		,,	1·89	,,
* ,, ,, . .	1,524		30		,,	2·18	,,
* ,, ,, . .	1,778		35		,,	2·39	,,
* ,, ,, . .	2,032		40		,,	2·71	,,

DERRICKS, SCOTCH

	tonne	ton			
* Hand, jib up to 18·3 m or up to 60 ft.. .	1·52	1½	,,	0·86	Hour
* ,, ,, ,, ,, ,, .	2·03	2	,,	0·92	,,
* ,, ,, ,, ,, ,, .	3·05	3	,,	1·29	,,
* ,, jib 18·6 m to 21·4 m or 61 ft. to 70 ft.	3·05	3	,,	1·41	,,
Extra for hand slew gear . . .			,,	0·27	,,
* Hand, jib up to 18·3 m or up to 60 ft.. .	5·08	5	,,	1·57	,,
* ,, jib 18·6 m to 21·4 m or 61 ft. to 70 ft. .	5·08	5	,,	1·76	,,
* Extra for hand slew gear . . .			,,	0·36	,,
* ,, ,, 3 ton/tonne bogies . .			set of 3	0·38	,,
* ,, ,, 5 ton/tonne bogies . .			,,	0·44	,,
All electric (single motor) jib 30·5 m or 100 ft .	3·05	3	Each	3·57	,,
Ditto, jib 30·8 m to 36·6 m or 101 ft. to 120 ft.	3·05	3	,,	3·86	,,
Extra per additional motor . . .			,,	1·09	,,
* Extra for plain bogies heavy duty 1·44 m or 4 ft. 8½ in. gauge.			set of 3	0·59	,,
All electric (single motor), jib up to 24·4 m or up to 80 ft.	5·08	5	Each	4·07	,,
Ditto, jib 24·7 m. to 30·5 m. or 81 ft. to 100 ft. .	5·08	5	,,	4·40	,,
Ditto, jib 30·8 m to 36·6 m or 101 ft. to 120 ft.	5·08	5	,,	4·55	,,
Extra per additional motor . . .			,,	1·22	,,
* Extra for plain bogies heavy duty 1·44 m or 4 ft. 8½ in. gauge.			set of 3	0·98	,,
All electric (single motor), jib up to 24·4 m or up to 80 ft.	10·20	10	Each	5·62	,,
Ditto, jib 24·7 m to 30·5 m or 81 ft. to 100 ft.	10·20	10	,,	6·01	,,

CIVIL ENGINEERING INDUSTRY

DERRICKS, SCOTCH – *continued*

Description			Unit	Hire rate £	Period
	tonne	tons			
Ditto, jib 30·8 m to 36·6 m or 101 ft. to 120 ft.	10·20	10	Each	6·54	Hour
* Extra for extended jib length . . .			,,	0·52	,,
Extra per additional motor . . .			,,	1·52	,,
* Extra for plain bogies heavy duty 1·44 m. or 4 ft. 8½ in. gauge.			set of 3	1·43	,,
All electric (single motor), jib up to 24·4 m or up to 80 ft.	15·24	15	Each	6·95	,,
Ditto, jib 24·7 m to 30·5 m or 81 ft. to 100 ft. .	15·24	15	,,	7·42	,,
Ditto, jib 30·8 m to 36·6 m or 101 ft. to 120 ft. .	15·24	15	,,	8·07	,,
* Extra for extended jib length . . .			,,	0·63	,,
Extra per additional motor . . .			,,	1·61	,,
* Extra for plain bogies heavy duty 1·44 m or 4 ft. 8½ in. gauge.			set of 3	1·61	,,
All electric (single motor), jib up to 24·4 m or up to 80 ft.	20·32	20	Each	8·43	,,
Ditto, jib 24·7 m to 30·5 m or 81 ft. to 100 ft. .	20·32	20	,,	8·99	,,
Ditto, jib 30·8 m or 101 ft. to 120 ft. .	20·32	20	,,	9·72	,,
* Extra for extended jib length . . .			,,	0·77	,,
Extra per additional motor . . .			,,	1·87	,,
* Extra for plain bogies heavy duty 1·44 m or 4 ft. 8½ in. gauge.			set of 3	2·05	,,
All electric (single motor), jib up to 24·4 m or up to 80 ft.	30·48	30	Each	10·83	,,
Ditto, jib 24·7 m to 30·5 m or 81 ft. to 100 ft. .	30·48	30	,,	11·07	,,
Ditto, jib 30·8 m to 36·6 m or 101 ft. to 120 ft.	30·48	30	,,	11·35	,,
* Extra for extended jib length . . .			,,	0·85	,,
Extra per additional motor . . .			,,	2·02	,,
* Extra for plain bogies heavy duty 1·44 m or 4 ft. 8½ in. gauge.			set of 3	3·44	,,
* Extra to power driven crane or derrick	tonne	ton			
Indicator Up to 5·08 or 5 capacity .	—		Each	0·20	,,
* Ditto Over 5·08 5 ,, .	—		,,	0·22	,,
* Ditto, Bowsill (or stringer)					
Up to 10·20 10 ,, .	—		,,	0·31	,,
* Ditto Up to 20·32 20 ,, .	—		,,	0·40	,,
* Ditto Up to 30·48 30 ,, .	—		,,	0·50	,,
Ditto, self travel gear	—			Equal to item for Extra per additional motor for each capacity:	
* Ditto, Kentledge weights	—		Per ton/ tonne	0·63	Week

DIVING GEAR

Rates to be negotiated

CIVIL ENGINEERING INDUSTRY

DUMPERS

Description		Unit	Hire rate £	Period
Small Dumper (site use only – *excluding* Tax, Insurance and extra cost of D.E.R.V.) (manual gravity tipping) (2-wheel drive)	Maker's Rated Capacity S.A.E. Heaped cu. m or cu. yd.			
	0·51 ⅝	Each	1·02	Hour
Ditto „ „	0·62 ¾	„	1·20	„
Ditto „ „	0·70 ⅞	„	1·39	„
Ditto „ „	0·80 1	„	1·65	„
Ditto „ „	1·02 1¼	„	1·83	„
Ditto „ „	1·20 1½	„	2·04	„
Ditto „ „	1·30 1⅝	„	2·09	„
Ditto „ „	1·40 1¾	„	2·39	„
Ditto (4 wheel drive)	0·80 1	„	2·02	„
Ditto „ „	1·02 1¼	„	2·43	„
Ditto „ „	1·20 1½	„	2·49	„
Ditto „ „	1·30 1⅝	„	2·51	„
Ditto „ „	1·70 2¼	„	3·19	„
Ditto (hydraulic tipping) (2-wheel drive)				
Ditto „ „	0·51 ⅝	„	1·15	„
Ditto „ „	0·62 ¾	„	1·36	„
Ditto „ „	0·70 ⅞	„	1·50	„
Ditto „ „	0·80 1	„	1·90	„
Ditto „ „	1·02 1¼	„	2·11	„
Ditto „ „	1·20 1½	„	2·30	„
Ditto „ „	1·30 1⅝	„	2·40	„
Ditto „ „	1·40 1¾	„	2·50	„
Ditto (4 wheel drive)	1·02 1¼	„	2·55	„
Ditto „ „	1·20 1½	„	2·74	„
Ditto „ „	1·30 1⅝	„	2·80	„
Ditto „ „	1·40 1¾	„	3·20	„
Ditto „ „	1·50 1⅞	„	3·35	„
Ditto „ „	1·60 2	„	3·40	„
Ditto „ „	1·70 2¼	„	3·81	„
Ditto (hydraulic tipping) (4 wheel drive)				
	2·30 3	„	4·32	„
Ditto „ „	2·50 3¼	„	4·69	„
Ditto „ „	2·75 3½	„	4·93	„
Ditto „ „	3·80 5	„	7·21	„
Ditto (high discharge) (2 wheel drive)				
	0·80 1	„	2·20	„
Ditto „ „	1·02 1¼	„	2·50	„
Ditto „ „	1·20 1½	„	2·70	„
Ditto „ „	1·30 1⅝	„	2·84	„
Ditto „ „	1·40 1¾	„	3·00	„
Ditto „ „	1·60 2	„	3·18	„
Ditto (4 wheel drive)	0·80 1	„	2·40	„
Ditto „ „	1·02 1¼	„	2·84	„
Ditto „ „	1·30 1⅝	„	3·01	„
Ditto „ „	1·50 1⅞	„	3·67	„
Ditto „ „	1·60 2	„	4·07	„
Ditto „ „	1·70 2¼	„	4·20	„
Ditto (turntable side tipping) (2 wheel drive)				
	0·71 ⅞	„	2·14	„
Ditto „ „	0·80 1	„	2·20	„
Ditto „ „	1·02 1¼	„	2·68	„
Ditto „ „	1·22 1½	..	3·07	„
Ditto „ „	1·30 1⅝	..	3·45	„

CIVIL ENGINEERING INDUSTRY

	Description	Unit	Hire rate £	Period

DUMPERS – *continued*
Small Dumper (site use only, *excluding* Tax, Insurance and extra cost of D.E.R.V.)

Maker's Rated Capacity S.A.E. Heaped

	cu. m	or	cu. yd.	Unit	Hire rate £	Period
Ditto (4 wheel drive) . .	1·02		1¼	Each	3·23	Hour
Ditto ,, ,, .	1·22		1½	,,	3·48	,,
Ditto ,, ,, .	1·30		1⅝	,,	3·83	,,
Ditto ,, ,, .	1·50		1⅞	,,	4·26	,,
Ditto ,, ,, .	1·70		2¼	,,	5·00	,,

up to and including cu. m or cu. yd.

	cu. m		cu. yd.		Hire rate	
Dump truck – rear dump .	10·5		13·8	,,	13·11	,,
,, ,, ,, .	12·0		15·7	,,	18·03	,,
,, ,, ,, .	14·0		18·4	,,	21·79	,,
,, ,, ,, .	16·0		21·0	,,	24·22	,,
,, ,, ,, .	18·4		24·1	,,	25·59	,,
,, ,, ,, .	21·4		28·0	,,	29·94	,,
,, ,, ,, .	22·5		29·5	,,	32·02	,,
,, ,, ,, .	24·0		31·4	,,	35·36	,,
,, ,, ,, .	32·0		41·9	,,	44·81	,,
,, ,, ,, .	44·0		57·6	,,	49·77	,,
Dump truck – articulated or trailer type	6·10		8·0	,,	9·22	,,
,, ,, ,, .	7·60		10·0	,,	12·64	,,
,, ,, ,, .	11·00		14·4	,,	17·15	,,
,, ,, ,, .	12·60		16·5	,,	20·38	,,
,, ,, ,, .	14·00		18·4	,,	24·14	,,
,, ,, ,, .	16·50		21·6	,,	27·17	,,
,, ,, ,, .	21·00		27·5	,,	31·01	,,

EXCAVATORS

Rope operated, full circle slew, crawler mounted, with single equipment (dragline) .

Maker's Rated Dragline Capacity

	cu. m	or	cu. yd.		Hire rate	
	0·50		⅝	,,	7·57	,,
,, ,, ,, .	0·60		¾	,,	9·19	,,
,, ,, ,, .	0·70		⅞	,,	10·22	,,
,, ,, ,, .	0·80		1	,,	11·12	,,
,, ,, ,, .	1·00		1¼	,,	13·15	,,
,, ,, ,, .	1·20		1½	,,	14·89	,,
,, ,, ,, .	1·40		1¾	,,	16·77	,,
,, ,, ,, .	1·50		2	,,	20·29	,,
,, ,, ,, .	2·00		2½	,,	26·63	,,
,, ,, ,, .	2·30		3	,,	30·39	,,
,, ,, ,, .	2·50		3¼	,,	32·70	,,
,, ,, ,, .	2·70		3½	,,	37·42	,,

Rope operated, full circle slew, crawler mounted, with single equipment (face shovel) .

Maker's Rated Face Shovel Capacity

	cu. m	or	cu. yd.		Hire rate	
	0·60		¾	,,	11·63	,,
,, ,, ,, .	0·70		⅞	,,	12·93	,,
,, ,, ,, .	1·00		1¼	,,	16·21	,,
,, ,, ,, .	1·40		1¾	,,	21·35	,,
,, ,, ,, .	1·50		2	,,	23·61	,,
,, ,, ,, .	2·50		3¼	,,	38·46	,,

CIVIL ENGINEERING INDUSTRY

	Description	Unit	Hire rate £	Period
EXCAVATORS – *continued*				
	Maker's Rated Standard Backacter Bucket Capacity			
Hydraulic, full circle slew, crawler mounted, with single equipment	cu. m or cu. yd.			
	0·42 ½	Each	9·89	Hour
Ditto (backacter) . .	0·50 ⅝	,,	10·46	,,
,, ,, ,, .	0·60 ¾	,,	11·61	,,
,, ,, ,, .	0·70 ⅞	,,	12·71	,,
,, ,, ,, .	0·80 1	,,	14·69	,,
,, ,, ,, .	0·90 1⅛	,,	16·68	,,
,, ,, ,, .	1·00 1¼	,,	18·64	,,
,, ,, ,, .	1·20 1½	,,	22·61	,,
,, ,, ,, .	1·40 1¾	,,	25·94	,,
,, ,, ,, .	1·50 2	,,	29·04	,,
	Maker's Rated Standard Shovel Bucket Capacity			
Hydraulic, full circle slew, crawler mounted, with single equipment	cu. m or cu. yd.			
	0·80 1	,,	16·35	,,
Ditto (shovel) . . .	1·30 1⅝	,,	25·77	,,
,, ,, ,, .	1·50 2	,,	30·28	,,
,, ,, ,, .	1·80 2⅜	,,	33·05	,,
,, ,, ,, .	2·50 3¼	,,	38·40	,,
	Maker's Rated Standard Bucket Capacity			
Hydraulic, full circle slew, wheeled type, with single equipment	cu. m or cu. yd.			
	0·42 ½	,,	9·44	,,
,, ,, ,, .	0·50 ⅝	,,	10·71	,,
,, ,, ,, .	0·60 ¾	,,	12·25	,,
,, ,, ,, .	0·70 ⅞	,,	12·93	,,
,, ,, ,, .	0·80 1	,,	14·17	,,
,, ,, ,, .	1·00 1¼	,,	16·37	,,
,, ,, ,, .	1·40 1¾	,,	20·00	,,
Extra for hydraulic breaker .	—	,,	6·47	,,
	Maker's Rated Loader Bucket Capacity			
Hydraulic, offset or centre post, half circle slew wheeled, dual purpose (back hoe/loader)	cu. m or cu. yd.			
	0·52 ⅝	,,	4·67	,,
,, ,, ,, .	0·60 ¾	,,	5·40	,,
,, ,, ,, .	0·70 ⅞	,,	5·77	,,
,, ,, ,, .	0·80 1	,,	5·87	,,
,, ,, ,, .	0·90 1⅛	,,	6·03	,,
,, ,, ,, .	1·00 1¼	,,	6·22	,,
,, ,, ,, .	1·20 1½	,,	6·51	,,
,, ,, ,, .	1·30 1⅝	,,	6·65	,,
Excavator mats	sq. m or sq. ft.			
* Light, thickness 150 mm or 6 in.	3·3 36	,,	0·52	,,
* Heavy, thickness 250 mm or 9/10 in.	3·3 36	,,	0·71	,,

CIVIL ENGINEERING INDUSTRY

	Description	Unit	Hire rate £	Period
GENERATING SETS AND TRANSFORMERS	**kVA**			
Generating set . . .	1	Each	0·64	Hour
,, ,,	1½	,,	0·80	,,
,, ,,	2	,,	0·93	,,
,, ,,	3	,,	1·04	,,
,, ,,	4	,,	1·25	,,
,, ,,	5	,,	1·49	,,
,, ,,	6¼	,,	1·58	,,
,, ,,	10	,,	2·11	,,
,, ,,	12½	,,	2·36	,,
,, ,,	16¼	,,	2·74	,,
,, ,,	20	,,	2·96	,,
,, ,,	23½	,,	3·30	,,
,, ,,	31	,,	3·62	,,
,, ,,	40	,,	4·04	,,
,, ,,	50	,,	4·98	,,
,, ,,	70	,,	6·57	,,
,, ,,	80	,,	7·55	,,
,, ,,	90	,,	8·57	,,
,, ,,	100	,,	8·89	,,
,, ,,	120	,,	10·48	,,
,, ,,	156	,,	12·19	,,
,, ,,	210	,,	14·06	,,
,, ,,	280	,,	17·04	,,
,, ,,	330	,,	20·22	,,
* Transformer, stationary .	1	,,	0·07	,,
* ,, ,,	2½	,,	0·14	,,
* ,, ,,	5	,,	0·28	,,
* ,, ,,	7½	,,	0·36	,,
* ,, ,,	10	,,	0·43	,,
* ,, ,,	12½	,,	0·55	,,
* ,, ,,	15	,,	0·63	,,
* ,, ,,	35	,,	0·75	,,
* ,, ,,	50	,,	0·85	,,
* ,, ,,	60	,,	0·95	,,
* ,, ,,	100	,,	1·29	,,
* ,, ,,	200	,,	1·71	,,
* ,, ,,	300	,,	2·20	,,
* ,, ,,	500	,,	2·80	,,

	Tower or mast height			
	Up to m or ft.			
Mobile lighting unit, 2 light tungsten halogen incl. 1½ kVA generator	4·9 16	,,	1·37	,,
Ditto, including 2½ kVA generator	7·6 25	,,	2·68	,,
Ditto, 4 light mercury vapour including 6¼ kVA generator .	15·3 50	,,	4·05	,,
Ditto, 4 light tungsten halogen including 7½ kVA generator .	18·3 60	·,	5·19	,,

CIVIL ENGINEERING INDUSTRY

HAULAGE	Description	Unit	Hire rate £	Period
* Trailer flat (towed with knock on brakes) site use only *excluding* Tax and Insurance .	Up to and including tonnes or tons			
	2·03 2	Each	3·76	Day
* „ „ „ „	3·05 3	„	4·91	„
* „ „ „ „	3·55 3½	„	5·10	„
	Maximum Lifting Capacity (forward loading) kg or cwt.			
Fork lift truck (yard type) .	1,016 20	„	2·36	Hour
„ „ „	1,270 25	„	2·41	„
„ „ „	1,524 30	„	2·60	„
„ „ „	2,032 40	„	3·56	„
„ „ „	2,540 50	„	3·98	„
Fork lift truck (rough terrain type) (2 wheel drive) . .	1,016 or 20	„	2·65	„
„ „ „	1,270 25	„	2·81	„
„ „ „	1,524 30	„	3·11	„
„ „ „	2,032 40	„	3·75	„
„ „ „	2,540 50	„	4·53	„
„ „ „	3,048 60	„	5·13	„
Fork lift truck (rough terrain type) (4 wheel drive) . .	1,524 30	„	3·39	„
„ „ „	2,032 40	„	4·18	„
„ „ „	2,540 60	„	5·02	„
„ „ „	3,048 60	„	5·81	„
„ „ „	3,556 70	„	6·81	„
„ „ „	4,572 90	„	8·56	„
„ „ „	5,580 110	„	10·21	„
	Belt width m or in. Length m or ft.			
Conveyor, mobile (including loading hopper) . . .	0·30 12 3·66 12	„	1·10	„
„ „ „	0·30 12 5·50 18	„	1·21	„
„ „ „	0·30 12 7·32 24	„	1·33	„
„ „ „	0·41 16 3·66 12	„	1·45	„
„ „ „	0·41 16 5·50 18	„	1·70	„
„ „ „	0·41 16 7·32 24	„	1·78	„
„ „ „	0·41 16 9·14 30	„	1·92	„
„ „ „	0·61 24 4·27 14	„	2·63	„
„ „ „	0·61 24 6·01 20	„	2·82	„
„ „ „	0·61 24 12·80 42	„	4·06	„
„ „ „	0·61 24 18·23 60	„	7·42	„
Extra for elevating type .	—	„	20%	„

HOISTS

Cantilever platform and centre slung scaffold tower type (complete with safety device and overwind limits):
Goods including winch, cage, platform, mast and guide rails up to:

	kg or cwt.			
16·5 m or 54 ft. .	508 10	„	1·42	„
Ditto, 16·5 m or 54 ft. .	672 15	„	1·87	„
Ditto, 15·6 m or 51 ft. .	1,016 20	„	2·27	„
Ditto, 15·6 m or 51 ft. .	1,524 30	„	2·94	„
Ditto, 15·6 m or 51 ft. .	2,032 40	„	3·87	„

CIVIL ENGINEERING INDUSTRY

	Description		Unit	Hire rate £	Period
HOISTS – *continued*					
Extra for additional mast sections:	kg	or cwt.			
* 1·68 m or 5·5 ft. .	508	10	Each	0·03	Hour
* Ditto, 3·35 m or 11 ft. .	508	10	,,	0·04	,,
* Ditto, 2·75 m or 9 ft. .	762	15	,,	0·05	,,
* Ditto, 5·50 m or 18 ft. .	762	15	·,	0·07½	,,
* Ditto, 2·75 m or 9 ft. .	1,016	20	,,	0·05½	,,
* Ditto, 5·50 m or 18 ft. .	1,016	20	,,	0·09	,,
* Ditto, 1·83 m or 6 ft. .	1,524	30	,,	0·07	,,
* Ditto, 2·75 m or 9 ft. .	1,524	30	,,	0·08	,,
* Ditto, 5·50 m or 18 ft. .	1,524	30	,,	0·12½	,,
* Ditto, 1·83 m or 6 ft. .	2,032	40	,,	0·13½	,,
* Ditto, 3·66 m or 12 ft. .	2,032	40	,,	0·15	,,
* Ditto, 5·50 m or 18 ft. .	2,032	40	,,	0·20	,,
* Extra for additional tubular side guides					
* 2·59 m or 8·5 ft. .	762	or 15	,,	0·05	,,
* Ditto, 2·75 m or 9 ft. .	1,016	20	,,	0·05½	,,
* Ditto, 1·83 m or 6 ft. .	1,524	30	,,	0·07	,,
* Ditto, 3·66 m or 12 ft. .	1,524	30	,,	0·13	,,
* Ditto, 3·35 m or 11 ft. .	2,032	40	,,	0·15	,,
* Ditto, 5·50 m or 18 ft. .	2,032	40	,,	0·20	,,
Goods (rack and pinion) incl. cage tower and guides . . top landing height – 50 m or 164 ft.	762	15	,,	2·24	,,
* Extra for additional tower section	1·68 m or 5·5 ft.		,,	0·04½	,,
* Extra for landing locking safety gate	—		,,	0·03	,,
* Extra for bottom gate and side screens . . .	—		,,	0·06	,,
* Extra for landing electric locking safety gate . .	—		,,	0·10	,,
* Extra for bottom electric locking safety gate and side screens	—		,,	0·16	,,
Passenger (rack and pinion) including cage tower and guides, 2 motor electric top landing height:	single cage				
60 m or 197 ft. . .	12 men		,,	6·08	,,
Ditto, 90 m or 295 ft. . .	12		,,	7·36	,,
Ditto, 60 m or 197 ft. . .	20		,,	6·36	,,
Ditto, 90 m or 295 ft. . .	20		,,	7·64	,,
Passenger (rack and pinion) including cages, tower and guides, 2 motor electric top landing height	twin cage				
60 m or 197 ft. . .	2 × 12 men		,,	9·44	,,
Ditto, 90 m or 295 ft. . .	2 × 12		,,	11·05	,,
Ditto, 60 m or 197 ft. . .	2 × 20		,,	10·12	,,
Ditto, 90 m or 295 ft. . .	2 × 20		,,	11·55	,,
Extra for additional tower * section	per m or 3·3 ft. single cage		,,	0·04½	,,
* ,, ,, ,,	,, ,, twin cage		,,	0·06	,,

CIVIL ENGINEERING INDUSTRY

	Description	Unit	Hire rate £	Period
HOISTS – *continued*				
* Extra for landing gate (electrically and mechanically interlocking) . . .	single cage	Each	0·32	Hour
* „ „ „	twin cage	„	0·65	„
* Mobile Hoist with 1 length of mast lending height:				
6·4 m or 21 ft. . .	305 kg or 6 cwt	„	1·20	„
Ditto, 7·3 m or 24 ft. . .	508 10	„	1·56	„
* Extra for additional mast sections				
2·86 m or 9·25 ft. .	305 kg or 6 cwt.	„	0·04½	„
* Ditto, 2·75/3·35 m or 9/11 ft.	508 10	„	0·06	„
* Extra for locking safety gate	—	„	0·03	„
* Extra for 3 side floorgate enclosure	—	„	0·08½	„
Scaffold hoist . . .	254 kg or 5 cwt.	„	0·59	„

	tonne	ton			
LIFTING AND JACKING GEAR					
* Shear legs, steel, 15 ft. .	Up to 1 or 1		„	0·95	Day
* Pipe gantry . . .	3 3		„	2·67	„
* Chain Blocks . . .	1 0·98		„	1·69	„
* „ „ . . .	2 1·97		„	2·08	„
* „ „ . . .	3·2 3·15		„	2·28	„
* „ „ . . .	4 3·94		„	2·48	„
* „ „ . . .	5 4·92		„	2·73	„
* „ „ . . .	6·3 6·20		„	3·10	„
* „ „ . . .	8 7·87		„	3·76	„
* „ „ . . .	10 9·84		„	5·20	„
* „ „ . . .	16 15·75		„	10·16	„
* „ „ . . .	20 19·68		„	11·92	„
	tonnes	tons			
* Screw Jack . . .	Not exceeding 10·20 or	10	„	0·23	„
* Hydraulic Jack (hand operated)	„ 5·08	5	„	0·44	„
* „ „ „ .	„ 10·20	10	„	0·70	„
* „ „ „ .	„ 15·24	15	„	0·87	„
* „ „ „ .	„ 20·32	20	„	1·10	„
* „ „ „ .	„ 25·40	25	„	1·25	„
* „ „ „ .	„ 35·56	35	„	1·53	„
* „ „ „ .	„ 50·80	50	„	2·50	„
* „ „ „ .	„ 101·60	100	„	3·55	„
* Ratchet Jack . . .	„ 5·08	5	„	0·64	„
* „ „ „ .	„ 10·20	10	„	1·05	„
* „ „ „ .	„ 15·24	15	„	1·67	„
* „ „ „ .	„ 20·32	20	„	1·71	„

	Safe working load				
	Lifting	Pulling			
	cwt. tonne	cwt. tonne			
* Lifting and pulling machine	15 0·75	15 0·75	„	0·55	„
* „ „ „ .	32 1·6	32 1·6	„	0·94	„
* „ „ „ .	3 ton/tonne	5 ton/tonne	„	1·42	„

CIVIL ENGINEERING INDUSTRY

Description		Unit	Hire rate £	Period
LOCOS AND RAILWAY EQUIPMENT				
Loco – 0·61 metre or 2 ft. 0 in. gauge electric including battery . . . . Draw bar pull 150 kg normal or 330 lb.		Each	1·73	Hour
* 2nd battery . . .	—	,,	0·79	,,
* Battery charger for above .	—	,,	0·21	,,
Loco – 0.61 metre or 2 ft. 0 in. gauge electric including battery . . . . Draw bar pull 273 kg normal or 600 lb.		,,	2·43	,,
* 2nd battery . . .	—	,,	0·90	,,
* Battery charger for above .	—	,,	0·22	,,
Loco – 0·61 metre or 2 ft. 0 in. gauge electric including battery . . . . Draw bar pull 318 kg normal or 700 lb.		,,	2·52	,,
* 2nd battery . . .	—	,,	0·93	,,
* Battery charger for above .	—	,,	0·55	,,
Loco – 0·61 metre or 2 ft. 0 in. gauge electric including battery . . . . Draw bar pull 454 kg normal or 1,000 lb.		,,	2·74	,,
* 2nd battery . . .	—	,,	0·96	,,
* Battery charger for above .	—	,,	0·60	,,
Loco – 0·61 metre or 2 ft. 0 in. gauge electric including battery . . . . Draw bar pull 907 kg normal or 2000 lb.		,,	6·04	,,
* 2nd battery . . .	—	,,	3·13	,,
* Battery charger for above .	—	,,	0·65	,,
Loco – 0·61 metre or 2 ft. 0 in. gauge diesel . . . 21 hp weight 2·29 tonnes 2¼ ton		,,	2·17	,,

	kg/metre or lb./yd. rail		per 10 lin. yd. or 9·2 m	
* Track – 0·61 metre or 2 ft.0 in. gauge . . .	7·4	15		0·29	Day
* ,, ,, ,, .	12·4	25	,,	0·35	,,
* ,, ,, ,, .	17·4	30	,,	0·39	,,
* Track 1·44 metre or 4 ft. 8½ in. gauge	29·6	60	,,	0·72	,,
* ,, ,, ,, ,,	52·2	90	,,	0·97	,,
* Extra over track for points or turnout:	Up to				
0·61 metre or 2 ft. 0 in. gauge	17·4	30	Each	0·80	,,
* 1·44 metre or 4 ft. 8½ in. gauge	29·6	60	,,	4·18	,,
* ,, ,, ,,	52·2	90	,,	5·22	,,
	Up to				
* Jim crow . . .	17·4	30	,,	0·27	,,
,, ...	52·2	90	,,	0·43	,,
	Capacity				
	cu. m	cu. ft.			
* Skip side tipping 'U' shaped	0·43 or	16	,,	0·93	,,
* ,, ,, ,, .	0·57	20	,,	1·09	,,
* ,, ,, ,, .	0·85	30	,,	1·20	,,

CIVIL ENGINEERING INDUSTRY

Description	Unit	Hire rate £	Period

LORRIES, VANS, ETC.

Lorry, ordinary (site use and public highway use – *including* Tax, Insurance and extra cost of petrol or D.E.R.V.)   .   .   .

Plated Gross Vehicle Weight

		tonnes	or	tons			
		3·56		3·5	Each	3·20	Hour
,,	,,	,, •		4·57	4·5	,,	3·55	,,

Description	tonnes		tons	Unit	Hire rate	Period
	3·56		3·5	Each	3·20	Hour
	4·57		4·5	,,	3·55	,,
	5·69		5·6	,,	3·65	,,
	6·60		6·5	,,	4·23	,,
	7·62		7·5	,,	4·37	,,
	8·74		8·6	,,	4·52	,,
	9·65		9·5	,,	4·66	,,
	10·16		10	,,	4·90	,,
	12·50		12·3	,,	5·56	,,
	13·72		13·5	,,	5·93	,,
	14·53		14·3	,,	6·50	,,
	16·26		16	,,	7·07	,,
	22·35		22	,,	8·42	,,
	24·38		24	,,	9·03	,,
	26·42		26	,,	10·50	,,
	28·45		28	,,	11·80	,,
	30·48		30	,,	12·26	,,

The "Lorry, ordinary" rows (shown with `,,` ditto marks for Description) correspond to:

Extra for crane attachment

maximum load

tonne	or	ton			
1·02	or	1	,,	0·36	,,
2·04		2	,,	0·44	,,
2·53		2½	,,	0·56	,,

Lorry tipper (site use and public highway use – *including* Tax, Insurance and extra cost of petrol or D.E.R.V.)   .   .   .

Plated Gross Vehicle Weight

tonnes	tons			
6·60	6·5	,,	5·19	,,
8·74	8·6	,,	5·51	,,
9·65	9·5	,,	5·65	,,
13·72	13·5	,,	7·72	,,
16·26	16	,,	9·04	,,
22·35	22	,,	10·75	,,
24·48	24	,,	11·83	,,
26·42	26	,,	13·50	,,
28·45	28	,,	14·37	,,
30·48	30	,,	14·89	,,

Extra for 2-way Tipping .

—	,,	0·53	,,

Extra for 3-way Tipping .

	,,	0·70	,,

Van or similar utility vehicle, ditto .   .   .

Carrying Capacity

tonne		cwt.			
0·25	or	5	,,	1·79	,,
0·35		7	,,	2·24	,,
0·50		10	,,	2·69	,,
0·60		12	,,	3·02	,,

tonnes	tons			
1·02	1·0	,,	3·05	,,
1·18	1·15	,,	3·47	,,
1·27	1·25	,,	3·54	,,
1·63	1·60	,,	3·62	,,
1·78	1·75	,,	3·72	,,
2·54	2·5	,,	4·23	,,
3·56	3·5	,,	5·04	,,
4·57	4·5	,,	5·58	,,
5·59	5·5	,,	5·81	,,

CIVIL ENGINEERING INDUSTRY

Description	Unit	Hire rate £	Period
LORRIES, VANS, ETC. *– continued*			

Passenger/goods, cross country, ditto . . .

Description	Unit	Hire rate £	Period
Wheelbase 88 in. or 2·24 m	Each	3·57	Hour
Wheelbase 109 in. or 2·77 m	,,	3·85	,,

Station wagon, ditto .

7 Seater	,,	3·72	,,
10 Seater	,,	4·10	,,
12 Seater	,,	4·20	,,

Personnel carrier/coach/ bus, ditto . . .

9/13 Seater	,,	3·75	,,
14/17 Seater	,,	4·35	,,

Road/sweeper/cleaner self propelled . . .

Hopper Capacity	Payload	Brush diameter Main	Side			
40 cu. ft.	20 cu. ft.	27½ in.	24 in.	,,	3·26	,,
or		or				
1·13 cu. m	0·57 cu. m	0·70 m	0·61 m	,,	4·75	,,
56 cu. ft	40 cu. ft.	14 in.	24 in.			
or		or				
1·59 cu. m	1·13 cu. m	0·36 m	61 m			

MONORAIL

	Struck Capacity cu. m		cu. ft.		Hire rate	Period
Power Wagon, manual tipper .	0·50	or	17·5	,,	2·56	,,
,, ,, hydraulic tipper	0·415		14·5	,,	2·99	,,
* Trailer Wagon, manual tipper	0·50		17·5	,,	0·88	,,
* ,, ,, hydraulic tipper	0·415		14·5	,,	1·16	,,
* Skip, side discharge . .	0·30		10·0	,,	0·27	,,
* ,, bottom discharge .	0·30		10·0	,,	0·34	,,
* Monorail, straight, including fittings . . .	3·66 m or 12 ft. 0 in. lengths			Per length	0·16	,,
* ,, ,, ,,	1·83 m or 6 ft. 0 in. lengths			,,	0·14	,,
* ,, ,, ,,	3·00 m or 10 ft. 0 in. lengths			,,	0·30	,,
* ,, 12 ft. radius curved, including fittings . . .	1·83 m or 6 ft. 0 in. lengths			,,	0·29	,,
* ,, other radii ditto .	1·83 m or 6 ft. 0 in. lengths			,,	0·37	,,
* Points or turnout .	—			Per set	1·16	Day
* Trolley transporter for power wagon	—			Each	0·58	,,
* Automatic Stop . . .	—			,,	0·14	,,
* Buffers . . .	—			,,	0·13	,,
,, heavy duty spring loaded	—			,,	0·25	,,
* Stand, standard . .	—			,,	0·11	,,
* ,, high level . .	0·76 m or 2 ft. 6 in.			,,	0·22	,,
* ,, ,, . .	1·37 m or 4 ft. 6 in.			,,	0·28	,,
* ,, ,, . .	1·68 m or 5 ft. 6 in.			,,	0·31	..

CIVIL ENGINEERING INDUSTRY

	Description	Unit	Hire rate £	Period
OFFICES, STORES, SHEDS, ETC.				
* Offices on site with usual fittings, i.e. heating equipment, desk, tables, chairs, stools, plan chest, etc. (including heating and lighting) . .	Timber sectional (excluding insurance)	Per 100 sq. ft. or 9·3 sq. m floor area	8·14	Week
Messroom on site with usual fittings, heating equipment, counter, tables, forms, etc. (excluding all kitchen equipment) (including heating and lighting)	Ditto	,,	8·28	,,
* Stores on site with usual fittings, i.e. counter, desk, chair, racks, shelves, bins (including heating and lighting)	Ditto	,,	6·65	,,
* Mobile office (caravan type) (excluding Tax and Insurance) (including heating and lighting) . . .	Up to 4·57 m or 15 ft. long	Each	21·00	,,
* Mobile office (caravan type) (excluding Tax and Insurance) (including heating and lighting) . . .	Up to 7·62 m or 25 ft. long	,,	28·92	,,
* Watchman's Hut . . .	—	,,	1·41	,,
* Latrine (including consumables) .	single	,,	5·50	,,
Men's shelter, including tarpaulins .	—	,,	3·94	,,
Proprietary Prefabricated Units .	Rates for special types of prefabricated and mobile buildings owned by main contractors to be based on those charged by proprietary firms, plus 12%			
PAINT SPRAYING MACHINES				
* Paint spraying machine, 1 gun type with 7·6 m or 25 ft. lengths of air and fluid hose and 9·0 litre or 2 gallon pressure tank.	(Excluding compressor)	Each	0·33	Hour
Ditto (power driven) . .	(Electric or petrol)	,,	1·29	,,
* Paint spraying machine, 2 gun type with 15·3 m or 50 ft. lengths of air and fluid hose and 22·7 litre or 5 gallon pressure tank	(Excluding compressor)	,,	0·55	,,
Ditto (power driven) . .	(Electric, petrol or diesel)	,,	1·39	,,

PILING PLANT
(Excluding Boiler or Compressor)

	Hammer weight			
* Piling Hammer, double-acting (steam or air)	kg	lb.		
	66	145	,,	0·73	,,
* ,, ,, ,, .	155	343	,,	1·02	,,
* ,, ,, ,, .	305	675	,,	1·30	,,
* ,, ,, ,, .	1,143	2,520	,,	2·13	,,
* ,, ,, ,, .	2,177	4,800	,,	2·87	,,
* ,, ,, ,, .	3,007	6,630	,,	3·61	,,
* ,, ,, ,, .	3,220	7,100	,,	6·41	,,
* ,, ,, ,, .	4,922	10,850	,,	9·81	,,
* ,, ,, ,, .	6,350	14,000	,,	12·42	,,

CIVIL ENGINEERING INDUSTRY

Description		Unit	Hire rate £	Period

PILING PLANT – *continued*

Ram weight

	kg	cwt.			
* Single acting Hammer (suitable for use with channel leaders) .	1,524	or	30	Each	2·40	Hour
* ,, ,, ,, ,, .	2,032		40	,,	2·50	,,
* ,, ,, ,, ,, .	2,540		50	,,	3·62	,,
* ,, ,, ,, ,, .	3,048		60	,,	4·06	,,
* ,, ,, ,, ,, .	4,064		80	,,	5·25	,,
* ,, ,, ,, ,, .	5,080		100	,,	6·53	,,
* ,, ,, ,, ,, .	6,096		120	,,	7·10	,,

Dead weight

	kg	cwt.			
* Drop Hammer (bare) (suitable for use with channel leaders)	508	or	10	,,	0·32	,,
* ,, ,, ,, ,, .	762		15	,,	0·38	,,
* ,, ,, ,, ,, .	1,016		20	,,	0·48	,,
* ,, ,, ,, ,, .	1,524		30	,,	0·63	,,
* ,, ,, ,, ,, .	2,032		40	,,	0·71	,,
* ,, ,· ,, ,, .	2,540		50	,,	0·77	,,
* ,, ,, ,, ,, .	3,048		60	,,	0·83	,,
* ,, ,, ,, ,, .	4,064		80	,,	0·95	,,

Dead weight

	kg	cwt.			
* Drop Hammer (bare) (suitable for use with tubular leaders fitted with leather guides and rubber inserts) . . .	1,524	or	30	,,	1·29	,,
* ,, ,, ,, ,, .	2,032		40	,,	1·47	,,
* ,, ,, ,, ,, .	2,540		50	,,	1·59	,,
* ,, ,, ,, ,, .	3,048		60	,,	1·71	,,
* ,, ,, ,, ,, .	4,064		80	,,	2·30	,,

* Internal drop Hammer (for use with cased piles):

Internal diameter of cased pile

Hammer weight

mm	in.	kg		cwt.			
254	10 .	762	or	15	,,	0·39	,,
* ,, 305	12 .	1,270		25	,,	0·47	,,
* ,, 305	12 .	1,778		35	,,	0·57	,,
* ,, 356	14 .	2,032		40	,,	0·61	,,
* ,, 356	14 .	2,794		55	,,	0·70	,,
* ,, 406	16 .	2,540		50	,,	0·64	,,
* ,, 406	16 .	3,556		70	,,	0·92	,,
* ,, 457	18 .	3,048		60	,,	0·83	,,
* ,, 457	18 .	4,064		80	,,	1·10	,,
* ,, 508	20 .	4,064		80	,,	1·10	,,
* ,, 508	20 .	5,588		110	,,	1·32	,,
* ,, 559	22 .	5,080		100	,,	1·21	,,

Hammer weight

	kg		lb.			
* Extractor Gear . . .	292	or	643	,,	0·49	,,
* ,, ,, . . .	1,295		2,856	,,	0·68	,,
* ,, ,, . . .	2,849		6,280	,,	1·00	,,

Ram weight

	kg		lb.			
* Extractor	91	or	200	,,	1·95	,,
* ,, ...	228		500	,,	2·59	,,
* ,, ...	363		800	,,	3·60	,,

CIVIL ENGINEERING INDUSTRY

PILING PLANT – *continued*	Description			Unit	Hire rate £	Period
	Hammer weight					
	kg	or	lb.			
* Heavy Duty Extractor .	1,701	or	3,750	Each	5·69	Hour
* „ „ . .	2,994		6,600	„	6·96	„
„ „ . .	4,582		10,100	„	7·91	„
* Flexible Reinforced Rubber Hose (for compressed air)	**Diameter**					
	mm	or	in.			
9·15m or 30 ft. length .	25		1	„	1·52	Day
* Ditto, 18·30 „ 60 „ .	25		1	„	1·98	„
* Ditto, 9·15 „ 30 „ .	32		1¼	„	1·79	„
* Ditto, 18·30 „ 60 „ .	32		1¼	„	2·37	„
* Ditto, 9·15 „ 30 „ .	38		1½	„	1·98	„
* Ditto, 18·30 „ 60 „ .	38		1½	„	2·64	„
* Ditto, 9·15 „ 30 „ .	51		2	„	2·46	„
* Ditto, 18·30 „ 60 „ .	51		2	„	3·31	„
* Ditto, 9·15 „ 30 „ .	64		2½	„	3·65	„
* Ditto, 18·30 „ 60 „ .	60		2½	„	4·71	„
N.B. Flexible Armoured Steam Hose 2½ times above rates.						
Diesel Hammer (including complete set of guiding equipment)	**Hammer or ram piston weight**					
	kg	or	lb.			
	500		1,100	„	6·09	Hour
„ „ „ .	907		2,000	„	7·80	„
„ „ „ .	1,247		2,750	„	9·00	„
„ „ „ .	1,270		2,800	„	9·20	„
„ „ „ .	1,361		3,000	„	9·50	„
„ „ „ .	1,503		3,310	„	10·00	„
„ „ „ .	1,815		4,000	„	11·30	„
„ „ „ .	2,200		4,850	„	13·09	„
„ „ „ .	2,268		5,000	„	13·20	„
„ „ „ .	2,500		5,510	„	13·60	„
„ „ „ .	2,994		6,600	„	14·92	„
„ „ „ .	3,457		7,620	„	21·18	„
„ „ „ .	4,083		9,500	„	25·20	„
„ „ „ .	4,600		10,150	„	26·04	„
* Pile Helmet for Pile . . 203 mm ×	203 mm or	8 in. ×	8 in.	„	0·28	„
* „ „ . . 254	254	10	10	„	0·43	„
* „ „ . . 305	305	12	12	„	0·58	„
* „ „ . . 356	356	14	14	„	0·60	„
* „ „ . . 381	381	15	15	„	0·64	„
* „ „ . . 406	406	16	16	„	0·67	„
* „ „ . . 457	457	18	18	„	0·68	„
* „ „ . . 508	508	20	20	„	0·80	„
(Plastic Dollies or equivalent to be paid for in addition)						
*Hanging Leaders Channel type (for use with drop hammer)	**Length of jib**					
	m		ft.			
* „ „ „ .	12·2	or	40	„	2·67	„
* „ „ „ .	13·7		45	„	2·92	„
„ „ „ .	15·3		50	„	3·04	„

CIVIL ENGINEERING INDUSTRY

	Description					Unit	Hire rate £	Period

PILING PLANT – *continued*

	Crane Boom		Nominal Length					
	m	or	ft.	m	or	ft.			
* Hanging Leaders, rectangular section 0·61 m × 0·61 m or 2 ft. × 2 ft. (for use with drop or diesel Hammers) . .	7·6	25	12·2	40	Each	4·24	,,
* ,, ,, ,, .	9·2	30	15·3	50	,,	4·48	,,
* ,, ,, ,, .	9·2	30	18·3	60	,,	5·02	,,
* ,, ,, ,, .	12·2	40	21·4	70	,,	5·33	,,
* ,, ,, ,, .	15·3	50	24·4	80	,,	5·82	,,

	m	or	ft.	m	or	ft.		
* Hanging Leaders, rectangular section 0·84 m × 0·84 m or 2 ft. 9 in. × 2 ft. 9 in. . .	12·2	40	18·3	60	,,	7·07	,,
* ,, ,, ,, .	12·2	40	21·4	70	,,	7·33	,,
* ,, ,, ,, .	15·3	50	24·1	79	,,	7·65	,,
* ,, ,, ,, .	18·3	60	29·9	98	,,	8·57	,,
* ,, ,, ,, .	21·4	70	34·5	113	,,	8·68	,,
* ,, ,, ,, .	24·4	80	36·6	120	,,	9·80	,,

PILING

Temporary steel piling and steel trench sheeting to be charged as a material. The residual value of recovered steel piling and steel trench sheeting to be the subject of special agreement.

PIPE BENDING EQUIPMENT

	tonne	or	ton			
* Pipe Winch . . .	0·51		½	,,	0·19	,,
* ,, . . .	1·02		1	,,	0·25	,,
* ,, . . .	2·03		2	,,	0·34	,,
* ,, . . .	3·05		3	,,	0·47	,,

	mm		dia. tube in.			
* Pipe Bending Machine (hand operated) single stage . .	Up to 51	or	2	,,	0·10	,,
* Ditto, two stage. . .	51		2	,,	0·13	,,
* Ditto, two stage. . .	76		3	,,	0·17	,,
* Ditto, two stage. . .	152		4	,,	0·40	,,
Ditto (power operated)						
* Ditto, two stage. . .	51		2	,,	0·75	,,
* Ditto, two stage. . .	76		3	,,	1·16	,,
* Ditto, two stage. . .	102		4	,,	1·67	,,
* Ditto, two stage. . .	152		6	,,	1·84	,,

CIVIL ENGINEERING INDUSTRY

PUMPS, PORTABLE	Description		Unit	Hire rate per day £	Extra Per working hour £
(up to 25 ft. suction head, exclusive of all hoses)	mm	or in.			
* Semi rotary (hand)	19	¾	Each	0·09	—
* ,, ,, ...	25	1	,,	0·10	—
* Diaphragm (hand)	51	2	,,	0·25	—
* ,, ,, ...	76	3	,,	0·36	—
Single diaphragm	51	2	,,	1·58	0·31
,, ,, ...	76	3	,,	2·52	0·42
,, ,, ...	102	4	,,	2·88	0·47
Double diaphragm	38	1½	,,	1·80	0·29
,, ,, ...	51	2	,,	3·62	0·38
,, ,, ...	76	3	,,	3·66	0·43
,, ,, ...	102	4	,,	4·40	0·62
Self priming centrifugal . . .	38	1½	,,	0·78	0·17
,, ,, ,, . . .	51	2	,,	1·61	0·40
,, ,, ,, . . .	76	3	,,	2·78	0·58
,, ,, ,, . . .	102	4	,,	4·66	0·78
,, ,, ,, . . .	152	6	,,	8·10	1·08
Sludge and Sewage	76	3	,,	6·26	0·85
,, ,, ...	102	4	,,	9·76	1·26
,, ,, ...	152	6	,,	13·33	1·96
,, ,, ...	204	8	,,	28·50	3·38
* Sump pneumatic (excluding compressor)	51	2	,,	1·93	0·12
* ,, ,, ,, ,, ,, .	64	2½	,,	2·03	0·13
* ,, ,, ,, ,, ,, .	76	3	,,	2·06	0·14
Sump electric	38	1½	,,	0·69	0·22
,, ,, ...	51	2	,,	0·88	0·24
,, ,, ...	64	2½	,,	1·38	0·34
,, ,, ...	76	3	,,	1·81	0·46
,, ,, ...	102	4	,,	2·11	0·70
Sump submersible	51	2	,,	2·39	0·27
,, ,, ...	76	3	,,	3·39	0·40
,, ,, ...	102	4	,,	6·10	0·64
,, ,, ...	152	6	,,	12·69	1·68

CIVIL ENGINEERING INDUSTRY

	Description	Unit	Hire rate £	Period
PUMPING EQUIPMENT				
Pump hoses, flexible, suction or delivery, including couplings, valve and strainer	Diameter	Per 25 ft. or 7·62 m		
	mm or in.	length		
* Suction	38 1½	length	0·72	Day
* „	51 2	„	0·84	„
* „	76 3	„	1·30	„
* „	102 4	„	2·26	„
* „	152 6	„	4·00	„
* „	204 8	„	5·50	„
* Delivery	38 1½	„	0·64	„
* „	51 2	„	0·76	„
* „	76 3	„	1·14	„
* „	102 4	„	1·67	„
* „	152 6	„	3·40	„
* „	204 8	„	4·52	„
* Additional lengths of hose	above sizes	pro rata		„
* Steel pipe suction or delivery, including flanges, bolts and joint rings (excluding valve and strainer)	Up to 76 mm or 3 in.	Per 6 ft. or 1·83 m	0·03	„
* „ „ „ .	152 6	„	0·08	„
* „ „ „ .	204 8	„	0·17	„
* Bend to be charged as 1·83 m or 6 ft. length . . .	—			
* Valve to be charged as 4 No./ 1·83 m or 6 ft. lengths . .	—			
* Steel pipe suction or delivery, screwed and socketed joints (excluding valve and strainer) .	Up to 51 or 2 in.	Per 12 ft. or 3·66 m	0·01½	„
* „ „ „ .	102 4	„	0·04	„
* „ „ „ .	152 6	„	0·10	„
* Bend to be charged as 1·83 m or 6 ft. length . . .	—			
* Valve to be charged as 2 No./ 3·66 m or 12 ft. lengths .	—			
RAMMERS AND COMPACTORS	Weight up to			
Vibro compactor or vibration rammer	kg or lb.			
	60 132	Each	0·39	Hour
* „ „ „ .	70 154	„	0·50	„
* „ „ „ .	78 172	„	0·56	„
* „ „ „ .	114 252	„	0·62	„
* „ „ „ .	200 441	„	0·90	„
Vibrating plate compactor .	78 172	„	0·55	„
„ „ „ .	90 199	„	0·60	„
„ „ „ .	120 265	„	0·87	„
„ „ „ .	135 298	„	0·95	„
„ „ „ .	176 385	„	1·05	„
„ „ „ .	220 485	„	1·20	„
„ „ „ .	310 684	„	1·49	„
„ „ „ .	470 1,037	„	2·03	„
„ „ „ .	494 1,090	„	2·10	„
„ „ „ .	520 1,147	„	2·18	„
Trench compactor .	265 594	„	1·34	„
Jumping rammer including trolley . . .	100 220	„	0·81	„

CIVIL ENGINEERING INDUSTRY

	Description	Unit	Hire rate £	Period
ROLLERS				
* Hand	—	Each	0·25	Day
	Unballasted weight up to and including			
Road Deadweight (steel 3 wheel/3 roll) Diesel . .	tonnes or tons 2·54 2½	„	2·60	Hour
„ „ „ .	4·06 4	„	3·60	„
„ „ „ .	6·10 6	„	4·62	„
„ „ „ .	8·13 8	„	5·49	„
„ „ „ .	10·20 10	„	6·06	„
„ „ „ .	12·70 12½	„	6·22	„
Ditto (tandem) Diesel . .	8·13 8	„	4·93	„
Rubber tyred, self propelled, 9 wheeled	7·11 7	„	8·31	„
Rubber tyred, self propelled, 7 wheeled	9·14 9	„	11·75	„
* Rubber tyred, trailer type .	8·13 8	„	2·76	„
* „ „ „ .	11·18 11	„	3·63	„
	Maker's weight Roll width up to up to			
Vibratory pedestrian operated (single roller) . . .	kg cwt. mm in. 550 or 11 712 or 28	„	0·97	„
„ „ „ .	680 13·5 184 32	„	1·20	„
Vibratory pedestrian operated (twin roller) . . .	650 13 611 24	„	1·50	„
„ „ „ .	950 19 762 30	„	1·77	„
„ „ „ .	1,300 26 915 36	„	2·18	„
„ „ „ .	1,750 35 970 38	„	2·43	„
	Up to and including H.P. kg or cwt. mm or in.			
Towed, Vibratory trailer .	1,460 28¾ 1,372 54 9	„	1·78	„
„ „ „	6,150 121 1,905 75 36	„	3·59	„
„ „ „	8,800 173 1,830 72 44	„	5·05	„
„ „ „	11,700 230 1,905 75 58	„	7·49	„
„ „ „	13,300 262 2,083 82 106	„	9·20	„
	Up to and including H.P. tonne ton m in.			
Self-propelled vibratory tandem (seated control) .	1·07 or 1¼ 0·80 or 31½ 6·25	„	1·56	„
„ „ „	2·03 2 1·02 40 10	„	3·22	„
„ „ „	3·05 3 1·12 44 18	„	5·87	„
„ „ „	6·10 6 36	„	9·75	„
„ „ „	7·11 7 50	„	10·48	„
„ „ „	8·63 8½ 38	„	11·80	„
„ „ „	12·20 12 67	„	13·89	„
Ditto, single roll, rubber tyred driving wheels . . .	8·13 8 120	„	14·00	„
„ „ „ .	10·20 10 125	„	15·75	„
* Scarifier (working time) extra over Roller including sharpening tines				
* „ „ „ .	1 Tine	„	1·80	„
* „ „ „ .	2 Tine	„	2·62	„
„ „ „ .	3 Tine Hydraulic	„	3·90	„
Compactor sheeps foot self-propelled Tamping foot heavy duty wheeled . . .	Maker's rated flywheel H.P. 170	„	16·49	„
„ „ „ .	300	„	30·92	„
„ „ „ .	400	„	38·04	„

CIVIL ENGINEERING INDUSTRY

	Description	Unit	Hire rate £	Period
SAWS MECHANICAL	Guide bar up to			
Chain Saw (1 man operated) .	0·31 m or 12 in.	Each	0·38	Hour
,, ,, ,, .	0·41 16	,,	0·47	,,
,, ,, ,, .	0·53 21	,,	0·58	,,
,, ,, ,, .	0·64 25	,,	0·67	,,
,, ,, ,, .	0·79 31	,,	0·75	,,
,, ,, ,, .	0·94 37	,,	0·83	,,
,, ,, ,, .	1·12 44	,,	0·91	,,
Ditto (2 man operated) .	1·27 50	,,	1·00	,,
,, ,, ,, .	1·52 60	,,	1·12	,,
,, ,, ,, .	2·29 90	,,	1·40	,,
	Saw diameter			
	m or in.			
Portable Saw Bench . .	0·26 10	,,	0·42	,,
,, ,, . .	0·31 12	,,	0·47	,,
,, ,, . .	0·41 16	,,	0·56	,,
,, ,, . .	0·46 18	,,	0·67	,,
,, ,, . .	0·51 20	,,	0·78	,,
,, ,, . .	0·61 24	,,	0·90	,,
,, ,, . .	0·66 26	,,	0·96	,,
,, ,, . .	0·76 30	,,	1·24	,,
Band Saw	—	,,	1·21	,,

		Per		
SCAFFOLDING				
* Tubular steel . . .	51 mm or 2 in. dia. Nominal	lin. ft.	0·0039	Week
* ,, alloy . . .	51 2 ,, ,, ,,	,,	0·0083	,,
* Putlog steel . . .	1·52 m or 5 ft. long	Each	0·026	,,
* ,, alloy . . .	1·52 5 ,,	,,	0·050	,,
* ,, steel . . .	1·83 6 ,,	,,	0·034	,,
* ,, alloy . . .	1·83 6 ,,	,,	0·060	,,
* Fitting steel, single, double or swivel coupler, joint pin, fixed base plate . . .	—	,,	0·012	,,
* Ditto, Adjustable Base Plate	—	,,	0·036	,,
* Ditto, Hop up Bracket .	—	,,	0·097	,,
* Ditto, Split Head trestle folding type/adjustable .	m or ft. in. to m or ft. in. 0·48 1 7 0·84 2 9	,,	0·37	,,
* ,, ,, ,, .	0·76 2 6 1·37 4 6	,,	0·42	,,
* ,, ,, ,, .	1·07 3 6 1·83 6 0	,,	0·50	,,
* ,, ,, ,, .	1·37 4 6 2·44 8 0	,,	0·60	,,
* Castor	—	,,	0·25	,,
* ,, rubber tyred . .	—	,,	0·35	,,
* ,, nylon tyred . .	—	,,	0·67	,,
* Jenny Wheel including 30·5 m or 100 ft. rope . .	254 mm or 10 in. dia.	,,	1·30	,,
* ,, ,, ,, .	305 12	,,	1·36	,,
* Board	—	,,	0·24	,,

Rates for special items of prefabricated scaffold units owned by the main contractors to be similar to those charged by Proprietary Firms, plus 12%.
NOTE: The rate for scaffold boards is subject to an increase of 5%.

CIVIL ENGINEERING INDUSTRY

Description	Unit	Hire rate £	Period

SHORING, PLANKING AND STRUTTING

Description		Unit	Hire rate £	Period
* Baulk timber, use and waste (excluding nails, dogs, wedges, etc., to be charged in addition as consumables) . . (Minimum charge per 0·028 cu. m or 1 cu. ft. £1·12)		1 cu. ft. or 0·028 cu. m.	0·062	Day
—				
* Timber for planking and strutting use and waste (excluding nails, dogs, wedges, etc., to be charged in addition as consumables) . . (Minimum charge per 0·028 cu. m or 1 cu. ft. £1·02)		”	0·085	”
—				
* Adjustable Steel Strut, extending	457–711 mm or 1 ft. 6 in.–2 ft. 4 in.	Each	0·26	Week
* ” ” ”	686mm–1·09 m or 2 ft. 3 in.–3 ft. 7 in.	”	0·30	”
* ” ” ”	1·04–1·70 m or 3 ft. 5 in.–5 ft. 7 in.	”	0·32	”

SHUTTERING

Description		Unit	Hire rate £	Period
* Steel Shutters, all types. Rates for items of steel shuttering to be based on those chargeable by Proprietary Firms. (Wedges, keys and consumables to be added) plus 12% . . .		100 lin. ft. or 30·48 m		
* Steel road forms . .	152 mm or 6 in.		0·84	Day
* ” ” . .	203 mm or 8 in.	”	1·06	”
* ” ” . .	203 mm or 8 in. (heavy section)	”	1·42	”
* ” ” . .	203 mm or 8 in. (heavy section with rail)	”	1·95	”
* ” ” . .	254 mm 10 in. (heavy section with rail)	”	2·36	”
* ” ” . .	305 mm 21 in. (heavy section with rail)	”	3·25	”
* Telescopic steel floor centre, extending inverted triangular plate type . . .	Span 1·22–1·83 m or 4–6 ft.	Each	0·44	Week
* ” ” ”	1·83–2·74 m 6–9 ft.	”	0·56	”
* ” ” ”	2·44–3·66 m 8–12 ft.	”	0·61	”
* ” ” ”	2·74–4·88 m 9–16 ft.	”	0·75	”
* Telescopic steel floor centre, extending lattice girder type .	Span 2·77 m or 9 ft. 1 in.	”	0·45	”
* ” ” ”	up to 4·17 m 13 ft. 8 in.	”	0·54	”
* ” ” ”	4·93 m 16 ft. 1 in.	”	0·58	”
* ” ” ”	5·56 m 18 ft. 3 in.	”	0·62	”
* Telescopic steel prop extending	1·04–1·83 m or 3 ft. 5 in.–6 ft. 3 in.	”	0·37	”
* ” ” ”	1·75–3·12 m. 5 ft. 9 in.–10 ft. 3 in.	”	0·48	”
* ” ” ”	1·98–3·35 m. 6 ft. 6 in.–11 ft.	”	0·49	”
* ” ” ”	2·44–3·96 m. 8 ft. 6 in.–13 ft.	”	0·53	”
* Column Shutter Clamps, extending	254–508 mm or 10 in.–1 ft. 8 in.	Per set	0·29	”
* ” ” ”	406–813 mm. 1 ft. 4 in.–2 ft. 8 in.	”	0·35	”
* ” ” ”	51 mm–1·02 m 2 ft–4 ft.	”	0·45	”

CIVIL ENGINEERING INDUSTRY

SHUTTERING – *continued*

Description	Unit	Hire rate £	Period

Beam Shutter Clamps, clamping width 4½ in. to 2 ft. 9½ in.

	Description	Unit	Hire rate £	Period
Beam Shutter Clamps, clamping width 4½ in. to 2 ft. 9½ in.	305 mm arm or 12 in. arm	Per set	0·40	Week
* „ „ „ .	457 mm arm 18 in. arm	„	0·45	„
* „ „ „ .	610 mm arm 24 in. arm	„	0·50	„
* Wall Shutter Clamps, concrete thickness . . .	102–305 mm or 4 in. to 1 ft.	„	0·40	„
* „ „ „	102–610 mm 4 in. to 2 ft.	„	0·45	„
* „ „ „	102–915 mm 4 in. to 3 ft.	„	0·53	„

Timber used for shuttering (excluding wedges, nails, screws, bolts, etc., to be charged in addition as consumables)

	Unit	Hire rate £	Period
* Rough (minimum charge per 0·028 cu. m or 1 cu. ft. £1·32) .	Per 1 cu. ft. or 0·028 cu. m	0·11	Day
* Wrot (minimum charge per 0·028 cu. m or 1 cu. ft £1·80) .	„	0·15	„

Plywood Sheeting (excluding nails, etc., to be charged in addition as consumables). (Excluding timber and steel or timber supports to be charged in addition.)

Douglas Fir .

		Unit	Hire rate £	Period
* Thickness ½ in. or 13 mm (minimum charge per 2·97 sq. m or 32 sq. ft. £4·08) . . .	Good 1 side	Per 32 sq. ft. or 2·97 sq. m	0·34	Day
* Thickness 13 mm or ½ in. (minimum charge per 2·97 sq. m or 32 sq. ft. £5·52) .	Good 2 sides	„	0·46	„
* Thickness 19 mm or ¾ in. (minimum charge per 2·97 sq. m or 32 sq. ft. £4·80) . .	Good 1 side	„	0·40	„
* Thickness 19 mm or ¾ in. (minimum charge per 2·97 sq. m or 32 sq. ft. £6·36) .	Good 2 sides	„	0·53	„
* Thickness 25 mm or 1 in. (minimum charge per 2·97 sq. m or 32 sq. ft. £8·40) . .	Good 1 side	„	0·70	„
* Thickness 25 mm or 1 in. (minimum charge per 2·97 sq. m or 32 sq. ft. £9·48) . . .	Good 2 sides	„	0·79	„

Plastic Faced

	Unit	Hire rate £	Period
* Thickness 13 mm or ½ in. (minimum charge per 2·97 sq. m or 32 sq. ft. £7·20) . . .	—	0·60	„
* Thickness 19 mm or ¾ in. (minimum charge per 2·97 sq. m or 32 sq. ft. £9·24) . . .	—	0·77	„

CIVIL ENGINEERING INDUSTRY

Description	Unit	Hire rate £	Period
SURVEYING INSTRUMENTS			
* Dumpy level and staff . —	Each	0·88	Day
* Quickset level and staff . —	,,	1·39	,,
* Engineer's Automatic and staff —	,,	1·89	,,
* Engineer's Precise parallel plate and staff . . . . —	,,	2·77	,,
* Theodolite, vernier reading to 20 seconds . . . . —	,,	2·43	,,
* Theodolite, microptic reading to 20 seconds . . . —	,,	4·54	,,
* Theodolite, microptic reading to 1 second . . . . —	,,	7·04	,,
* Ranging rod or Pole . . —	,,	0·04	,,
* Laser levelling device for pipe laying . . . . —	,,	8·20	,,
* Electronic distance measuring device including theodolite . —	,,	20·00	,,
TAR SPRAYING AND COLD EMULSION PAINT			
* Tar Boiler and Sprayer (excluding firing) hand operated. 120 gallons or 545 litres	,,	0·55	Hour
ditto, power operated . . 250 gallons or 1,136 litres	,,	1·00	,,
* Gritter (attached to lorry) extra over lorry . . . —	,,	0·45	,,
* Cold Emulsion Sprayer hand operated . . . . Up to 40 gallons or 182 litres	,,	1·48	,,
ditto, power operated . . Up to 300 gallons or 1,363 litres	,,	4·95	,,
TOOLS, PNEUMATIC			
(excluding compressor) Compressor tool (including sharpening) with up to and including 50 ft. of hose.			
* Breaker including steels . —	,,	1·06	,,
* Light pneumatic pick including steels . . . . —	,,	1·02	,,
* Pneumatic clay Spade including blade . . . . —	,,	1·00	,,
* Compressor Tool Silencer or Muffler . . . . —	,,	0·06	,,
* Chipping/Scaling/Caulking-hammer . . . . —	,,	0·10	,,
Hand-held Rock Drill without Drill Rods or Detachable Bits:			
* Light weight . . . 16–20 kg or 35–43 lb.	,,	0·29	,,
* Middle weight . . . 20–24 kg 45–52½ lb.	,,	0·44	,,
* Heavy duty . . . 25–32 kg. 55–69½ lb.	,,	0·52	,,
(Drill Rods and Bits for Hand-held Rock Drills to be paid for in addition as consumable stores.)			
* Additional hoses . . Per 15·3 m or 50 ft.	,,	0·06	,,
* Drill (excluding consumables)* —	,,	0·18	,,
* Reversible drill, ditto . —	,,	0·32	,,

CIVIL ENGINEERING INDUSTRY

	Description	Unit	Hire rate £	Period
TOOLS, PNEUMATIC – *continued*				
* Grinder, ditto . . .	76 mm or 3 lb. light weight	Each	0·06	Hour
* Ditto, ditto . . .	177 mm or 7 lb. heavy duty	„	0·17½	„
* Sander (excluding pad and disc) ditto	—	„	0·20	„
* Riveting hammer, ditto	—	„	0·08	„
* Chain Saw up to 0·58 m or 23 in., ditto	—	„	0·42	„
* Pneumatic paint scraper tool	—	„	0·13	„
Polisher	—	„	0·10	„
*Consumables to be charged in addition				

		diameter		Unit	Hire rate	Period
TOOLS, PORTABLE ELECTRIC (excluding generator or power) (consumables, to be charged in addition)		mm	in.			
* Drill		6 or	¼	„	0·08	„
* „		10	⅜	„	0·10	„
* „		13	½	„	0·13	„
* „		19	¾	„	0·21	„
* „		32	1¼	„	0·30	„
* Extra for Stand . . .		—		„	0·05	„
* Morse Taper . . .		—		„	0·18	„
* Bench grinder and pedestal .		152	6	„	0·16	„
* „ „ „ .		180	7	„	0·16½	„
* „ „ „ .		204	8	„	0·21	„
* „ „ „ .		250	10	„	0·31	„
* Angle Grinder . . .		180	7	„	0·17	„
* „ „ . . .		204	8	„	0·17½	„
* „ „ . . .		230	9	„	0·18	„
* Extra for Stand . . .		—		„	0·04	„
* Portable Grinder . .		102	4	„	0·11½	„
* „ „ . .		152	6	„	0·13½	„
* Sander		180	7	„	0·20	„
* „		230	9	„	0·21	„
* Polisher/Sander . . .		—		„	0·12½	„
* Electric Breaker . . .		—		„	0·49	„
* Electric Hammer . .		—		„	0·32	„
* Electric Hammer Kit . .		29	1⅛	„	0·30	„
* „ „ . .		51	2	„	0·35	„
* Extra for power to be added to above items where applicable.		—		„	30%	„

CIVIL ENGINEERING INDUSTRY

	Description	Unit	Hire rate £	Period
TRACTORS, SCRAPERS, Etc.	Maker's Rated Flywheel horse power			
Tractor (Crawler) . . .	Up to 51·4	Each	5·90	Hour
,, ,, . . .	,, 59·4	,,	8·74	,,
,, ,, . . .	,, 70·0	,,	9·44	,,
,, ,, . . .	,, 82·4	,,	11·18	,,
,, ,, . . .	,, 96·4	,,	11·74	,,
,, ,, . . .	,, 112·9	,,	13·17	,,
,, ,, . . .	,, 132·4	,,	15·30	,,
,, ,, . . .	,, 154·9	,,	17·99	,,
,, ,, . . .	,, 181·4	,,	21·77	,,
,, ,, . . .	,, 211·9	,,	24·59	,,
,, ,, . . .	,, 249·9	,,	29·00	,,
,, ,, . . .	,, 294·9	,,	33·57	,,
,, ,, . . .	,, 344·9	,,	38·96	,,
,, ,, . . .	,, 410·0	,,	44·81	,,
Tractor (crawler) with bull or angle dozer (hydraulically or winch operated) . . .	Up to 51·4	,,	6·62	,,
,, ,, ,, .	,, 59·4	,,	9·91	,,
,, ,, ,, .	,, 70·0	,,	11·18	,,
,, ,, ,, .	,, 82·4	,,	12·22	,,
,, ,, ,, .	,, 96·4	,,	13·21	,,
,, ,, ,, .	,, 112·9	,,	14·67	,,
,, ,, ,, .	,, 132·4	,,	16·85	,,
,, ,, ,, .	,, 154·9	,,	18·29	,,
,, ,, ,, .	,, 181·4	,,	24·12	,,
,, ,, ,, .	,, 211·9	,,	28·35	,,
,, ,, ,, .	,, 249·9	,,	32·39	,,
,, ,, ,, .	,, 294·9	,,	37·74	,,
,, ,, ,, .	,, 344·9	,,	41·95	,,
,, ,, ,, .	,, 410·0	,,	49·28	,,
	S.A.E. rated capacity cu. metre cu. yd.			
Tractor loading Shovel (Crawler)	0·60 or ¾	,,	6·37	,,
,, ,, ,, .	0·80 1	,,	8·64	,,
,, ,, ,, .	1·00 1¼	,,	10·66	,,
,, ,, ,, .	1·20 1½	,,	12·88	,,
,, ,, ,, .	1·30 1⅝	,,	13·50	,,
,, ,, ,, .	1·40 1¾	,,	14·11	,,
,, ,, ,, .	1·60 2	,,	16·00	,,
,, ,, ,, .	1·80 2¼	,,	17·85	,,
,, ,, ,, .	2·00 2½	,,	19·05	,,
,, ,, ,, .	2·10 2¾	,,	21·29	,,
,, ,, ,, .	3·50 4½	,,	34·78	,,
Tractor loading Shovel (Crawler) with Back Hoe Equipment . . .	0·60 ¾	,,	7·44	,,
,, ,, ,, .	0·80 1	,,	9·93	,,
,, ,, ,, .	0·90 1⅛	,,	10·97	,,
,, ,, ,, .	1·00 1¼	,,	11·13	,,
,, ,, ,, .	1·30 1⅝	,,	12·02	,,

CIVIL ENGINEERING INDUSTRY

	Description		Unit	Hire rate £	Period
TRACTORS, SCRAPERS, Etc. – *continued*	S.A.E. Rated Capacity				
Tractor loading Shovel	cu. metre	cu. yd.			
with 4 in 1 attachment .	0·60 or	¾	Each	6·79	Hour
,, ,, ,, .	0·80	1	,,	9·49	,,
,, ,, ,, .	1·00	1¼	,,	11·20	,,
,, ,, ,, .	1·20	1½	,,	13·63	,,
,, ,, ,, .	1·40	1¾	,,	14·71	,,
,, ,, ,, .	1·60	2	,,	17·68	,,
,, ,, ,, .	1·80	2¼	,,	18·81	,,
,, ,, ,, .	2·00	2½	,,	20·41	,,
,, ,, ,, .	2·10	2¾	,,	22·85	,,
Tractor loading Shovel (Crawler) with additional	Heaped capacity up to cu. metre or	cu. yd.			
hydraulic mounted ripper .	0·60	¾	,,	6·79	,,
,, ,, ,, .	0·80	1	,,	8·97	,,
,, ,, ,, .	1·00	1¼	,,	11·15	,,
,, ,, ,, .	1·20	1½	,,	13·22	,,
,, ,, ,, .	1·40	1¾	,,	15·07	,,
,, ,, ,, .	1·50	1⅞	,,	16·43	,,
,, ,, ,, .	1·60	2	,,	16·90	,,
,, ,, ,, .	1·80	2¼	,,	18·61	,,
Tractor loading Shovel (wheeled) with 4 wheel drive .	0·80	1	,,	7·13	,,
,, ,, ,, .	1·00	1¼	,,	8·50	,,
,, ,, ,, .	1·20	1½	,,	9·40	,,
,, ,, ,, .	1·30	1⅝	,,	10·76	,,
,, ,, ,, .	1·40	1¾	,,	11·31	,,
,, ,, ,, .	1·60	2	,,	11·50	,,
,, ,, ,, .	1·80	2¼	,,	13·48	,,
,, ,, ,, .	2·00	2½	,,	14·07	,,
,, ,, ,, .	2·10	2¾	,,	15·58	,,
,, ,, ,, .	2·30	3	,,	16·72	,,
,, ,, ,, .	2·70	3½	,,	18·54	,,
,, ,, ,, .	3·10	4	,,	22·15	,,
,, ,, ,, .	3·50	4½	,,	23·68	,,
,, ,, ,, .	4·70	6	,,	31·50	,,
Tractor loading Shovel (wheeled) with 4 wheel drive –	S.A.E. Rated Capacity cu. metre or	cu. yd.			
Articulated	1·00	1¼	,,	9·96	,,
,, ,, ,, .	1·20	1½	,,	10·17	,,
,, ,, ,, .	1·40	1¾	,,	11·80	,,
,, ,, ,, .	1·50	1⅞	,,	12·36	,,
,, ,, ,, .	1·60	2	,,	12·98	,,
,, ,, ,, .	1·80	2¼	,,	14·88	,,
,, ,, ,, .	2·00	2½	,,	15·55	,,
,, ,, ,, .	2·10	2¾	,,	16·56	,,
,, ,, ,, .	2·30	3	,,	18·50	,,
,, ,, ,, .	2·70	3½	,,	20·72	,,
,, ,, ,, .	3·10	4	,,	23·80	,,
,, ,, ,, .	3·50	4½	,,	25·63	,,
,, ,, ,, .	3·85	5	,,	28·46	,,
,, ,, ,, .	5·00	6½	,,	36·76	,,

CIVIL ENGINEERING INDUSTRY

	Description		Unit	Hire rate £	Period
TRACTORS, SCRAPERS, Etc. – *continued*					
Tractor loading Shovel (wheeled) with 4 wheel drive – Articulated with 4 in 1 attachment	S.A.E. Rated Capacity cu. metre or cu. yd.				
	1·00	$1\frac{1}{4}$	Each	10·58	Hour
,, ,, ,, .	1·40	$1\frac{3}{4}$	,,	13·31	,,
,, ,, ,, .	1·60	2	,,	14·02	,,
,, ,, ,, .	2·10	$2\frac{3}{4}$	,,	18·34	,,
,, ,, ,, .	2·70	$3\frac{1}{2}$	,,	22·31	,,
,, ,, ,, .	5·40	7	,,	42·69	,,
,, ,, ,, .	7·65	10	,,	60·88	,,
Tractor loading Shovel (wheeled) with 2 wheel drive .	0·30	$\frac{3}{8}$	,,	4·13	,,
,, ,, ,, .	0·40	$\frac{1}{2}$	,,	4·51	,,
,, ,, ,, .	0·50	$\frac{5}{8}$	,,	4·67	,,
,, ,, ,, .	0·60	$\frac{3}{4}$	,,	5·13	,,
,, ,, ,, .	0·70	$\frac{7}{8}$	,,	5·48	,,
,, ,, ,, .	0·80	1	,,	6·07	,,
,, ,, ,, .	0·90	$1\frac{1}{8}$	,,	6·54	,,
,, ,, ,, .	1·00	$1\frac{1}{4}$	,,	7·24	,,
,, ,, ,, .	1·10	$1\frac{3}{8}$	,,	7·87	,,
,, ,, ,, .	1·20	$1\frac{1}{2}$	,,	8·42	,,
Tractor loading Shovel (wheeled) 2 wheel drive with additional back hoe . .	0·60 or	$\frac{3}{4}$	,,	6·82	,,
,, ,, ,, .	0·80	1	,,	7·94	,,
,, ,, ,, .	0·90	$1\frac{1}{8}$	,,	8·50	,,
	Blade width up to metres or feet	Flywheel H.P. up to			
Motor Grader . . .	3·20	10·5	70	,,	7·02	,,
,, ,, . . .	3·20	10·5	101	,,	10·94	,,
,, ,, . . .	3·20	10·5	120	,,	13·07	,,
,, ,, . . .	3·70	12	138	,,	14·46	,,
,, ,, . . .	3·70	12	150	,,	15·13	,,
,, ,, . . .	3·70	12	180	,,	16·84	,,
,, ,, . . .	3·70	12	201	,,	18·51	,,
,, ,, . . .	3·70	12	210	,,	19·14	,,
,, ,, . . .	3·70	12	253	,,	22·23	,,
,, ,, . . .	4·00	13	170	,,	16·93	,,
,, ,, . . .	4·00	13	201	,,	19·00	,,
,, ,, . . .	4·00	13	230	,,	21·02	,,
,, ,, . . .	4·00	13	253	,,	22·30	,,
,, ,, . . .	4·70	14	163	,,	16·42	,,
,, ,, . . .	4·70	14	180	,,	17·85	,,
,, ,, . . .	4·70	14	210	,,	19·90	,,
,, ,, . . .	4·70	14	230	,,	21·30	,,
,, ,, . . .	4·70	14	240	,,	22·00	,,
	Up to and including H.P.				
Wheeled Tractor (rubber tyred)	50		,,	2·62	,,
,, ,, ,, .	62		,,	3·59	,,
,, ,, ,, .	75		,,	5·62	,,
,, ,, ,, .	94		,,	7·24	,,
,, ,, ,, .	102		,,	7·60	,,
,, ,, ,, .	116		,,	9·03	,,
,, ,, ,, .	126		,,	10·80	,,
,, ,, ,, .	138		,,	12·55	,,

CIVIL ENGINEERING INDUSTRY

Description		Unit	Hire rate £	Period
TRACTORS, SCRAPERS, Etc. – *continued*				
	Heaped capacity up to cu. metres or cu. yd.			
* Scraper, tractor drawn (when hired with tractor) . .	3·0 4·0	Each	2·55	Hour
* „ „ „ .	5·0 6·6	„	3·10	„
* „ „ „ .	8·0 10·5	„	3·75	„
* „ „ „ .	11·0 14·4	„	4·30	„
* „ „ „ .	17·0 22·0	„	4·90	„
* „ „ „ .	20·0 26·2	„	5·16	„
Motorized Scraper (rubber tyred) (single engine) . .	15·3 20	„	37·09	„
„ „ „ .	16·1 21	„	40·70	„
„ „ „ .	23·0 30	„	51·59	„
„ „ „ .	33·6 44	„	74·12	„
„ „ „ .	41·3 54	„	80·00	„
Motorized Scraper (rubber tyred) (twin engine) . .	15·3 20	„	47·49	„
„ „ „ .	24·5 32	„	64·80	„
„ „ „ .	33·6 44	„	89·03	„
Motorized elevating Scraper (rubber tyred) (single engine) .	7·3 9½	„	28·54	„
„ „ „ .	8·5 11	„	34·29	„
„ „ „ .	12·3 16	„	39·36	„
„ „ „ .	17·6 23	„	46·68	„
„ „ „ .	26·8 35	„	64·04	„
Ditto (twin engine) . .	20·7 27	„	62·36	„
„ „ „ .	26·0 34	„	78·40	„
Machine mounted percussion breaker	—	„	22·96	„
TRENCHERS	Flywheel H.P.			
Chain bucket or wheel type .	up to 12	„	2·09	„
„ „ „ .	13 to 20	„	3·34	„
„ „ „ .	21 to 39	„	4·54	„
„ „ „ .	40 to 50	„	4·89	Day
„ „ „ .	55 to 58	„	6·50	„
„ „ „ .	61 to 70	„	7·03	„
„ „ „ .	76 to 85	„	8·30	„
„ „ „ .	90 to 110	„	9·63	„
WATER SUPPLY	(Exclusive of supporting structure) litres or gallons			
* Water storage tank . .	1,136 l. or 250 g.	Each	0·40	Day
* „ „ .	2,272 l. or 500 g.	„	0·61	„
* „ „ .	2,276 l. or 501 g. to 4,543 l. or 1,000 g.	„	1·34	„
* „ „ .	4,548 l. or 1,001 g. to 9,086 l. or 2,000 g.	„	2·24	„
* Fuel storage tank .	2,272 l. or 500 g.	„	0·77	„
* „ „ .	2,276 l. or 501 g. to 4,543 l. or 1,000 g.	„	1·60	„
* „ „ .	4,548 l. or 1,001 g. to 9,086 l. or 2000, g.	„	2·61	„
* Water storage tank trailer (rubber tyred) . .	1,136 l. or 250 g.	„	3·91	„
* „ „ „ .	1,140 l. or 251 g. to 2,272 l. or 500 g.	„	5·90	„
Water tanker mobile – self propelled	4,543 l. or 1,000 g.	„	4·38	Hour
„ „ „ .	4,548 l. or 1,001 g. to 6,814 l. or 1,500 g.	„	4·50	„
* Water dandy or barrow .	—	„	0·38	Day

CIVIL ENGINEERING INDUSTRY

Description		Unit	Hire rate £	Period

WELDING AND CUTTING SETS

Description		Unit	Hire rate £	Period
Oxy-acetylene cutting and welding set inclusive of oxygen and acetylene (excluding underwater equipment) . . .	—	Each	3·83	Hour
Welding set, diesel . .	300 amp. Single operator	,,	2·16	,,
(exclusive of electrodes to be charged in addition)				
,, ,, ,,	480 amp. Double operator	,,	5·25	,,
Ditto, transformer electric .	300 amp. Single operator	,,	1·40	,,
* Hand screen . . .	—	,,	0·15	Day
* Helmet	—	,,	0·22	,,
* Standard kit for 300 amp. set	—	,,	1·06	,,

WINCHES

Description	Steady pull tonnes tons	Drop Hammer duty tonnes tons	Unit	Hire rate £	Period
Double drum steam friction winch	3·05 or 3	2·03 or 2	,,	4·71	Hour
,, ,, ,, .	4·60 4½	3·05 3	,,	6·27	,,
Ditto, diesel friction winch .	3·05 3	2·03 2	,,	5·00	,,
,, ,, ,, .	4·60 4½	3·05 3	,,	6·08	,,

Description	Steady pull	Drop Hammer duty tonne ton	Unit	Hire rate £	Period
Light type double drum diesel winch		1·02 or 1	,,	2·56	,,
,, ,, ,, ,, .		2·03 2	,,	2·78	,,
Ditto, portable single drum diesel winch		2·54 2½	,,	2·37	,,
Ditto, electric winch . .		2·54 2½	,,	1·97	,,
Ditto, double drum diesel winch		4·06 4	,,	3·32	,,
* Ditto, hand operated crab .		1·02 1	,,	0·18	,,
* ,, ,, ,, .		2·03 2	,,	0·21	,,
* ,, ,, ,, .		3·05 3	,,	0·29	,,
* ,, ,, ,, .		5·08 5	,,	0·37	,,
* Ditto, hand operated portable rubber tyred	1·52 tonnes 1½ tons		,,	0·35	,,
Ditto mounted on lorry . . (power ex lorry engine) . .	direct lift Wheeled load on ramp 1·02 tonne 1 ton 3·05 tonnes 3 tons		,,	0·36	,,
Ditto (power ex lorry battery).	1·52 tonne 1½ tons 3·05 tonnes 3 tons		,,	0·50	,,
N.B. Special rope requirements be charged extra.					

MISCELLANEOUS PLANT AND CONSUMABLE STORES

Description		Unit	Hire rate £	Period
* Air testing machine for drains	—	,,	0·58	Day
* Anvil	89 kg or 196 lb.	,,	0·27	,,
* Fencing Chestnut . .	per 30·5 metre or 100 lin. ft.	,,	0·23	,,
* Fencing pole and trestle or similar	per 6·1 metre or 20 lin. ft.	,,	0·23	,,
* Firing for pipe jointer . .	—	,,	0·32	Hour
* Ditto, for blacksmith's hearth	—	,,	0·64	,,
* ,, ,, ,, ,,	—	,,	1·60	,,
* Firing for watchman . .	—	,,	0·32	,,

CIVIL ENGINEERING INDUSTRY

	Description	Unit	Hire rate £	Period
MISCELLANEOUS PLANT AND CONSUMABLE STORES–*continued*				
* Fire devil (watchman's) .	—	Each	0·07	Day
* Fire devil with lowering hook, lead pot and ladle. . .	—	,,	0·16	,,
* Forge (blacksmith's) . .	—	,,	0·26	,,
* Road barrier . .	standard 2·44 to 3·05 m or 8 to 10 ft.	Each sq. yd. or 0·83 sq. m	0·39	,,
* Tarpaulin . . .	—		0·03	,,
* Tilley Lamp (flood) . .	—	Each	1·17	Night
Traffic signals . . .	Fixed time portable	,,	55·45	Week
,, ,, . . .	vehicle activated portable with detector loops	,,	72·82	,,
	ditto with radar detectors	,,	78·72	,,
* Watchman's lamp incl. paraffin	—	,,	0·37	Night
	cu. m/min. or cu. ft./min.			
Space Heater . . .	5·66 200	Each	0·52	Hour
,, ,, . . .	11·32 400	,,	0·83	,,
,, ,, . . .	25·47 900	,,	1·25	,,
,, ,, . . .	28·30 1,000	,,	1·37	,,
,, ,, . . .	38·20 1,350	,,	1·70	,,
,, ,, . . .	48·11 1,700	,,	2·02	,,
,, ,, . . .	50·94 1,800	,,	2·12	,,
,, ,, . . .	56·60 2,000	,,	2·19	,,
Brickwork and masonry saw (excluding abrasive discs) .	H.P. Blade dia.			
	2 305 mm or 12 in.	,,	0·44	,,
,, ,, ,,	5·5 356 mm or 14 in.	,,	0·74	,,
,, ,, ,,	7·5 457 mm or 18 in.	,,	1·37	,,
* Road traffic warning cones .	—	,,	0·06	Day
		Per 0·093 m or 1 sq ft.		
* Crossing plates (steel sheets)	thickness 20mm. or ¾ in.		0·12	,,
* Flashing traffic warning lamp unit	—	Each	0·60	,,
* Flashing traffic warning lamp unit including bollard or tripod	—	,,	0·75	,,
* Pendant barrier marker including fencing pins . . .	26 m or 85 ft. cord length	,,	0·46	,,
		Per 9·1 m or 30 lin. ft.		
* Drainage rods incl. accessories	102 to 152 mm or 4 to 6 in.	Each	0·40	,,
* Ditto extra per rod . .	0·91 m or 3 ft.	,,	0·02½	,,
* Drain stopper (expanding) .	Up to 102 mm or 4 in. dia.	,,	0·09	,,
* ,, ,, ,,	152 6	,,	0·13	,,
* ,, ,, ,,	305 12	,,	0·23	,,
* ,, ,, ,,	457 18	,,	0·52	,,
* ,, ,, ,,	610 24	,,	1·07	,,
* ,, ,, ,,	762 30	,,	1·98	,,
* ,, ,, ,,	914 36	,,	3·06	,,
* ,, ,, ,,	1,220 48	,,	4·34	,,
	Water extraction per 24 hours			
Dehumidifier . . .	68 litres 15 gallons	,,	3·36	,,
,,	91 20	,,	4·48	,,
,, . . .	160 35	,,	6·16	,,
Ditto with temperature control	160 35	,,	9·60	,,
* Ladders – all types and lengths	—		Rates to be negotiated	

CIVIL ENGINEERING INDUSTRY

4. SUPPLEMENTARY CHARGES

Notes and Conditions

(1) Cost of free transport provided by contractors for workmen to and from the site to be charged at cost plus $12\frac{1}{2}\%$.

(2) *Subcontractors*

    (i) *Ordinary Subcontractors*

All subcontractors' accounts, other than labour only subcontractors or plant hire to be charged at full amount of invoice (without deduction of any cash discounts not exceeding $2\frac{1}{2}\%$) plus 10%.

    (ii) *Plant Hire*

    (*a*) for plant hire, fuel, oil and grease, insurances, transport, etc., to be charged a full amount of invoice (without deduction of any cash discounts not exceeding $2\frac{1}{2}\%$) plus $12\frac{1}{2}\%$, to which should be added consumables where supplied by the contractor;

    (iii) *Labour Subcontractors*

    (*a*) for driver or operative, full amount of invoice plus 64%.

Cost of labour to be dealt with as in Section 1 Note 3. Any minor items invoiced, other than labour, to be dealt with in accordance with Note 2(ii) above.

(4) The cost of internal transport on a site to be charged in addition at the appropriate daywork rates for labour, plant hire, etc.

(5) The net cost of operating welfare facilities to be charged by the contractor plus $12\frac{1}{2}\%$.

(6) The cost of additional insurance premiums for abnormal contract work or special site conditions to be charged at cost plus $12\frac{1}{2}\%$.

(7) The cost of watching and lighting specially necessitated by daywork is to be paid for separately.

(8) The charges referred to above are as at date of publication and will be subject to periodic review.

5. VALUE ADDED TAX

Notes and Conditions

(1) Value Added Tax has not been included in any of the rates in this schedule but will be chargeable if payable to H M Customs and Excise by the contractor.

Fees for Professional Services

Extracts from the scales of fees for architects, quantity surveyors and consulting engineers are given together with extracts from the Greater London Council (General Powers) Act 1965, showing the fees for district surveyors. These extracts are reproduced by kind permission of the bodies concerned. Attention is drawn to the fact that the full scales are not reproduced here and that the extracts are given for guidance only. The full authorized scales should be studied before concluding any agreement and the reader should ensure that the fees quoted here are still current at the time of reference.

ARCHITECTS' FEES

QUANTITY SURVEYORS' FEES

CONSULTING ENGINEERS' FEES

DISTRICT SURVEYORS' FEES

ARCHITECTS' FEES

LAST PRINTING OCTOBER 1977, REVISED OCTOBER 1979

INTRODUCTION

These Conditions of Engagement are for the mutual benefit of clients and architects. They determine the minimum fees for which members of the RIBA may undertake work and describe the professional services which clients may expect in return.

Members of the RIBA are governed by the Charters, Bylaws and Code of Professional Conduct of the Royal Institute, which determine their relationships with the public and their professional colleagues.

It is the duty of members to uphold and apply the Conditions of Engagement adopted by the Royal Institute. The engagement of a member shall therefore be in accordance with these Conditions and the fees and charges herein shall apply. However, higher fees and charges may be appropriate depending on the circumstances of the work, in which case these shall be agreed between architect and client when the former is engaged.

Members may not work speculatively nor compete with one another in respect of percentage fees or time charges. Where a prospective client is considering the engagement of one of a number of firms the members concerned may give guidance on the engagement of architects but shall not submit estimates of fees for competitive purposes.

Members whose engagements have been confirmed may give provisional forecasts of the cost of their professional services, quoting the appropriate percentage fees on any constructional cost limits supplied by the client and the estimated time charges on all other work. The services to be provided should always be clearly stated, and estimates of time charges should indicate the time likely to be spent by principals and staff at various salary levels.

Members working outside the United Kingdom shall apply these Conditions wherever they are recognized. Where they are not recognized, members shall apply the locally recognized scale or, in the absence of such a scale, may determine their own fees. Where members are likely to be in competition with their fellow members abroad they are advised to consult the RIBA.

Any value added tax chargeable on the services of the architect is chargeable to the client at the appropriate rate current at the time the tax is charged in accordance with Clause 1.16 of these Conditions. Clients who are taxable persons under the Finance Act 1972 will be able to recover such input tax from Customs and Excise.

ARCHITECTS' FEES

PART 1: GENERAL

1.00. This part deals with general conditions of engagement and will apply irrespective of the nature or extent of services to be provided.

1.1. REMUNERATION

1.10. The services normally provided by an architect in studying his client's needs, advising him, preparing, directing and co-ordinating design and inspecting work executed under a building contract are described in Part 2: Normal Services. Other services that an architect may provide are described in Part 4: Other Services.

1.11. The Normal Services for a building project are divided into a sequence of Work Stages A to H through which the architect's work progresses, augmented by services which vary widely in nature and extent with the circumstances of the project.

1.12. Fees for Work Stages C to H are generally calculated as a percentage of the total construction cost of the works, as described in Part 3 of these Conditions.

1.13. Fees for Work Stages A and B and for those other services which are likely to vary widely in nature and extent are charged additionally on a time basis, as described in Part 5 of these Conditions.

1.14. In exceptional circumstances, any service normally charged on a percentage fee may, by prior written agreement between architect and client, be charged on a time basis, in which case the architect shall notify the RIBA.

1.15. The minimum fees and charges described in these Conditions may not be sufficient in all circumstances, in which case higher fees and charges may be agreed between the client and architect when the architect is commissioned.

1.16. The amount of any value added tax on the services of the architect arising under the Finance Act 1972 (or any statutory modification or re-enactment thereof) shall be chargeable to the client in addition to the amount of the architect's fees and charges calculated in accordance with Part 3, Part 5 and Part 6 of these Conditions.

1.2. CONSULTANTS

1.20. Normal Services do not include quantity surveying, town planning, civil, structural, mechanical, electrical or heating and ventilating engineering or similar consultants' services. Where the provision of such services is within the competence of the architect's own office or where they are provided by consultants in association with the architect, fees shall be in accordance with the scales of fees of the appropriate professional bodies, but all time charges shall be in accordance with Part 5 of these Conditions.

1.21. Where the services of more than one profession are provided by a single firm or consortium, fees shall be the sum of the appropriate fees for the individual professional services rendered.

1.22. The architect will advise on the need for independent consultants and will be responsible for the direction and integration of their work but not for the detailed design, inspection and performance of the work entrusted to them.

1.23. Independent consultants and quantity surveyors should be nominated or approved by the architect in agreement with the client. They should be appointed and paid by the client.

ARCHITECTS' FEES

1.3. RESPONSIBILITIES

1.30. The architect must have the authority of his client before initiating any service or Work Stage.

1.31. The architect shall not make any material alteration, addition to or omission from the approved design without the knowledge and consent of the client, except if found necessary during construction for constructional reasons in which case he shall inform the client without delay.

1.32. The architect shall inform the client if he has reason to believe the total authorized expenditure or contract period are likely to be materially varied.

1.33. The architect shall advise on the selection and appointment of the contractor and shall make such periodic visits to the site as he considers necessary to inspect generally the progress and quality of the work and to determine in general if the work is proceeding in accordance with the contract documents.

1.34. The architect shall not be responsible for the contractor's operational methods, techniques, sequences or procedures, nor for safety precautions in connection with the work, nor shall he be responsible for any failure by the contractor to carry out and complete the work in accordance with the terms of the building contract between the client and the contractor.

1.4. SPECIALIST SUB-CONTRACTORS AND SUPPLIERS

1.40. The architect may recommend that specialist sub-contractors and suppliers should design and execute any part of the work. He will be responsible for the direction and integration of their design, and for general inspection of their work in accordance with Stage H of the Normal Services, but not for the detailed design or performance of the work entrusted to them.

1.5. COPYRIGHT

1.50. The provisions of this Section shall apply without prejudice to the architect's lien on drawings against unpaid fees.

1.51. In accordance with the provisions of the Copyright Act 1956, copyright in all drawings and in the work executed from them, except drawings and works for the Crown, will remain the property of the architect unless otherwise agreed.

1.52. Where an architect has completed Stage D or where an architect provides detail design in Stage E F G the client unless otherwise agreed shall, on payment or tender of any fees due to the architect, be entitled to reproduce the design by proceeding to execute the project, but only on the site to which the design relates.

1.53. Where an architect has not completed Stage D or where he and his client have agreed that Clause 1.52 shall not apply, the client may not reproduce the design by proceeding to execute the project without the consent of the architect and payment of any additional fee that may be agreed in exchange for the architect's consent.

1.54. The architect shall not unreasonably withhold his consent under Clause 1.53 but where his services are limited to making and negotiating Town Planning consents he may withhold his consent unless otherwise determined by an arbitrator appointed in accordance with Clause 7.50.

1.6. INSPECTION

1.60. During his on-site inspections made in accordance with Clause 1.33 the architect

ARCHITECTS' FEES

shall endeavour to guard the client against defects and deficiencies in the work of the contractor, but shall not be required to make exhaustive or continuous inspections to check the quality or quantity of the work.

1.61. Where frequent or constant inspection is required a clerk or clerks of works should be employed. He shall be nominated or approved by the architect and be under the architect's direction and control. He may be appointed and paid by the client or employed by the architect.

1.62. Where the need for frequent or constant on-site inspection by the architect is agreed to be necessary, a resident architect shall be appointed by the architect.

1.63. Where the architect employs a resident architect or a clerk or clerks of works he shall be reimbursed by the client in accordance with either Part 5 or Part 6 of these Conditions.

1.7. DELAY AND CHANGES IN INSTRUCTIONS

1.70. Extra work and expense caused in any Stage resulting from delay in receiving instructions, delays in building operations, changes in the client's instructions, phased contracts, bankruptcy or liquidation of the contractor or any other cause beyond the control of the architect, shall be additionally charged on a time basis.

1.8. FEES FOR HOSPITAL WORKS

1.80. The Health Ministers have asked Hospital Boards to apply these scales to their commissions and have advised the Boards that claims from architects to re-negotiate terms for long-running commissions, from some suitable future break-point and where a substantial amount of work remains to be done, should be considered.

1.81. For certain hospitals, defined as 'hospital projects let as a single commission by a client representing a University and a Hospital Board' with total construction cost of £7,000,000 or more, the minimum fee shall be 6%. Such projects are to be assessed as a whole, irrespective of any phases into which they might be divided for building.

1.9. SIGNING BUILDINGS

1.90. 'Where the architect has been commissioned for and is completing the Normal Services he shall be entitled at his own expense, or may be required by the client at the client's expense, to sign the building by inscription or otherwise on a suitable and reasonably visible part of the permanent fabric of the building.'

PART 2: NORMAL SERVICES

2.00. This part describes the services normally provided by an architect for a building project. The fees for Work Stages C to H are generally charged on a percentage basis as described in Part 3 of these Conditions. Stage C begins where the architect's brief has been determined in sufficient detail. Fees otherwise, including work in Stages A and B to determine the architect's brief, are charged additionally on a time basis as described in Part 5 of these Conditions. Initial consultations may be given free of charge.

ARCHITECTS' FEES

2.1. WORK STAGES

2.10. Work Stages charged on a *time* basis:

A *Inception*

Receiving an initial statement of requirements, outlining possible courses of action, and advising on the need for a quantity surveyor and consultants. Determining the brief in sufficient detail for subsequent Stages to begin.

B *Feasibility Studies*

Undertaking a preliminary technical appraisal of a project sufficient to enable the client to decide whether and in what form to proceed, and making town planning inquiries or application for outline town planning approval. Such an appraisal may include an approximation of the cost of meeting the client's requirements, a statement on the need for consultants, an outline timetable and a suggested contract procedure.

2.11. Work stages normally charged on a *percentage* basis:

C *Outline proposals*

Analysing the client's requirements and where necessary instructing the quantity surveyor and consultants. Preparing, describing and illustrating outline proposals, including an approximation of the cost of meeting them. Informing the client of any major decisions which are needed and receiving any amended instructions.

D *Scheme design*

Preparing in collaboration with the quantity surveyor, and consultants if appropriate, a scheme design consisting of drawings, and outline specification sufficient to indicate spatial arrangements, materials and appearance. Presenting a report on the scheme, the estimated cost and timetable for the project, for the client's approval.

E F G *Detailed design, production drawings, specifications and bills of quantities*

Completing a detailed design, incorporating any design work done by consultants, nominated sub-contractors and suppliers. Carrying out cost checks as necessary. Obtaining quotations and other information from nominated sub-contractors and suppliers. Preparing production drawings and specification of materials and workmanship required. Supplying information necessary for the preparation of bills and quantities, if any.

H *Tender action to completion*

Obtaining and advising on tenders and preparing and advising on the contract and the appointment of the contractor. Supplying information to the contractor, arranging for him to take possession of the site and examining his programme. Making periodic visits to the site as described in Clause 1.33; issuing certificates and other administrative duties under the contract. Accepting the building on behalf of the client, providing scale drawings showing the main lines of drainage and obtaining drawings of other services as executed, and giving initial guidance on maintenance.

2.2. DEVELOPMENT STUDIES

To be charged on a time basis.

2.20. Services where a client's initial statement of requirements in Stage A requires a

ARCHITECTS' FEES

special service (such as operational research) before consideration of the brief and development of outline proposals as described in Stage C can begin.

2.3. DEVELOPMENT PLANS

To be charged on a time basis.

2.30. Preparing development plans for any large building or complex of buildings which will be carried out in phases over a number of years.

2.31. Preparing a layout only, or preparing a layout for a greater area than that which is to be developed immediately.

2.4. SITES AND BUILDINGS

To be charged on a time basis.

2.40. Advising on the selection and suitability of sites, conducting negotiations concerned with sites or buildings, making measured surveys, taking levels and preparing plans of sites and buildings or existing buildings.

2.41. Making inspections, preparing reports or giving general advice on the condition of premises.

2.42. Work in connection with soil investigations.

2.5. CONSTRUCTIONAL RESEARCH

To be charged on a time basis.

2.50. Research where the development of a scheme design in Stage D involves special constructional research, including the design, construction or testing of prototype buildings or models.

2.6. NEGOTIATIONS

To be charged on a time basis.

2.60. Exceptional negotiations such as those arising from applications for Town Planning, Building Byelaw, Building Act or Building Regulations approvals.

2.61. Providing information, making all applications other than those covered by the Normal Services, such as those including applications for licences, negotiations in connection with party walls and grant aids.

2.62. Submission to the Royal Fine Art Commission and town planning appeals.

2.7. SPECIAL DRAWINGS

To be charged on a time basis.

2.70. Preparing any special drawings, models or technical information specially for the use of the client, or for Town Planning, Byelaw and Building Regulations approvals; for negotiations with ground landlords, adjoining owners, public authorities, licensing authorities, mortgagors and others.

2.8. FURNISHINGS AND WORKS OF ART

To be charged on a time basis.

2.80. Advising on the selection and suitability of loose furniture, fittings and soft furnishings, on the commissioning or selection of works of art, obtaining tenders and supervising their installation.

2.9. APPROVALS IN THE NORMAL SERVICES

2.90. Except in Scotland, Stages C to E F G of the Normal Services include the duty

ARCHITECTS' FEES

of making and negotiating applications for Town Planning consents, Building Byelaw, Building Act and Building Regulations approvals, as appropriate. All work in connection with these applications will not necessarily be included in any particular Stage.

2.91. In Scotland, the Normal Services cover the duty of preparing drawings and technical information necessary for submission of applications for licences, Town Planning and Building (Scotland) Act approvals as appropriate. The actual completion of the application and its presentation to the appropriate Court is not part of the architect's responsibility.

PART 3: FEES FOR THE NORMAL SERVICES

3.00. This part describes how the percentage fees for the Normal Services are calculated and may be varied, and when they and other charges are due. Percentage fees are based on the total construction cost of the works and on the issue of the final certificate shall be re-calculated on the actual total construction cost.

3.1. TOTAL CONSTRUCTION COST

3.10. The total construction cost shall be the cost, as certified by the architect, of all works (including site works) executed under his direction, subject to the following conditions:

3.101. The total construction cost shall include the cost of all work designed or supervised by consultants which the architect is responsible for directing and co-ordinating in accordance with Clause 1.22, irrespective of whether such work is carried out under separate building contracts for which the architect may not be responsible. The architect shall be informed of the cost of any such separate contracts.

3.102. The total construction cost shall not include nominated subcontractor's design fees for work on which consultants would otherwise have been employed. Where such fees are not known, the architect shall estimate a reduction from the total construction cost.

3.103. For the purpose of calculating the appropriate fees, the total construction cost shall include the actual or estimated cost of any work executed which is excluded from the contract but otherwise designed by the architect.

3.104. The total construction cost shall include the cost of built-in furniture and equipment. Where the cost of any special equipment is excluded from the total construction cost, the architect shall charge for work in connection with such items on a time basis.

3.105. Where appropriate the cost of old materials used in the work shall be calculated as if they were new.

3.106. Where any material, labour or carriage are supplied by a client who is not the builder, the cost shall be estimated by the architect as if they were supplied by the builder and included in the total cost.

3.107. Where the client is the builder, a statement of the ascertained gross cost of the works may be used in calculating the total construction cost of the works. In the absence of such a statement, the architect's own estimate shall be used. In both a statement of the ascertained gross cost and an architect's estimate there shall be included an allowance for the builder's profit and overheads.

ARCHITECTS' FEES

3.11. The fee for any part of the work omitted on the client's instruction shall be calculated in accordance with Section 3.5 of these Conditions.

3.2. NEW WORKS

3.20. Fees for new works generally are shown in Table 1 on page 83.

3.3. WORKS TO EXISTING BUILDINGS

3.30. Higher percentages are chargeable for works to existing buildings and are shown in Table 2 on page 83.

3.31. The percentage in Table 2 will not necessarily be sufficient for alterations to all buildings, especially those of historic importance, and higher fees may be appropriate.

3.32. Where extensions to existing buildings are substantially independent, fees may be as for new works, but the fee for those sections of works which marry existing buildings to the new shall be charged separately at the fee in Table 2 applicable to an independent commission of similar value.

3.4. REPETITION

3.40. Where a building is repeated for the same client fees for the superstructures excluding all work below the top of ground floor slabs may be reduced as follows:

3.401. On all except the first three of any houses of the same design.

3.402. On all except the first, i.e. the prototype, of all other building types to the same design.

3.41. Where a single building incorporates a number of identical compartments such as floors in multi-storey or complete structural bays in single-storey buildings, fees may be reduced on all identical compartments in excess of 10 provided that the building does not otherwise attract fee reductions and that it is completed in a single contract.

3.42. Reductions shall not be made for repeated individual dwelling units in multi-storey housing schemes but such schemes may qualify for fee reductions under Sub-clause 3.402 or Clause 3.41.

3.43. Reductions in accordance with Clauses 3.40 and 3.41 shall be made by waiving either the fee for either Stages D and E F G of the Normal Services where a complete design can be re-used without modification other than the handing of plans, or Stage E F G where a complete design can be re-used with only minor modification.

3.44. The handing of a plan shall not constitute a modification.

3.45. The total construction cost of the works shall be taken first and the fee for normal or partial services calculated thereon. The appropriate reduction shall then be applied to the cost of the repeated superstructures or sections and the result deducted from the full fee.

3.46. Screen walls and outbuildings and garages shall be excluded from the construction cost of works on which fees are waived unless they are included in the type drawings and specifications.

3.47. The fees for work in Stage H of the Normal Services shall not be reduced for repetitive works or repeated buildings, and any additional work arising out of repetition shall be charged on a time basis.

ARCHITECTS' FEES

3.5. PARTIAL SERVICES

3.50. Where for any reason the architect provides only part of the Normal Services described in Part 2 of these Conditions he shall be entitled to commensurate remuneration, and his fees and charges shall be calculated as follows:

3.501. Where an architect completes the work described in any of Stages C to E F G he shall be entitled to the appropriate proportion of the full percentage fee for the service in accordance with Table 3.

3.502. Where an architect is commissioned to undertake only the work described in Stage H, whether in whole or part, fees shall be on a time basis.

3.503. Where an architect originally engaged to provide the Normal Services does part only of the work described in Stage H, he shall be entitled to not less than the percentage fee otherwise due to him under Clause 3.61.

3.504. Where an architect provides part only of the services described in Stages C to E F G, fees for service in any Stage which is incomplete shall be on a time basis, except by prior written agreement in accordance with Clause 3.51.

3.505. Where an architect has previously completed the work described in Stages C to E F G on a commission which has been abandoned under the terms of Part 7 of these Conditions and the commission is resumed within two years, fees for the work in Stage H shall be on a percentage basis. Where the commission is resumed after two years Sub-Clause 3.502 will apply.

3.51. Where work done by a client results in the omission of part of Stages C to H described in Part 2 of these Conditions or a sponsored constructional method is used, a commensurate reduction in fees may be made by prior written agreement, provided each such agreement specifies in sufficient detail the work to be done by the client which would otherwise have formed part of the Stages provided by the architect, and is either made in accordance with the RIBA Memorandum on the application of this Clause or is approved by the RIBA.

3.52. All percentage fees for partial service shall be based on the architect's current estimate of the total construction cost of the work. Such estimates may be based on an accepted tender or, subject to Clause 3.53, on the lowest of unaccepted tenders.

3.53. Where partial service is provided in respect of works for which the executed cost is not known and no tender has been accepted, percentage fees shall be based either on the architect's estimated total construction cost or the most recent cost limit agreed with the client, whichever is the lower.

3.6. MODE AND TIME OF PAYMENT

3.60. On completion of each Stage of Stages C to H of the Normal Services described in Part 2 of these Conditions, the appropriate proportion of the full percentage fee calculated on the current estimated construction cost of the works, plus any other fee and out of pocket expenses which have accrued, shall be due for payment.

3.61. Notwithstanding Clause 3.60, fees in respect of Stages E F G and H shall be due for payment in instalments proportionate to the drawings and other work completed or value of the works certified from time to time.

ARCHITECTS' FEES

3.62. Alternatively, the architect and client may arrange for interim payment of fees and charges during all Stages of the work, including payment during Stage H by instalment other than those related to the value of the works certified from time to time.

3.63. On the issue of the final certificate the final instalment of all fees and other charges shall then be due for payment.

NOTE: *An alternative method of calculating fees based on accepted tender costs for jobs on fluctuations (or variations of price) contracts and fixed fee contracts for the Department of the Environment was published in a practice note in the RIBA Journal in November 1975.*

PERCENTAGE FEES FOR THE NORMAL SERVICES

Minimum charges are laid down in Tables 1 and 2 so that a fee shall not be less than the fee for works having a lower construction cost.

TABLE 1: NEW WORKS

Total construction cost	Minimum % rate	Minimum charges for work stages completed up to and including:			
		H	EFG	D	C
Up to £2,500	10·0	—	—	—	—
£2,500–£8,000	8·5	£250	£187·50	£87·50	£37·50
£8,000–£14,000	7·5	£680	£510·00	£238·00	£102·00
£14,000–£25,000	6·5	£1,050	£787·50	£367·50	£157·50
£25,000–£750,000	6·0	£1,625	£1,218·75	£568·75	£243·75
*£750,000–£1,750,000	5·75	£45,000	£33,750·00	£15,750·00	£6,750·00
*£1,750,000 and over	5·5	£100,625	£75,468·75	£35,218·75	£15,093·75

* Does not apply to certain hospitals (see Clause 1.81) or to works for which the fee is reduced for repetition as provided in Section 3.4. In those cases the minimum fee for works having a total construction cost of £25,000 and over shall be 6%.

TABLE 2: WORKS TO EXISTING BUILDINGS

Total construction cost	Minimum % rate	Minimum charges for work stages completed up to and including:			
		H	EFG	D	C
Up to £2,500	13·0	—	—	—	—
£2,500–£8,000	12·5	£325	£243·75	£113·75	£48·75
£8,000–£14,000	12·0	£1,000	£750·00	£350·00	£150·00
£14,000–£25,000	11·0	£1,680	£1,260·00	£588·00	£252·00
£25,000 and over	10·0	£2,750	£2,062·50	£962·50	£412·50

TABLE 3: APPORTIONMENT OF FEES BETWEEN STAGES OF SERVICE

On completion of each Stage of the Normal Services described in Part 2 of these Conditions, the following proportions of the cumulative fee shown in Tables 1 and 2 are payable:

Work stage	Proportion of fee	Cumulative total
C	15%	15%
D	20%	35%
E F G	40%	75%
H	25%	100%

ARCHITECTS' FEES

PART 4: OTHER SERVICES

4.00. This part describes other services which may be provided by the architect. Unless otherwise stated fees for these services shall be on a time basis.

4.1. TOWN PLANNING

4.10. Fees for town planning work shall be in accordance with the Professional Charges of the Royal Town Planning Institute, except that all layouts shall be charged on a time basis and all time charges shall be in accordance with Part 5 of these Conditions.

4.2. QUANTITY SURVEYING, VALUING AND SURVEYING

4.20. Fees for preparing bills of quantities, valuing works executed where no quantity surveyor is employed, valuation of properties and other surveying work not described elsewhere in these Conditions, shall be in accordance with the Professional Charges of the Royal Institution of Chartered Surveyors.

4.3. GARDEN AND LANDSCAPE DESIGN

4.30. Fees for garden and landscape design executed under separate contracts shall be in accordance with the Scale of Professional Charges to the Institute of Landscape Architects.

4.4. BUILDING SURVEYS AND STRUCTURAL INVESTIGATIONS

4.40. Preparing schedules of dilapidations and negotiating them on behalf of landlords or tenants.

4.41. Fees for taking particulars on site, preparing specifications and/or schedules for repairs or restoration work and inspecting their execution shall be charged on a time basis or, by prior agreement between architect and client, on a percentage basis. Where percentage fees are charged the rates shown in Table 2 will normally be appropriate.

4.42. Making structural investigations, the limits of which shall be clearly defined and agreed in writing, such as are necessary to ascertain whether or not there are defects in the walls, roof, floors and drains of a building which may materially affect its life and value.

4.5. SEPARATE TRADES CONTRACTS

4.50. Where there are separate contracts for each trade the fees shall be determined by prior written agreement and shall not be less than 20 per cent higher than the fee for Stages C to H of the Normal Services.

4.6. INTERIOR DESIGN, SHOPFITTING AND FURNITURE DESIGN

4.60. Fees may be charged on a percentage or time basis for the following work. Where percentage fees are charged, rates up to double those shown in Table 1 will normally be appropriate:

ARCHITECTS' FEES

4.601. Special services, including the provision of special sketch studies, detailed advice on the selection of furniture, fittings and soft furnishings and inspection of making up such furnishings for interior design work executed under a special building contract or subcontract or a contract separate from that for other works on which the architect may be employed.

4.602. Works of a special quality, such as special shopfitting, fronts and interiors, exhibition design and similar works, including both the remodelling of existing shops and the design of new units, both independently and within the shell of an existing building, irrespective of whether the architect is employed for shopfitting design only or the work forms part of a general building contract.

4.61. Where all shopfitting drawings are provided by specialist subcontractors the fee shall be as for the Normal Services described in Part 2 of these Conditions.

4.62. For the design of special items of furniture and fittings for limited production only, i.e. not more than 49 off, the percentage fee shall be either 15% of the total production cost or calculated on a time basis.

4.63. Payment for the design of mass-produced items of furniture may be by royalty, or by time charges and sale of copyright. Fees for the design of prototypes shall be either on a time basis or an advance on royalties.

4.7. BUILDING SYSTEMS AND COMPONENTS

4.70. For the development of building systems, percentage fees on the total production cost may be agreed specially. Otherwise, fees shall be either on a time basis or an advance on royalties.

4.71. Payment for the design of mass-produced building components may be by royalty, or by time charges and sale of copyright. Fees for development work in connection with the design of prototypes shall be either on a time basis or an advance on royalties.

4.72. Where an architect recommends to an independent client the use of a building system or components on which he is receiving royalties, the client shall be so informed. The total construction cost shall not be reduced but the architect may reduce his fees to the extent of the royalties received.

4.8. LITIGATION AND ARBITRATION

4.80. For qualifying to give evidence, settling proofs, conferences with solicitors and counsel, attendance in court or at arbitration or town planning inquiries, or before other tribunals, for services in connection with litigation, and for arbitration, fees shall be on a time basis.

4.81. Time charges shall be in accordance with Part 5.

4.9. CONSULTANCY

4.90. For acting as consultant architect, fees shall be on a time basis.

4.91. Where an architect is retained to provide consultancy or other services on a regular or intermittent basis, annual retention fees may be charged, and where appropriate may be merged with subsequent percentage fees or time charges.

Fees for Professional Services

ARCHITECTS' FEES

PART 5: TIME CHARGES

5.00. Time charges are based on hourly rates for principals and other operational staff. In assessing the rate at which time should be charged, all relevant factors should be considered, including the complexity of the work, the qualifications, experience and responsibility of the architect, and the character of any negotiations.

5.1. HOURLY RATES

5.10. The hourly rate for principals shall be agreed.

5.11. The minimum hourly rate for architectural and other operational staff, including resident architects and clerks of works not appointed and paid direct by the client, shall be 15 pence per hour for each £100 of gross annual salary, which shall include bonus payments and the employer's share of other overheads such as national insurance and occupational pension schemes.

5.2. TRAVELLING TIME

5.20. Where work is being charged on a time basis, travelling time shall be charged in accordance with Section 5.1 of these Conditions.

5.21. Where work charged on a percentage fee is at such a distance that an exceptional amount of time is spent in travelling, additional charges may be made by prior written agreement.

ARCHITECTS' FEES

PART 6: OUT OF POCKET EXPENSES

6.00. In addition to the fees under any other part of these Conditions, the architect shall be reimbursed for all reasonable out of pocket expenses actually and properly incurred in connection with the commission. Such expenses include the following:

6.1. DRAWING AND DOCUMENTS

6.10. Printing, reproduction or purchase costs of all documents, drawings, maps, models, photographs, and other records, including all those used in communication between architect, client, quantity surveyor, consultants and contractors, and for inquiries to contractors, subcontractors and suppliers, notwithstanding any obligation on the part of the architect to supply such documents to those concerned, except that contractors and suppliers will pay for any prints additional to those to which they are entitled under the contract.

6.2. HOTEL AND TRAVELLING EXPENSES

6.20. Hotel and travelling expenses, including mileage allowances for cars at recognized rates, and other similar disbursements.

6.3. DISBURSEMENTS

6.30. All payments made on behalf of the client, including expenses incurred in advertising for tenders, clerks of works, and other resident site staff, including the time and expenses of interviewers and reasonable expenses for interviewees.

6.31. Fees and other charges for specialist professional advice, including legal advice, which have been incurred by the architect with the specific authority of the client.

6.32. Postage and telephone charges incurred by the architect may be charged by prior written agreement.

6.4. COMPOUNDING OF EXPENSES

6.40. By prior written agreement, expenses may be estimated or standardized in whole or part, or compounded for an increase in the percentage fee.

ARCHITECTS' FEES

PART 7: TERMINATIONS, ABANDONED WORKS AND INTERPRETATION

7.00. An engagement entered into between the architect and the client may be terminated at any time by either party on the expiry of reasonable notice, when the architect shall be entitled to remuneration in accordance with Section 3.5.

7.1. ABANDONED COMMISSIONS

7.10. Where the construction of works is cancelled or postponed on the client's instructions, or the architect is instructed to stop work indefinitely at any time, the commission may be deemed to be abandoned and fees for partial service shall be due.

7.11. Notwithstanding Clause 7.10, if instructions necessary for the architect to continue work are not received from the client six months after such instructions were requested, the commission shall be deemed to have been abandoned.

7.12. Where a commission is abandoned or any part of the works is omitted at any time before completion, fees for partial service in respect of the whole or part of the works shall be charged for all service provided with due authority.

7.2. RESUMED COMMISSIONS

7.20. If a commission which has been abandoned is resumed without substantial alteration within six months, any fees paid under Section 7.1 shall rank solely as payments on account toward the total fees payable on the execution of the works and calculated on their total construction cost.

7.21. Where a commission which has been abandoned is resumed at any time with substantial alteration or is resumed after six months, any fees paid under Section 7.1 above shall be regarded as final payment for the service originally rendered. The resumed commission shall then be deemed separate, and fees charged in accordance with Section 3.5 of these Conditions.

7.22. All additional work arising out of a commission which is resumed in accordance with Clause 7.20 shall be charged on a time basis.

7.3. INTERPRETATION

7.30. Any question arising out of these Conditions may be referred in writing by architect or client to the RIBA for advice provided always that any difference or dispute between them is determined in accordance with either Clause 7.40 or 7.50.

7.4. DISPUTES

7.40. Any difference or dispute on the application of these Conditions to fees charged by a member of the RIBA may by agreement between the parties be referred to the RIBA for an opinion, provided always that such opinion is sought on a joint statement of undisputed facts and the parties undertake to accept it as final.

ARCHITECTS' FEES

7.5. ARBITRATION

7.50. Where any difference or dispute arising out of these Conditions cannot be resolved in accordance with Clause 7.40, it shall be referred to the arbitration of a person to be agreed between the parties, or, failing agreement within 14 days after either party has given to the other a written request to concur in the appointment of an arbitrator, a person to be nominated at the request of either party by the President of the Institute of Arbitrators, provided that in a difference or dispute arising out of the provisions of Section 1.5 the arbitrator shall, unless otherwise agreed, be a chartered architect.

QUANTITY SURVEYORS' FEES

Scale 37 itemized scale of professional charges for quantity surveying services for building works issued by the Royal Institution of Chartered Surveyors. The scale is recommended and not mandatory.

EFFECTIVE FROM MARCH 1980

1.0. GENERALLY

1.1. The fees are in all cases exclusive of travelling and other expenses (for which the actual disbursement is recoverable unless there is some prior arrangement for such charges) and of the cost of reproduction of bills of quantities and other documents, which are chargeable in addition at net cost.

1.2. The fees are in all cases exclusive of services in connection with the allocation of the cost of the works for purposes of calculating value added tax for which there shall be an additional fee based on the time involved (see paras. 19.1 and 19.2).

1.3. If any of the materials used in the works are supplied by the employer or charged at a preferential rate, then the actual or estimated market value thereof shall be included in the amounts upon which fees are to be calculated.

1.4. The fees are in all cases exclusive of preparing a specification of the materials to be used and the works to be done, but the fees for preparing bills of quantities and similar documents do include for incorporating preamble clauses describing the materials and workmanship (from instructions given by the architect and/or consulting engineer).

1.5. If the quantity surveyor incurs additional costs due to exceptional delays in building operations or any other cause beyond the control of the quantity surveyor then the fees may be adjusted by agreement between the employer and the quantity surveyor to cover the reimbursement of these additional costs.

1.6. The fees and charges are in all cases exclusive of value added tax which will be applied in accordance with legislation.

1.7. Copyright in bills of quantities and other documents prepared by the quantity surveyor is reserved to the quantity surveyor.

CONTRACTS BASED ON BILLS OF QUANTITIES: PRE-CONTRACT SERVICES

2.0. BILLS OF QUANTITIES

2.1. **Basic scale**

For preparing bills of quantities and examining tenders received and reporting thereon.

(*a*) *Category A:* Banks; churches and similar places of religious worship; clubs; council offices; court-houses; crematoria; houses; libraries; old people's homes not exceeding two storeys; petrol service stations; police stations; public houses and inns; sports pavilions; teaching laboratories and industrial laboratories of a like nature; theatres; town halls; universities, polytechnics and colleges of further education (other than halls of residence and hostels); and the like.

QUANTITY SURVEYORS' FEES

Value of work £	Category A fee £	£
Up to 75,000	115+3·0% (Minimum fee £600)	
75,000– 150,000	2,365+2·3% on balance over	75,000
150,000– 300,000	4,090+1·8% on balance over	150,000
300,000– 750,000	6,790+1·5% on balance over	300,000
750,000–1,500,000	13,540+1·2% on balance over	750,000
1,500,000–3,000,000	22,540+1·1% on balance over	1,500,000
Over 3,000,000	39,040+1·0% on balance over	3,000,000

(b) *Category B:* Ambulance and fire stations; canteens; church halls; cinemas; community centres; departmental stores; enclosed sports stadia and swimming baths; halls of residence; hospitals; hostels; industrial laboratories other than those included in Category A; motels; offices other than those included in Categories A and C; old people's homes exceeding two storeys; railway stations; recreation centres; residential hotels; restaurants; schools; self-contained flats and maisonettes not exceeding two storeys (including old people's accommodation of this nature); shops; telephone exchanges; and the like, being works of a simpler character than those in Category A or works with some element of repetition.

Value of work £	Category B fee £	£
Up to 75,000	105+2·8% (Minimum fee £560)	
75,000– 150,000	2,205+2·0% on balance over	75,000
150,000– 300,000	3,705+1·5% on balance over	150,000
300,000– 750,000	5,955+1·1% on balance over	300,000
750,000–1,500,000	10,905+1·0% on balance over	750,000
1,500,000–3,000,000	18,405+0·9% on balance over	1,500,000
Over 3,000,000	31,905+0·8% on balance over	3,000,000

(c) *Category C:* Factories; self-contained flats and maisonettes exceeding two storeys (including old people's accommodation of this nature); garages; multi-storey car parks; offices with open floor space to be finished to tenants' requirements (tenants' requirements to be Category A); open-air sports stadia and swimming baths; warehouses; and the like, being works containing little internal detail or works with a large amount of repetition.

NOTE: *When a building contract comprises a number of dissimilar buildings, with a resultant decrease in the amount of repetition, then the fees set out below may be adjusted by agreement between the employer and the quantity surveyor.*

QUANTITY SURVEYORS' FEES

Value of work £	Category C fee
Up to 75,000	90+2·5% (Minimum fee £500)
75,000– 150,000	1,965+1·8% on balance over 75,000
150,000– 300,000	3,315+1·2% on balance over 150,000
300,000– 750,000	5,115+0·9% on balance over 300,000
750,000–1,500,000	9,165+0·8% on balance over 750,000
1,500,000–3,000,000	15,165+0·7% on balance over 1,500,000
Over 3,000,000	25,655+0·6% on balance over 3,000,000

(*d*) The scales of fees for preparing bills of quantities (paras. 2.1 (*a*) to (*c*) are overall scales based upon the inclusion of all provisional and prime cost items, subject to the provision of para. 2.1 (*g*). When work normally included in a building contract is the subject of a separate contract for which the quantity surveyor has not been paid fees under any other clause hereof, the value of such work shall be included in the amount upon which fees are charged.

(*e*) Fees shall be calculated upon the accepted tender for the whole of the work subject to the provisions of para. 2.6. In the event of no tender being accepted, fees shall be calculated upon the basis of the lowest original bona fide tender received. In the event of no such tender being received, the fees shall be calculated upon a reasonable valuation of the works based upon the original bills of quantities.

N O T E : *In the foregoing context 'bona fide tender' shall be deemed to mean a tender submitted in good faith without major errors of computation and not subsequently withdrawn by the tenderer.*

(*f*) In calculating the amount upon which fees are charged the total of any credits and the totals of any alternative bills shall be aggregated and added to the amount described above. The value of any omission or addition forming part of an alternative bill shall not be added unless measurement or abstraction from the original dimension sheets was necessary.

(*g*) Where the value of the air conditioning, heating, ventilating and electrical services included in the tender documents together exceeds 25% of the amount calculated as described in paras. 2.1 (*d*) and (*e*), then, subject to the provisions of para. 2.2, no fee is chargeable on the amount by which the value of these services exceeds the said 25%. In this context the term 'value' excludes general contractor's profit, attendance, builder's work in connection with the services, preliminaries and any similar additions.

(*h*) When a contract comprises buildings which fall into more than one category, the fee shall be calculated as follows:

   (i) The amount upon which fees are chargeable shall be allocated to the categories of work applicable and the amounts so allocated expressed as percentages of the total amount upon which fees are chargeable.

   (ii) Fees shall then be calculated for each category on the total amount upon which fees are chargeable.

   (iii) The fee chargeable shall then be calculated by applying the percentages of work in each category to the appropriate total fee and adding the resultant amounts.

QUANTITY SURVEYORS' FEES

(*j*) When a project is the subject of a number of contracts then, for the purpose of calculating fees, the values of such contracts shall not be aggregated but each contract shall be taken separately and the scale of charges (paras. 2.1 (*a*) to (*h*)) applied as appropriate.

(*k*) Where the quantity surveyor is specifically instructed to provide cost planning services the fee calculated in accordance with paras. 2.1 (*a*) to (*j*) shall be increased by a sum calculated in accordance with the following table and based upon the same value of work as that upon which the aforementioned fee has been calculated:

Categories A & B; (as defined in paras. 2.1 (*a*) and (*b*)).

Value of work £	Fee £	£
Up to 300,000	0·7%	
300,000–1,500,000	2,100+0·4% on balance over 300,000	
1,500,000–3,000,000	6,900+0·35% on balance over 1,500,000	
Over 3,000,000	12,150+0·3% on balance over 3,000,000	

Category C: (as defined para. 2.1 (*c*))

Value of work £	Fee £	£
Up to 300,000	0·5%	
300,000–1,500,000	1,500+0·3% on balance over 300,000	
1,500,000–3,000,000	5,100+0·25% on balance over 1,500,000	
Over 3,000,000	8,850+0·2% on balance over 3,000,000	

2.2. Air conditioning, heating, ventilating and electrical services

(*a*) Where bills of quantities are prepared by the quantity surveyor for the air conditioning, heating, ventilating and electrical services there shall be a fee for these services (which shall include examining tenders received and reporting thereon), in addition to the fee calculated in accordance with para. 2.1, as follows:

Value of work £	Additional fee £	£
Up to 60,000	2·5%	
60,000– 120,000	1,500+2·25% on balance over 60,000	
120,000– 240,000	2,850+2·0% on balance over 120,000	
240,000– 375,000	5,250+1·75% on balance over 240,000	
375,000– 500,000	7,612+1·25% on balance over 375,000	
Over 500,000	9,175+1·15% on balance over 500,000	

(*b*) The values of such services, whether the subject of separate tenders or not, shall be aggregated and the total value of work so obtained used for the purpose of calculating the additional fee chargeable in accordance with para. (*a*).

(Except that when more than one firm of consulting engineers is engaged on the design of these services, the separate values for which each such firm is responsible shall be aggregated and the additional fees charged shall be calculated independently on each such total value so obtained.)

QUANTITY SURVEYORS' FEES

(c) Fees shall be calculated upon the accepted tender for the whole of the air conditioning, heating, ventilating and electrical services for which bills of quantities have been prepared by the quantity surveyor. In the event of no tender being accepted, fees shall be calculated upon the basis of the lowest original bona fide tender received. In the event of no such tender being received, the fees shall be calculated upon a reasonable valuation of the services based upon the original bills of quantities.

NOTE: *In the foregoing context 'bona fide tender' shall be deemed to mean a tender submitted in good faith without major errors of computation and not subsequently withdrawn by the tenderer.*

(d) When cost planning services are provided by the quantity surveyor for air conditioning, heating, ventilating and electrical services (or for any part of such services) there shall be an additional fee based on the time involved (see paras. 19.1 and 19.2). Alternatively the fee may be on a lump sum or percentage basis agreed between the employer and the quantity surveyor.

NOTE: *The incorporation of figures for air conditioning, heating, ventilating and electrical services provided by the consulting engineer is deemed to be included in the quantity surveyor's services under para. 8.*

2.3. Works of alteration

On works of alteration or repair, or on those sections of the works which are mainly works of alteration or repair, there shall be a fee of 1·0% in addition to the fee calculated in accordance with paras. 2.1 and 2.2.

2.4. Works of redecoration and associated minor repairs

On works of redecoration and associated minor repairs, there shall be a fee of 1·5% in addition to the fee calculated in accordance with paras. 2.1 and 2.2.

2.5. Bills of quantities prepared in special forms

Fees calculated in accordance with paras. 2.1, 2.2, 2.3, and 2.4 include for the preparation of bills of quantities on a normal trade basis. If the employer requires additional information to be provided in the bills of quantities or the bills to be prepared in an elemental, operational or similar form, then the fee may be adjusted by agreement between the employer and the quantity surveyor.

2.6. Reduction of tenders

(a) When cost planning services have been provided by the quantity surveyor and a tender, when received, is reduced before acceptance and if the reductions are not necessitated by amended instructions of the employer or by the inclusion in the bills of quantities of items which the quantity surveyor has indicated could not be contained within the approved estimate, then in such a case no charge shall be made by the quantity surveyor for the preparation of bills of reductions and the fee for the preparation of the bills of quantities shall be based on the amount of the reduced tender.

(b) When cost planning services have not been provided by the quantity surveyor and if a tender, when received, is reduced before acceptance, fees are to be calculated upon the amount of the unreduced tender. When the preparation of bills of reductions is required, a fee is chargeable for preparing such bills of reductions as follows:

QUANTITY SURVEYORS' FEES

(*i*) 2·0% upon the gross amount of all omissions requiring measurement or abstraction from original dimension sheets.

(*ii*) 3·0% upon the gross amount of all additions requiring measurement.

(*iii*) 0·5% upon the gross amount of all remaining additions.

NOTE: *The above scale for the preparation of bills of reductions applies to work in all categories.*

2.7. Generally

If the works are substantially varied at any stage or if the quantity surveyor is involved in an excessive amount of abortive work, then the fees shall be adjusted by agreement between the employer and the quantity surveyor.

3.0. NEGOTIATING TENDERS

3.1. (*a*) For negotiating and agreeing prices with a contractor:

Value of work £	Fee £	£
Up to 75,000	0·5%	
75,000– 300,000	375+0·3% on balance over 75,000	
300,000– 600,000	1,050+0·2% on balance over 300,000	
Over 600,000	1,650+0·1% on balance over 600,000	

(*b*) The fee shall be calculated on the total value of the works as defined in paras. 2.1 (*d*), (*e*), (*f*), (*g*) and (*j*).

(*c*) For negotiating and agreeing prices with a contractor for air conditioning, heating, ventilating and electrical services there shall be an additional fee as para. 3.1 (*a*) calculated on the total value of such services as defined in para. 2.2 (*b*).

CONSULTATIVE SERVICES AND PRICING BILLS OF QUANTITIES

4.1. Consultative services

Where the quantity surveyor is appointed to prepare approximate estimates, feasibility studies or submissions for the approval of financial grants or similar services, then the fee shall be based on the time involved (see paras. 19.1 and 19.2) or alternatively, on a lump sum or percentage basis agreed between the employer and the quantity surveyor.

4.2. Pricing bills of quantities

(*a*) For pricing bills of quantities, if instructed, to provide an estimate comparable with tenders, the fee shall be one-third ($33\frac{1}{3}$%) of the fee for negotiating and agreeing prices with a contractor, calculated in accordance with paras. 3.1 (*a*) and (*b*).

(*b*) For pricing bills of quantities, if instructed, to provide an estimate comparable with tenders for air conditioning, heating, ventilating and electrical services the fee shall be one-third ($33\frac{1}{3}$%) of the fee calculated in accordance with para. 3.1 (*c*).

CONTRACTS BASED ON BILLS OF QUANTITIES: POST-CONTRACT SERVICES

Alternative Scales (I and II) for post-contract services are set out below to be used at the quantity surveyor's discretion by prior agreement with the employer.

QUANTITY SURVEYORS' FEES

5.0. ALTERNATIVE I: OVERALL SCALE OF CHARGES FOR POST-CONTRACT SERVICES

5.1. If the quantity surveyor appointed to carry out the post-contract services did not prepare the bills of quantities then the fees in paras. 5.2 and 5.3 shall be increased to cover the additional services undertaken by the quantity surveyor.

5.2. **Basic scale**

For taking particulars and reporting valuations for interim certificates for payments on account to the contractor, preparing periodic assessments of anticipated final cost and reporting thereon, measuring and making up bills of variations including pricing and agreeing totals with the contractor, and adjusting fluctuations in the cost of labour and materials if required by the contract.

(*a*) *Category A:* Banks; churches and similar place of religious worship; clubs; council offices; court-houses; crematoria; houses; libraries; old people's homes not exceeding two storeys; petrol service stations; police stations; public houses and inns; sports pavilions; teaching laboratories and industrial laboratories of a like nature; theatres; town halls; universities, polytechnics and colleges of further education (other than halls of residence and hostels); and the like.

Value of work £	Category A fee £	£
Up to 75,000	75+2·0% (Minimum fee £400)	
75,000– 150,000	1,575+1·7% on balance over	75,000
150,000– 300,000	2,850+1·6% on balance over	150,000
300,000– 750,000	5,250+1·3% on balance over	300,000
750,000–1,500,000	11,100+1·2% on balance over	750,000
1,500,000–3,000,000	20,100+1·1% on balance over	1,500,000
Over 3,000,000	36,600+1·0% on balance over	3,000,000

(*b*) *Category B:* Ambulance and fire stations; canteens; church halls; cinemas; community centres; departmental stores; enclosed sports stadia and swimming baths; halls of residence; hospitals; hostels; industrial laboratories other than those included in Category A; motels; offices other than those included in Categories A and C; old people's homes exceeding two storeys; railway stations; recreation centres; residential hotels; restaurants; schools; self-contained flats and maisonettes not exceeding two storeys (including old people's accommodation of this nature); shops; telephone exchanges; and the like, being works of a simpler character than those in Category A or works with some element of repetition.

Value of work £	Category B fee £	£
Up to 75,000	75+2·0% (Minimum fee £400)	
75,000– 150,000	1,575+1·7% on balance over	75,000
150,000– 300,000	2,850+1·5% on balance over	150,000
300,000– 750,000	5,100+1·1% on balance over	300,000
750,000–1,500,000	10,050+1·0% on balance over	750,000
1,500,000–3,000,000	17,550+0·9% on balance over	1,500,000
Over 3,000,000	31,050+0·8% on balance over	3,000,000

QUANTITY SURVEYORS' FEES

(c) *Category C.* Factories; self-contained flats and maisonettes exceeding two storeys (including old people's accommodation of this nature); garages; multi-storey car parks; offices with open floor space to be finished to tenants' requirements (tenants' requirements to be Category A); open-air sports stadia and swimming baths; warehouses; and the like, being works containing little internal detail or works with a large amount of repetition.

N O T E : *When a building contract comprises a number of dissimilar buildings, with a resultant decrease in the amount of repetition, then the fees set out below may be adjusted by agreement between the employer and the quantity surveyor.*

Value of work £	Category C fee £	£
Up to 75,000	60+1·6% (Minimum fee £320)	
75,000– 150,000	1,260+1·5% on balance over	75,000
150,000– 300,000	2,385+1·4% on balance over	150,000
300,000– 750,000	4,485+1·1% on balance over	300,000
750,000–1,500,000	9,435+0·9% on balance over	750,000
1,500,000–3,000,000	16,185+0·8% on balance over	1,500,000
Over 3,000,000	28,185+0·7% on balance over	3,000,000

(d) The scales of fees for post-contract services (paras. 5.2 (a) to (c) are overall scales based upon the inclusion of all nominated sub-contractors' and nominated suppliers' accounts, subject to the provision of para. 5.2 (g). When work normally included in a building contract is the subject of a separate contract for which the quantity surveyor has not been paid fees under any other clause hereof, the value of such work shall be included in the amount on which fees are charged.

(e) Fees shall be calculated upon the basis of the account for the whole of the work, subject to the provisions of para. 5.3.

(f) In calculating the amount on which fees are charged the total of any credits is to be added to the amount described above.

(g) Where the value of air conditioning, heating, ventilating and electrical services included in the tender documents together exceeds 25% of the amount calculated as described in paras. 5.2 (d) and (e) above, then, subject to the provisions of para. 5.3, no fee is chargeable on the amount by which the value of these services exceeds the said 25%. In this context the term 'value' excludes general contractors' profit, attendance, builders work in connection with the services, preliminaries and any other similar additions.

(h) When a contract comprises buildings which fall into more than one category, the fee shall be calculated as follows:

    (i) The amount upon which fees are chargeable shall be allocated to the categories of work applicable and the amounts so allocated expressed as percentages of the total amount upon which fees are chargeable.

    (ii) Fees shall then be calculated for each category on the total amount upon which fees are chargeable.

    (iii) The fee chargeable shall then be calculated by applying the percentages of work in each category to the appropriate total fee and adding the resultant amounts.

QUANTITY SURVEYORS' FEES

(*j*) When a project is the subject of a number of contracts then, for the purposes of calculating fees, the values of such contracts shall not be aggregated but each contract shall be taken separately and the scale of charges (paras. 5.2 (*a*) to (*h*), applied as appropriate.

(*k*) When the quantity surveyor is required to prepare valuations of materials or goods off site, an additional fee shall be charged based on the time involved (see paras. 19.1 and 19.2).

(*l*) The basic scale for post-contract services includes for a simple routine of periodically estimating final costs. When the employer specifically requests a cost monitoring service which involves the quantity surveyor in additional or abortive measurement an additional fee shall be charged based on the time involved (see paras. 19.1 and 19.2), or alternatively on a lump sum or percentage basis agreed between the employer and the quantity surveyor.

(*m*) The above overall scales of charges for post contract services assume normal conditions when the bills of quantities are based on drawings accurately depicting the building work the employer requires. If the works are materially varied to the extent that substantial remeasurement is necessary then the fee for post-contract services shall be adjusted by agreement between the employer and the quantity surveyor.

5.3. Air conditioning, heating, ventilating and electrical services

(*a*) Where final accounts are prepared by the quantity surveyor for the air conditioning, heating, ventilating and electrical services there shall be a fee for these services, in addition to the fee calculated in accordance with para. 5.2, as follows:

Value of work £	Additional fee £		£
Up to 60,000	2·0%		
60,000– 120,000	1,200+1·6%	on balance over 60,000	
120,000– 500,000	2,160+1·25%	on balance over 120,000	
500,000–2,000,000	6,910+1·0%	on balance over 500,000	
Over 2,000,000	21,910+0·9%	on balance over 2,000,000	

(*b*) The values of such services, whether the subject of separate tenders or not, shall be aggregated and the total value of work so obtained used for the purpose of calculating the additional fee chargeable in accordance with para. (*a*).
(Except that when more than one firm of consulting engineers is engaged on the design of these services the separate values for which each such firm is responsible shall be aggregated and the additional fee charged shall be calculated independently on each such total value so obtained.)

(*c*) The scope of the services to be provided by the quantity surveyor under para. (*a*) above shall be deemed to be equivalent to those described for the basic scale for post-contract services.

(*d*) When the quantity surveyor is required to prepare periodic valuations of materials or goods off site, an additional fee shall be charged based on the time involved (see paras. 19.1 and 19.2).

(*e*) The basic scale for post-contract services includes for a simple routine of periodically estimating final costs. When the employer specifically requests a

QUANTITY SURVEYORS' FEES

cost monitoring service which involves the quantity surveyor in additional or abortive measurement an additional fee shall be charged based on the time involved (see paras. 48 and 49), or alternatively on a lump sum or percentage basis agreed between the employer and the quantity surveyor.

(*f*) Fees shall be calculated upon the basis of the account for the whole of the air conditioning, heating, ventilating and electrical services for which final accounts have been prepared by the quantity surveyor.

6.0. ALTERNATIVE II: SCALE OF CHARGES FOR SEPARATE STAGES OF POST-CONTRACT SERVICES

6.1. If the quantity surveyor appointed to carry out the post-contract services did not prepare the bills of quantities then the fees in paras. 6.2 and 6.3 shall be increased to cover the additional services undertaken by the quantity surveyor.

NOTE: *The scales of fees in paras. 6.2 and 6.3 apply to work in all categories (including air conditioning, heating, ventilating and electrical services).*

6.2. Valuations for interim certificates

(*a*) For taking particulars and reporting valuations for interim certificates for payments on account to the contractor.

Total of valuations £	Fee £	£
Up to 150,000	+0·5%	
150,000– 500,000	750+0·4% on balance over	150,000
500,000–3,000,000	2,150+0·3% on balance over	500,000
Over 3,000,000	9,650+0·2% on balance over	3,000,000

NOTES: *1 Subject to note 2 below, the fees are to be calculated on the total of all interim valuations (i.e. the amount of the final account less only the net amount of the final valuation).*

2 When consulting engineers are engaged in supervising the installation of air conditioning, heating, ventilating and electrical services and their duties include reporting valuations for inclusion in interim certificates for payments on account in respect of such services, then valuations so reported shall be excluded from any total amount of valuations used for calculating fees.

(*b*) When the quantity surveyor is required to prepare valuations of materials or goods off site, an additional fee shall be charged based on the time involved (see paras. 19.1 and 19.2).

6.3. Preparing accounts of variations upon contracts

For measuring and making up bills of variations including pricing and agreeing totals with the contractor:

(*a*) An initial lump sum of £300 shall be payable on each contract.

(*b*) 2·0% upon the gross amount of omissions requiring measurement or abstraction from the original dimension sheets.

(*c*) 3·0% upon the gross amount of additions requiring measurement and upon dayworks.

(*d*) 0·5% upon the gross amount of remaining additions which shall be deemed to include all nominated sub-contractors' and nominated suppliers' accounts which do not involve measurement or checking of quantities but only checking against lump sum estimates.

QUANTITY SURVEYORS' FEES

(e) 3·0% upon the aggregate of the amounts of the increases and/or decreases in the cost of labour and materials in accordance with any fluctuations clause in the conditions of contract, except where a price adjustment formula applies.

(f) On contracts where fluctuations are calculated by the use of a price adjustment formula method the following scale shall be applied to the account for the whole of the work:

Value of work £	Fee £	£
Up to 150,000	150+0·5%	
150,000–500,000	900+0·3% on balance over 150,000	
Over 500,000	1,950+0·1% on balance over 500,000	

(g) When consulting engineers are engaged in supervising the installation of air conditioning, heating, ventilating and electrical services and their duties include for the adjustment of accounts and pricing and agreeing totals with the sub-contractors for inclusion in the measured account, then any totals so agreed shall be excluded from any amounts used for calculating fees.

6.4. Cost monitoring services

The fee for providing all approximate estimates of final cost and/or a cost monitoring service shall be based on the time involved (see paras. 19.1 and 19.2), or alternatively on a lump sum or percentage basis agreed between the employer and the quantity surveyor.

7.0. BILLS OF APPROXIMATE QUANTITIES, INTERIM CERTIFICATES AND FINAL ACCOUNTS

7.1. Basic scale

For preparing bills of approximate quantities suitable for obtaining competitive tenders which will provide a schedule of prices and a reasonably close forecast of the cost of the works, but subject to complete remeasurement, examining tenders and reporting thereon, taking particulars and reporting valuations for interim certificates for payments on account to the contractor, preparing periodic assessments of anticipated final cost and reporting thereon, measuring and preparing final account, including pricing and agreeing totals with the contractor and adjusting fluctuations in the cost of labour and materials if required by the contract:

(a) *Category A.* Banks; churches and similar places of religious worship; clubs; council offices; court-houses; crematoria; houses; libraries; old people's homes not exceeding two storeys; petrol service stations; police stations; public houses and inns; sports pavilions; teaching laboratories and industrial laboratories of a like nature; theatres; town halls; universities, polytechnics and colleges of further education (other than halls of residence and hostels); and the like.

QUANTITY SURVEYORS' FEES

Value of work	*Category A fee*	
£	£	£
Up to 75,000	190+5·0% (Minimum fee £1,000)	
75,000– 150,000	3,940+4·0% on balance over	75,000
150,000– 300,000	6,940+3·4% on balance over	150,000
300,000– 750,000	12,040+2·8% on balance over	300,000
750,000–1,500,000	24,640+2·4% on balance over	750,000
1,500,000–3,000,000	42,640+2·2% on balance over	1,500,000
Over 3,000,000	75,640+2·0% on balance over	3,000 000

(b) *Category B.* Ambulance and fire stations; canteens; church halls; cinemas; community centres; departmental stores; enclosed sports stadia and swimming baths; halls of residence; hospitals; hostels; industrial laboratories other than those included in Category A; motels; offices other than those included in Categories A and C; old people's homes exceeding two storeys; railway stations; recreation centres; residential hotels; restaurants; schools; self-contained flats and maisonettes not exceeding two storeys (including old people's accommodation of this nature); shops; telephone exchanges; and the like, being works of a simpler character than those in Category A or works with some element of repetition.

Value of work	*Category B fee*	
£	£	£
Up to 75,000	180+4·8% (Minimum fee £960)	
75,000– 150,000	3,780+3·7% on balance over	75,000
150,000– 300,000	6,555+3·0% on balance over	150,000
300,000– 750,000	11,055+2·2% on balance over	300,000
750,000–1,500,000	20,955+2·0% on balance over	750,000
1,500,000–3,000,000	35,955+1·8% on balance over	1,500,000
Over 3,000,000	62,955+1·6% on balance over	3,000,000

(c) *Category C.* Factories; self-contained flats and maisonettes exceeding two storeys (including old people's accommodation of this nature); garages; multi-storey car parks; offices with open floor space to be finished to tenants' requirements (tenants' requirements to be Category A); open air sports stadia and swimming baths; warehouses; and the like, being works containing little internal detail or works with a large amount of repetition.

NOTE: *When a building contract comprises a number of dissimilar buildings, with a resultant decrease in the amount of repetition, then the fees set out below may be adjusted by agreement between the employer and the quantity surveyor.*

Value of work	*Category C fee*	
£	£	£
Up to 75,000	150+4·1% (Minimum fee £820)	
75,000– 150,000	3,225+3·3% on balance over	75,000
150,000– 300,000	5,700+2·6% on balance over	150,000
300,000– 750,000	9,600+2·0% on balance over	300,000
750,000–1,500,000	18,600+1·7% on balance over	750,000
1,500,000–3,000,000	31,350+1·5% on balance over	1,500,000
Over 3,000,000	53,850+1·3% on balance over	3,000,000

QUANTITY SURVEYORS' FEES

(*d*) The scales of fees for pre-contract and post-contract services (paras. 7.1 (*a*) to (*c*)) are overall scales based upon the inclusion of all nominated sub-contractors' and nominated suppliers' accounts, subject to the provision of para. 7·1 (*g*). When work normally included in a building contract is the subject of a separate contract for which the quantity surveyor has not been paid fees under any other clause hereof, the value of such work shall be included in the amount on which fees are charged.

(*e*) Fees shall be calculated upon the basis of the account for the whole of the work, subject to the provisions of para. 7.2.

(*f*) In calculating the amount on which fees are charged the total of any credits is to be added to the amount described above.

(*g*) Where the value of air conditioning, heating, ventilating and electrical services included in tender documents together exceeds 25% of the amount calculated as described in paras. 7.1 (*d*) and (*e*), then, subject to the provisions of para. 7.2 no fee is chargeable on the amount by which the value of these services exceeds the said 25%. In this context the term 'value' excludes general contractors' profit, attendance, builders' work in connection with the services, preliminaries and any other similar additions.

(*h*) When a contract comprises buildings which fall into more than one category, the fee shall be calculated as follows:

    (*i*) The amount upon which fees are chargeable shall be allocated to the categories of work applicable and the amount so allocated expressed as percentages of the total amount upon which fees are chargeable.

    (*ii*) Fees shall then be calculated for each category on the total amount upon which fees are chargeable.

    (*iii*) The fee chargeable shall then be calculated by applying the percentages of work in each category to the appropriate total fee and adding the resultant amounts.

(*j*) When a project is the subject of a number of contracts then, for the purposes of calculating fees, the values of such contracts shall not be aggregated but each contract shall be taken separately and the scale of charges (paras. 7.1 (*a*) to (*h*)) applied as appropriate.

(*k*) Where the quantity surveyor is specifically instructed to provide cost planning services, the fee calculated in accordance with paras. 7.1 (*a*) to (*j*) shall be increased by a sum calculated in accordance with the following table and based upon the same value of work as that upon which the aforementioned fee has been calculated:

Categories A & B: (as defined in paras. 7.1 (*a*) and (*b*))

Value of work £	Fee £	£
Up to　　300,000	0·7%	
300,000–1,500,000	2,100+0·4% on balance over	300,000
1,500,000–3,000,000	6,900+0·35% on balance over	1,500,000
Over　　3,000,000	12,150+0·3% on balance over	3,000,000

QUANTITY SURVEYORS' FEES

Category C: (as defined in para. 7.1 (*c*))

Value of work £	*Fee* £	£
Up to 300,000	0·5%	
300,000–1,500,000	1,500+0·3% on balance over 300,000	
1,500,000–3,000,000	5,100+0·25% on balance over 1,500,000	
Over 3,000,000	8,850+0·2% on balance over 3,000,000	

(*l*) When the quantity surveyor is required to prepare valuations of materials or goods off site, an additional fee shall be charged based on the time involved (see paras. 19.1 and 19.2).

(*m*) The basic scale for post-contract services includes for a simple routine of periodically estimating final costs. When the employer specifically requests a cost monitoring service which involves the quantity surveyor in additional or abortive measurement an additional fee shall be charged based on the time involved (see paras. 19.1 and 19.2), or alternatively on a lump sum or percentage basis agreed between the employer and the quantity surveyor.

7.2. Air conditioning, heating, ventilating and electrical services

(*a*) Where bills of approximate quantities and final accounts are prepared by the quantity surveyor for the air conditioning, heating, ventilating and electrical services there shall be a fee for these services in addition to the fee calculated in accordance with para. 7.1 as follows:

Value of work	*Additional fee*
Up to 60,000	4·5%
60,000– 120,000	2,700+3·85% on balance over 60,000
120,000– 240,000	5,010+3·25% on balance over 120,000
240,000– 375,000	8,910+3·0% on balance over 240,000
375,000– 500,000	12,960+2·5% on balance over 375,000
500,000–2,000,000	16,085+2·15% on balance over 500,000
Over 2,000,000	48,335+2·05% on balance over 2,000,000

(*b*) The value of such services, whether the subject of separate tenders or not, shall be aggregated and the total value of work so obtained used for the purpose of calculating the additional fee chargeable in accordance with para. (*a*).

(Except that when more than one firm of consulting engineers is engaged on the design of these services, the separate values for which each such firm is responsible shall be aggregated and the additional fees charged shall be calculated independently on each such total value so obtained.)

(*c*) The scope of the services to be provided by the quantity surveyor under para. (*a*) above shall be deemed to be equivalent to those described for the basic scale for pre-contract and post-contract services.

(*d*) When the quantity surveyor is required to prepare valuations of materials or goods off site, an additional fee shall be charged based on the time involved (see paras. 19.1 and 19.2).

QUANTITY SURVEYORS' FEES

(*e*) The basic scale for post-contract services includes for a simple routine of periodically estimating final costs. When the employer specifically requests a cost monitoring service, which involves the quantity surveyor in additional or abortive measurement, an additional fee shall be charged based on the time involved (see paras. 19.1 and 19.2), or alternatively on a lump sum or percentage basis agreed between the employer and the quantity surveyor.

(*f*) Fees shall be calculated upon the basis of the account for the whole of the air conditioning, heating, ventilating and electrical services for which final accounts have been prepared by the quantity surveyor.

(*g*) When cost planning services are provided by the quantity surveyor for air conditioning, heating, ventilating and electrical services (or for any part of such services) there shall be an additional fee based on the time involved (see paras. 19.1 and 19.2) or alternatively on a lump sum or percentage basis agreed between the employer and the quantity surveyor.

NOTE: *The incorporation of figures for air conditioning, heating, ventilating and electrical services provided by the consulting engineer is deemed to be included in the quantity surveyor's services under para. 7.1.*

7.3. Works of alteration

On works of alteration or repair, or on those sections of the work which are mainly works of alteration or repair, there shall be a fee of 1·0% in addition to the fee calculated in accordance with paras. 7.1 and 7.2.

7.4. Works of redecoration and associated minor repairs

On works of redecoration and associated minor repairs, there shall be a fee of 1·5% in addition to the fee calculated in accordance with paras. 7.1 and 7.2.

7.5. Bills of quantities and/or final accounts prepared in special forms

Fees calculated in accordance with paras. 7.1, 7.2, 7.3, and 7.4 include for the preparation of bills of quantities and/or final accounts on a normal trade basis. If the employer requires additional information to be provided in the bills of quantities and/or final accounts or the bills and/or final accounts to be prepared in an elemental, operational or similar form, then the fee may be adjusted by agreement between the employer and the quantity surveyor.

7.6. Reduction of tenders

(*a*) When cost planning services have been provided by the quantity surveyor and a tender, when received, is reduced before acceptance and if the reductions are not necessitated by amended instructions of the employer or by the inclusion in the bills of approximate quantities of items which the quantity surveyor has indicated could not be contained within the approved estimate, then in such a case no charge shall be made by the quantity surveyor for the preparation of bills of reductions and the fee for the preparation of bills of approximate quantities shall be based on the amount of the reduced tender.

QUANTITY SURVEYORS' FEES

(*b*) When cost planning services have not been provided by the quantity surveyor and if a tender, when received, is reduced before acceptance, fees are to be calculated upon the amount of the unreduced tender. When the preparation of bills of reductions is required, a fee is chargeable for preparing such bills of reductions as follows:

(*i*) 2·0% upon the gross amount of all omissions requiring measurement or abstraction from original dimension sheets.
(*ii*) 3·0% upon the gross amount of all additions requiring measurement.
(*iii*) 0·5% upon the gross amount of all remaining additions.

N O T E : *The above scale for the preparation of bills of reductions applies to work in all categories.*

7.7. Generally

If the works are substantially varied at any stage or if the quantity surveyor is involved in an excessive amount of abortive work, then the fees shall be adjusted by agreement between the employer and the quantity surveyor.

8.0. NEGOTIATING TENDERS

8.1. (*a*) For negotiating and agreeing prices with a contractor:

Value of work £	*Fee* £	£
Up to 75,000	0·5%	
75,000–300,000	375+0·3% on balance over	75,000
300,000–600,000	1,650+0·2% on balance over	300,000
Over 600,000	1,650+0·1% on balance over	600,000

(*b*) The fee shall be calculated on the total value of the works as defined in paras. (*d*), (*e*), (*f*), (*g*) and (*j*).
(*c*) For negotiating and agreeing prices with a contractor for air conditioning, heating, ventilating and electrical services there shall be an additional fee as para. 8.1 (*a*) calculated on the total value of such services as defined in para. 7.2 (*b*).

9.0. CONSULTATIVE SERVICES AND PRICING BILLS OF APPROXIMATE QUANTITIES

9.1. Consultative services

Where the quantity surveyor is appointed to prepare approximate estimates, feasibility studies or submissions for the approval of financial grants or similar services, then the fee shall be based on the time involved (see paras. 19.1 and 19.2) or alternatively, on a lump sum or percentage basis agreed between the employer and the quantity surveyor.

9.2. Pricing bills of approximate quantities

For pricing bills of approximate quantities, if instructed, to provide an estimate comparable with tenders, the fees shall be the same as for the corresponding services in paras. 4.2 (*a*) and (*b*).

QUANTITY SURVEYORS' FEES

10.0. INSTALMENT PAYMENTS

10.1. For the purpose of instalment payments the fee for the preparation of bills of approximate quantities only shall be the equivalent of forty per cent (40%) of the fees calculated in accordance with the appropriate sections of paras. 7.1 to 7.5 and the fee for providing cost planning services shall be in accordance with the appropriate sections of para. 7.1 (*k*); both fees shall be based on the total value of the bills of approximate quantities ascertained in accordance with the provisions of para. 2.1 (*e*).

10.2. In the absence of agreement to the contrary, fees shall be paid by instalments as follows:

(*a*) Upon acceptance by the employer of a tender for the works the above defined fees for the preparation of bills of approximate quantities and for providing cost planning services.

(*b*) In the event of no tender being accepted, the aforementioned fees shall be paid within three months of completion of the bills of approximate quantities.

(*c*) The balance by instalments at intervals to be agreed between the date of the first certificate and one month after final certification of the contractor's account.

10.3. In the event of the project being abandoned at any stage other than those covered by the foregoing, the proportion of fee payable shall be by agreement between the employer and the quantity surveyor.

11.0. SCHEDULES OF PRICES

11.1. The fee for preparing, pricing and agreeing schedules of prices shall be based on the time involved (see paras. 19.1 and 19.2). Alternatively, the fee may be on a lump sum or percentage basis agreed between the employer and the quantity surveyor.

12.0. COST PLANNING AND APPROXIMATE ESTIMATES

12.1. The fee for providing cost planning services or for preparing approximate estimates shall be based on the time involved (see paras. 19.1 and 19.2). Alternatively, the fee may be on a lump sum or percentage basis agreed between the employer and the quantity surveyor.

CONTRACTS BASED ON SCHEDULES OF PRICES: POST-CONTRACT SERVICES

13.0. FINAL ACCOUNTS

13.1. **Basic scale**

(*a*) For taking particulars and reporting valuations for interim certificates for payments on account to the contractor, preparing periodic assessments of anticipated final cost and reporting thereon, measuring and preparing final account including pricing and agreeing totals with the contractor, and adjusting fluctuations in the cost of labour and materials if required by the contract, the fee shall be equivalent to sixty per cent (60%) of the fee calculated in accordance with paras. 7.1 (*a*) to (*j*).

QUANTITY SURVEYORS' FEES

(*b*) When the quantity surveyor is required to prepare valuations of materials or goods off site, an additional fee shall be charged on the basis of the time involved (see paras. 19.1 and 19.2).

(*c*) The basic scale for post-contract services includes for a simple routine of periodically estimating final costs. When the employer specifically requests a cost monitoring service, which involves the quantity surveyor in additional or abortive measurement, an additional fee shall be charged based on the time involved (see paras. 19.1 and 19.2), or alternatively on a lump sum or percentage basis agreed between the employer and the quantity surveyor.

13.2. Air conditioning, heating, ventilating and electrical services

Where final accounts are prepared by the quantity surveyor for the air conditioning, heating, ventilating and electrical services there shall be a fee for these services, in addition to the fee calculated in accordance with para. 13.1, equivalent to sixty per cent (60%) of the fee calculated in accordance with paras. 7.2 (*a*) to (*f*).

13.3. Works of alterations

On works of alteration or repair, or on those sections of the work which are mainly works of alteration or repair, there shall be a fee of 1·0% in addition to the fee calculated in accordance with paras. 13.1 and 13.2.

13.4. Works of redecoration and associated minor repairs

On works of redecoration and associated minor repairs, there shall be a fee of 1·5% in addition to the fee calculated in accordance with paras. 13.1 and 13.2.

13.5. Final accounts prepared in special forms

Fees calculated in accordance with paras. 38, 39, 40 and 41 include for the preparation of final accounts on a normal trade basis. If the employer requires additional information to be provided in the final accounts or the accounts to be prepared in an elemental, operational or similar form, then the fee may be adjusted by agreement between the employer and the quantity surveyor.

14.0. COST PLANNING

14.1.
The fee for providing a cost planning service shall be based on the time involved (see paras. 19.1 and 19.2). Alternatively, the fee may be on a lump sum or percentage basis agreed between the employer and the quantity surveyor.

15.0. ESTIMATES OF COST

15.1.
For preparing an approximate estimate, calculated by measurement, of the cost of work, and, if required under the terms of the contract, negotiating, adjusting and agreeing the estimate:

Value of work £	Fee £		£
Up to 15,000	1·25%		
15,000– 75,000	187+1·0%	on balance over	15,000
75,000–300,000	787+0·75%	on balance over	75,000
Over 300,000	2,475+0·5%	on balance over	300,000

QUANTITY SURVEYORS' FEES

(*b*) The fee shall be calculated upon the total of the approved estimates.

16.0. FINAL ACCOUNTS

16.1. (*a*) For checking prime costs, reporting for interim certificates for payments on account to the contractor and preparing final accounts:

Value of work £		Fee £		£
Up to	15,000	2·5%		
15,000–	75,000	375+2·0%	on balance over	15,000
75,000–	300,000	1,575+1·5%	on balance over	75,000
Over	300,000	4,950+1·25%	on balance over	300,000

(*b*) The fee shall be calculated upon the total of the final account with the addition of the value of credits received for old materials removed and less the value of any work charged for in accordance with para. 16.1 (*c*).

(*c*) On the value of any work to be paid for on a measured basis, the fee shall be 3%.

(*d*) When the quantity surveyor is required to prepare valuations of materials or goods off site, an additional fee shall be charged based on the time involved (see paras. 19.1 and 19.2).

(*e*) The above charges do not include the provision of checkers on the site. If the quantity surveyor is required to provide such checkers an additional charge shall be made by arrangement.

17.0. COST REPORTING AND MONITORING SERVICES

17.1. The fee for providing cost reporting and/or monitoring services (e.g. preparing periodic assessments of anticipated final costs and reporting thereon) shall be based on the time involved (see paras. 19.1 and 19.2) or alternatively, on a lump sum or percentage basis agreed between the employer and the quantity surveyor.

18.0. ADDITIONAL SERVICES

18.1. For additional services not normally necessary, such as those arising as a result of the termination of a contract before completion, liquidation, fire damage to the buildings, services in connection with arbitration, litigation and investigation of the validity of contractors' claims, services in connection with taxation matters and all similar services where the employer specifically instructs the quantity surveyor, the charges shall be in accordance with paras. 19.1 and 19.2.

19.0. TIME CHARGES

19.1. (*a*) For consultancy and other services performed by a principal, a fee by arrangement according to the circumstances including the professional status and qualifications of the quantity surveyor.

(*b*) When a principal does work which would normally be done by a member of staff, the charge shall be calculated as para. 19.2 below.

QUANTITY SURVEYORS' FEES

19.2. (*a*) For services by a member of staff, the charges for which are to be based on the time involved, such charges shall be calculated on the hourly cost of the individual involved plus 145%.

(*b*) A member of staff shall include a principal doing work normally done by an employee (as para. 19.1 (*b*) above), technical and supporting staff, but shall exclude secretarial staff or staff engaged upon general administration.

(*c*) For the purpose of para 19.2 (*b*) above, a principal's time shall be taken at the rate applicable to a senior assistant in the firm.

(*d*) The supervisory duties of a principal shall be deemed to be included in the addition of 145% as para. 19.2 (*a*) above and shall not be charged separately.

(*e*) The hourly cost to the employer shall be calculated by taking the sum of the annual cost of the member of staff of:

(*i*) Salary and bonus but excluding expenses;

(*ii*) Employer's contributions payable under any Pension and Life Assurance Schemes;

(*iii*) Employer's contributions made under the National Insurance Acts, the Redundancy Payments Act and any other payments made in respect of the employee by virtue of any statutory requirements; and

(*iv*) Any other payments or benefits made or granted by the employer in pursuance of the terms of employment of the member of staff;

and dividing by 1,650.

19.3. The foregoing Time Charges under paras. 19.1 and 19.2 are intended for use where other paragraphs of the Scale (not related to Time Charges) form a significant proportion of the overall fee. In all other cases an increased time charge may be agreed.

20.0. INSTALMENT PAYMENTS

20.1. In the absence of agreement to the contrary, payments to the quantity surveyor shall be made by instalments by arrangement between the employer and the quantity surveyor.

QUANTITY SURVEYORS' FEES

Scale 36 inclusive scale of professional charges for quantity surveying services for building works issued by the Royal Institution of Chartered Surveyors. The scale is recommended and not mandatory.

EFFECTIVE FROM MARCH 1980

1.0. GENERALLY

1.1. This scale is for use when an inclusive scale of professional charges is considered to be appropriate by mutual agreement between the employer and the quantity surveyor.

1.2. This scale does not apply to civil engineering works, housing schemes financed by local authorities and the Housing Corporation and housing improvement work for which separate scales of fees have been published.

1.3. The fees cover quantity surveying services as may be required in connection with a building project irrespective of the type of contract from initial appointment to final certification of the contractor's account such as:

(*a*) Budget estimating; cost planning and advice on tendering procedures and contract arrangements.

(*b*) Preparing tendering documents for main contract and specialist sub-contracts; examining tenders received and reporting thereon or negotiating tenders and pricing with a selected contractor and or sub-contractors.

(*c*) Preparing recommendations for interim payments on account to the contractor; preparing periodic assessments of anticipated final cost and reporting thereon; measuring work and adjusting variations in accordance with the terms of the contract and preparing final account, pricing same and agreeing totals with the contractor.

(*d*) Providing a reasonable number of copies of bills of quantities and other documents; normal travelling and other expenses. Additional copies of documents, abnormal travelling and other expenses (e.g. in remote areas or overseas) and the provision of checkers on site shall be charged in addition by prior arrangement with the employer.

1.4. If any of the materials used in the works are supplied by the employer or charged at a preferential rate, then the actual or estimated market value thereof shall be included in the amounts upon which fees are to be calculated.

1.5. If the quantity surveyor incurs additional costs due to exceptional delays in building operations or any other cause beyond the control of the quantity surveyor then the fees may be adjusted by agreement between the employer and the quantity surveyor to cover the reimbursement of these additional costs.

1.6. The fees and charges are in all cases exclusive of value added tax which will be applied in accordance with legislation.

1.7. Copyright in bills of quantities and other documents prepared by the quantity surveyor is reserved to the quantity surveyor.

2.0. INCLUSIVE SCALE

2.1. The fees for the services outlined in para. 3, subject to the provision of para. 9, shall be as follows:

QUANTITY SURVEYORS' FEES

(a) *Category A.* Banks; churches and similar places of religious worship; clubs; council offices; court-houses; crematoria; houses; libraries; old people's homes not exceeding two storeys; petrol service stations; police stations; public houses and inns; sports pavilions; teaching laboratories and industrial laboratories of a like nature; theatres; town halls; universities, polytechnics and colleges of further education (other than halls of residence and hostels); and the like.

Value of work £	*Category A fee* £	£
Up to 75,000	190+6·0% (Minimum fee £1,200)	
75,000– 150,000	4,690+5·0% on balance over	75,000
150,000– 300,000	8,440+4·3% on balance over	150,000
300,000– 750,000	14,890+3·4% on balance over	300,000
750,000–1,500,000	30,190+3·0% on balance over	750,000
1,500,000–3,000,000	52,690+2·8% on balance over	1,500,000
Over 3,000,000	94,690+2·4% on balance over	3,000,000

(b) *Category B.* Ambulance and fire stations; canteens; church halls; cinemas; community centres; departmental stores; enclosed sports stadia and swimming baths; halls of residence; hospitals; hostels; industrial laboratories other than those included in Category A; motels; offices other than those included in Categories A and C; old people's homes exceeding two storeys; railway stations; recreation centres; residential hotels; restaurants; schools; self-contained flats and maisonettes not exceeding two storeys (including old people's accommodation of this nature); shops; telephone exchanges; and the like, being works of a simpler character than those in Category A or works with some element of repetition.

Value of work £	*Category B fee* £	£
Up to 75,000	180+5·8% (Minimum fee £1,160)	
75,000– 150,000	4,530+4·7% on balance over	75,000
150,000– 300,000	8,055+3·9% on balance over	150,000
300,000– 750,000	13,905+2·8% on balance over	300,000
750,000–1,500,000	26,505+2·6% on balance over	750,000
1,500,000–3,000,000	46,005+2·4% on balance over	1,500,000
Over 3,000,000	82,005+2·0% on balance over	3,000,000

(c) *Category C.* Factories; self-contained flats and maisonettes exceeding two storeys (including old people's accommodation of this nature); garages; multi-storey car parks; offices with open floor space to be finished to tenants' requirements (tenants' requirements to be Category A); open-air sports stadia and swimming baths; warehouses; and the like, being works containing little internal detail or works with a large amount of repetition.

NOTE : *When a building contract comprises a number of dissimilar buildings, with a resultant decrease in the amount of repetition, then the fees set out below may be adjusted by agreement between the employer and the quantity surveyor.*

QUANTITY SURVEYORS' FEES

Value of work £	Category C fee £	£
Up to 75,000	150+4·9% (Minimum fee £980)	
75,000– 150,000	3,825+4·1% on balance over 75,000	
150,000– 300,000	6,900+3·3% on balance over 150,000	
300,000– 750,000	11,850+2·5% on balance over 300,000	
750,000–1,500,000	23,100+2·2% on balance over 750,000	
1,500,000–3,000,000	39,600+2·0% on balance over 1,500,000	
Over 3,000,000	69,600+1·6% on balance over 3,000,000	

(*d*) Fees shall be calculated upon the total of the final account for the whole of the work including all nominated sub-contractors' and nominated supplier's accounts. When work normally included in a building contract is the subject of a separate contract for which the quantity surveyor has not been paid fees under any other clause hereof, the value of such work shall be included in the amount upon which fees are charged.

(*e*) When a contract comprises buildings which fall into more than one category, the fee shall be calculated as follows:

  (*i*) The amount upon which fees are chargeable shall be allocated to the categories of work applicable and the amounts so allocated expressed as percentages of the total amount upon which fees are chargeable.

  (*ii*) Fees shall then be calculated for each category on the total amount upon which fees are chargeable.

  (*iii*) The fee chargeable shall then be calculated by applying the percentages of work in each category to the appropriate total fee and adding the resultant amounts.

  (*iv*) A consolidated percentage fee applicable to the total value of the work may be charged by prior agreement between the employer and the quantity surveyor. Such a percentage shall be based on this scale and on the estimated cost of the various categories of work and calculated in accordance with the principles stated above.

(*f*) When a project is the subject of a number of contracts then, for the purpose of calculating fees, the values of such contracts shall not be aggregated but each contract shall be taken separately and the scale of charges (paras. 2.1 (*a*) to (*e*)) applied as appropriate.

2.2. Air conditioning, heating, ventilating and electrical services

(*a*) When the services outlined in para. 1.3 are provided by the quantity surveyor for the air conditioning, heating, ventilating and electrical services there shall be a fee for these services in addition to the fee calculated in accordance with para. 2.1 as follows:

Value of work £	Additional fee £	£
Up to 60,000	5·0%	
60,000– 120,000	3,000+4·7% on balance over 60,000	
120,000– 240,000	5,820+4·0% on balance over 120,000	
240,000– 375,000	10,620+3·6% on balance over 240,000	
375,000– 500,000	15,480+3·0% on balance over 375,000	
500,000–2,000,000	19,230+2·7% on balance over 500,000	
Over 2,000,000	59,730+2·4% on balance over 2,000,000	

QUANTITY SURVEYORS' FEES

(*b*) The value of such services, whether the subject of separate tenders or not, shall be aggregated and the total value of work so obtained used for the purpose of calculating the additional fee chargeable in accordance with para. (*a*). (Except that when more than one firm of consulting engineers is engaged on the design of these services, the separate values for which each such firm is responsible shall be aggregated and the additional fees charged shall be calculated independently on each such total value so obtained.)

(*c*) Fees shall be calculated upon the basis of the account for the whole of the air conditioning, heating, ventilating and electrical services for which bills of quantities and final accounts have been prepared by the quantity surveyor.

2.3. Works of alteration

On works of alteration or repair, or on those sections of the work which are mainly works of alteration or repair, there shall be a fee of 1·0% in addition to the fee calculated in accordance with paras. 2.1 and 2.2.

2.4. Works of redecoration and associated minor repairs

On works of redecoration and associated minor repairs, there shall be a fee of 1·5% in addition to the fee calculated in accordance with paras. 2.1 and 2.2.

2.5. Generally

If the works are substantially varied at any stage or if the quantity surveyor is involved in an excessive amount of abortive work, then the fees shall be adjusted by agreement between the employer and the quantity surveyor.

3.0. ADDITIONAL SERVICES

3.1. For additional services not normally necessary, such as those arising as a result of the termination of a contract before completion, liquidation, fire damage to the buildings, services in connection with arbitration, litigation and investigation of the validity of contractors' claims, services in connection with taxation matters and all similar services where the employer specifically instructs the quantity surveyor, the charges shall be in accordance with para. 4.1 below.

4.0. TIME CHARGES

4.1. (*a*) For consultancy and other services performed by a principal, a fee by arrangement according to the circumstances including the professional status and qualifications of the quantity surveyor.

(*b*) When a principal does work which would normally be done by a member of staff, the charge shall be calculated as para. 4.2 below.

4.2. (*a*) For services by a member of staff, the charges for which are to be based on the time involved, such charges shall be calculated on the hourly cost of the individual involved plus 145%.

(*b*) A member of staff shall include a principal doing work normally done by an employee (as para. 4.1 (*b*) above), technical and supporting staff, but shall exclude secretarial staff or staff engaged upon general administration.

QUANTITY SURVEYORS' FEES

(*c*) For the purpose of para. 4.2 (*b*) above, a principal's time shall be taken at the rate applicable to a senior assistant in the firm.

(*d*) The supervisory duties of a principal shall be deemed to be included in the addition of 145% as para. 4.2 (*a*) above and shall not be charged separately.

(*e*) The hourly cost to the employer shall be calculated by taking the sum of the annual cost of the member of staff of:

    (*i*) Salary and bonus but excluding expenses;

    (*ii*) Employer's contributions payable under any Pension and Life Assurance Schemes;

    (*iii*) Employer's contributions made under the National Insurance Acts, the Redundancy Payments Act and any other payments made in respect of the employee by virtue of any statutory requirements; and

    (*iv*) Any other payments or benefits made or granted by the employer in pursuance of the terms of employment of the member of staff;

and dividing by 1,650.

5.0. INSTALMENT PAYMENTS

5.1. In the absence of agreement to the contrary, fees shall be paid by instalments as follows:

(*a*) Upon acceptance by the employer of a tender for the works, one half of the fee calculated on the amount of the accepted tender.

(*b*) The balance by instalments at intervals to be agreed between the date of the first certificate and one month after final certification of the contractor's account.

5.2. (*a*) In the event of no tender being accepted, one half of the fee shall be paid within three months of completion of the tender documents. The fee shall be calculated upon the basis of the lowest original bona fide tender received. In the event of no tender being received, the fee shall be calculated upon a reasonable valuation of the works based upon the tender documents.

N O T E : *In the foregoing context 'bona fide tender' shall be deemed to mean a tender submitted in good faith without major errors of computation and not subsequently withdrawn by the tenderer.*

(*b*) In the event of the project being abandoned at any stage other than those covered by the foregoing, the proportion of fee payable shall be by agreement between the employer and the quantity surveyor.

CONSULTING ENGINEERS' FEES

CONDITIONS OF ENGAGEMENT – DECEMBER 1970 PRINTING WITH MAY 1977 AMENDMENTS

1. PREAMBLE

The following paragraphs describe the scope of professional services provided by the consulting engineer and give general advice about his appointment.

The Association of Consulting Engineers has drawn up standard Conditions of Engagement to form the basis of the agreement between the client and the consulting engineer, for five different types of appointment as hereafter described. Each of the standard Conditions of Engagement is preceded by a recommended Memorandum of Agreement. These two documents, taken together, constitute the recommended form of Agreement in each case.

When the standard Conditions of Engagement are not so used, it is in the interests of both the client and the consulting engineer that there should be an exchange of letters defining the duties which the consulting engineer is to perform and the terms of payment.

All fees and charges set out in this Agreement and Schedule are exclusive of Value Added Tax, the amount of which, at the rate and in the manner prescribed by law, shall be paid by the Client to the Consulting Engineer. Where Value Added Tax is chargeable on Disbursements and Out-of-pocket Expenses, this will be based upon the VAT-exclusive cost of such outgoings.

2. REPORTS AND ADVISORY WORK

For reports, and for advisory work, the services required from the consulting engineer will usually comprise one or more of the following:

(*a*) Investigating and advising on a project and submitting a report thereon. The consulting engineer may be asked to examine alternatives; review all technical aspects; make an economic appraisal of costs and benefits; draw conclusions and make recommendations.

(*b*) Inspecting an existing structure (e.g. a reservoir or a building or an installation) and reporting thereon. If the client requires continuing advice on maintenance or operation of an existing project, the consulting engineer may be appointed to make periodic visits.

(*c*) Making a special investigation of an engineering problem and reporting thereon.

(*d*) Making valuations of plant and undertakings.

Payment for the services provided under Items (*a*) to (*d*) above should normally be on a time basis, as described in Section 10. It is further recommended that the conditions governing an appointment under Item (*a*) above should be based on:

Schedule 1 – Conditions of Engagement for Report and Advisory Work.

3. DESIGN AND SUPERVISION OF CONSTRUCTION

When the client has approved the proposals recommended in the report (see Section 2 (*a*) above) and has decided to proceed with the construction of engineering works or the installation of plant it is normal practice for the consulting engineer who prepared or assisted in the preparation of the Report to be appointed for the subsequent stages of the work or the relevant part thereof. The consulting engineer will assist the client in obtaining the requisite approval, then prepare the designs and tender documents to enable competitive tenders to be obtained or orders to be placed, and will be responsible for the technical control and administration of the construction of the works.

CONSULTING ENGINEERS' FEES

It is recommended that the agreement for this type of appointment should be based on the most appropriate of the following:

Schedule 2 – Conditions of Engagement for the Design and Supervision of Civil, Mechanical and Electrical Works.

Schedule 3 – Conditions of Engagement for the Design and Supervision of Structural Engineering Work in Buildings and Other Structures, where an Architect is Appointed by the Client.

Schedule 4 – Conditions of Engagement for the Design and Supervision of Engineering Systems (formerly referred to as 'Engineering Services') in Buildings and Other Projects.

4. VARIATION IN TERMS OF PAYMENT FOR DESIGN AND SUPERVISION OF CONSTRUCTION

The Association of Consulting Engineers believes that, when the circumstances are normal and the design work is of average complexity, competent and responsible engineering services cannot be provided at a level of remuneration lower than that represented by the scales of percentage charges shown in the standard A.C.E. Conditions of Engagement. Variation in these terms may, however, be necessary in special circumstances. For example, the following would justify an increase in the charge for services:

(a) The design work is of an unusually complex character.

(b) The Works are to be constructed by means of an abnormally large number of separate contracts.

(c) A substantial proportion of the project involves alterations or additions to existing structures, plant or services.

(d) The completion of the project is retarded through circumstances over which the consulting engineer has no control – e.g. by the client in order to spread the aggregate cost over an extended period. If the project is to be built in stages, the definition of 'Works' in the standard A.C.E. Conditions of Engagement should apply separately to each stage.

The following would justify a reduction in the charge for services:

(e) The design work is unusually simple or repetitive in character.

(f) The fact that part of the normal services set out in the applicable standard A.C.E. Conditions of Engagement has already been provided by the consulting engineer (e.g. included in a project report, as described in Section 2) and has been paid for by the client. In such a case an appropriate reduction in the charge for services should be agreed between the client and the consulting engineer to take account only of those earlier services, directly relevant to the Works and not to alternative schemes, which would otherwise have to be performed as part of the normal services.

6. PARTIAL SERVICES

When the client wishes to appoint the consulting engineer for partial services only, and not for both design and supervision of construction, it is important that both parties recognize the limitation which such an appointment places upon the professional responsibility of the consulting engineer who cannot be held liable for matters that are outside his control and knowledge.

CONSULTING ENGINEERS' FEES

The terms of reference for the appointment should be carefully drawn up and the relevant A.C.E. Conditions of Engagement should be adapted to suit the scope of services required.

Professional charges for partial services are usually best calculated on the time basis described in Section 10, but may, in suitable cases, be a commensurate part of the percentage charge for normal services shown in the standard Conditions of Engagement.

Examples of partial service are described in the two succeeding Sections 7 and 8.

7. INSPECTION SERVICES

The inspection of materials and plant during manufacture or on site is usually required during the construction stage of a project. Consequently, the arrangements for this service are described in the standard Conditions of Engagement referred to in Section 3. If, however, the client wishes to engage the consulting engineer to provide only inspection services, the professional charges of the consulting engineer may be either a percentage of the cost of the materials to be inspected, or on a time basis as described in Section 10.

8. INDUSTRIALIZED BUILDING

A client who has decided to employ a building contractor for the design and construction of an industrialized building may require the services of a consulting engineer for advice on the selection of a contractor, for checking the design and for supervising construction. It is recommended that the agreement for this type of appointment should be based on:

Schedule 5 – Conditions of Engagement for Structural Engineering Services in Connection with Industrialized Building.

9. ACTING AS ARBITRATOR, UMPIRE OR EXPERT WITNESS

A consulting engineer may be appointed to act as arbitrator or umpire, or be required to attend as an expert witness at Parliamentary Committees, Courts of Law, Arbitrations or Official Inquiries.

Payment for any of these services should be on the basis of a lump sum retainer plus time charges and expenses as described in Section 10, not less than three hours per day being chargeable for attendance, however short, either before or after a mid-day adjournment.

10. PAYMENT ON A TIME BASIS

When it is impossible to estimate in advance the duration and extent of the consulting engineer's services, neither a lump sum payment alone nor a percentage of the estimated construction cost would normally be a fair basis of remuneration. The most satisfactory and equitable method of payment in these cases makes allowance for the actual time occupied in providing the services required, and comprises the following elements, as applicable:

10.1. A charge in the form of hourly rate(s) for the services of a partner or a consultant of the firm. The hourly rate will depend upon his standing, the nature of the work and any special circumstances. Alternatively a lump sum fee may be charged instead of the said hourly rate(s).

10.2. A charge which covers technical and supporting staff salary and other payroll costs actually incurred by the consulting engineer, together with a fair proportion of his

CONSULTING ENGINEERS' FEES

overhead costs, plus an element of profit. This charge is most conveniently calculated by applying a multiplier to the salary cost and then adding the net amount of other payroll costs. For technical and supporting staff working in or based on the consulting engineer's office, a multiplier of 2·5 is normally applicable. The major part of the multiplier is attributable to the consulting engineer's overheads which may include, *inter alia*, the following indirect costs and expenses:

(a) rent, rates and other expenses of up-keep of his office, its furnishings, equipment and supplies;
(b) insurance premiums other than those recovered in the payroll cost;
(c) administrative, accounting, secretarial and financing costs;
(d) the expense of keeping abreast of advances in engineering;
(e) the expense of preliminary arrangements for new or prospective projects;
(f) loss of productive time of technical staff between assignments.

10.3. A charge for use of a computer or other special equipment.
10.4. Disbursements made by the consulting engineer which can be directly identified with the work.

When calculating amounts chargeable on a time basis, a consulting engineer is entitled to include time spent by partners, consultants and the technical and supporting staff in travelling in connection with the performance of the services. The time spent by secretarial staff or by staff engaged on general accountancy or administration duties in the consulting engineer's office is not chargeable unless otherwise agreed.

11. APPOINTMENTS OUTSIDE THE UNITED KINGDOM

The standard A.C.E. Conditions of Engagement are suitable for appointments in the United Kingdom. For work outside the United Kingdom the arrangement and wording of the clauses will in general be found suitable, but modifications and supplementary clauses may be needed to suit the circumstances and locality of the work.

GENERAL CONDITIONS AND SCALE OF CHARGES FOR WORK CARRIED OUT UNDER SCHEDULE 1

1. Definitions

'Salary Cost' means the annual salary including bonuses of any person employed by the consulting engineer, divided by 1,600 (being deemed to be the average annual total of effective working hours of an employee) and multiplied by the number of working hours spent by such person in performing any of the services in respect of which payment under this Agreement is to be made to the consulting engineer upon the basis of salary cost. For the purposes of this definition the annual salary of a person employed by the consulting engineer for a period less than a full year shall be calculated pro rata to such person's salary (including bonuses) for such lesser period.

CONSULTING ENGINEERS' FEES

'Other Payroll Cost' means the annual amount of all contributions and payments made by the consulting engineer on behalf of or in respect of a person employed by him for staff pension and life assurance schemes, and also for National Health Insurance, Graduated Pension Fund, Selective Employment Tax and for any other tax, charge, levy, impost or payment of any kind whatsoever which the consulting engineer at any time during the performance of this Agreement is obliged by law to make on behalf of or in respect of such person, divided by 1,600 (being deemed to be the average annual total of effective working hours of an employee) and multiplied by the number of working hours spent by such person in performing any of the services in respect of which payment under this Agreement is to be made to the consulting engineer upon the basis of Other Payroll Cost. For the purposes of this definition the annual amount of all contributions and payments made by the consulting engineer on behalf of or in respect of a person employed by him for a period less than a full year shall be calculated pro rata to the amount of such contributions and payments for such lesser period.

OBLIGATIONS OF CONSULTING ENGINEER

6. Normal services

The services to be provided by the consulting engineer shall comprise:

(a) all or any of the services stated in the Appendix to the Memorandum of Agreement, and

(b) advising the client as to the need for the client to be provided with additional services in accordance with Clause 7.

7. Additional services not included in normal services

7.1. As services additional to those specified in Clause 6, the consulting engineer shall, if so requested by the client, provide any of the services specified in Clause 7.2 and provide or take all reasonable steps to arrange for the provision of any of the services specified in Clause 7.3.

7.2. (a) Carrying out work consequent upon a decision by the client to seek parliamentary powers.

(b) Carrying out work in connection with any application by the client for any order, sanction, licence, permit or other consent, approval or authorization necessary to enable the Task to proceed.

(c) Carrying out work arising from the failure of the client to award a contract in due time.

(d) Carrying out work consequent upon any assignment of a contract by the

CONSULTING ENGINEERS' FEES

contractor or upon the failure of the contractor properly to perform any contract or upon delay by the client in fulfilling his obligations under Clause 8 or in taking any other step necessary for the due performance of the Task.

(e) Advising the client upon and carrying out work following the taking of any step in or towards any litigation or arbitration relating to the Task.

(f) Carrying out work in conjunction with others employed to provide any of the services specified in Clause 7.3.

7.3. (a) Specialist technical advice on any abnormal aspects of the Task.

(b) Architectural, legal, financial and other professional services.

(c) Services in connection with the valuation, purchase, sale or leasing of lands and the obtaining of wayleaves.

(d) The carrying out of marine, air and land surveys, and the making of model tests or special investigations.

7.4. The consulting engineer shall obtain the prior agreement of the client to the arrangements which he proposes to make on the client's behalf for the provision of any of the services specified in Clause 7.3. The client shall be responsible to any person or persons providing such services for the cost thereof.

OBLIGATIONS OF CLIENT

8. Information to be supplied to the consulting engineer

8.1. The client shall supply to the consulting engineer without charge and within a reasonable time all necessary and relevant data and information in the possession of the client and shall give such assistance as shall reasonably be required by the consulting engineer in the performance of the Task.

8.2. The client shall give his decision on all sketches, drawings, reports, recommendations, tender documents and other matters properly referred to him for decision by the consulting engineer in such reasonable time as not to delay or disrupt the performance by the consulting engineer of the Task.

9. Payment for services

9.1. In respect of services provided by the consulting engineer under Clauses 6 and 7, the client shall pay the consulting engineer:

(a) For technical and supporting staff working in or based on the consulting engineer's office: Up to a maximum of Salary Cost times 2·5, plus Other Payroll Cost.

(b) For technical and supporting staff, working as field staff, in or based on any field office established in pursuance of Clause 10: Salary Cost times the multiplier specified in Clause 4 (a) of the Memorandum of Agreement plus Other Payroll Cost.

(c) The fee specified in Clause 4 (b) of the Memorandum of Agreement, which shall be deemed to cover the services of Partners and Consultants of the firm but not their expenses, which are reimbursable separately under Clause 11 (c).

(d) A reasonable charge for the use of a computer or other special equipment which charge shall, if possible, be agreed between the client and the consulting engineer before the work is put in hand.

CONSULTING ENGINEERS' FEES

9.2. Time spent by technical and supporting staff in connection with the use of a computer or other special equipment, including the development and writing of programmes and the operation of the computer in trial and final runs, and time spent by technical and supporting staff in travelling in connection with the Task, shall be chargeable.

9.3. Unless otherwise agreed between the client and the consulting engineer, the consulting engineer shall not be entitled to any payment in respect of time spent by secretarial staff or by staff engaged on general accountancy or administration duties in the consulting engineer's office.

10. Payment for field staff facilities

The client shall be responsible for the cost of providing such field office accommodation, furniture, telephones, equipment and transport as shall be reasonably necessary for the use of field staff, and for the reasonable running costs of such necessary field office accommodation and other facilities, including those of stationery, telephone calls, telegrams and postage. Unless otherwise agreed between the client and the consulting engineer, the consulting engineer shall arrange for the provision of field office accommodation and facilities for the use of field staff.

11. Disbursements

The client shall reimburse the consulting engineer in respect of all the consulting engineer's disbursements properly made in connection with:

(*a*) Printing, reproduction and purchase of all documents, drawings, maps and records.

(*b*) Telegrams and telephone calls other than local.

(*c*) Travelling, hotel expenses and other similar disbursements.

(*d*) Advertising for tenders and for field staff.

(*e*) The provision of additional services to the client pursuant to Clause 7.4.

GENERAL CONDITIONS AND SCALE OF CHARGES FOR WORK CARRIED OUT UNDER SCHEDULE 2

1. Definitions

'Salary Cost' means the annual salary including bonuses of any person employed by the consulting engineer, divided by 1,600 (being deemed to be the average annual total of effective working hours of an employee) and multiplied by the number of working hours spent by such person in performing any of the services in respect of which payment under this Agreement is to be made to the consulting engineer upon the basis of Salary Cost. For the purposes of this definition the annual salary of a person employed by the consulting engineer for a period less than a full year shall be calculated pro rata to such person's salary (including bonuses) for such lesser period.

CONSULTING ENGINEERS' FEES

'Other Payroll Cost' means the annual amount of all contributions and payments made by the consulting engineer on behalf of or in respect of a person employed by him for staff pension and life assurance schemes and also for National Health Insurance, Graduated Pension Fund, Selective Employment Tax and for any other tax, charge, levy, impost or payment of any kind whatsoever which the consulting engineer at any time during the performance of this Agreement is obliged by law to make on behalf of or in respect of such person, divided by 1,600 (being deemed to be the average annual total of effective working hours of an employee) and multiplied by the number of working hours spent by such person in performing any of the services in respect of which payment under this Agreement is to be made to the consulting engineer upon the basis of Other Payroll Cost. For the purposes of this definition the annual amount of all contributions and payments made by the consulting engineer on behalf of or in respect of a person employed by him for a period less than a full year shall be calculated pro rata to the amount of such contributions and payments for such lesser period.

OBLIGATIONS OF CONSULTING ENGINEER

6. Normal services

6.1. *Design Stage I*

The services to be provided by the consulting engineer at this stage shall comprise all or any of the following as may be necessary in the particular case:

(a) Investigating data and information relevant to the Works which are reasonably accessible to the consulting engineer, and considering any reports relating to the Works which have either been previously prepared by the consulting engineer or else prepared by others and made available to the consulting engineer by the client.

(b) Making any normal topographical survey of the proposed site of the Works which may be necessary to supplement the topographical information already available to the consulting engineer.

(c) Advising the client on the need to carry out any geotechnical investigations which may be necessary to supplement the geotechnical information already available to the consulting engineer, arranging for such investigations when authorized by the client, certifying the amount of any payments to be made by the client to the persons or firms carrying out such investigations under the consulting engineer's direction and advising the client on the results of such investigations.

(d) Advising the client on the need for arrangements to be made, in accordance with Clause 7, for the carrying out of special surveys, special investigations

CONSULTING ENGINEERS' FEES

or model tests, and advising the client of the results of any such surveys, investigations or tests carried out.

(e) Consulting any architect appointed by the client in connection with the architectural treatment of the Works.

(f) Preparing such documents as are reasonably necessary to enable the client to consider the consulting engineer's general proposals for the construction of the Works in the light of the investigations carried out by him at this stage, and to enable the client to apply for approval in principle of the execution of the Works in accordance with such proposals.

6.2. *Design Stage II*

The services to be provided by the consulting engineer at this stage shall comprise all or any of the following as may be necessary in the particular case:

(a) Preparing designs and tender drawings in connection with the Works.

(b) Advising as to the appropriate conditions of contract to be incorporated in any contract to be made between the client and a contractor.

(c) Preparing such specifications, schedules and bills of quantities as may be necessary to enable the client to obtain tenders or otherwise award a contract for carrying out the Works.

(d) Advising the client as to the suitability for carrying out the Works of persons and firms tendering and further as to the relative merits of tenders, prices and estimates received for carrying out the Works.

As soon as the consulting engineer shall have submitted advice to the client upon tenders, his services at this stage shall be complete.

6.3. *Construction Stage*

The consulting engineer shall not accept any tender in respect of the Works unless the client gives him instructions in writing to do so, and any acceptance so made by the consulting engineer on the instructions of the client shall be on behalf of the client. The services to be provided by the consulting engineer at this stage shall include all or any of the following as may be necessary in the particular case:

(a) Advising on the preparation of formal contract documents relating to accepted tenders for carrying out the Works or any part thereof.

(b) Inspecting and testing during manufacture and installation such electrical and mechanical materials, machinery and plant supplied for incorporation in the Works as are usually inspected and tested by consulting engineers, and arranging and witnessing acceptance tests.

(c) Advising the client on the need for special inspection or testing other than that referred to in sub-clause (b).

(d) Advising the client on the appointment of site-staff in accordance with Clause 8.

(e) Preparing bar bending schedules and any further designs and drawings which may be necessary.

(f) Examining the contractor's proposals.

(g) Making such visits to site as the consulting engineer shall consider necessary to satisfy himself as to the performance of any site-staff appointed pursuant to Clause 8, and to satisfy himself that the Works are executed generally

CONSULTING ENGINEERS' FEES

according to contract and otherwise in accordance with good engineering practice.

(*h*) Giving all necessary instructions to the contractor, provided that the consulting engineer shall not without the prior approval of the client give any instructions which in the opinion of the consulting engineer are likely substantially to increase the cost of the Works unless it is not in the circumstances practicable for the consulting engineer to obtain such prior approval.

(*i*) Issuing certificates for payment to the contractor.

(*j*) Performing any services which the consulting engineer may be required to carry out under any contract for the execution of the Works, including where appropriate the supervision of any specified tests and of the commissioning of the Works, provided that the consulting engineer may decline to perform any services specified in a contract the terms of which have not initially been approved by the consulting engineer.

(*k*) Delivering to the client on the completion of the Works such records and manufacturers' manuals as are reasonably necessary to enable the client to operate and maintain the Works.

(*l*) Deciding any dispute or difference arising between the client and the contractor and submitted to the consulting engineer for his decision, provided that this service shall not extend to advising the client following the taking of any step in or towards any arbitration or litigation in connection with the Works.

6.4. *General*

Without prejudice to the preceding provisions of this clause, the consulting engineer shall from time to time as may be necessary advise the client as to the need for the client to be provided with additional services in accordance with Clause 7.

7. Additional services not included in normal services

7.1. As services additional to those specified in Clause 6, the consulting engineer shall, if so requested by the client, provide any of the services specified in Clause 7.2 and provide or take all reasonable steps to arrange for the provision of any of the services specified in Clause 7.3.

7.2. (*a*) Preparing any report or additional contract documents required for consideration of proposals for the carrying out of alternative works.

(*b*) Carrying out work consequent upon a decision by the client to seek parliamentary powers.

(*c*) Carrying out work in connection with any application already made by the client for any order, sanction, licence, permit or other consent, approval or authorization necessary to enable the Works to proceed.

(*d*) Carrying out work arising from the failure of the client to award a contract in due time.

(*e*) Preparing details for shop fabrication of ductwork, metal or plastic frameworks.

CONSULTING ENGINEERS' FEES

(*f*) Carrying out work consequent upon any assignment of a contract by the contractor or upon the failure of the contractor properly to perform any contract or upon delay by the client in fulfilling his obligations under Clause 9 or in taking any other step necessary for the due performance of the Works.

(*g*) Advising the client and carrying out work following the taking of any step in or towards any litigation or arbitration relating to the Works.

(*h*) Carrying out work in conjunction with others employed to provide any of the services specified in Clause 7.3.

(*i*) Carrying out such other additional services, if any, as are specified in Clause 8 of the Memorandum of Agreement.

7.3. (*a*) Specialist technical advice on any abnormal aspects of the Works.

(*b*) Architectural, legal, financial and other professional services.

(*c*) Services in connection with the valuation, purchase, sale or leasing of lands and the obtaining of wayleaves.

(*d*) The carrying out of marine and air surveys, and land surveys other than those referred to in Clause 6, and the making of model tests or special investigations.

(*e*) The carrying out of special inspections or tests advised by the consulting engineer under sub-clause 6.3 (*c*).

7.4. The consulting engineer shall obtain the prior agreement of the client to the arrangements which he proposes to make on the client's behalf for the provision of any of the services specified in Clause 7.3. The client shall be responsible to any person or persons providing such services for the cost thereof.

8. Supervision on site

8.1. If in the opinion of the consulting engineer the nature of the Works, including the carrying out of any geotechnical investigation pursuant to Clause 6.1, warrants full-time or part-time engineering supervision on site, the client shall not object to the appointment of such suitably qualified technical and clerical site-staff as the consulting engineer shall consider reasonably necessary to enable such supervision to be carried out.

8.2. Persons appointed pursuant to the previous sub-clause shall be employed either by the consulting engineer or, if the client and the consulting engineer shall so agree, by the client directly, provided that the client shall not employ any person as a member of the site-staff unless the consulting engineer has first selected or approved such person as suitable for employment.

8.3. The terms of service of all site-staff to be employed by the consulting engineer shall be subject to the approval of the client, which approval shall not be unreasonably withheld.

8.4. The client shall procure that the contracts of employment of site-staff employed by the client shall stipulate that the person employed shall in no circumstances take or act upon instructions other than those of the consulting engineer.

8.5. Where any of the services specified in Clause 6.3. are performed by site-staff employed by the client, the consulting engineer shall not be responsible for any failure on the part of such staff properly to comply with any instructions given by the consulting engineer.

CONSULTING ENGINEERS' FEES

OBLIGATIONS OF CLIENT

9. Information to be supplied to the consulting engineer

9.1. The client shall supply to the consulting engineer, without charge and within a reasonable time, all necessary and relevant data and information in the possession of the client and shall give such assistance as shall reasonably be required by the consulting engineer in the performance of his services under this Agreement.

9.2. The client shall give his decision on all sketches, drawings, reports, recommendations, tender documents and other matters properly referred to him for decision by the consulting engineer in such reasonable time as not to delay or disrupt the performance by the consulting engineer of his services under this Agreement.

10. Payment for normal services

10.1. *Payment depending upon the actual cost of the Works*

10.1.1. The sum payable by the client to the consulting engineer for his services under Clause 6 shall be calculated as follows:

(*a*) The Works shall first be classified into one or more of the following classes as shall be appropriate –

Class 1: Structural work in reinforced concrete, prestressed concrete, steel and other metals.

Class 2: Buildings including engineering systems associated with buildings, but excluding Class 1 work.

Class 3: Civil engineering including geotechnical investigation, but excluding Class 1 and Class 2 work.

Class 4: Mechanical and electrical plant and equipment.

(*b*) The cost of each relevant class of work shall next be calculated, and

(*c*) The sum payable by the client to the consulting engineer shall then be calculated and shall be an amount or the sum of the amounts calculated in respect of each relevant class of work in accordance with the Scales of Charges set out in Clause 10.1.3 where the cost of the class of work is not less than £10,000, or in accordance with the Scale of Charges in Clause 11 or otherwise as may be agreed between the client and the consulting engineer, where the cost of the class of work is less than £10,000.

10.1.2. The cost of work shall be calculated in accordance with Clause 19. Where the Works have been classified in accordance with Clause 10.1.1 into more than one class, there shall be attributed to each class an appropriate portion of any 'General or Preliminary Items' included in the total cost of the Works, so that the total cost of all classes of work shall equal the total cost of the Works. In the classification of the Works, Class 1 work shall be taken to include concrete, reinforcement, prestressing tendons and anchorages, formwork, inserts and all labours.

10.1.3. The Scales of Charges (suitable for work of average complexity) referred to in Clause 10.1.1 are as follows:

CONSULTING ENGINEERS' FEES

Cost of class of work		Charge for Class 1 work	Charge for Class 2, Class 3 or Class 4 work
	£		
On the first	10,000	15%	11%
On the next	15,000	13%	9%
On the next	25,000	11½%	7½%
On the next	50,000	9¾%	6½%
On the next	100,000	8¼%	6%
On the next	300,000	7¼%	5½%
On the next	500,000	6½%	5%
On the next	1,000,000	6¼%	4½%
On the next	2,000,000	6%	4¼%
On the next	4,000,000	5¾%	4%
On the remainder		5¼%	3¾%

The charge for Class 1 work can also be calculated conveniently from the appropriate line of the following table:

Cost of Class 1 work		Charge		
	£	£		£
From 10,000–	25,000	1,500 + 13% of balance over	10,000
25,000–	50,000	3,450 + 11½% of balance over	25,000
50,000–	100,000	6,325 + 9¾% of balance over	50,000
100,000–	200,000	11,200 + 8¼% of balance over	100,000
200,000–	500,000	19,450 + 7¼% of balance over	200,000
500,000–1,000,000		41,200 + 6½% of balance over	500,000
1,000,000–2,000,000		73,700 + 6¼% of balance over	1,000,000
2,000,000–4,000,000		136,200 + 6% of balance over	2,000,000
4,000,000–8,000,000		256,200 + 5¾% of balance over	4,000,000
Over 8,000,000		486,200 + 5¼% of balance over	8,000,000

The charge for Class 2, Class 3 or Class 4 work can also be calculated conveniently from the appropriate line of the following table:

Cost of Class 2, Class 3 or Class 4 work		Charge		
	£	£		£
From 10,000–	25,000	1,100 + 9% of balance over	10,000
25,000–	50,000	2,450 + 7½% of balance over	25,000
50,000–	100,000	4,325 + 6½% of balance over	50,000
100,000–	200,000	7,575 + 6% of balance over	100,000
200,000–	500,000	13,575 + 5½% of balance over	200,000
500,000–1,000,000		30,075 + 5% of balance over	500,000
1,000,000–2,000,000		55,075 + 4½% of balance over	1,000,000
2,000,000–4,000,000		100,075 + 4¼% of balance over	2,000,000
4,000,000–8,000,000		185,075 + 4% of balance over	4,000,000
Over 8,000,000		345,075 + 3¾% of balance over	8,000,000

CONSULTING ENGINEERS' FEES

10.1.4. If the client decides to have the Works constructed in more than one phase and as a consequence the services which it may be necessary for the consulting engineer to perform under Clause 6 have to be undertaken by the consulting engineer separately in respect of each phase, then the provisions of this payment clause shall apply separately to each phase and as if the expression 'the Works' as used in this clause meant, in the case of each phase, the work comprised in that phase.

10.2. *Payment of a fixed sum*

The sum payable by the client to the consulting engineer for his services under Clause 6 shall be the sum specified in Clause 5 (*a*) of the Memorandum of Agreement, provided that the consulting engineer shall, in addition to the said sum, be paid in accordance with Clause 11.2 for any services of the kind specified in Clause 6.3 (1) which it is necessary for him to provide.

10.3. *Payment on the basis of Salary Cost times multiplier, plus Other Payroll Cost plus fee*

10.3.1. In respect of services provided by the consulting engineer under the following clauses:

Clause 6	Normal Services
Clause 7	Additional Services
Clause 12	Computer, etc.
Clause 13.3	Site Visits
Clause 15	Alterations, etc.
Clause 16	Works Damaged, etc.

the client shall pay the consulting engineer:

(*a*) Technical and supporting staff Salary Cost times the multiplier specified in Clause 5 (*b*) of the Memorandum of Agreement, plus Other Payroll Cost.

(*b*) The fee specified in Clause 5 (*c*) of the Memorandum of Agreement, which shall be deemed to cover the services of partners and consultants of the firm but not their expenses which are reimbursable separately under Clause 14 (*c*).

(*c*) Any charge for the use of a computer or other special equipment payable under Clause 12 (*d*).

10.3.2. Time spent by technical and supporting staff in travelling in connection with the Works shall be chargeable on the above basis.

10.3.3. Unless otherwise agreed between the client and the consulting engineer, the consulting engineer shall not be entitled to any payment in respect of time spent by secretarial staff or by staff engaged on general accountancy or administration duties in the consulting engineer's office.

10.3.4. The consulting engineer shall submit to the client at the time of submission of the monthly accounts referred to in Clause 20 such supporting data as may be agreed between the client and the consulting engineer.

CONSULTING ENGINEERS' FEES

11. Payment for additional services

11.1. In respect of additional services provided by the consulting engineer under Clause 7, the client shall, subject to Clause 10.3, pay the consulting engineer in accordance with the Scale of Charges set out in Clause 11.2.

11.2. Scale of Charges:

    (a) Partners and consultants: At the hourly rate or rates specified in Clause 5 (*d*) of the Memorandum of Agreement.

    (b) Technical and supporting staff: Up to a maximum of Salary Cost times 2·5, plus Other Payroll Cost.

    (c) Time spent by partners, consultants, technical and supporting staff in travelling in connection with the Works shall be chargeable on the above basis.

    (d) Unless otherwise agreed between the client and the consulting engineer, the consulting engineer shall not be entitled to any payment in respect of time spent by secretarial staff or by staff engaged on general accountancy or administration duties in the consulting engineer's office.

12. Payment for use of computer or other special equipment

Where the client has agreed to pay the consulting engineer:

    (a) in accordance with Clause 10.3 and the consulting engineer decides to use a computer or other special equipment in the carrying out of any of his services, or

    (b) for his services under Clause 6 in accordance with Clause 10.1 or 10.2 and the consulting engineer decides to use a computer or other special equipment in carrying out any additional services in accordance with Clause 7 or is expressly required by the client to use a computer or other special equipment in the carrying out of his services under Clause 6.

the client shall, unless otherwise agreed between the client and the consulting engineer, pay the consulting engineer:

    (c) for the time spent in connection with the use of a computer or other special equipment, including the development and writing of programmes and the operation of the computer in trial and final runs, in accordance with Clause 10.3 when applicable and otherwise in accordance with the Scale of Charges set out in Clause 11.2, and

    (d) a reasonable charge for the use of the computer or other special equipment, which charge shall, if possible, be agreed between the client and the consulting engineer before the work is put in hand.

13. Payment for site supervision

13.1. In addition to any other payment to be made by the client to the consulting engineer under this agreement, the client shall

    (a) reimburse the consulting engineer in respect of all salary and wage payments made by the consulting engineer to site-staff employed by the consulting engineer pursuant to Clause 8 and in respect of all other expenditure incurred by the consulting engineer in connection with the selections engagement and employment of site-staff, and

CONSULTING ENGINEERS' FEES

(*b*) pay to the consulting engineer a sum calculated at 7% of the amounts payable to the consulting engineer under the preceding sub-clause in respect of head office overhead costs incurred on site-staff administration,

provided that in lieu of payments under (*a*) and (*b*) above the client and the consulting engineer may agree upon inclusive monthly or other rates to be paid by the client to the consulting engineer for each member of site-staff employed by the consulting engineer.

13.2. The client shall also in all cases be responsible for the cost of providing such local office accommodation, furniture, telephones, equipment and transport as shall be reasonably necessary for the use of site-staff appointed pursuant to Clause 8, and for the reasonable running costs of such necessary local office accommodation and other facilities, including those of stationery, telephone calls, telegrams and postage. Unless otherwise agreed between the client and the consulting engineer, the consulting engineer shall arrange, whether through the contractor or otherwise, for the provision of local office accommodation and facilities for the use of site-staff.

13.3. In cases where the consulting engineer has thought it proper that site-staff should not be appointed, or where the necessary site-staff is not available at site due to sickness or any other cause, the consulting engineer shall, subject to Clause 10.3, be paid in accordance with the Scale of Charges set out in Clause 11.2 for site visits which would have been unnecessary but for the absence or non-availability of site-staff.

14. Disbursements

The client shall in all cases reimburse the consulting engineer in respect of all the consulting engineer's disbursements properly made in connection with:

(*a*) Printing, reproduction and purchase of all documents, drawings, maps and records.

(*b*) Telegrams and telephone calls other than local.

(*c*) Travelling, hotel expenses and other similar disbursements.

(*d*) Advertising for tenders and for site-staff.

(*e*) The provision of additional services to the client pursuant to Clause 7.4.

The client may, however, by agreement between himself and the consulting engineer make to the consulting engineer a lump sum payment or payment of a sum calculated as a percentage charge on the cost of the Works in satisfaction of his liability to the consulting engineer in respect of the consulting engineer's disbursements.

15. Payment for alteration or modification to design

If after the completion by the consulting engineer of his services under Clause 6.1, or where the client has agreed to make payment to the consulting engineer in accordance with Clause 10.2 at any time after the date of this Agreement, any design whether completed or in progress or any specification, drawing or other document prepared in whole or in part by the consulting engineer shall require to be modified or revised by reason of instructions received by the consulting engineer from the client, or by reason of circumstances which could not reasonably have been fore-

CONSULTING ENGINEERS' FEES

seen, then the client shall make additional payment to the consulting engineer for making any necessary modifications or revisions and for any consequential reproduction of documents. Subject to Clause 10.3, and unless otherwise agreed between the client and the consulting engineer, the additional sum to be paid to the consulting engineer shall be calculated in accordance with the Scale of Charges set out in Clause 11.2, and shall also include any appropriate reimbursements in accordance with Clause 14.

16. Payment when Works are damaged or destroyed

If at any time before completion of the Works any part of the Works or any materials, plant or equipment whether incorporated in the Works or not shall be damaged or destroyed, the client shall make additional payment to the consulting engineer in respect of any expenses incurred or additional work required to be carried out by the consulting engineer as a result of such damage or destruction. Subject to Clause 10.3, the amount of such additional payment shall be calculated in accordance with the Scale of Charges set out in Clause 11.2, and shall also include any appropriate reimbursements in accordance with Clause 14.

19. Cost of the Works

19.1. The cost of the Works or any part thereof shall be deemed to include:

(a) The cost to the client of the Works however incurred, including any payments (before deduction of any liquidated damages or penalties payable by the contractor to the client) made by the client to the contractor by way of bonus, incentive or ex-gratia payments, or in settlement of claims.

(b) A fair valuation of any labour, materials, manufactured goods, machinery or other facilities provided by the client, and of the full benefit accruing to the contractor from the use of construction plant and equipment belonging to the client which the client has required to be used in the execution of the Works.

(c) The market value, as if purchased new, of any second-hand materials, manufactured goods and machinery incorporated in the Works.

(d) The cost of geotechnical investigations (Clause 6.1 (c)).

(e) A fair proportion of the total cost to the client of any work in connection with the provision or diversion of public utilities systems which is carried out, other than by the contractor, under arrangements made by the consulting engineer. The said fair proportion shall be assessed with reference to the costs incurred by the consulting engineer in making such arrangements.

19.2. The cost of the Works shall not include:

(a) Administration expenses incurred by the client.

(b) Costs incurred by the client under this Agreement.

(c) Interest on capital during construction, and the cost of raising moneys required for carrying out the construction of the Works.

(d) Cost of land and wayleaves.

CONSULTING ENGINEERS' FEES

GENERAL CONDITIONS AND SCALE OF CHARGES FOR WORK CARRIED OUT UNDER SCHEDULE 3

1. Definitions

'Salary Cost'
means the annual salary including bonuses of any person employed by the consulting engineer, divided by 1,600 (being deemed to be the average annual total of effective working hours of an employee) and multiplied by the number of working hours spent by such person in performing any of the services in respect of which payment under this Agreement is to be made to the consulting engineer upon the basis of Salary Cost. For the purposes of this definition the annual salary of a person employed by the consulting engineer for a period less than a full year shall be calculated pro rata to such person's salary (including bonuses) for such lesser period.

'Other Payroll Cost'
means the annual amount of all contributions and payments made by the consulting engineer on behalf of or in respect of a person employed by him for staff pension and life assurance schemes, and also for National Health Insurance, Graduated Pension Fund, Selective Employment Tax and for any other tax, charge, levy, impost or payment of any kind whatsoever which the consulting engineer at any time during the performance of this Agreement is obliged by law to make on behalf of or in respect of such person, divided by 1,600 (being deemed to be the average annual total of effective working hours of an employee) and multiplied by the number of working hours spent by such person in performing any of the services in respect of which payment under this Agreement is to be made to the consulting engineer upon the basis of Other Payroll Cost. For the purposes of this definition the annual amount of all contributions and payments made by the consulting engineer on behalf of or in respect of a person employed by him for a period less than a full year shall be calculated pro rata to the amount of such contributions and payments for such lesser period.

OBLIGATIONS OF CONSULTING ENGINEER

6. Normal services

6.1. *Preliminary or sketch plan stage*
The services to be provided by the consulting engineer at this stage shall comprise all or any of the following as may be necessary in the particular case:

CONSULTING ENGINEERS' FEES

(*a*) Investigating data and information relating to the project and relevant to the Works which are reasonably accessible to the consulting engineer, and considering any reports relating to the Works which have either been prepared by the consulting engineer or else prepared by others and made available to the consulting engineer by the client.

(*b*) Advising the client on the need to carry out any geotechnical investigations which may be necessary to supplement the geotechnical information already available to the consulting engineer, arranging for such investigations when authorized by the client, certifying the amount of any payments to be made by the client to the persons or firms carrying out such investigations under the consulting engineer's directions, and advising the client on the results of such investigations.

(*c*) Advising the client on the need for arrangements to be made, in accordance with Clause 7, for the carrying out of special surveys, special investigations or model tests, and advising the client of the results of any such surveys, investigations or tests carried out.

(*d*) Consulting any local or other authorities on matters of principle in connection with the structural design of the Works.

(*e*) Providing sufficient structural information to enable the architect to produce his sketch plans.

6.2. *Tender Stage*

The services to be provided by the consulting engineer at this stage shall include all or any of the following as may be necessary in the particular case, for the purpose of enabling tenders to be obtained:

(*a*) Developing the design of the Works in collaboration with the architect and preparing calculations, drawings and specifications of the Works to enable a Bill of Quantities to be prepared.

(*b*) Advising on conditions of contract relevant to the Works and forms of tender and invitations to tender as they relate to the Works.

(*c*) Consulting any local or other authorities in connection with the structural design of the Works, and preparing typical details and typical calculations.

(*d*) Advising the client as to the suitability for carrying out the Works of persons and firms tendering.

6.3. *Working drawing stage*

The services to be provided by the consulting engineer at this stage shall include all or any of the following as may be necessary in the particular case:

(*a*) Advising the client as to the relative merits of tenders, prices and estimates received for carrying out the Works.

(*b*) Preparing such calculations and details relating to the Works as may be required for submission to any appropriate authority.

(*c*) Preparing any further designs, specifications and drawings, including bar bending schedules, necessary for the information of the contractor to enable him to carry out the Works, but excepting the preparation of any shop details relating to the Works or any part thereof.

(*d*) Checking shop details for accuracy, general dimensions and adequacy of members and connections.

CONSULTING ENGINEERS' FEES

6.4. *Construction stage*

The consulting engineer shall not accept any tender in respect of the Works unless the client gives him instructions in writing to do so, and any acceptance so made by the consulting engineer on the instructions of the client shall be on behalf of the client. The services to be provided by the consulting engineer at this stage shall include all or any of the following as may be necessary in the particular case:

(*a*) Advising on the preparation of formal contract documents relating to accepted tenders for carrying out the Works or any part thereof.

(*b*) Advising the client on the need for special inspections or tests.

(*c*) Advising the client and the architect on the appointment of site-staff in accordance with Clause 8.

(*d*) Examining the contractor's proposals.

(*e*) Making such visits to site as the consulting engineer shall consider necessary to satisfy himself as to the performance of any site-staff appointed pursuant to Clause 8, and to satisfy himself that the Works are executed generally according to contract and otherwise in accordance with good engineering practice.

(*f*) Giving all necessary instructions to the contractor, provided that the consulting engineer shall not without the prior approval of the client give any instructions which in the opinion of the consulting engineer are likely substantially to increase the cost of the Works unless it is not in the circumstances practicable for the consulting engineer to obtain such prior approval.

(*g*) Advising on certificates for payment to the contractor.

(*h*) Performing any services which the consulting engineer may be required to carry out under any contract for the execution of the Works, including where appropriate the supervision of any specified tests, provided that the consulting engineer may decline to perform any services specified in a contract the terms of which have not initially been approved by the consulting engineer.

(*i*) Delivering to the client on the completion of the Works such records as are reasonably necessary to enable the client to operate and maintain the Works.

(*j*) Assisting in settling any dispute or difference which may arise between the client and the contractor, provided that this service shall not extend to advising the client following the taking of any step in or towards any arbitration or litigation in connection with the Works.

6.5. *General*

Without prejudice to the preceding provisions of this clause, the consulting engineer shall from time to time as may be necessary advise the client as to the need for the client to be provided with additional services in accordance with Clause 7.

7. Additional services not included in normal services

7.1. As services additional to those specified in Clause 6, the consulting engineer shall, if so requested by the client, provide any of the services specified in Clause 7.2 and provide or take all reasonable steps to arrange for the provision by others of any of the services specified in Clause 7.3.

CONSULTING ENGINEERS' FEES

7.2. (*a*) Preparing any report or additional contract documents required for consideration of proposals for the carrying out of alternative works.

(*b*) Carrying out work consequent upon a decision by the client to seek parliamentary powers.

(*c*) Carrying out work in connection with any application already made by the client for any order, sanction, licence, permit or other consent, approval or authorization necessary to enable the Works to proceed.

(*d*) Carrying out work arising from the failure of the client to award a contract in due time.

(*e*) Preparing preliminary estimates.

(*f*) Preparing details for shop fabrication of ductwork, metal or plastic frameworks.

(*g*) Checking and advising upon any part of the project not designed by the consulting engineer.

(*h*) Preparing interim or other reports or detailed valuations, including estimates or cost analyses based on measurement or forming an element of a cost planning service.

(*i*) Carrying out work consequent upon any assignment of a contract by the contractor or upon the failure of the contractor properly to perform any contract or upon delay by the client in fulfilling his obligations under Clause 9 or in taking any other step necessary for the due performance of the Works.

(*j*) Advising the client upon and carrying out work following the taking of any step in or towards any litigation or arbitration relating to the Works.

(*k*) Carrying out work in conjunction with others employed to provide any of the services specified in Clause 7.3.

(*l*) Carrying out such other additional services, if any, as are specified in Clause 6 of the Memorandum of Agreement.

7.3. (*a*) Specialist technical advice on any abnormal aspects of the Works.

(*b*) Legal, financial and other professional services.

(*c*) Services in connection with the valuation, purchase, sale or leasing of lands and the obtaining of wayleaves.

(*d*) The surveying of sites or existing works.

(*e*) Investigation of the nature and strength of existing works and the making of model tests or special investigations.

(*f*) The carrying out of special inspections or tests advised by the consulting engineer under Clause 6.4 (*b*).

7.4. The consulting engineer shall obtain the prior agreement of the client to the arrangements which he proposes to make on the client's behalf for the provision of any of the services specified in Clause 7.3. The client shall be responsible to any person or persons providing such services for the cost thereof.

8. Supervision on site

8.1. If in the opinion of the consulting engineer the nature of the Works warrants full-time or part-time engineering supervision on site, the client shall not object to the appointment of such suitably qualified technical and clerical site-staff as

CONSULTING ENGINEERS' FEES

the consulting engineer shall consider reasonably necessary to enable such supervision to be carried out.

8.2. Persons appointed pursuant to the previous sub-clause shall be employed either by the consulting engineer or, if the client and the consulting engineer shall so agree, by the client directly, provided that the client shall not employ any person as a member of the site-staff unless the consulting engineer has first selected or approved such person as suitable for employment.

8.3. The terms of service of all site-staff to be employed by the consulting engineer shall be subject to the approval of the client, which approval shall not be unreasonably withheld.

8.4. The client shall procure that the contracts of employment of site-staff employed by the client shall stipulate that the person employed shall in no circumstances take or act upon instructions other than those of the consulting engineer.

8.5. Where any of the services specified in Clause 6.4 are performed by site-staff employed by the client, the consulting engineer shall not be responsible for any failure on the part of such staff properly to comply with any instructions given by the consulting engineer.

8.6. Where a clerk of works nominated by the architect is charged with the supervision of engineering works on site, his selection and appointment shall be subject to the approval of the consulting engineer. In respect of engineering works, such clerk of works shall take instructions solely from the consulting engineer, who shall inform the architect of all such instructions.

OBLIGATIONS OF CLIENT

9. Information to be supplied to the consulting engineer

9.1. The client shall supply to the consulting engineer, without charge, and within a reasonable time, all necessary and relevant data and information in the possession of the client and shall give such assistance as shall reasonably be required by the consulting engineer in the performance of his services under this Agreement. The information to be provided by the client to the consulting engineer shall include:

(a) All such drawings as may be necessary to make the client's or the architect's requirements clear, including plans and sections of all buildings (to a scale of not less than 1 to 100) and essential details (to a scale of not less than 1 to 25) together with site plans (to a scale of not less than 1/1,250) and levels.

(b) Copies of all contract documents, including a priced bill of quantities relating to those parts of the project which are relevant to the Works.

(c) Copies of all variation orders and supporting documents relating to other parts of the project.

9.2. The client shall give his decision on all sketches, drawings, reports, recommendations, tender documents and other matters properly referred to him for decision by the consulting engineer in such reasonable time as not to delay or disrupt the performance by the consulting engineer of his services under this Agreement.

CONSULTING ENGINEERS' FEES

10. Payment for normal services

10.1. The sum payable by the client to the consulting engineer for his services under Clause 6 shall be an amount or the sum of the amounts calculated as follows:

(*a*) On the total cost of the Works, including reinforced concrete but excluding load-bearing brickwork or blockwork, a charge calculated in accordance with the Scale of Charges set out in Clause 10.3 where the said cost is not less than £10,000, or in accordance with the Scale of Charges set out in Clause 11.2 or otherwise as may be agreed between the client and the consulting engineer, where the said cost is less than £10,000.

(*b*) On the cost of reinforced concrete, a further charge of 3%.

(*c*) On the cost of load-bearing brickwork or blockwork, a charge of $3\frac{1}{2}$%.

10.2. The cost of work shall be calculated in accordance with Clause 20; reinforced concrete work shall include concrete, reinforcement, prestressing tendons and anchorages, formwork, inserts and all labours, and a due proportion of 'General or Preliminary Items'; load-bearing brickwork or blockwork shall include facing brickwork or blockwork and all other brickwork or blockwork necessary for the stability of the load-bearing walls, together with all labours and materials normally associated with the construction of load-bearing walls, and a due proportion of 'General or Preliminary Items'.

10.3. The Scale of Charges (suitable for work of average complexity) referred to in Clause 10.1 is as follows:

Cost of the Works		Charge
	£	
On the first	10,000	11%
On the next	15,000	9%
On the next	25,000	$7\frac{1}{2}$%
On the next	50,000	$6\frac{1}{2}$%
On the next	100,000	6%
On the remainder		5%

The charge can also be calculated conveniently from the appropriate line of the following table:

Cost of the Works £	Charge £		£
From 10,000– 25,000	1,100 + 9%	of balance over	10,000
25,000– 50,000	2,450 + $7\frac{1}{2}$%	of balance over	25,000
50,000–100,000	4,325 + $6\frac{1}{2}$%	of balance over	50,000
100,000–200,000	7,575 + 6%	of balance over	100,000
Over 200,000	13,575 + 5%	of balance over	200,000

10.4. If the client decides to have the Works constructed in more than one phase and as a consequence the services which it may be necessary for the consulting engineer to perform under Clause 6 have to be undertaken by the consulting engineer separately in respect of each phase, then the provisions of this pay-

CONSULTING ENGINEERS' FEES

ment clause shall apply separately to each phase and as if the expression 'the Works' as used in this clause meant, in the case of each phase, the work comprised in that phase.

11. Payment for additional services

11.1. In respect of additional services provided by the consulting engineer under Clause 7, the client shall pay the consulting engineer in accordance with the Scale of Charges set out in Clause 11.2.

11.2. Scale of Charges:

(*a*) Partners and consultants: At the hourly rate or rates specified in Clause 4 of the Memorandum of Agreement.

(*b*) Technical and supporting staff: Up to a maximum of Salary Cost times 2·5, plus Other Payroll Cost.

(*c*) Time spent by partners, consultants, technical and supporting staff in travelling in connection with the Works shall be chargeable on the above basis.

(*d*) Unless otherwise agreed between the client and the consulting engineer, the consulting engineer shall not be entitled to any payment in respect of time spent by secretarial staff or by staff engaged on general accountancy or administration duties in the consulting engineer's office.

12. Payment for quantity surveying services

If the client requires the consulting engineer to carry out quantity surveying services in respect of the Works designed by him the client shall pay the consulting engineer in respect of such services in accordance with the current scale of professional charges for quantity surveying services published by the Royal Institution of Chartered Surveyors.

13. Payment for use of computer or other special equipment

Where the consulting engineer decides to use a computer or other special equipment in carrying out any additional services in accordance with Clause 7 or is expressly required by the client to use a computer or other special equipment in the carrying out of his services under Clause 6, the client shall, unless otherwise agreed between the client and the consulting engineer, pay the consulting engineer:

(*a*) for the time spent in connection with the use of a computer or other special equipment, including the development and writing of programmes and the operation of the computer in trial and final runs, in accordance with the Scale of Charges set out in Clause 11.2, and

(*b*) a reasonable charge for the use of the computer or other special equipment, which charge shall, if possible, be agreed between the client and the consulting engineer before the work is put in hand.

14. Payment for site supervision

14.1. In addition to any other payment to be made by the client to the consulting engineer under this Agreement, the client shall:

CONSULTING ENGINEERS' FEES

(*a*) reimburse the consulting engineer in respect of all salary and wage payments made by the consulting engineer to site-staff employed by the consulting engineer pursuant to Clause 8 and in respect of all other expenditure incurred by the consulting engineer in connection with the selection, engagement and employment of site-staff, and

(*b*) pay to the consulting engineer a sum calculated at 7% of the amounts payable to the consulting engineer under the preceding sub-clause in respect of head office overhead costs incurred on site-staff administration,

provided that in lieu of payments under (*a*) and (*b*) above the client and the consulting engineer may agree upon inclusive monthly or other rates to be paid by the client to the consulting engineer for each member of site-staff employed by the consulting engineer.

14.2. The client shall also in all cases be responsible for the cost of providing such local office accommodation, furniture, telephones, equipment and transport as shall be reasonably necessary for the use of site-staff appointed pursuant to Clause 8, and for the reasonable running costs of such necessary local office accommodation and other facilities, including those of stationery, telephone calls, telegrams and postage. Unless otherwise agreed between the client and the consulting engineer, the consulting engineer shall arrange, whether through the contractor or otherwise, for the provision of local office accommodation and facilities for the use of site-staff.

14.3. In cases where the consulting engineer has thought it proper that site-staff should not be appointed, or where the necessary site-staff is not available at site due to sickness or any other cause, the consulting engineer shall be paid in accordance with the Scale of Charges set out in Clause 11.2 for site visits which would have been unnecessary but for the absence or non-availability of site-staff.

15. Disbursements

The client shall in all cases reimburse the consulting engineer in respect of all the consulting engineer's disbursements properly made in connection with:

(*a*) Printing, reproduction and purchase of all documents, drawings, maps and records.

(*b*) Telegrams and telephone calls other than local.

(*c*) Travelling, hotel expenses and other similar disbursements.

(*d*) Advertising for tenders and for site-staff.

(*e*) The provision of additional services to the client pursuant to Clause 7.4.

The client may, however, by agreement between himself and the consulting engineer make to the consulting engineer a lump sum payment or payment of a sum calculated as a percentage charge on the cost of the Works in satisfaction of his liability to the consulting engineer in respect of the consulting engineer's disbursements.

16. Payment for alteration or modification to design

If after the completion by the consulting engineer of his services under Clause 6.1, any design whether completed or in progress or any specification, drawing or other document prepared in whole or in part by the consulting engineer shall require to be

CONSULTING ENGINEERS' FEES

modified or revised by reason of instructions received by the consulting engineer from the client or from the architect, or by reason of circumstances which could not reasonably have been foreseen, then the client shall make additional payment to the consulting engineer for making any necessary modifications or revisions and for any consequential reproduction of documents. Unless otherwise agreed between the client and the consulting engineer, the additional sum to be paid to the consulting engineer shall be calculated in accordance with the Scale of Charges set out in Clause 11.2, and shall also include any appropriate reimbursements in accordance with Clause 15.

17. Payment when works are damaged or destroyed

If at any time before completion of the Works any part of the Works or any materials, plant or equipment whether incorporated in the Works or not shall be damaged or destroyed, the client shall make additional payment to the consulting engineer in respect of any expenses incurred or additional work required to be carried out by the consulting engineer as a result of such damage or destruction. The amount of such additional payment shall be calculated in accordance with the Scale of Charges set out in Clause 11.2 and shall also include any appropriate reimbursements in accordance with Clause 15.

20. Cost of the Works

20.1. The cost of the Works or any part thereof shall be deemed to include:

(a) The cost to the client of the Works however incurred, including any payments (before deduction of any liquidated damages or penalties payable by the contractor to the client) made by the client to the contractor by way of bonus, incentive or ex-gratia payments, or in settlement of claims.

(b) The cost of such operations as are necessary to enable the Works to be carried out, including all excavations and supports thereof, filling, shoring, pumping and other operations for the control of water, and the cost of preliminary and general items in the proportion that the cost of the Works bears to the total cost of the project.

(c) A fair valuation of any labour, materials, manufactured goods, machinery or other facilities provided by the client, and of the full benefit accruing to the contractor from the use of construction plant and equipment belonging to the client which the client has required to be used in the execution of the Works.

(d) The market value, as if purchased new, of any second-hand materials, manufactured goods and machinery incorporated in the Works.

(e) The cost of geotechnical investigations (Clause 6.1 (b)).

20.2. The cost of the Works shall not include:

(a) Administration expenses incurred by the client.

(b) Costs incurred by the client under this Agreement.

(c) Interest on capital during construction, and the cost of raising moneys required for carrying out the construction of the Works.

(d) Cost of land and wayleaves.

CONSULTING ENGINEERS' FEES
GENERAL CONDITIONS AND SCALE OF CHARGES
FOR WORK CARRIED OUT UNDER SCHEDULE 4

1. Definitions

'Salary Cost'

means the annual salary including bonuses of any person employed by the Consulting Engineer, divided by 1600 (being deemed to be the average annual total of effective working hours of an employee) and multiplied by the number of working hours spent by such person in performing any of the services in respect of which payment under this Agreement is to be made to the Consulting Engineer upon the basis of Salary Cost. For the purposes of this definition the annual salary of a person employed by the Consulting Engineer for a period less than a full year shall be calculated pro rata to such person's salary (including bonuses) for such lesser period.

'Other Payroll Cost'

means the annual amount of all contributions and payments made by the Consulting Engineer on behalf of or in respect of a person employed by him for staff pension and life assurance schemes and also for National Health Insurance, Graduated Pension Fund, Selective Employment Tax and for any other tax, charge, levy, impost or payment of any kind whatsoever which the Consulting Engineer at any time during the performance of this Agreement is obliged by law to make on behalf of or in respect of such person, divided by 1600 (being deemed to be the average annual total of effective working hours of an employee) and multiplied by the number of working hours spent by such person in performing any of the services in respect of which payment under this Agreement is to be made to the Consulting Engineer upon the basis of Other Payroll Cost. For the purposes of this definition the annual amount of all contributions and payments made by the Consulting Engineer on behalf of or in respect of a person employed by him for a period less than a full year shall be calculated pro rata to the amount of such contributions and payments for such lesser period.

'Tender Drawings'

means drawings prepared by the Consulting Engineer in sufficient detail to enable those tendering to interpret correctly the design for the Works and to submit competitive tenders for the execution of the Works.

'Detail Drawings'

means drawings, additional to the Tender Drawings, specially prepared by the Consulting Engineer at the request of the Architect or the Client.

'Co-ordination Drawings'

means drawings prepared by the Consulting Engineer

CONSULTING ENGINEERS' FEES

	or the Contractor, showing the inter-relation of two or more engineering systems.
'Installation Drawings'	means drawings prepared by the Contractor for approval by the Consulting Engineer, showing details of the Contractor's proposals for the execution of the Works.
'Builder's Work Drawings'	means drawings normally prepared by the Contractor for approval by the Consulting Engineer, showing details of work of a structural nature which is required to be carried out by a builder or other party to facilitate the execution of the Works.
'Record Drawings'	means drawings normally prepared by the Contractor for approval by the Consulting Engineer, showing clearly the general scheme and details of the Works as completed.

OBLIGATIONS OF CONSULTING ENGINEER

6. Normal services

6.1 *Design Stage I*

The services to be provided by the Consulting Engineer at this stage shall comprise all or any of the following as may be necessary in the particular case:

(*a*) Investigating data and information relating to the Project and relevant to the Works which are reasonably accessible to the Consulting Engineer, and considering any reports relating to the works which have either been prepared by the Consulting Engineer or else prepared by others and made available to the Consulting Engineer by the Client.

(*b*) Advising the Client on the need for arrangements to be made, in accordance with Clause 7, for the carrying out of special surveys, special investigations or model tests, and advising the Client of the results of any such surveys, investigations or tests carried out.

(*c*) Consulting any local or other authorities on matters of principle in connection with the design of the Works.

(*d*) Consulting the Architect or any other professional adviser appointed by the Client in connection with the Project.

(*e*) Providing sufficient preliminary information and approximate estimates (based on unit volume, unit surface area or similar bases of estimation) regarding the Works to enable the Client or the Architect to prepare architectural sketch plans and budget estimates for the Project.

6.2 *Design Stage II*

The services to be provided by the Consulting Engineer at this stage shall comprise all or any of the following as may be necessary in the particular case:

(*a*) Preparing designs, Tender Drawings and specifications for the Works in such detail as may be necessary to enable competitive tenders for the execution of the Works to be obtained.

CONSULTING ENGINEERS' FEES

(b) Providing outline information as to plant rooms, chimneys, air-conditioning and ventilation ducts, main service ducts and other similar elements incorporated in the building structure, and providing information as to the approximate weights of items of heavy plant and equipment which are to be incorporated in the Works.

(c) Advising on conditions of contract relevant to the Works and forms of tender and invitations to tender as they relate to the Works.

(d) Advising the Client as to the suitability for carrying out the Works of persons and firms tendering, and further as to the relative merits of tenders, prices and estimates received for carrying out the Works.

As soon as the Consulting Engineer shall have submitted advice to the Client upon tenders, his services at this stage shall be complete.

6.3 *Installation Stage*

The Consulting Engineer shall not accept any tender in respect of the Works unless the Client gives him instructions in writing to do so, and any acceptance so made by the Consulting Engineer on the instructions of the Client shall be on behalf of the Client. The services to be provided by the Consulting Engineer at this stage shall include all or any of the following as may be necessary in the particular case:

(a) Advising on the preparation of formal contract documents relating to accepted tenders for carrying out the Works or any part thereof.

(b) Advising the Client on the appointment of site-staff in accordance with Clause 10.

(c) Providing the Contractor with such further information as is necessary in the opinion of the Consulting Engineer, to enable the Installation Drawings to be prepared.

(d) Examining the Contractor's proposals.

(e) Making such visits to site as the Consulting Engineer shall consider necessary to satisfy himself as to the performance of any site-staff appointed pursuant to Clause 10, and to satisfy himself that the Works are executed generally according to his designs and specifications and otherwise in accordance with good engineering practice.

(f) Giving all necessary instructions to the Contractor, provided that the Consulting Engineer shall not without the prior approval of the Client give any instructions which are in his opinion likely substantially to increase the cost of the Works unless it is not in the circumstances practicable for the Consulting Engineer to obtain such prior approval.

(g) Advising the Client or the Architect as to the need to vary any part of the Project for a reason or reasons relating to the Works.

(h) Approving the Contractor's commissioning procedures and performance tests, and inspecting the Works on completion.

(i) Advising on interim valuations, issuing certificates for payment to the Contractor where appropriate and advising on the settlement of the Contractor's final accounts.

(j) Performing any services which the Consulting Engineer may be required to carry out under any document which he has prepared relating to the Works. The Consulting Engineer may decline to perform any services specified in a

CONSULTING ENGINEERS' FEES

contract the terms of which have not initially been expressly approved by him in writing.

(*k*) Delivering to the Client on the completion of the Works copies of Record Drawings, the Contractor's operating instructions, and, where appropriate, certificates of works tests.

(*l*) Assisting in settling any dispute or difference which may arise between the Client and the Contractor, provided that this service shall not extend to advising the Client following the taking of any step in or towards any arbitration or litigation in connection with the Works.

6.4 *General*

Without prejudice to the preceding provisions of this clause, the Consulting Engineer shall from time to time as may be necessary advise the Client as to the need for the Client to be provided with additional services in accordance with Clause 7.

7. Additional services not included in normal services

7.1 As services additional to those specified in Clause 6, the Consulting Engineer shall, if so requested by the Client, provide any of the services specified in Clause 7.2 and provide or take all reasonable steps to arrange for the provision of any of the services specified in Clause 7.3.

7.2 (*a*) Preparing any report or additional contract documents required for consideration of proposals for the carrying out of alternative works.

(*b*) Carrying out work consequent upon a decision by the Client to seek parliamentary powers.

(*c*) Carrying out work in connection with any application already made by the Client for any order, sanction, licence, permit or other consent, approval or authorization necessary to enable the Works to proceed.

(*d*) Carrying out work arising from the failure of the Client to award a contract in due time.

(*e*) Carrying out special cost investigations or detailed valuations, including estimates or cost analyses based on measurement or forming an element of a cost planning service.

(*f*) Carrying out surveys, including surveys of existing installations.

(*g*) Negotiating and arranging for the provision or diversion of utility services.

(*h*) Negotiating any contract or sub-contract with a Contractor selected otherwise than by competitive tendering, including checking and agreeing quantities and nett costs of materials and labour, arithmetical checking and agreeing added percentages to cover overheads and profit.

(*i*) Preparing Detail Drawings, Co-ordination Drawings, Builder's Work Drawings, Record Drawings or any detailed Schedules.

(*j*) Preparing details for shop fabrication of ductwork, metal or plastic frameworks.

(*k*) Checking and advising upon any part of the Project not designed by the Consulting Engineer.

(*l*) Inspecting or witnessing the testing of materials or machinery during manufacture.

CONSULTING ENGINEERS' FEES

(*m*) Carrying out commissioning procedures or performance tests.

(*n*) Carrying out work consequent upon any assignment of the contract by the Contractor or upon the failure of the Contractor properly to perform any contract or upon delay by the Client in fulfilling his obligations under Clause 11 or in taking any other step necessary for the due performance of the Works.

(*o*) Providing manuals and other documents describing the design, operation and maintenance of the Works.

(*p*) Advising the Client and carrying out work following the taking of any step in or towards any litigation or arbitration relating to the Works.

(*q*) Carrying out work in conjunction with others employed to provide any of the services specified in Clause 7.3.

(*r*) Carrying out such other additional services, if any, as are specified in Clause 6 of the Memorandum of Agreement or made necessary by RIBA 'Plan of Work' or other special procedures required by the Client or the Architect.

7.3 (*a*) Specialist technical advice on any abnormal aspects of the Works.

The services provided by the Consulting Engineer under Clause 6 will include the provision of all expert technical advice and skills which are normally required for the class of work for which the Consulting Engineer's services are engaged.

(*b*) Legal, financial, architectural, structural and other professional services.

(*c*) Services in connection with the valuation, purchase, sale or leasing of lands and the obtaining of wayleaves.

(*d*) Investigation of the nature and strength of existing works and the making of model tests or special investigations.

7.4 The Consulting Engineer shall obtain the prior agreement of the Client to the arrangements which he proposes to make on the Client's behalf for the provision of any of the services specified in Clause 7.3. The Client shall be responsible to any person or persons providing such services for the cost thereof.

8. Bills of quantities

8.1 The Consulting Engineer shall advise the Client on the need for preparing Bills of Quantities in respect of the Works before the invitation of tenders therefor.

8.2 If the Client so requests, the Consulting Engineer shall

(*a*) prepare detailed Bills of Quantities for the Works if the design of the Project is sufficiently advanced to enable him so to do, or

(*b*) prepare Bills of Approximate Quantities for the Works, if the design of the Project is not sufficiently advanced to enable him to prepare detailed Bills of Quantities.

8.3 In any case in which the Consulting Engineer has prepared Bills of Approximate Quantities, he shall subsequently correct the quantities and reprice the Bills in consultation with the Contractor.

CONSULTING ENGINEERS' FEES

9. Variations

9.1 The Consulting Engineer shall initiate all necessary variation orders in connection with the Works.

9.2 The Consulting Engineer shall measure or assess the extent of all variations in the Works and shall negotiate, and if possible agree with the Contractor, the value thereof. The Consulting Engineer shall also check all relevant entries in the Contractor's interim and final accounts.

9.3 If, however, the Consulting Engineer and the Contractor agree to remeasure the Works on completion for the purposes of Clause 8.3, Clause 9.2 shall not apply.

9.4 The Consulting Engineer shall check and approve the Contractor's assessments of the value of all fluctuations (increases and/or decreases) in the cost of labour and materials.

10. Supervision on site

10.1 If in the opinion of the Consulting Engineer the nature of the Works warrants full-time or part-time engineering supervision on site, the Client shall not object to the appointment of such suitably qualified technical and clerical site-staff as the Consulting Engineer shall consider reasonably necessary to enable such supervision to be carried out.

10.2 Persons appointed pursuant to the previous sub-clause shall be employed either by the Consulting Engineer or, if the Client and the Consulting Engineer shall so agree, by the Client directly, provided that the Client shall not employ any person as a member of the site-staff unless the Consulting Engineer has first selected or approved such person as suitable for employment.

10.3 The terms of service of all site-staff to be employed by the Consulting Engineer shall be subject to the approval of the Client, which approval shall not be unreasonably withheld.

10.4 The Client shall procure that the contracts of employment of site-staff employed by the Client shall stipulate that the person employed shall in no circumstances take or act upon instructions other than those of the Consulting Engineer.

10.5 Where any of the services specified in Clause 6.3 are performed by site-staff employed by the Client, the Consulting Engineer shall not be responsible for any failure on the part of such staff properly to comply with any instructions given by the Consulting Engineer.

OBLIGATIONS OF CLIENT

11. Information to be supplied to the consulting engineer

11.1 The Client shall supply to the Consulting Engineer, without charge and within a reasonable time, all necessary and relevant data and information in the possession of the Client and shall give such assistance as shall reasonably be required by the Consulting Engineer in the performance of his services under this Agreement. The information to be provided by the Client to the Consulting Engineer shall include:

CONSULTING ENGINEERS' FEES

(a) All such drawings as may be necessary to make the Client's or the Architect's requirements clear, including plans and sections of all buildings (to a scale of not less than 1 to 100) and essential details (to a scale of not less than 1 to 25) together with site plans (to a scale of not less than 1/1,250) and levels.

(b) Copies of all contract documents, variation orders and supporting documents relating to those parts of the Project which are relevant to the Works.

11.2 The Client shall give his decision on all sketches, drawings, reports, recommendations, tender documents and other matters properly referred to him for decision by the Consulting Engineer in such reasonable time as not to delay or disrupt the performance by the Consulting Engineer of his services under this Agreement.

12. Payment for normal services

12.1 The sum payable by the Client to the Consulting Engineer for his services under Clause 6 shall be calculated in accordance with the Scale of Charges set out in Clause 12.2 where the cost of the Works is not less than £10,000, or in accordance with the Scale of Charges set out in Clause 13.2, or otherwise as may be agreed between the Client and the Consulting Engineer, where the cost of the Works is less than £10,000.

12.2 The scale of Charges (suitable for work of average complexity) referred to in Clause 12.1 is as follows:

Cost of the Works		Charge
On the first	£ 10,000	11%
On the next	£ 15,000	9%
On the next	£ 25,000	8%
On the next	£ 50,000	$7\frac{1}{2}$%
On the next	£100,000	7%
On the next	£800,000	$6\frac{3}{4}$%
On the remainder		$6\frac{1}{2}$%

The charge can also be calculated conveniently from the appropriate line of the following table:

Cost of the Works		Charge	
From £ 10,000 to £ 25,000	*£ 1,100 + 9%*	*of balance over £ 10,000*	
£ 25,000 to £ 50,000	*£ 2,450 + 8%*	*of balance over £ 25,000*	
£ 50,000 to £ 100,000	*£ 4,450 + 7½%*	*of balance over £ 50,000*	
£ 100,000 to £ 200,000	*£ 8,200 + 7%*	*of balance over £ 100,000*	
£ 200,000 to £1,000,000	*£15,200 + 6¾%*	*of balance over £ 200,000*	
Over £1,000,000	*£69,200 + 6½%*	*of balance over £1,000,000*	

12.3 If the Client decides to have the Works constructed in more than one phase and as a consequence the services which it may be necessary for the Consulting Engineer to perform under Clause 6 have to be undertaken by the Consulting Engineer separately in respect of each phase, then the provisions of

CONSULTING ENGINEERS' FEES

this payment clause shall apply separately to each phase and as if the expression 'the Works' as used in this clause meant, in the case of each phase, the work comprised in that phase.

13. Payment for additional services

13.1 In respect of additional services provided by the Consulting Engineer under Clause 7, the Client shall pay the Consulting Engineer in accordance with the Scale of Charges set out in the next sub-clause.

13.2 *Scale of Charges:*

(*a*) Partners and Consultants: At the hourly rate or rates specified in Clause 4 of the Memorandum of Agreement.

(*b*) Technical and supporting staff: Up to a maximum of Salary Cost times 2.5, plus Other Payroll Cost.

(*c*) Time spent by Partners, Consultants, technical and supporting staff in travelling in connection with the Works shall be chargeable on the above basis.

(*d*) Unless otherwise agreed between the Client and the Consulting Engineer, the Consulting Engineer shall not be entitled to any payment in respect of time spent by secretarial staff or by staff engaged on general accountancy or administration duties in the Consulting Engineer's office.

14. Payment for bills of quantities

14.1 For services provided under Clause 8 the Consulting Engineer shall be paid an additional amount according to the procedure adopted, as follows:

(*a*) Where Bills of Quantities are prepared, either by the Consulting Engineer or by a Quantity Surveyor appointed by the Client to provide services in respect of the Works, the Consulting Engineer shall be paid for providing the additional information required for preparing the Bill of Quantities, at the following rates:

	Detailed Bills of Quantities	*Bills of Approximate Quantities*
On the cost of the Works in the Bills		
On the first £500,000	1%	$\frac{1}{2}\%$
On the excess thereafter	$\frac{3}{4}\%$	$\frac{3}{8}\%$

(*b*) For preparing Bills of Quantities as described in Clauses 8.2(*a*) and 8.2(*b*), the Consulting Engineer shall be paid at the following rates:

	Detailed Bills of Quantities	*Bills of Approximate Quantities*
On the cost of the Works in the Bills and subject to Clause 14.1(*c*)		
On the first £500,000	$1\frac{1}{2}\%$	$\frac{3}{4}\%$
On the excess thereafter	$1\frac{1}{4}\%$	$\frac{5}{8}\%$

In the case of Approximate Bills of Quantities the Consulting Engineer shall be paid a further amount of $1\frac{1}{2}\%$ on the cost of the Works, for correcting and repricing the Bills.

CONSULTING ENGINEERS' FEES

(c) On any Agreements concluded after 8th March 1977 the amount payable to the Consulting Engineer shall be increased by $12\frac{1}{2}\%$ of the sum as calculated in Clause 14.1(b).

14.2 Payments made in accordance with either Clause 14.1 or Clause 14.2 shall be calculated on the total cost of the works in the Bills, including all provisional and prime cost items therein.

14.3 Unless and until the Bills are fully accurately priced, the cost of the works in the Bills shall be the total of the tender recommended for acceptance or, where no tender has been received, shall be the Consulting Engineer's best estimate of the said cost.

15. Payment for variations

15.1 (a) For his services under Clause 9, the Consulting Engineer shall be paid an additional amount of $2\frac{1}{2}\%$ on the value of additional works and of fluctuations (increases and/or decreases) in the cost of labour and materials, and $1\frac{1}{2}\%$ on the value of omitted works.

(b) On any Agreements concluded after 8th March 1977, the amount payable to the Consulting Engineer shall be increased by $12\frac{1}{2}\%$ of the sum as calculated in Clause 15.1(a). Where measurement of the completed Works is carried out for the purposes of Clause 8.3, the preceding sub-clause shall not apply.

16. Payment for use of computer or other special equipment

Where the Consulting Engineer decides to use a computer or other special equipment in carrying out any additional services in accordance with Clause 7 or is expressly required by the Client to use a computer or other special equipment in the carrying out of his services under Clause 6, the Client shall, unless otherwise agreed between the Client and the Consulting Engineer, pay the Consulting Engineer

(a) for the time spent in connection with the use of a computer or other special equipment, including the development and writing of programmes and the operation of the computer in trial and final runs, in accordance with the Scale of Charges set out in Clause 13.2, and

(b) a reasonable charge for the use of the computer or other special equipment, which charge shall, if possible, be agreed between the Client and the Consulting Engineer before the work is put in hand.

17. Payment for site supervision

17.1 In addition to any other payment to be made by the Client to the Consulting Engineer under this Agreement, the Client shall

(a) reimburse the Consulting Engineer in respect of all salary and wage payments made by the Consulting Engineer to site-staff employed by the Consulting Engineer pursuant to Clause 10 and in respect of all other expenditure incurred by the Consulting Engineer in connection with the selection, engagement and employment of site-staff, and

CONSULTING ENGINEERS' FEES

(*b*) pay to the Consulting Engineer a sum calculated at 7 per cent of the amounts payable to the Consulting Engineer under the preceding sub-clause in respect of head office overhead costs incurred on site-staff administration.

provided that in lieu of payments under (*a*) and (*b*) above the Client and the Consulting Engineer may agree upon inclusive monthly or other rates to be paid by the Client to the Consulting Engineer for each member of site-staff employed by the Consulting Engineer.

17.2 The Client shall also in all cases be responsible for the cost of providing such local office accommodation, furniture, telephones, equipment and transport as shall be reasonably necessary for the use of site-staff appointed pursuant to Clause 10, and for the reasonable running costs of such necessary local office accommodation and other facilities, including those of stationery, telephone calls, telegrams and postage. Unless otherwise agreed between the Client and the Consulting Engineer, the Consulting Engineer shall arrange, whether through the Contractor or otherwise, for the provision of local office accommodation and facilities for the use of site-staff.

17.3 In cases where the Consulting Engineer has thought it proper that site-staff should not be appointed, or where the necessary site-staff is not available at site due to sickness or any other cause, the Consulting Engineer shall be paid in accordance with the Scale of Charges set out in Clause 13.2 for site visits which would have been unnecessary but for the absence or non-availability of site-staff.

18. Disbursements

The Client shall in all cases reimburse the Consulting Engineer in respect of all the Consulting Engineer's disbursements properly made in connection with:

(*a*) Printing, reproduction and purchase of all documents, drawings, maps and records.
(*b*) Telegrams and telephone calls other than local.
(*c*) Travelling, hotel expenses and other similar disbursements.
(*d*) Advertising for tenders and for site-staff.
(*e*) The provision of additional services to the Client pursuant to Clause 7.4.

The Client may, however, by agreement between himself and the Consulting Engineer make to the Consulting Engineer a lump sum payment or payment of a sum calculated as percentage charge on the cost of the Works in satisfaction of his liability to the Consulting Engineer in respect of the Consulting Engineer's disbursements.

19. Payment for alteration or modification to design

If after the completion by the Consulting Engineer of his services under Clause 6.1 any design whether completed or in progress or any specification, drawing or other document prepared in whole or in part by the Consulting Engineer shall require to be modified or revised by reason of instructions received by the Consulting Engineer from the Client or from the Architect, or by reason of circumstances which could not reasonably have been

CONSULTING ENGINEERS' FEES

foreseen, then the Client shall make additional payment to the Consulting Engineer for making any necessary modifications or revisions and for any consequential reproduction of documents. Unless otherwise agreed between the Client and the Consulting Engineer, the additional sum to be paid to the Consulting Engineer shall be calculated in accordance with the Scale of Charges set out in Clause 13.2, and shall also include any appropriate reimbursements in accordance with Clause 18.

20. Payment when works are damaged or destroyed

If at any time before completion of the Works any part of the Works or any materials, plant or equipment whether incorporated in the Works or not shall be damaged or destroyed, the Client shall make additional payment to the Consulting Engineer in respect of any expenses incurred or additional work required to be carried out by the Consulting Engineer as a result of such damage or destruction. The amount of such additional payment shall be calculated in accordance with the Scale of Charges set out in Clause 13.2, and shall also include any appropriate reimbursements in accordance with Clause 18.

23. Cost of the works

23.1 The cost of the Works or any part thereof shall be deemed to include:

(a) The cost to the Client of the Works however incurred, including any payments (before deduction of any liquidated damages or penalties payable by the Contractor to the Client) made by the Client to the Contractor by way of bonus, incentive or ex-gratia payments, or in settlement of claims.

(b) Where the Works are carried out as a sub-contract or sub-contract awarded under a main contract, the allowances made in the main contract to cover attendance and profit relating to the Works, together with the cost of items of builder's work required in connection with the Works, and a part of the cost of the preliminary and general items included in the main contract being the proportion that the cost of the sub-contract or sub-contracts bears to the total cost of the main contract.

(c) A fair proportion of the cost of any chimneys, and any air-conditioning or ventilation ducts and their insulation, forming part of the building structure. The said fair proportion shall be assessed with reference to the Consulting Engineer's estimate of the cost of independent construction.

(d) A fair valuation of any labour, materials, manufactured goods, machinery or other facilities provided by the Client, and of the full benefit accruing to the Contractor from the use of construction plant and equipment belonging to the Client which the Client has required to be used in the execution of the Works.

(e) The market value, as if purchased new, of any second-hand materials, manufactured goods and machinery incorporated in the Works.

23.2 The cost of the Works shall not include:

(a) Administration expenses incurred by the Client.

(b) Costs incurred by the Client under this Agreement.

(c) Interest on capital during construction, and the cost of raising moneys required for carrying out the construction of the Works.

(d) Cost of land and wayleaves.

DISTRICT SURVEYORS' FEES

FEES PAYABLE BY BUILDER, OWNER OR OCCUPIER TO COUNCIL IN RESPECT OF SERVICES RENDERED BY DISTRICT SURVEYOR

(*A*) In respect of buildings, structures or works to which the provisions of the London Building Acts and any by-laws made in pursuance of those Acts apply (except buildings, structures and works exempt by virtue of section 149 (Buildings exempt from provisions of Parts III and IV, etc.) of this Act) –

	£
When the cost is £100 or less –	
the sum of	10·00
When the cost exceeds £100 but does not exceed £1,000 –	
(*i*) the sum of	10·00
plus (*ii*) for every £100 (or fractional part thereof) by which the cost exceeds £100 –	
the sum of	2·25
When the cost exceeds £1,000 but does not exceed £5,000 – . . .	
(*i*) the sum of	30·25
plus (*ii*) for every £100 (or fractional part thereof) by which the cost exceeds £1,000 –	
the sum of	0·75
When the cost exceeds £5,000 but does not exceed £1,000,000 –	
(*i*) the sum of	60.25
plus (*ii*) for every £100 (or fractional part thereof) by which the cost exceeds £5,000 –	
the sum of	0·45
When the cost exceeds £1,000,000 but does not exceed £2,000,000 –	
(*i*) the sum of	4,537·75
plus (*ii*) for every £100 (or fractional part thereof) by which the cost exceeds £1,000,000 –	
the sum of	0·30
When the cost exceeds £2,000,000 –	
(*i*) the sum of	7,537·75
plus (*ii*) for every £100 (or fractional part thereof) by which the cost exceeds £2,000,000 –	
the sum of	0·20

Provided that –

(1) when the work is done as the result of a notice served under section 62 (Certification of dangerous structures) of this Act without the necessity of a complaint being made to a magistrates' court and the cost thereof does not exceed £15 no fee shall be payable in respect thereof; and

(2) when the work is done as the result of a notice served under the said section 62 or an order of a magistrates' court and the cost thereof exceeds £15 the fee payable shall be

DISTRICT SURVEYORS' FEES

reduced by the amount of the fee payable under item 2 of paragraph (*i*) of heading (*b*) of Part I of Schedule 1 to this Act for an inspection and report as to the completion of the works when such inspection is coincident with any other inspection made by the district surveyor in connection with his supervision of work under the London Building Acts and any by-laws made in pursuance of those Acts.

(*B*) In respect of structures to which Part IV (Special and temporary buildings and structures) of this Act applies the same amount as for a building to which heading (*A*) of this Part of this schedule applies.

Provided that this paragraph shall not apply in any case in which the local authority, being a London borough council, is the authority to grant consent under the said Part IV for the setting up, erection or retention of the structure.

(*C*) In respect of public buildings the said amount as for a building to which heading (*A*) of this Part of this schedule applies with the addition of 50% of the amount of such fee.

RULES

1. Any fees payable in respect of works to a party wall comprise the fees payable in respect of both sides of the wall.

2. No fee shall be payable in respect of the fixing of a chimney pot.

3. No fee shall be payable in respect of the repairing of a chimney top unless the top has been pulled down to a greater extent than 12 in.

4. No fees shall be payable in respect of the repairing of a parapet unless the parapet has been pulled down to a greater extent than 12 in.

Building Costs and Tender Prices Index

The tables which follow show the changes in building costs and tender prices since 1970.

To avoid confusion it is essential that the terms 'building costs' and 'tender price' are clearly defined and understood.

'Building costs' are the costs actually incurred by the builder in the course of his business the major ones being those for labour and materials.

'Tender price' is the price for which a builder offers to erect a building. This includes 'building costs' but also takes into account market considerations such as the availability of labour and materials and the prevailing economic situation. This means that in 'boom' periods when there is a surfeit of building work to be done 'tender prices' may increase at a greater rate than 'building costs' whilst in a period when work is scarce 'tender prices' may actually fall when building costs are rising.

Building costs

This table reflects the fluctuations, since 1970 in wages and materials costs to the builder. In compiling the table the proportion of labour to material has been assumed to be 40:60. The wages element has been assessed from a contract wages sheet revalued for each variation in labour costs whilst the changes in the cost of materials have been based on the indices prepared by the Department of Trade and Industry. No allowance has been made for changes in productivity, plus rates or hours worked which may occur in particular conditions and localities.

1970 = 100

Year	First quarter	Second quarter	Third quarter	Fourth quarter	Annual average
1970	97	99	101	103	100
1971	105	109	110	111	109
1972	112	113	119	132	119
1973	134	136	143	147	140
1974	153	163	172	176	166
1975	189	199	215	218	205
1976	224	233	251	257	241
1977	264	271	280	282	274
1978	284	289	306	311	298
1979	317	325	360	369	343
1980	378	388			

NOTE: *The above indices, up to the first quarter of 1975, are arithmetic transformations of those which appeared in the 101st edition of this price book recalculated to a base of 1970. Indices published in earlier editions with a 1956 base may be converted to a base of 1970 by dividing by 1·87.*

Tender Prices

This table reflects the changes in tender prices since 1970. It indicates the level of pricing contained in the lowest tenders for new work in the London area (over £30,000 in value) compared with a common base.

1970 = 100

Year	First quarter	Second quarter	Third quarter	Fourth quarter	Annual average
1970	98	98	103	104	100
1971	109	112	116	121	115
1972	133	138	147	161	145
1973	176	190	207	223	199
1974	234	242	238	233	237
1975	243	249	236	241	242
1976	233	236	247	249	241
1977	253	254	263	265	259
1978	272	280	303	335	298
1979	343	352	385	403	371
1980	432	476			

Prices

This part of the book contains the following sections:

Market Prices of Materials

The prices given, unless otherwise stated, include the cost of delivery to sites in the Inner London area. Prices should be adjusted in cases where high transport costs may be incurred or where small quantities are involved.

Prices are given for the units in which the materials are sold and this means that both imperial and metric terms are used.

In general, the prices given are those in force at April/May 1980 and later information may be given in the 'Stop Press', but readers are advised to check them against current quotations.

All prices quoted are exclusive of Value Added Tax.

EXCAVATION AND EARTHWORK

	Unit	£
Removing excavated material from site, not over 11 miles		
Hand loaded	m³	4·90
Machine loaded	"	2·88
Removing rubbish from site by container	"	7·40 plus container hire charge of approx. £0·40 per day
Hardcore	"	2·49
Ashes	"	6·54
Pulverised fuel ash	"	5·25
Granular fill	tonne	5·10

CONCRETE WORK

READY MIXED CONCRETE

The following are average prices which will vary depending on size of order and site conditions affecting discharge

Ballast concrete using coarse aggregate to B.S. 882 of the size shown

	Unit	Minimum cube strength at 28 days, N/mm²	Normal setting cement to B.S. 12	Rapid hardening cement to B.S. 12	Sulphate resisting cement
			£	£	£
10 mm	m³	21·00	31·10	31·60	32·65
10 ,,	,,	26·00	32·20	32·80	33·75
20 ,,	,,	16·00	29·45	30·00	30·80
20 ,,	,,	21·00	30·00	30·45	31·25
20 ,,	,,	26·00	31·00	31·65	32·75
20 ,,	,,	28·00	31·30	31·85	32·75
20 ,,	,,	31·50	31·40	32·00	32·90
40 ,,	,,	21·00	30·00	30·45	31·35
40 ,,	,,	26·00	31·00	31·55	32·60
Lightweight concrete using sintered pulverized fuel ash aggregate	,,	21·00	37·15	—	—
,, ,, ,, ,,	,,	26·00	38·70	—	—
,, ,, ,, ,,	,,	28·00	39·00	—	—

CONCRETE AGGREGATES

	Unit	£
Concrete aggregates to B.S. 882		
40 mm all-in aggregate	tonne	5·10
20 ,, ,, ,,	,,	5·10
40 ,, shingle	,,	5·10
20 ,, ,,	,,	5·10
10 ,, ,,	,,	5·10
Sharp sand	,,	5·10

LIGHTWEIGHT AGGREGATES

	Unit	£
Aglite		
Medium or fine, in bulk	,,	20·24
Medium or fine, in bags	,,	33·17
Lytag		
Granular, 13–8 mm	m³	10·91
Granular, 8–5 mm	,,	10·91
Fine, 5 mm down	,,	11·49

CONCRETE WORK

CEMENT		Unit	£
Portland cement			
Normal-setting quality to B.S. 12			
In bulk to silos	. . . .	tonne	38·22
In bags .	. . . . . .	,,	40·94
Rapid-hardening quality to B.S. 12	. . . .	,,	43·10
Sulphate resisting cement	. . . . .	,,	47·14
Aluminous cement to B.S. 915	. . . . .	,,	93·08

JOINT FILLERS, SEALERS AND WATERSTOPS

			Thickness		
		10	12·5	20	25
	Unit	mm	mm	mm	mm
		£	£	£	£
Bitumen impregnated joint filler					
75 mm wide	m	0·21	0·24	0·38	0·45
100 mm ,, ...	,,	0·24	0·27	0·43	0·51
150 mm ,, ...	,,	0·35	0·39	0·63	0·75
175 mm ,, ...	,,	0·38	0·41	0·67	0·80
200 mm ,, ...	,,	0·42	0·46	0·76	0·90

Polyethylene sealant backer and joint strip

		£
Cord		
12 mm diameter	15 metres	1·86
18 mm ,, ...	,,	2·98
25 mm ,, ...	,,	4·04
38 mm ,, ...	10 metres	6·16
Strip		
6 × 12 mm	15 metres	0·54
6 × 18 mm	,,	0·86
6 × 25 mm	,,	1·16
12 × 18 mm	,,	1·54
12 × 25 mm	,,	2·10
12 × 50 mm	,,	4·38
Polysulphide sealant, gun grade, grey or black . .	litre	6·23
Butyl mastic sealing compound	,,	2·75
Hot applied rubber bitumen sealing compound . .	Kg	0·59
Bituminous joint sealing compound	litre	1·38
Primer for ditto	,,	2·05
Waterstops		
Rubber, 150 mm wide with centre bulb . .	m	5·35
,, ,, ,, ,, with plain web . .	,,	5·05
,, 230 mm ,, with centre bulb .	,,	8·56
,, ,, ,, ,, with plain web .	,,	6·52
P.V.C., 140 mm wide eyeleted, with centre bulb .	,,	1·56
,, 190 ,, ,, eyeleted, plain web .	,,	1·89
,, 240 ,, ,, eyeleted, with centre bulb	,,	2·68

		Three way	Four way	Flat mitre	Edge mitre
Intersections for water stops		£	£	£	£
Rubber					
150 mm wide with centre bulb . .	No.	14·54	15·85	13·25	13·25
,, ,, ,, with plain web . .	,,	13·23	14·54	12·10	12·10
230 mm ,, with centre bulb . .	,,	16·89	18·07	15·70	15·70
,, ,, ,, with plain web . .	,,	15·85	16·89	14·55	14·55
P.V.C.					
140 mm wide, eyeleted, with centre bulb	,,	4·99	6·08	2·55	2·24
190 ,, ,, eyeleted, plain web .	,,	4·73	6·01	3·22	2·16
240 ,, ,, eyeleted, with centre bulb	,,	7·23	8·88	4·82	3·15

Sports Halls and Swimming Pools

A design and briefing guide

G. A. Perrin, Perrin Associates

'. . . . lavishly illustrated with black and white photographs, colour prints and numerous plans of UK and European halls and pools . . . . The book's format is such as to provide a valuable reference tool for architects and recreation managers during the pre-contract research and discussion period.

The sections include primary and secondary briefs, with design and technical guides and feasibility studies. There is also a management guide setting out objectives, staffing and other related matters.'

Local Government News

1980   222 pages
Hardback: 0 419 11140 9   £16.50

E & F N Spon Ltd

The technical division of Associated Book Publishers Ltd., 11 New Fetter Lane, London EC4P 4EE

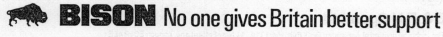

CONCRETE WORK

	Unit	B.S. 4449 £	B.S. 4461 £

STEEL REINFORCING RODS
(HOME PRODUCED)

Steel reinforcing rods cut to length, bent, bundled and labelled in the following diameters in quantities over 20 tonnes

	Unit	B.S. 4449 £	B.S. 4461 £
40 mm	tonne	234·64	242·64
32 mm	,,	233·58	241·58
25 mm	,,	231·43	239·43
20 mm	,,	231·24	239·24
16 mm	,,	235·84	243·84
12 mm	,,	263·86	271·86
10 mm	,,	274·36	282·36
8 mm	,,	283·40	291·40
6 mm	,,	291·72	299·72

Extra for quantities of:

	Unit	£
15 to under 20 tonnes	,,	2·00
10 to under 15 tonnes	,,	4·00
5 to under 10 tonnes	,,	6.00
Under 5 tonnes	,,	8·00

STEEL WIRE MESH FABRIC

Steel wire mesh fabric to B.S. 4483 in standard widths in quantities over 5 tonnes

	Unit	£
Square		
B.S. Ref. A 393, 6·16 kg/m²	m²	1·83
,, ,, A 252, 3·95 ,,	,,	1·18
,, ,, A 193, 3·02 ,,	,,	0·90
,, ,, A 142, 2·22 ,,	,,	0·68
,, ,, A 98, 1·54 ,,	,,	0·47
Structural		
B.S. Ref. B 1131, 10·90 ,,	,,	3·46
,, ,, B 785, 8·14 ,,	,,	2·43
,, ,, B 503, 5·93 ,,	,,	1·77
,, ,, B 385, 4·53 ,,	,,	1·35
,, ,, B 283, 3·73 ,,	,,	1·12
,, ,, B 196, 3·05 ,,	,,	0·91
Long		
B.S. Ref. C 785, 6·72 ,,	,,	2·01
,, ,, C 636, 5·55 ,,	,,	1·65
,, ,, C 503, 4·34 ,,	,,	1·29
,, ,, C 385, 3·41 ,,	,,	1·02
,, ,, C 283, 2·61 ,,	,,	0·78
Hy-rib reinforcement		
Ref. 2411, 5·71 Kg/m²	,,	3·89
,, 2611, 4·02 ,,	,,	3·27

PRECAST CONCRETE

Precast concrete 21 N/mm², 20 mm aggregate finished fair on exposed faces.

	Unit	£
Lintels reinforced with mild steel rods		
114 × 150 mm with one 12 mm rod	m	3·90
150 × 150 mm with two 12 mm rods	,,	4·20
225 × 150 mm ,, 12 mm ,,	,,	6·25
225 × 225 mm ,, 16 mm ,,	,,	9·40
Coping, weathered, with sinkings at each end for mortices		
175 × 75 mm	,,	3·80
325 × 75 mm	,,	5·10

CONCRETE WORK

	Unit	£

PRECAST CONCRETE – *continued*

Duct covers cast in short lengths reinforced with steel fabric weighing
2·22 kg/m²

	Unit	£
50 mm thick	m²	13·75
75 mm „	„	14·80
100 mm „	„	16·60
Concrete and glass pavement lights in precast panels for		
Pedestrian traffic	„	56·00
Vehicular ditto	„	74·50
Concrete and glass roof lights in precast panels		
Single glazed	„	56·00
Double glazed	„	84·80

FLOOR POTS

Hollow clay floor pots to B.S. 3921

	Unit	£
300 × 300 × 75 mm	1000	166·21
300 × 300 × 100 mm	„	193·60
300 × 300 × 125 mm	„	221·05
300 × 300 × 150 mm	„	246·10
300 × 300 × 175 mm	„	275·66
300 × 300 × 200 mm	„	303·23
300 × 300 × 225 mm	„	357·91
300 × 300 × 250 mm	„	395·17
Standard filler tiles		
300 × 75 × 15 mm	Per 1000	170·56
300 × 100 × 15 mm	multiples	287·05
300 × 125 × 15 mm	of 10 tiles	346·81

WATER-PROOFERS AND D.P.C. MEMBRANES

	Unit	£
Colemanoid No. 1	litre	0·49
Pudlo	kg	0·33
Cementone No. 2 powder	„	0·61
Sealocrete powder	„	0·56
Sealocrete Double-Strength Premix	litre	0·46
Aston Evo Plast	„	0·49
Synthaprufe	„	0·61
Tretol 202T/200T	„	0·36

CONCRETE WORK

	Unit	£
ANTI-FROST ADDITIVES		
Sealocrete Double-strength Premix	litre	0·46
Colemanoid No. 1	,,	0·49
Tretol Anti-freezer	,,	0·32
CEMENT RETARDER		
Sika	,,	0·61
BUILDING PAPER		
Sisalkraft		
Subsoil grade	100 m²	21·70
Standard grade	,,	34·95
Ibeco waterproof kraft in rolls 36 in. × 100 yd.		
No. 40 to B.S. 1521, Class B, Grade B.2	Roll	6·87
No. 60 to B.S. 1521, Class B, Grade B.1	,,	9·81
No. 80 to B.S. 4016, breather type	,,	14·38
POLYTHENE SHEETING		
Polythene sheeting		
0·07 mm thick, 65 mu gauge	100 m²	5·40
0·13 mm thick, 125 mu gauge	,,	9·35
0·26 mm thick, 250 mu gauge	,,	18·70
Visqueen sheeting		
1200 Super, 0·30 mm thick	,,	26·90
2000 T, 0·46 mm thick	,,	38·00
CLIPS, GUARDS, INSERTS AND ANCHORS		
50 mm Bulldog floor clips	1000	51·81
Acoustic Bulldog clips	,,	341·55
Galvanized steel plate column guards in 1 m lengths, ex works:		
75 × 75 × 3 mm	No.	4·95
75 × 75 × 4·5 mm	,,	6·68
Mild steel cast in sockets with nailing flange		
6 mm diameter thread	100	24·05
12 mm ,, ,, ...	,,	41·45
24 mm ,, ,, ...	,,	173·30
Coated white nylon steel corner guard, ex works		
75 × 75 × 1·5 mm in 1000 mm lengths	No.	6·24
75 × 75 × 1·5 mm in 1500 mm lengths	,,	9·66
18 gauge galvanized steel dovetail masonry slot in 10 ft. lengths . .	100 lin ft.	14·00
Galvanized steel anchors		
Standard brick, 14 g., 4 in. projection	100	6·20
Standard stone, ⅛ in. thick, 3½ in. projection	,,	7·80
Standard faience, ⅛ in. thick, 1¾ in. projection	,,	5·70
Slotted metal inserts, galvanized finish, 2·6 mm thick complete with end caps		
and foam filling		
75 mm long	No.	0·60
100 mm long	,,	0·65
150 mm long	,,	0·75
6 metres long	,,	13·90
Nuts with springs for slotted metal inserts		
6 mm diameter	100	13·30
10 mm diameter	,,	17·90
12 mm diameter	,,	18·20

BRICKWORK AND BLOCKWORK

	Unit	£
CEMENT		
Portland cement	tonne	40·94
White Portland cement	„	79·32
SAND		
Sand for brickwork to B.S. 1200	„	4·26
LIME		
White hydrated lime to B.S. 890 for non-hydraulic calcium lime .	„	44·65
MORTAR PLASTICIZER		
Mortar plasticizer	litre	0·41
BRICKS		
Fletton bricks, 215 × 102·5 × 65 mm		
Plain	1000	44·24
Keyed	„	44·38
Extra for strapak	„	3·00
Stock bricks, 215 × 102·5 × 65 mm		
2nd hard	„	153·43
Mild	„	109·93
Old English Stock	„	111·43
Warnham pressed Class, 5, 215 × 102·5 × 65 mm . . .	„	88·73
Southwater engineering, 215 × 102·5 × 65 mm		
Class A	„	159·23
Class B	„	121·23
Single bullnose	„	333·23
Double bullnose	„	393·23
Lingfield Engineering wirecuts Class 7, 215 × 102·5 × 65 mm .	„	95·33
Phorpres facing bricks, 215 × 102·5 × 65 mm		
Rustic	„	52·34
Tudors	„	55·26
Sandfaced	„	56·01
Chiltern	„	56·16
Golden buffs	„	55·79
Dapple light	„	57·56
Brindle	„	57·06
Milton Buff Ridgefaced	„	57·73
Regency	„	63·23
Windsor	„	61·46

Third Edition

Building Design Evaluation

Costs-in-use

P A Stone

This book provides a complete guide for accurate assessment of costs-in-use which are essential for the evaluation and benefit for all building and development plans.

The third edition, now in paperback as well, has been produced in the context of changes in technology, ever-rising costs in labour and materials and growing relative scarcity of some of those materials, particularly energy. The tables of discount rates have been further extended and an additional appendix on building costs and prices included.

1980   256 pages
Hardback: 0 419 11720 2   £10.00
Paperback: 0 419 11660 5   £5.75

E & F N Spon Ltd
technical division of Associated Book Publishers Ltd, 11 New Fetter Lane, London EC4P 4EE

Methuen Inc
733 Third Avenue
New York
NY 10017

In line with your aims...

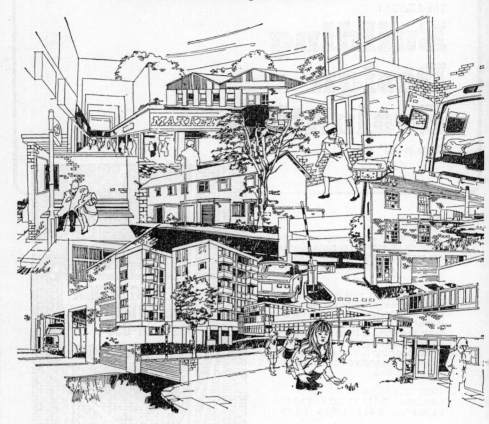

LBC facings and commons.
Quality, value and appearance central to good building everywhere.

BRICKWORK AND BLOCKWORK

BRICKS – *continued*	Unit	£
Tonbridge facing bricks, 215 × 102·5 × 65 mm Hand-made multicoloured	1000	186·18
Dorking multicoloured stocks 215 × 102·5 × 65 mm	,,	121·38
Midhurst White facing bricks to B.S. 187, 190 × 90 × 65 mm		
Calcium silicate		
Class 3	,,	60·70
,, 4	,,	64·90
Flint silicate		
Class 3	,,	70·70
Bariwall radiation shielding bricks, 250 × 120 × 65 mm	,,	900·00
Special jointing mortar	Tonne	75·00

BUILDING BLOCKS

	Unit	75 mm £	90 mm £	100 mm £	140 mm £	*Thickness* 150 mm £	190 mm £	215 mm £	230 mm £	305 mm £
Precast concrete blocks, lytag aggregate, 440 × 215 mm										
Cellular Hollow	m²	2·11	—	3·05	3·88	—	—	5·30	—	
Solid	,,	2·50	—	3·12	4·66	—	6·10	6·85	—	
Aglite blocks, 440 × 215 mm										
Hollow	,,	—	—	—	—	4·25	—	6·20	—	
Solid	,,	2·00	—	2·48	—	4·53	—	6·66	—	
Celcon blocks, 450 × 225 mm	,,	2·57	3·03	3·37	4·72	5·05	6·40	7·24	7·75	
Forticrete grey granite blocks, 400 × 200 mm										
Hollow	,,	—	—	—	6·18	—	8·20	—	—	
Solid	,,	—	4·61	—	7·63	—	10·71	—	—	
Hemelite blocks, 450 × 225 mm										
Cored or cellular	,,	—	—	3·10	3·90	—	—	—	—	
Solid	,,	2·20	2·68	2·79	3·90	4·03	5·29	5·92	5·92	
Lignacite blocks, 450 × 225 mm										
Cored		—	—	3·73	—	6·02	—	—	—	
Solid	,,	2·84	—	3·80	—	—	—	—	—	
Thermalite blocks, 450 × 225 mm										
Standard	,,	2·45	2·93	3·12	4·56	4·89	6·19	7·01	—	9·94
Smooth face	,,	2·73	3·27	3·48	5·09	5·45	6·91	7·82	—	
Hollow clay blocks, keyed both sides										
300 × 225 mm	1000	122·11	—	149·44	—	—	—	—	—	
338 × 150 mm	,,	—	—	—	—	205·95	—	—	—	

BRICKWORK AND BLOCKWORK

	Unit	£
DAMP-PROOFING MATERIALS		
Cold rolled copper strip to B.S. 743 0·56 mm thick		
114 mm wide	m	1·01
229 mm ,,	,,	2·05
Damp-course slates to B.S. 743		
14 in. × 4½ in.	100	25·33
14 in. × 9 in.	,,	46·00
Bituminous damp-course to B.S. 743		
Hessian base	m²	1·21
Fibre base	,,	0·84
Lead-lined hessian base	,,	3·10
Lead-lined fibre base	,,	3·05
Asbestos damp-course to B.S. 743		
Standard	,,	1·25
Lead-lined	,,	3·42
Pitch polymer damp-course	,,	1·76
Polythene	,,	0·43
Bitumen polythene standard grade	,,	2·04
SUNDRIES		
Wall-ties, twisted type		
Metal to BS 1243		
Galvanized iron, ⅛ in.	100	5·55
Stainless steel, ⅛ in.	,,	26·50
Copper alloy, ⅛ in.	,,	53·00
Polypropylene	,,	2·44
Wall-ties, butterfly type		
Metal to BS 1243		
Galvanized iron, 10 s.w.g.	,,	4·05
Stainless steel, 10 s.w.g.	,,	10·05
Copper, 10 s.w.g.	,,	20·00
24 gauge metal mesh reinforcement in 270 ft. coils		
2½ in. wide	Coil	6·11
4½ in. ,,	,,	10·78
Mastics		
General purpose mastic in cartridge	No.	0·96
Bituminous mastic in cartridge	,,	0·57
Air brick, galvanized cast iron, light		
220 × 70 mm	,,	1·35
220 × 145 mm	,,	2·60
220 × 220 mm	,,	3·75
Air brick, red terracotta		
220 × 70 mm	,,	0·60
220 × 145 mm	,,	1·10
220 × 220 mm	,,	2·35
Air brick, fibrous plaster		
220 × 70 mm	,,	0·34
220 × 145 mm	,,	0·40
220 × 220 mm	,,	0·50
Moler flue bricks		
230 × 114 × 64 mm	1000	226·00
230 × 114 × 76 mm	,,	264·00
Setting powder for flue bricks	tonne	167·00

BRICKWORK AND BLOCKWORK

	Unit	£
SUNDRIES – *continued*		
Firebricks		
2½ in.	100	42·00
3 in.	,,	47·00
Fireclay, loose	tonne	80·00
Cramps, 1 × ⅛ in., galvanized hoop iron, 10 in. girth . .	100	15·85
Plain creasing tiles	1000	167·00
Plain hand-made sand-faced roofing tiles, 11 × 7 in. . .	,,	316·00

MASONRY

	Unit	£
NATURAL STONE		
Stone in blocks		
Monks Park	m³	115·00
Portland, building quality depending on size and quantity . . .	,,	66·00
		to
		144·00
Doulting	,,	109·00
Ancaster Hardwhite	,,	179·00
Ancaster Weatherbed	,,	179·00
NATURAL SLATE		
Slate slabs for shelving, etc., fine rubbed one face, ex works		
25 mm thick	m²	53·00
38 mm thick	,,	79·00
50 mm thick	,,	105·00
CAST STONE		
Cast stone to B.S. 1217 to match Portland stone. (*Prices are for fair quantities of any one section*)		
Sill, plain sunk, weathered, and throated and grooved for water bar in approximately 1·2 m lengths		
229 × 76 mm	m	4·33
279 × 76 mm	,,	4·63
Extra for stoolings for 50 × 114 mm jambs	No.	0·22
Coping, weathered 1 m lengths with sinkings at each end for mortices		
178 × 76 mm	m	3·55
330 × 76 mm	,,	4·78
Saddleback coping, once throated		
178 × 76 mm	,,	4·00
330 × 76 mm	,,	5·13
Extra for external angle stone to 76 mm coping	No.	1·46

Smoke Control in Fire Safety Design

E G Butcher and A C Parnell

Foreword by **Kenneth L Holland,** CBE, H.M. Chief
Inspector of fire services for England and Wales

In recent years, the proportion of fire
deaths due to smoke and toxic fumes has
increased alarmingly. Smoke control has
consequently become a most important
consideration in designing for fire safety;
this book sets out to explain its basic
principles.

The opening chapters deal with smoke
quality and quantity and the factors
governing its spread through a building.
Later chapters consider specific problems
such as smoke control in the actual fire
area; escape routes without complete
structural protection; and the use of
pressurization to exclude smoke from fully
protected escape routes. The text is
enhanced by numerous figures and
tables, clearly illustrating the discussion.

1979   192 pages   illustrated
Hardback: 0 419 11190 5   £14.00

E & F N Spon Ltd
The technical division
of Associated Book Publishers
Ltd, 11 New Fetter Lane,
London EC4P 4EE

ROOFING

The following are intended to represent prices quoted to a general contractor and are not necessarily consistent with the 'Prices for Measured Work', the latter being based upon prices for roofing supplied and fixed by specialists.

ASBESTOS CEMENT SLATES

		Unit	£
Asbestos cement slates, blue			
500 × 250 mm		1000	325·38
600 × 300 mm		,,	451·66
600 × 350 mm		,,	678·36

NOTE: *The above are subject to a surcharge and carriage charge on orders under £1000 in value.*

NATURAL SLATES

		Unit	£
Welsh slates to B.S. 680			
255 × 150 mm 0·55 tonnes per thousand		1000	154·00
255 × 205 mm 0·70 ,, ,, ,,		,,	169·25
255 × 255 mm 0·85 ,, ,, ,,		,,	197·00
305 × 150 mm 0·65 ,, ,, ,,		,,	168·30
305 × 255 mm 1·00 ,, ,, ,,		,,	262·35
405 × 255 mm 1·30 ,, ,, ,,		,,	393·00
510 × 255 mm 1·60 ,, ,, ,,		,,	654·75
610 × 305 mm 2·25 ,, ,, ,,		,,	1098·00
Westmorland green slates, random sizes at (computed covering capacity, 22 sq. yd. per ton at 3 in. lap)		Tonne	402·00
Blue plain ridge or hip tiles 450 mm		No.	1·20

	Unit	B.S. 473 Concrete tiles £	B.S. 402 Broseley or Staffordshire machine-made tiles £	B.S. 402 Best hand-made sand-faced tiles £
PLAIN TILES				
Plain tiles	1000	94·20	103·80	202·85
Eaves tiles	,,	94·20	100·65	200·18
Tile and half	,,	183·20	202·60	402·35
Angle tiles	100	60·52	101·07	123·10
Valley tiles	,,	60·52	101·07	123·10
Half-round or segmental ridge tile 300 mm long	,,	91·52	99·35	120·35
Bonnet hip tile	,,	60·52	101·07	123·10

ROOFING

	Unit	£
SINGLE LAP TILES		
Lincolnshire red sandfaced pantiles, 13¼ × 9½ in.	1000	311·25
Double roll verge tiles	,,	881·25
Concrete patent interlocking tiles, 413 × 330 mm to B.S. 550	,,	248·80
Marley Modern tiles, 413 × 330 mm	,,	274·60
UNDERLAYS AND SLATERS' FELTS		
Fine granule surfaced felt weighing 14 kg/10 m² roll	Roll	5·26
Reinforced felt weighing 22·5 Kg/15 m²	,,	12·76
ASBESTOS-CEMENT SHEETS		
Grey corrugated sheets (Big Six or similar)	m²	3·15
Lining panels to suit corrugated sheets	,,	2·67
Eaves filler piece	No.	2·40
Eaves filler and flashing piece	,,	2·28
Two-piece adjustable ridge capping	,,	4·61
As above, but ventilating type	,,	5·60
Two-piece adjustable north-light ridge capping	,,	4·14
Ridge finials	,,	2·84
Apron flashing piece	,,	2·41
Flashing piece for bottom of glazing	,,	2·40
Horizontal flashing piece	,,	6·55
Angle cover piece 2400 mm long	,,	6·66
Barge boards 2400 mm long	,,	6·66
Extra for Colourglaze treatment gloss finish to exterior surfaces of corrugated sheeting	m²	2·36

NOTE: *The above prices are subject to a surcharge and carriage charge on orders under £1000 in value.*

			Fire rating	
CORRUGATED PLASTICS TRANSLUCENT SHEETS	Unit	300 *grade* AB Class 3 £	205 *grade* AA Class 2 £	101 *grade* AA Class 1 and 0 £
Filon translucent corrugated sheets to suit the following profiles				
3 in. standard asbestos	m²	7·53	8·00	9·83
Big Six asbestos	,,	6·02	6·70	8·50
British Steel Long Rib				
900 roofing	,,	7·01	7·75	9·70
900 vertical	,,	7·01	7·75	9·70

ROOFING

	Unit	BS 1105 Type B £	High density £
ROOF DECKING			
Unreinforced wood-wool slabs			
50 mm thick	m²	2·50	3·11
75 mm „	„	3·67	3·34
100 mm „	„	5·07	5·27
Interlocking reinforced wood-wool slabs			
50 mm thick	„	6·95	7·49
75 mm „	„		12·62
100 mm „	„		13·18
Extra for			
Pre-felted quality (glass fibre)	„	1·44	1·44
Pre-screeded deck	„	0·62	0·62
Pre-screeded and proofed deck	„	0·95	0·95
Pre-felted deck plus pre-screeded soffit	„	3·52	3·52
SHEET METAL			
Sheet lead to B.S. 1178	100 kg		109·75
Aluminium sheet, 0·9 mm commercial quality	„		115·00
Zinc sheet, 0·80 mm	„		94·85
Copper sheet			
23 gauge	„		182·00
24 „	„		184·40
SUNDRIES			
Slate or tile battens Celcure or Tanalised finish			
19 × 38 mm	100 m		16·45
19 × 50 mm	„		20·47
25 × 50 mm	„		28·90
Copper nails	kg		7·56
Aluminium nails	„		4·50
Aluminium alloy roofing strip for felt roofing			
38 mm on face, 48 mm fixing arm	m		1·54
Angles	No.		1·64
38 mm on face, 102 mm fixing arm	m		1·90
Angles	No.		1·97
63·5 mm on face 63·5 mm fixing arm	m		1·90
Angles	No.		1·97
Aluminium alloy roofing strip for asphalt roofing			
76·2 mm on face 101·6 mm fixing arm	m		2·52
Angles	No.		2·67
47·6 mm on face 101·6 mm fixing arm	m		2·24
Angles	No.		2·03

Market Prices of Materials

CARPENTRY AND JOINERY

TIMBER		Unit	£
Softwood			
Carcassing quality			
Up to 4·80 m lengths	. . . .	m³	120·00
4·80–6 m lengths	. . . .	,,	125·00
6·30–6·90 m lengths	. . . .	,,	132·00
Extra for			
Stress grading, subject to specification and quality			
S2/M75 grade	. . . .	,,	18·00
S2/MSS ,,	. . . .	,,	14·00
S2/M50 ,,	. . . .	,,	12·00
S2/MGS ,,	. . . .	,,	8·00
Joinery quality			
Average Redwood	. . . .	,,	166·00
Top-grade Redwood	. . . . . .	,,	175·00
No. 2 clear and better Douglas Fir	. . . . .	,,	289·00
No. 2 clear and better Hemlock	. . . . .	,,	246·00
P.T. and G. and P.E. flooring to B.S. 1297, 150 mm nominal widths			
16 mm finish	. . . .	10 m²	28·00
19 mm finish	. . . .	,,	33·00
Hardwood, fair average specification, average prices, ex wharf			
African walnut	. . . .	m³	288·00
Afrormosia	. . . .	,,	435·00
Agba	. . . .	,,	254·00
Beech	. . . .	,,	270·00
Brazilian mahogany	. . . . . .	,,	328·00
Iroko	. . . .	,,	232·00
Keruing	. . . .	,,	155·00
Meranti	. . . .	,,	235·00
Obeche	. . . .	,,	186·00
Sapele	. . . .	,,	311·00
Utile	. . . .	,,	367·00
West African mahogany	. . . . . .	,,	295·00

BLOCKBOARD	16 mm	18 mm	22 mm	25 mm
Prices are per 10m² for full bundles ex wharf	£	£	£	£
Birch	48·50	51·10	58·70	65·30
Gaboon	56·30	59·35	67·35	75·70

LAMINBOARD		16 mm	
Prices are per 10 m² for full bundles ex wharf		£	
Birch		64·50	78·80	87·60

PLYWOOD	6/6·5 mm	9/9·5 mm	12/12·5 mm	18/19 mm	24/25 mm
Prices are for full bundles of 10 m² ex wharf	£	£	£	£	£
Russian Birch, internal quality, Grade BB .	20·70	29·00	37·75	56·15	—
French Gaboon. external quality, Grade B/BB	26·15	35·65	46·55	67·50	93·60
Finnish Birch, external quality Grade BB .	27·65	35·82	45·60	67·10	88·22
Douglas Fir, external quality					
Good one side	—	—	—	36·50	—
Good two sides	—	—	—	47·50	—
Regular sheathing, unsanded . . .	—	14·55	19·75	29·35	49·85
Select sheathing, unsanded . . .	—	17·50	22·75	38·20	—

CARPENTRY AND JOINERY

CHIPBOARD	Unit	Plain £	Melamine faced £
Wood chipboard			
12 mm thick .	10 m²	13·65	26·35
18 mm ,, .	,,	18·30	47·30
25 mm ,, .	,,	27·50	—
Square edge flooring		£	
18 mm thick .	,,	19·45	
22 mm ,, .	,,	25·40	
Tongued and grooved flooring			
18 mm thick .	,,	21·35	
22 mm ,, .	,,	28·00	

BUILDING BOARDS

See '*Finishings*'

SHEET INSULATION	Unit	£
Sisalation reflective insulation		
Single sided .	100 m²	54·20
Double sided .	,,	79·05
Expanded polystyrene boards		
Standard grade		
12 mm thick	m²	0·23
18 mm ,,	,,	0·34
25 mm ,,	,,	0·48
38 mm ,,	,,	0·73
50 mm ,,	,,	0·96
Fire-resisting grade		
12 mm thick	,,	0·31
18 mm ,,	,,	0·46
25 mm ,,	,,	0·64
38 mm ,,	,,	0·97
50 mm ,,	,,	1·27
Fibred building mat, unfaced		
60 mm thick	,,	1·18
80 mm ,,	,,	1·56
100 mm ,,	,,	1·91
Building roll		
60 mm thick	,,	1·48
80 mm ,,	,,	1·89
100 mm ,,	,,	2·28
Flanged building roll		
Paper faced		
60 mm thick	,,	1·58
80 mm ,,	,,	1·98
100 mm ,,	,,	2·38
Polythene faced		
60 mm thick	,,	1·70
80 mm ,,	,,	2·11
100 mm ,,	,,	2·50

CARPENTRY AND JOINERY

SUNDRIES	Unit	£
Timber connectors		
1½ in. single sided for ⅜ in. bolts	100	4·30
1½ in. double sided ,, ,, ,,	,,	5·20
2 in. single sided for ½ in. bolts	,,	4·85
2 in. double ,, ,, ,,	,,	5·50
2½ in. single ,, ,, ,,	,,	7·00
2½ in. double ,, ,, ,,	,,	7·80
3 in. single ,, ,, ,,	,,	9·00
3 in. double ,, ,, ,,	,,	9·35
Teco split ring connectors		
2½ in. diameter	No.	0·30
4 in. ,,	,,	0·58
Teco shear plate connectors		
2⅝ in. diameter	,,	0·30
4 in. ,,	,,	1·00
Tie down straps		
1·6 mm tie down straps		
660 mm long for,		
Joists	,,	0·42½
Plates	,,	0·33
Joist hangers		
3 mm welded galvanized mild steel joist hangers for building into brickwork or blockwork		
75 mm deep × 50 mm wide	,,	0·70
75 mm deep × 75 mm wide	,,	0·80
100 mm deep × 50 mm wide	,,	0·76
100 mm deep × 75 mm wide	,,	0·87
100 mm deep × 100 mm wide	,,	1·05
125 mm deep × 50 mm wide	,,	0·82
125 mm deep × 75 mm wide	,,	0·95
125 mm deep × 100 mm wide	,,	1·14
150 mm deep × 50 mm wide	,,	0·89
150 mm deep × 75 mm wide	,,	1·02
150 mm deep × 100 mm wide	,,	1·23
175 mm deep × 50 mm wide	,,	0·97
175 mm deep × 75 mm wide	,,	1·08
175 mm deep × 100 mm wide	,,	1·34
200 mm deep × 50 mm wide	,,	1·05
200 mm deep × 75 mm wide	,,	1·15
200 mm deep × 100 mm wide	,,	1·43
300 mm deep × 50 mm wide	,,	1·59
300 mm deep × 75 mm wide	,,	1·80
300 mm deep × 100 mm wide	,,	2·13
Adhesives		
Evo-stik Impact 528 general purpose impact adhesive	litre	2·25
Evo-stik 863 for expanded polystyrene	,,	1·41
Evo-stik 'Resin W' for woodwork	,,	2·16

JOINERY – GENERAL

	Unit	Softwood £	Mahogany £	Afrormosia £
PLAIN LININGS				
Tongued angles:				
32 × 100 mm	m	1·95	3·50	4·56
32 × 140 mm	,,	2·63	4·50	6·12
38 × 100 mm	,,	2·35	4·17	5·80
38 × 150 mm	,,	2·82	5·16	6·96
Crosstongued joints; tongued angles:				
38 × 230 mm	,,	4·75	8·37	10·95
PANELLED PARTITIONS				
Over 300 mm wide:				
32 mm frame, 9 mm plywood panels	m²	14·80	30·75	38·60
38 mm ,, ,, ,, ,,	,,	15·80	33·40	42·00
50 mm ,, 12 mm ,, ,,	,,	19·15	42·00	47·25
Moulding worked on the solid both sides; over 300 mm wide:				
32 mm frame, 9 mm plywood panels	,,	17·00	35·15	43·06
38 mm ,, ,, ,, ,,	,,	17·80	39·35	46·40
50 mm ,, 12 mm ,, ,,	,,	21·50	43·60	52·00
Mouldings planted on both sides; over 300 mm wide:				
32 mm frame, 9 mm plywoood panels	,,	17·12	41·00	49·00
38 mm ,, ,, ,, ,,	,,	18·00	44·85	52·10
50 mm ,, 12 mm ,, ,,	,,	21·60	48·85	57·00
Diminishing stiles; over 300 mm wide; upper portion open panels for glass divided into medium panes:				
32 mm frame, 9 mm plywood panels	,,	24·50	57·65	68·20
38 mm ,, ,, ,, ,,	,,	27·40	62·00	71·00
50 mm ,, 12 mm ,, ,,	,,	34·00	65·75	75·70

JOINERY – GENERAL

DOORS	Unit	Softwood 38 mm thick £	44 mm thick £	50 mm thick £
Matchboard doors; 50 mm framing, 32 mm braces, intermediate and bottom rails; 19 mm tongued, grooved and vee jointed both sides vertical filling one side:				
626 × 2040 mm	No.	—	—	36·00
726 × 2040 mm	,,	—	—	38·80
826 × 2040 mm	,,	—	—	42·00
Panelled doors; four plywood panels; mouldings worked on the solid both sides:				
626 × 2040 mm	,,	29·80	35·00	35·00
726 × 2040 mm	,,	32·75	38·70	38·70
826 × 2040 mm	,,	35·70	42·40	42·40
Panelled doors; one open panel for glass; mouldings worked on the solid one side; 19 × 13 mm beads one side:				
626 × 2040 mm	,,	20·70	21·10	21·10
726 × 2040 mm	,,	22·00	23·00	23·00
826 × 2040 mm	,,	23·15	26·25	26·25
Panelled doors; 250 mm intermediate rail; two open panels for glass; mouldings worked on the solid one side; 19 × 13 mm beads one side:				
626 × 2040 mm	,,	24·00	26·30	26·30
726 × 2040 mm	,,	25·65	28·30	28·30
826 × 2040 mm	,,	27·15	30·25	30·25
Rebates; beaded	m	0·47	0·47	0·47
Rounded edges or rounded heels	,,	0·45	0·45	0·45

	Unit	Mahogany 50 mm £	63 mm £	Afrormosia 50 mm £	63 mm £
Panelled doors; four panels; mouldings worked on the solid both sides:					
626 × 2040 mm	No.	60·00	68·00	91·00	105·00
726 × 2040 mm	,,	66·00	75·20	101·50	117·30
826 × 2040 mm	,,	72·00	82·50	112·00	130·00
Panelled doors; one open panel for glass; mouldings worked on the solid one side; 19 × 13 mm beads one side brass screws and cups:					
626 × 2040 mm	,,	45·50	51·50	59·00	67·00
726 × 2040 mm	,,	49·50	55·50	65·00	72·00
826 × 2040 mm	,,	52·50	58·00	67·00	77·00
Panelled doors; 250 mm intermediate rail; two open panels for glass; mouldings worked on the solid one side; 19 × 13 mm beads one side; brass screws and cups:					
626 × 2040 mm	,,	49·00	55·00	66·00	76·00
726 × 2040 mm	,,	53·00	58·50	69·00	84·00
826 × 2040 mm	,,	56·50	64·00	79·50	90·00
Panelled doors, 150 mm wide stiles in one width; 430 mm wide cross-tongued bottom rail; five panels raised and fielded one side; mouldings worked on the solid both sides:					
626 × 2040 mm	,,	123·00	134·00	159·00	173·00
726 × 2040 mm	,,	129·00	150·00	170·00	187·00
826 × 2040 mm	,,	139·00	158·00	290·00	203·00
Rebates: beaded	m	0·57	0·57	0·57	0·57
Rounded edges or rounded heels . . .	,,	0·57	0·57	0·57	0·57

JOINERY – GENERAL

	Unit	Softwood £	Mahogany £	Afrormosia £
CASEMENTS				
Rebated once; moulded:				
38 mm thick	m²	8·10	14·00	17·10
50 mm ,,	,,	9·60	18·23	21·00
Rebated once; moulded; divided into medium panes:				
38 mm thick	,,	11·80	17·00	22·15
50 mm ,,	,,	14·60	22·75	26·60

DOUBLE HUNG SASH WINDOWS

Cased frames of 100 × 25 mm inner linings, grooved once; 114 × 25 mm outer linings, grooved once; 125 × 38 mm head linings, rebated twice; 125 × 32 mm pulley stiles; rebated twice; grooved once; 150 × 13 mm back linings; 50 × 19 mm parting slips; 25 × 13 mm parting beads; 25 × 19 mm inside beads; 150 × 75 mm oak sill, sunk weathered twice, throated once; 50 mm thick sashes, rebated once, moulded, moulded horns; over 1·25 m² each; both sashes divided into medium

	Unit	Softwood £	Mahogany £	Afrormosia £
panes	,,	56·00	—	—
Ditto, but cased mullions	,,	59·40	—	—

DOOR FRAMES OR WINDOW FRAMES

Rebated once; rounded once:

	Unit	Softwood £	Mahogany £	Afrormosia £
100 × 32 mm	m	1·96	3·63	5·00
75 × 50 mm	,,	2·35	4·00	6·00
100 × 50 mm	,,	3·13	4·88	7·40
150 × 50 mm	,,	4·25	7·49	10·35
100 × 65 mm	,,	3·50	5·55	8·90
150 × 75 mm	,,	5·60	9·64	14·85

MULLIONS OR TRANSOMS

Rebated twice; rounded twice:

	Unit	Softwood £	Mahogany £	Afrormosia £
75 × 50 mm	,,	2·46	4·37	6·12
100 × 50 mm	,,	3·25	6·15	7·60
150 × 75 mm	,,	5·75	5·30	15·00

SILLS

Sunk weathered once, rebated once; grooved four times; rounded once:

	Unit	Softwood £	Mahogany £	Afrormosia £
100 × 75 mm	,,	4·34	7·54	10·72
150 × 75 mm	,,	5·75	9·80	15·41

JOINERY – GENERAL

	Unit	Softwood £	Mahogany £	Afrormosia £
SHELVES				
Butt joints; over 300 mm wide:				
19 mm thick	m²	9·10	15·50	22·20
25 mm „	„	10·60	18·65	27·40
32 mm „	„	13·25	22·20	33·75
Crosstongued joints; over 300 mm wide:				
19 mm thick	„	9·55	16·20	22·90
25 mm „	„	11·20	19·30	28·00
32 mm „	„	14·00	22·90	34·00
Slatted; 50 mm longitudinal slats at 75 mm centres:				
19 mm thick	„	6·25	12·55	16·15
25 mm „	„	7·40	15·20	20·00
32 mm „	„	8·75	18·70	25·00
In one width; 230 mm wide:				
19 mm thick	m	2·05	3·60	5·00
25 mm „	„	2·40	4·65	6·70
32 mm „	„	3·13	5·70	8·20
Shiplap joints:				
19 mm thick	„	1·66	2·00	2·12
25 mm „	„	1·66	2·00	2·12
32 mm „	„	1·66	2·00	2·12
WEATHER FILLETS				
Rebated once; moulded:				
50 × 25 mm	m	0·79	1·78	2·30
Ends; moulded	No.	0·68	0·86	0·86
STOP FILLETS				
38 × 13 mm	m	0·44	0·86	0·95
38 × 19 mm	„	0·54	0·97	1·14
38 × 25 mm	„	0·64	1·09	1·35
50 × 25 mm	„	0·79	1·31	1·50
GLAZING BEADS				
19 × 13 mm	„	0·47	0·72	0·80
25 × 13 mm	„	0·54	0·89	0·94
25 × 13 mm; with brass screws and cups	„	0·69	1·13	1·43
32 × 25 mm; „ „ „ „ „	„	0·86	1·46	1·56

JOINERY – GENERAL

WINDOW BOARDS	Unit	Softwood £	Mahogany £	Afrormosia £
150 mm wide; rebated once: rounded once:				
25 mm thick	m	1·79	3·17	4·83
32 mm ,, ...	,,	2·05	3·97	5·55
38 mm ,, ...	,,	2·53	4·76	6·85
Crosstongued joints; 300 mm wide; rebated once; rounded once:				
25 mm thick	,,	3·29	6·11	9·00
32 mm ,, ...	,,	4·10	7·65	10·65
38 mm ,, ...	,,	4·84	9·00	12·30
In one width; 230 mm wide; rebated once; rounded once:				
25 mm thick	,,	2·64	4·66	6·85
32 mm ,, ...	,,	3·37	5·75	8·45
38 mm ,, ...	,,	3·85	7·00	10·00

SKIRTINGS				
Splayed once or moulded:				
75 × 19 mm	,,	0·63	1·25	2·12
95 × 19 mm	,,	0·82	1·73	2·55
75 × 25 mm	,,	0·82	1·73	2·55
100 × 25 mm	,,	1·00	1·91	3·10

ARCHITRAVES				
Splayed once or moulded:				
64 × 13 mm	,,	0·39	0·83	1·59
38 × 19 mm	,,	0·39	0·82	1·52
75 × 25 mm	,,	0·86	1·85	2·91
100 × 25 mm	,,	1·15	2·13	3·50
150 × 25 mm	,,	1·50	2·80	4·35

GROUNDS				
Sawn; splayed:				
19 × 13 mm	,,	0·30	—	—

JOINERY – GENERAL

STANDARD DOORS	Unit	2′ 6″ × 6′ 6″ £	2′ 9″ × 6′ 6″ £
Panelled doors, 2 X G type; 44 mm finished thickness	No.	19·85	20·38
Panelled doors, 2 X GG type 44 mm finished thickness	„	19·15	19·60
Panelled doors, six panels raised and fielded, moulding worked on solid both sides, 44 mm finished thickness	„	—	104·00
Ledged and braced doors 25 mm ledger and braces, 19mm boarding.	„	24·50	27·00
Framed, ledged and braced doors 44 mm finished thickness . . .	„	26·45	28·90

		526 × 2040 mm £	626 × 2040 mm £	726 × 2040 mm £	826 × 2040 mm £
Flush doors					
Internal pattern					
Hardboard faced prepared for painting					
Honeycomb core					
40 mm finished thickness . .	„	6·85	6·95	7·05	7·15
Solid core					
44 mm finished thickness . .	„	17·90	18·00	18·15	18·25
Internal pattern lipped all edges					
Plywood faced					
Honeycomb core					
40 mm finished thickness . .	„	11·85	12·00	12·10	12·20
Solid core					
44 mm finished thickness . .	„	21·85	22·00	22·10	22·30
Sapele veneered hardboard faced					
Honeycomb core					
40 mm finished thickness . .	„	16·75	16·90	17·05	17·10
Solid core					
44 mm finished thickness . .	„	25·25	28·80	28·90	29·00
Fire check to BS. 476/8 30/20					
Sapele veneered hardboard faced					
Half hour check					
44 mm finished thickness . .	„	26·95	27·10	27·20	27·30
One hour resistant					
54 mm finished thickness . .	„	55·15	55·25	55·35	55·45

		826 × 2040 mm £	807 × 2000 mm £	907 × 2000 mm £
External pattern lipped all edges, 44 mm finished thickness, honeycomb core				
Plywood faced	„	—	19·15	18·90
				19·15
Plywood faced with standard glass opening	„	—	24·00	23·77
				24·00

JOINERY – GENERAL

STANDARD SOFTWOOD WINDOWS

Standard metric softwood windows without glazing bars, with 75 × 150 mm softwood sill, knotted and primed before delivery

*B.S. notation	Nominal size Width	Height								Unit	£
Casement type										No.	13·40
N 26 V	450 mm	750 mm	.	.	.	.	.	.	.		15·90
1 26 P	600 mm	750 mm	.	.	.	.	.	.	.	,,	15·95
1 36 V	600 mm	900 mm	.	.	.	.	.	.	.	,,	30·45
2 36 T	1200 mm	1050 mm	.	.	.	.	.	.	.	,,	31·60
2 40 T	1200 mm	1200 mm	.	.	.	.	.	.	.	,,	43·30
3 36 V	1800 mm	1050 mm	.	.	.	.	.	.	.	,,	45·00
3 40 V	1800 mm	1200 mm	.	.	.	.	.	.	.	,,	48·00
3 46 V	1800 mm	1350 mm	.	.	.	.	.	.	.	,,	
Top hung type										,,	14·65
TH 126	600 mm	750 mm	.	.	.	.	.	.	.	,,	15·65
TH 130	600 mm	900 mm	.	.	.	.	.	.	.	,,	26·20
TH 236	1200 mm	1050 mm	.	.	.	.	.	.	.	,,	27·35
TH 240	1200 mm	1200 mm	.	.	.	.	.	.	.	,,	28·80
TH 246 S	1200 mm	1350 mm	.	.	.	.	.	.	.	,,	30·90
TH 336	1800 mm	1050 mm	.	.	.	.	.	.	.	,,	32·00
TH 340	1800 mm	1200 mm	.	.	.	.	.	.	.	,,	34·75
TH 346 S	1800 mm	1350 mm	.	.	.	.	.	.	.	,,	40·00
TH 450 S	2400 mm	1500 mm	.	.	.	.	.	.	.	,,	

*Nearest equivalent imperial code

PIVOT WINDOWS

High performance pivot windows with 75 × 150 mm softwood sill, adjustable ventilators, weather stripping, knotted and primed before delivery

Nominal size Width	Height									Unit	£
900 mm	900 mm	.	.	.	.	.	.	.	.	No.	47·85
1200 mm	1050 mm	.	.	.	.	.	.	.	.	,,	51·80
1200 mm	1200 mm	.	.	.	.	.	.	.	.	,,	53·10
1200 mm	1350 mm	.	.	.	.	.	.	.	.	,,	61·30
1800 mm	1050 mm	.	.	.	.	.	.	.	.	,,	62·50
1800 mm	1200 mm	.	.	.	.	.	.	.	.	,,	64·10
1800 mm	1350 mm	.	.	.	.	.	.	.	.	,,	77·00
2400 mm	1500 mm	.	.	.	.	.	.	.	.		83·35

SASH WINDOWS

Double hung sash windows with glazing bars with 63 × 175 mm softwood sill, standard flush external linings, spiral spring balances, knotted and primed before delivery

Nominal size Width	Height									Unit	£
593 mm	1096 mm	.	.	.	.	.	.	.	.	No.	69·40
821 mm	1096 mm	.	.	.	.	.	.	.	.	,,	75·85
821 mm	1401 mm	.	.	.	.	.	.	.	.	,,	82·50
1050 mm	1401 mm	.	.	.	.	.	.	.	.	,,	91·25
1050 mm	1706 mm	.	.	.	.	.	.	.	.	,,	98·80
1617 mm	1401 mm	.	.	.	.	.	.	.	.		177·00

JOINERY – GENERAL

STANDARD HARDWOOD WINDOWS

Standard windows in Phillipine Mahogany, 45 × 140 mm sill, weather stripping, fittings and finished one coat primer stain

Side hung units

Width	Nominal size Height		Unit	£
600 mm	600 mm (a)	. . . .	No.	31·00
600 mm	900 mm (a)	. . . .	,,	35·00
1200 mm	1050 mm (b)	. . . .	,,	60·30
1200 mm	1200 mm (b)	. . . .	,,	63·60
1800 mm	1050 mm (b)	. . . .	,,	99·15
1800 mm	1200 mm (c)	. . . .	,,	104·60
2400 mm	1200 mm (c)	. . . .	,,	116·10

Top hung units

Width	Nominal size Height			£
600 mm	600 mm (a)	. . . .	,,	31·35
900 mm	600 mm (a)	. . . .	,,	40·35
1200 mm	1050 mm (a)	. . . .	,,	57·75
1200 mm	1200 mm (a)	. . . .	,,	62·80
1200 mm	1500 mm (a*)	. . . .	,,	68·70
1800 mm	1050 mm (b)	. . . .	,,	76·15
1800 mm	1200 mm (b)	. . . .	,,	82·45
1800 mm	1350 mm (b*)	. . . .	,,	86·70
2400 mm	1500 mm (b*)	. . . .	,,	107·25

(a) denotes one opening light.
(b) ,, one opening and one fixed light.
(c) ,, two opening light.
(*) ,, sub light.

JOINERY–GENERAL

STANDARD SOFTWOOD STAIRCASES

Softwood staircases; 25 mm treads with rounded nosing; 12 mm plywood risers; 32 mm strings, bullnosed bottom tread, 50 × 75 mm hardwood handrail, 2 No. 32 × 140 mm balustrade knee rails, 32 × 50 mm stiffeners, 100 × 100 mm newel posts, hardwood caps, left in the white.

	Unit	£
Straight flight; 838 mm wide; 2688 mm going; 2600 mm rise with two newel posts	No.	94·50
As last but with three top treads winding	,,	134·00
As last but in two short flights to quarter space landing and with three newel posts, no landing boards, carriages or bearers	,,	101·25
As last but balustrade formed of ex 32 mm square parana pine balusters	,,	101·25

STANDARD BALUSTRADE

Landing balustrade 3 m long; 50 × 75 mm hardwood handrail, 32 × 140 mm subrail and three lay rails	,,	35·50
Landing balustrade 3 m long; 50 × 75 mm hardwood handrail, ex 32 mm square parana pine balusters	,,	22·50

STORAGE CUPBOARDS

Cupboard units of high density chipboard faced both sides with laminated melamine, with backs and plinths. B.S. references are to pattern only.

Base units

B.S. ref.	Length	Height	Depth	Unit	£
5 B6	500 mm	900 mm	500 mm	No.	30·00
5 B6 D	500 mm	900 mm	500 mm	,,	41·00
10 B6	1000 mm	900 mm	500 mm	,,	55·00
10 B6 D	1000 mm	900 mm	500 mm	,,	72·50
5 B6	500 mm	900 mm	600 mm	,,	32·80
5 B6 D	500 mm	900 mm	600 mm	,,	44·00
10 B 6	1000 mm	900 mm	600 mm	,,	57·50
10 B6 D	1000 mm	900 mm	600 mm	,,	74·70

Wall units

B.S. ref.					
5 WT	500 mm	900 mm	300 mm	,,	29·50
6 WT	600 mm	900 mm	300 mm	,,	32·25
10 WT	1000 mm	900 mm	300 mm	,,	51·00
5 W	500 mm	600 mm	300 mm	,,	23·00
6 W	600 mm	600 mm	300 mm	,,	25·50
10 W	1000 mm	600 mm	300 mm	,,	42·00

Store units

B.S. ref.					
6 TS	600 mm	1950 mm	500 mm	,,	69·00

Sink units

B.S. ref.	Length	Height	Depth		
12 S6	1000 mm	900 mm	500 mm	,,	44·00
15 S6	1500 mm	900 mm	500 mm	,,	60·00
12 S6	1200 mm	900 mm	600 mm	,,	56·50
15 S6	1500 mm	900 mm	600 mm	,,	65·50

Plastic laminated worktops to suit base units

500 mm × 500 mm	,,	9·60
500 mm × 1000 mm	,,	14·60
600 mm × 500 mm	,,	10·00
600 mm × 1000 mm	,,	15·90

JOINERY – FITTINGS

	Unit	Softwood £	Iroko £	Teak £
BACKS, FRONTS OR SIDES				
Crosstongued joints; over 300 mm wide:				
25 mm thick	m²	13·65	25·65	39·00
DIVISIONS				
Crosstongued joints; over 300 mm wide:				
25 mm thick	,,	13·65	25·65	39·00
WORKTOPS				
Crosstongued joints; over 300 mm wide:				
25 mm thick	,,	13·65	25·65	39·00
DRAINING BOARDS				
Crosstongued joints; over 300 mm wide:				
25 mm thick	,,	—	29·00	42·00
Grooves; cross grain	m	—	0·51	0·51
Flutes; stopped	,,	—	0·62	0·62

FLUSH DOORS

Softwood skeleton core; 4 mm plywood facing both sides, fixed with adhesive under hydraulic pressure; 13 mm softwood lippings all edges:

	Unit	Softwood £		
450 × 750 × 35 mm thick	No.	11·00	—	—

	Unit	Softwood £	Iroko £	Mahogany £
BEARERS				
38 × 19 mm	m	0·61	1·27	1·27
50 × 25 mm	,,	0·71	1·60	1·60
50 × 50 mm	,,	1·33	3·02	3·02
75 × 50 mm	,,	1·79	4·00	4·00
BEARERS; FRAMED				
38 × 19 mm	,,	0·70	1·73	1·73
50 × 25 mm	,,	0·92	2·02	2·02
50 × 50 mm	,,	1·85	3·44	3·44
75 × 50 mm	,,	2·64	4·95	4·95
FRAMING TO BACKS, FRONTS OR SIDES				
38 × 19 mm	,,	0·70	1·73	1·73
50 × 25 mm	,,	0·92	2·02	2·02
50 × 50 mm	,,	1·85	3·44	3·44
75 × 50 mm	,,	2·64	4·95	4·95

JOINERY – FITTINGS

	Unit	Plywood £	Blockboard £
SHELVES			
Over 300 mm wide:			
19 mm thick	m²	17·50	14·55
25 mm „	„	22·40	17·40
BACKS, FRONTS OR SIDES			
Over 300 mm wide:			
6 mm thick	„	8·30	—
9 mm „	„	10·00	—
12 mm „	„	12·00	12·45
19 mm „	„	18·40	15·15
25 mm „	„	23·25	18·10
DIVISIONS			
Over 300 mm wide:			
6 mm thick	„	8·30	—
9 mm „	„	10·00	—
12 mm „	„	12·00	12·43
19 mm „	„	18·40	15·17
25 mm „	„	23·25	18·07
WORKTOPS			
Over 300 mm wide:			
19 mm thick	„	18·40	15·17
25 mm „	„	23·25	18·07
FLUSH DOORS			
13 mm hardwood lipping all edges:			
450 × 750 × 19 mm thick	No.	8·60	6·75
450 × 750 × 25 mm „	„	9·80	7·90
600 × 900 × 19 mm „	„	11·50	9·15
600 × 900 × 25 mm „	„	13·30	10·60

JOINERY – STAIRCASES

	Unit	Softwood £	Mahogany £	Oak £
BOARD LANDINGS				
Crosstongued joints; 75 × 50 mm sawn softwood bearers:				
25 mm thick	m²	12·50	23·40	54·00
32 mm „	„	14·60	25·30	63·00
TREADS AND RISERS				
Treads, crosstongued joints and risers; rounded nosings; tongued, grooved; glued and blocked together; and 175 × 50 mm sawn softwood carriage:				
25 mm treads; 19 mm risers	m²	16·00	30·00	67·20
32 mm „ 25 mm „	„	17·00	36·25	75·00
Ends; quadrant:				
25 mm treads; 19 mm risers	No.	7·48	20·65	33·35
32 mm „ 25 mm „	„	7·82	22·35	38·30
Ends; housed:				
25 mm treads; 19 mm risers	„	0·88	1·65	1·65
32 mm „ 25 mm „	„	0·88	1·65	1·65
Winders, crosstongued joints and risers in one width; rounded nosings; tongued, grooved, glued and blocked together; one 175 × 50 mm sawn softwood carriage:				
25 mm treads; 19 mm risers	m²	17·50	32·75	71·00
32 mm „ 25 mm „	„	18·50	40·25	79·00
Wide ends; housed:				
25 mm treads; 19 mm risers	No.	1·00	2·02	2·02
32 mm „ 25 mm „	„	1·00	2·02	2·02
Narrow ends; housed:				
25 mm treads; 19 mm risers	„	0·90	1·65	1·65
32 mm „ 25 mm „	„	0·90	1·65	1·65

JOINERY – STAIRCASES

	Unit	Softwood £	Mahogany £	Oak £
CLOSED STRINGS				
Crosstongued joints; 280 mm wide; rounded once; fixing with screws:				
32 mm thick	m	5·75	10·25	22·60
38 mm ,,	,,	7·45	12·00	26·15
50 mm ,,	,,	8·75	15·60	31·00
Ends; fitted:				
32 mm thick	No.	1·17	2·21	2·21
38 mm ,,	,,	1·17	2·21	2·21
50 mm ,,	,,	1·17	2·21	2·21
Ends; framed:				
32 mm thick	,,	1·85	2·55	2·55
38 mm ,,	,,	1·85	2·55	2·55
50 mm ,,	,,	1·85	2·55	2·55
Extra; short ramp:				
32 mm thick	,,	2·34	5·22	10·45
38 mm ,,	,,	2·34	6·26	10·75
50 mm ,,	,,	2·34	6·50	12·35
Extra; tongued heading joint				
32 mm thick	,,	1·05	2·00	2·00
38 mm ,,	,,	1·05	2·00	2·00
50 mm ,,	,,	1·05	2·52	2·52
Ramped; crosstongued joints; 280 mm wide; rounded once; fixing with screws				
32 mm thick	m	5·75	10·20	22·60
38 mm ,,	,,	7·45	12·10	26·15
50 mm ,,	,,	8·75	15·60	31·00
In one width; 230 mm wide; rounded twice:				
32 mm thick	,,	5·55	9·15	18·15
38 mm ,,	,,	6·60	10·75	21·00
50 mm ,,	,,	8·15	13·40	27·50
APRON LININGS				
In one width; 230 mm wide:				
19 mm thick	,,	2·50	4·55	10·00
25 mm ,,	,,	2·90	6·10	13·35
HANDRAILS				
Rounded:				
44 × 50 mm	m	1·70	2·90	5·90
63 × 75 mm	,,	2·90	5·30	11·25
75 × 100 mm	,,	4·10	8·85	16·40
Moulded:				
44 × 50 mm	,,	1·90	3·70	6·60
63 × 75 mm	,,	3·30	6·30	11·75
75 × 100 mm	,,	4·30	9·85	17·80
Ramped; rounded:				
44 × 50 mm	,,	5·40	8·20	11·45
63 × 75 mm	,,	7·00	14·30	21·30
75 × 100 mm	,,	10·00	19·18	30·15
Ramped; moulded:				
44 × 50 mm	,,	6·40	9·35	13·50
63 × 75 mm	,,	8·30	16·30	23·75
75 × 100 mm	,,	12·00	20·80	32·00

Market Prices of Materials

JOINERY – STAIRCASES

HANDRAILS – *continued*		Unit	Softwood £	Mahogany £	Oak £
Add for grooved once:					
44 × 50 mm	. . . . . .	m	0·26	0·39	0·39
63 × 75 mm	. . . . . .	,,	0·26	0·39	0·39
75 × 100 mm	. . . . . .	,,	0·26	0·39	0·39
Add for ends; framed:	.				
44 × 50 mm	. . . . . .	No.	0·88	1·90	1·90
63 × 75 mm	. . . . . .	,,	1·06	2·10	2·10
75 × 100 mm	. . . . . .	,,	1·06	2·10	2·10
Add for ends; framed on rake:					
44 × 50 mm	. . . . . .	,,	1·06	2·16	2·16
63 × 75 mm	. . . . . .	,,	1·41	2·30	2·30
75 × 100 mm	. . . . . .	,,	1·53	2·40	2·40
Add for mitres; handrail screws:					
44 × 50 mm	. . . . . .	,,	0·90	1·12	1·12
63 × 75 mm	. . . . . .	,,	1·06	1·26	1·27
75 × 100 mm	. . . . . .	,,	1·18	1·52	1·52
Add for heading joints; on rake; handrail screws:					
44 × 50 mm	. . . . . .	,,	1·88	2·28	2·28
63 × 75 mm	. . . . . .	,,	2·12	2·53	2·53
75 × 100 mm	. . . . . .	,,	2·23	2·78	2·78
BALUSTERS					
25 × 25 mm	. . . . . .	m	0·80	1·32	1·75
32 × 32 mm	. . . . . .	,,	0·97	1·60	2·00
Ends; housed:					
25 × 25 mm	. . . . . .	No.	0·50	0·90	0·90
32 × 32 mm	. . . . . .	,,	0·56	0·90	0·90
Ends; housed on rake:					
25 × 25 mm	. . . . . .	,,	0·90	1·14	1·14
32 × 32 mm	. . . . . .	,,	0·90	1·14	1·14
NEWELS					
75 × 75 mm	. . . . . .	m	3·32	6·35	12·55
100 × 100 mm	. . . . . .	,,	5·30	10·43	22·15
Ends; rounded:					
75 × 75 mm	. . . . . .	No.	1·24	1·52	1·52
100 × 100 mm	. . . . . .	,,	1·40	1·77	1·77
Newel caps; rounded; housed; fixing with adhesive:					
100 × 100 × 50 mm	. . . . . .	,,	3·30	4·65	4·65
125 × 125 × 50 mm	. . . . . .	,,	3·85	5·00	5·00

IRONMONGERY

BUTTS AND HINGES

	Unit	51 mm	64 mm	76 mm	102 mm	127 mm
		£	£	£	£	£
Steel butts:						
Medium . .	Pair	0·16	0·20	0·21	0·40	—
Strong . . .	,,	—	0·56	0·73	1·05	—
Extra strong . .	,,	—	—	1·79	2·36	—
Rising . . .	,,	—	—	1·10	1·79	—
Lift-off . . .	,,	—	—	1·05	1·50	—
Solid drawn brass butt hinges with steel pins and washers .	,,	—	—	2·20	4·25	5·80
Steel butt hinges with nylon washers and loose pins .	,,	—	—	1·26	1·80	—
Nylon butt hinges .	,,	—	0·36	0·42	0·66	—
Parliament hinges:						
Steel . . .	,,	—	—	1·75	2·95	3·89
Brass . . .	,,	—	—	—	6·54	—

		25 × 73 mm	32 × 82 mm	38 × 89 mm	51 × 108 mm
		£	£	£	£
Steel backflap hinges .	,,	0·27	0·30	0·37	0·51

		305 mm	406 mm	457 mm	607 mm
		£	£	£	£
Wrought iron hook and band hinge .	,,	3·00	3·85	4·90	7·60

		51 mm	64 mm	76 mm	
		£	£	£	
Double strap gate hinge for thickness of gate shown .	Set	8·00	10·85	10·85	

		230 mm	300 mm	375 mm	450 mm
		£	£	£	£
Steel tee hinges medium weight .	Pair	0·60	0·78	1·13	1·46

		76 mm	102 mm	127 mm	152 mm	178 mm
		£	£	£	£	£
Springe hinges:						
Single action . .	,,	7·65	7·70	9·15	10·90	14·50
Double action . .	,,	10·35	11·75	14·10	17·10	21·50

SLIDING DOOR GEAR

	Unit	*For doors maximum weight/height*		
		125 kg/2.7 m	365 kg/3.6 m	550 kg/4.2 m
		£	£	£
Galvanized steel top track . .	m	3·14	6·75	8·68
Floor channel, steel . . .	,,	2·90	4·30	4·30
Open side wall bracket . .	No.	1·08	1·26	4·05
Timber door hangers . .	,,	5·40	9·55	11·70
Metal door hangers . . .	,,	4·80	8·35	10·00
Timber door guides . .	,,	1·35	2·32	2·32
Metal door guides . .	,,	1·64	1·81	1·81
Bow handles . . .	,,	2·32	2·32	2·32
Flush handles . . .	,,	1·49	1·49	1·49
Drop bolt	,,	10·60	10·60	10·60
Locking bar . . .	,,	5·30	5·30	5·30
Padlock	,,	5·55	5·55	5·55
Door stop, rubber buffers . .	,,	6·70	6·70	6·70

IRONMONGERY

	Unit	£

DOOR CLOSERS

Overhead door closers for doors weighing up to approximately

45 kg	No.	33·70
70 kg	,,	41·90
90 kg	,,	55·50

Concealed overhead door closers for doors weighing up to approximately

50 kg	,,	59·90
80 kg	,,	66·60

Floor springs with hydraulic check action with shoe, adjustable top centre and loose box, for doors weighing approximately 90 kg

Single action	,,	74·50
Double action	,,	84·50

SECURITY DEVICES

Security hinge bolts	Pair	3·80
Padlock bar for double doors	No.	12·00
Close shackle padlock	,,	15·80
Window catch for wood casements with key	,,	2·00
Window lock for metal casements with key	,,	3·70
Door viewer	,,	3·70
Mortice bolt with key	,,	1·95

BOLTS

	Unit	102 mm £	152 mm £	203 mm £	254 mm £
Tower bolts, black japanned					
Straight	No.	0·46	0·60	0·87	1·22
Necked	,,	—	1·00	1·30	1·83
Barrel bolts, black japanned					
Straight	,,	—	1·00	1·30	1·60

	Unit	£
Monkey tail garage bolts, 381 mm	No.	3·93
Foot bolts, 171 mm, sheradized	,,	2·67
Flush lever action bolts, 200 mm		
Brass	,,	4·38
S.A.A.	,,	3·27
W.C. indicating bolts		
B.M.A.	,,	3·87
Chromium plated	,,	3·87
Automatic panic bolts, aluminium box, steel shoots and cross rail		
Single doors	,,	23·60
Double rebated doors	Set	30·00
Swing doors	,,	47·15

IRONMONGERY

	Unit	£
LOCKS AND LATCHES		
Norfolk latch	No.	1·30
Cupboard lock, 51 × 25 mm two lever, brass	"	1·10
Cylinder rim night latch, japanned case		
Chromium-plated cylinder	"	4·60
Brass cylinder	"	4·15
Cylinder rim night latch, automatic deadlocking, enamelled case		
Chromium-plated cylinder	"	10·80
Mortice latch, steel case, brass bolt	"	0·68
Rim latch, ditto	"	1·60
Mortice sash lock, three lever, steel case, brass striking plate	"	3·85
Mortice deadlock, ditto	"	3·10
Rim lock, enamelled case	"	1·60
Rim deadlock, four lever, enamelled case	"	4·60
Mortice budget lock and key	"	2·16

	Unit	Stainless Steel £	Silver Anodized Aluminium £	B.M.A. £
DOOR FURNITURE				
Set of lever mortice furniture	No.	—	4·76	3·76
Finger plate, 305 × 76 mm	"	2·56	0·60	4·10
Kicking plate				
749 × 152 mm	"	7·00	2·78	12·20
825 × 152 mm	"	7·60	3·10	12·90

	Unit	£
SUNDRIES		
50 mm stainless steel numerals	No.	0·40
Postal knocker, silver anodized	"	4·00
Centre door knobs, silver anodized	"	5·20
Door chain and bolt, brass	"	1·42
Hat and coat hooks, silver anodized	"	0·36
Brass Bales catch 64 × 25 mm	"	0·80
Cabin hooks, 152 mm, steel	"	0·27
Shelf brackets 203 × 152 mm grey finish	"	0·19
Adjustable metal shelving, White or grey finish.		
Wall uprights		
1219 mm long	"	3·03
2394 " "	"	4·33
Straight brackets		
229 mm	"	0·65
368 "	"	1·58
470 "	"	2·15
Metal shelves 1 m long		
267 mm wide	"	5·62
368 " "	"	7·15
470 " "	"	7·85

STRUCTURAL STEELWORK

	Unit	£

ROLLED STEEL JOISTS
(Figures in brackets are weights in kg/m)

	Unit	£
254 × 203 mm (81·85)	tonne	270·25
254 × 114 mm (37·20)	,,	245·75
203 × 152 mm (52·09)	,,	264·75
203 × 102 mm (25·33)	,,	238·25
178 × 102 mm (21·54)	,,	248·00
152 × 127 mm (37·20)	,,	239·00
152 × 89 mm (17·09)	,,	252·50
152 × 76 mm (17·86)	,,	235·75
127 × 114 mm (29·76)	,,	234·50
127 × 114 mm (26·79)	,,	234·50
127 × 76 mm (16·37)	,,	254·00
127 × 76 mm (13·36)	,,	254·00
114 × 114 mm (26·79)	,,	231·50
102 × 102 mm (23·06)	,,	233·00
102 × 64 mm (9·65)	,,	248·40
102 × 44 mm (7·44)	,,	250·40
89 × 89 mm (19·35)	,,	237·25
76 × 76 mm (12·65)	,,	272·25
76 × 76 mm (14·67)	,,	272·25

UNIVERSAL BEAMS
(Figures in brackets are available weights in kg/m)

	Unit	£
914 × 419 mm (343, 388)	,,	257·25
914 × 305 mm (201, 224, 253, 289)	,,	251·25
838 × 292 mm (176, 194, 226)	,,	248·75
762 × 267 mm (147, 173, 197)	,,	247·25
686 × 254 mm (125, 140, 152, 170)	,,	247·25
610 × 305 mm (149, 179, 238)	,,	247·25
610 × 229 mm (101, 113, 125, 140)	,,	244·75
533 × 210 mm (82, 92, 101, 109, 122)	,,	241·75
457 × 191 mm (67, 74, 82, 89, 98)	,,	229·25
457 × 152 mm (52, 60, 67, 74, 82)	,,	250·25
406 × 178 mm (54, 60, 67, 74)	,,	230·75

STRUCTURAL STEELWORK

	Unit	£

UNIVERSAL BEAMS – *continued*

	Unit	£
406 × 140 mm (39, 46)	tonne	246·55
356 × 171 mm (45, 51, 57, 67)	,,	234·55
356 × 127 mm (33, 39)	,,	245·55
305 × 165 mm (40, 46, 54)	,,	234·55
305 × 127 mm (37, 42, 48)	,,	237·05
305 × 102 mm (25, 28, 33)	,,	247·55
254 × 146 mm (31, 37, 43)	,,	233·55
254 × 102 mm (22, 25, 28)	,,	246·55
203 × 133 mm (25, 30)	,,	236·55

UNIVERSAL COLUMNS

(Figures in brackets are available weights in kg/m)

	Unit	£
356 × 406 mm (235, 287, 340, 393, 467, 551, 634)	,,	258·25
356 × 368 mm (129, 153, 177, 202)	,,	251·25
305 × 305 mm (97, 118, 137, 158, 198, 240, 283)	,,	235·75
254 × 254 mm (73, 89, 107, 132, 167)	,,	235·75
203 × 203 mm (46, 52, 60, 71, 86)	,,	234·55
152 × 152 mm (23, 30, 37)	,,	240·55

CHANNELS

(Figures in brackets are weights in kg/m)

	Unit	£
432 × 102 mm (65·54)	,,	245·75
381 × 102 mm (55·10)	,,	232·00
305 × 102 mm (46·18)	,,	233·00
305 × 89 mm (41·69)	,,	232·00
254 × 89 mm (35·74)	,,	234·50
254 × 76 mm (28·29)	,,	242·50
229 × 89 mm (32·76)	,,	236·00
229 × 76 mm (26·06)	,,	242·50
203 × 89 mm (29·78)	,,	238·75
203 × 76 mm (23·82)	,,	239·75
178 × 89 mm (26·81)	,,	241·25
178 × 76 mm (20·84)	,,	234·75
152 × 89 mm (23·84)	,,	234·75
152 × 76 mm (17·88)	,,	233·50
127 × 64 mm (14·90)	,,	233·50
76 × 38 mm (6·70)	,,	223·90
51 × 25 mm (4·56)	,,	228·90
51 × 25 mm (3·48)	,,	228·90

STRUCTURAL STEELWORK

ANGLES	Unit	£
30 × 30 × 5 mm thick	tonne	223·75
40 × 40 × 4 mm thick	,,	224·75
40 × 40 × 5 mm thick	,,	220·25
40 × 40 × 6 mm thick	,,	217·75
45 × 45 × 4 mm thick	,,	223·75
45 × 45 × 5 mm thick	,,	217·75
45 × 45 × 6 mm thick	,,	215·75
50 × 50 × 5 mm thick	,,	216·75
50 × 50 × 6 mm thick	,,	212·75
50 × 50 × 8 mm thick	,,	214·75
60 × 60 × 5 mm thick	,,	206·00
60 × 60 × 6 mm thick	,,	205·00
60 × 60 × 8 mm thick	,,	204·50
60 × 60 × 10 mm thick	,,	203·00
70 × 70 × 6 mm thick	,,	206·00
70 × 70 × 8 mm thick	,,	205·50
70 × 70 × 10 mm thick	,,	204·50
80 × 80 × 6 mm thick	,,	205·00
80 × 80 × 8 mm thick	,,	204·50
80 × 80 × 10 mm thick	,,	204·00
90 × 90 × 6 mm thick	,,	209·00
90 × 90 × 8 mm thick	,,	206·50
90 × 90 × 10 mm thick	,,	200·50
90 × 90 × 12 mm thick	,,	203·50
100 × 100 × 8 mm thick	,,	196·50
100 × 100 × 10 mm thick	,,	189·50
100 × 100 × 12 mm thick	,,	194·50
100 × 100 × 15 mm thick	,,	194·50
120 × 120 × 8 mm thick	,,	215·75
120 × 120 × 10 mm thick	,,	208·25
120 × 120 × 12 mm thick	,,	202·75
120 × 120 × 15 mm thick	,,	203·25
150 × 150 × 10 mm thick	,,	219·25
150 × 150 × 12 mm thick	,,	217·25
150 × 150 × 15 mm thick	,,	222·75
150 × 150 × 18 mm thick	,,	224·25
200 × 200 × 16 mm thick	,,	214·25
200 × 200 × 18 mm thick	,,	221·75
200 × 200 × 20 mm thick	,,	216·75
200 × 200 × 24 mm thick	,,	219·25
65 × 50 × 5 mm thick	,,	207·90
65 × 50 × 6 mm thick	,,	206·90
65 × 50 × 8 mm thick	,,	205·90
75 × 50 × 6 mm thick	,,	207·90
75 × 50 × 8 mm thick	,,	206·90
80 × 60 × 6 mm thick	,,	206·90
80 × 60 × 7 mm thick	,,	209·40
80 × 60 × 8 mm thick	,,	206·40
100 × 65 × 7 mm thick	,,	203·90
100 × 65 × 8 mm thick	,,	208·40
100 × 65 × 10 mm thick	,,	208·40
100 × 75 × 8 mm thick	,,	202·90
100 × 75 × 10 mm thick	,,	208·40
100 × 75 × 12 mm thick	,,	199·40

STRUCTURAL STEELWORK

ANGLES – *continued*	Unit	£
125 × 75 × 8 mm thick	tonne	207·40
125 × 75 × 10 mm thick	,,	199·40
125 × 75 × 12 mm thick	,,	210·40
150 × 75 × 10 mm thick	,,	216·75
150 × 75 × 12 mm thick	,,	213·25
150 × 75 × 15 mm thick	,,	216·25
150 × 90 × 10 mm thick	,,	221·25
150 × 90 × 12 mm thick	,,	215·75
150 × 90 × 15 mm thick	,,	218·75
200 × 100 × 10 mm thick	,,	222·75
200 × 100 × 12 mm thick	,,	219·25
200 × 100 × 15 mm thick	,,	232·25
200 × 150 × 12 mm thick	,,	219·75
200 × 150 × 15 mm thick	,,	218·75
200 × 150 × 18 mm thick	,,	218·75

TEES		
51 × 51 × 6 mm	,,	221·90
51 × 51 × 8 mm	,,	223·90
51 × 51 × 9 mm	,,	223·90
44 × 44 × 5 mm	,,	226·90
44 × 44 × 6 mm	,,	224·90

STRUCTURAL HOLLOW SECTIONS TO B.S. 4848: PART 2
The following are basis prices for one size and thickness delivered ex-works in 10 tonne loads

	Unit	Theoretical wt kg per m	B.S. 4360 Grade 43C £	Grade 50C £
Circular sections:				
26·9 × 3·2 mm thick	100 m	1·87	60·98	67·02
42·4 × 3·2 mm thick	,,	3·09	100·76	110·83
60·3 × 5·0 mm thick	,,	6·82	215·74	237·32
114·3 × 6·3 mm thick	,,	16·80	531·45	584·59
139·7 × 6·3 mm thick	,,	20·70	752·11	827·32
168·3 × 6·3 mm thick	,,	25·20	915·62	997·64
219·1 × 10·0 mm thick	,,	51·60	1995·37	2194·91
244·5 × 10·0 mm thick	,,	57·80	2235·13	2458·64
273·0 × 10·0 mm thick	,,	64·90	2509·68	2760·65
323·9 × 12·5 mm thick	,,	96·00	3712·32	4083·55
Rectangular and square sections:				
20 × 20 × 2·6 mm thick	,,	1·39	46·00	—
50 × 30 × 3·2 mm thick	,,	3·66	107·42	118·17
60 × 60 × 4·0 mm thick	,,	6·97	204·58	225·03
80 × 80 × 5·0 mm thick	,,	11·70	343·41	377·75
100 × 50 × 4·0 mm thick	,,	8·86	260·05	286·05
100 × 100 × 6·3 mm thick	,,	18·40	540·06	594·06
100 × 100 × 10·0 mm thick	,,	27·90	871·79	958·98
120 × 80 × 6·3 mm thick	,,	18·40	574·94	632·44
150 × 150 × 6·3 mm thick	,,	28·30	982·15	1129·48
200 × 100 × 8·0 mm thick	,,	35·40	1238·82	1424·64
200 × 200 × 10·0 mm thick	,,	59·30	2075·20	2386·47
300 × 200 × 12·5 mm thick	,,	92·60	3240·54	3726·59

METALWORK

EXTRUDED ALUMINIUM SECTIONS

Extruded aluminium sections to B.S. 1474:HE9-TF in random lengths

	Unit	Weight, kg/m	£
Flat bars			
10 × 6 mm	100 kg	0·165	223·50
13 × 6 mm	,,	0·219	223·50
19 × 6 mm	,,	0·339	223·50
25 × 13 mm	,,	0·894	222·00
39 × 10 mm	,,	0·985	222·00
Equal angles			
13 × 1·6 mm	,,	0·106	254·76
19 × 1·6 mm	,,	0·167	223·50
25 × 1·6 mm	,,	0·211	223·50
25 × 3 mm	,,	0·421	223·50
31 × 1·6 mm	,,	0·266	223·50
51 × 10 mm	,,	2·451	209·93
Tees			
19 × 19 × 3 mm	,,	0·301	223·50
25 × 13 × 3 mm	,,	0·310	223·50
25 × 32 × 3 mm	,,	0·475	223·50
38 × 38 × 6 mm	,,	1·228	209·93
Channels			
19 × 19 × 3 mm	,,	0·451	223·50
38 × 38 × 3 mm	,,	0·988	212·60
51 × 51 × 3 mm	,,	1·245	209·93

STEEL LINTELS

Galvanized steel lintels with internal key for plaster for standard cavity walls

	Unit	£
143 mm high		
600 mm long	No.	4·54
900 mm long	,,	6·80
1200 mm long	,,	9·07
1800 mm long	,,	14·09
219 mm high		
2250 mm long	,,	21·08
2850 mm long	,,	28·87
3300 mm long	,,	33·43
4200 mm long	,,	52·16

METALWORK

	Unit	Thickness 0·5 mm £	0·8 mm £
ALUMINIUM ALLOY SHEETS			
Profiled aluminium sheets, Trough type, 100 mm pitch, 26 mm deep, cover width 900 mm			
Mill finish	m	2·89	4·54
Stoved paint finish one side	,,	3·68	5·33
Trough type, 125·7 mm pitch, 38 mm deep, cover width 880 mm			
Mill finish	,,	—	4·80
Stoved paint finish one side	,,	—	5·59

	Unit	Min. £	Max. £
METAL WINDOWS			
Prices for metal windows vary, depending on quantity, and those given represent minimum and maximum.			
Aluminium windows, white acrylic finish			
VS2A vertical sliding windows glazed with and including 3 mm or 4 mm O.Q. sheet glass			
6 VV 9A 600 × 900 mm	No.	53·16	70·88
9 VV 11A 900 × 1100 mm	,,	65·37	87·16
12 VV 15A 1200 × 1500 mm	,,	78·80	105·08
HS 3A horizontal sliding windows glazed with and including 3 mm, 4 mm, or 5 mm O.Q. sheet glass			
18 HH 9A 900 × 1800 mm	,,	76·75	102·36
15 HH 13A 1300 × 1500 mm	,,	81·33	108·44
15 HH 9A 900 × 1500 mm	,,	70·92	94·56
HS3A horizontal sliding windows glazed with and including 11 mm sealed double glass units			
D24 HH 13A 1300 × 2400 mm	,,	170·88	227·84
D15 HH 7A 700 × 1500 mm	,,	96·80	129·08
Steel windows			
Module 100 windows			
Basic fixed lights			
6F2 600 × 200 mm	,,	2·34	3·12
12F2 1200 × 200 mm	,,	3·87	5·16
18F2 1800 × 200 mm	,,	5·22	6·96
6F9 600 × 900 mm	,,	3·87	5·16
12F9 1200 × 900 mm	,,	5·21	6·95
18F9 1800 × 900 mm	,,	7·80	10·40
6F15 600 × 1500 mm	,,	5·10	6·80
12F15 1200 × 1500 mm	,,	7·65	10·20
18F15 1800 × 1500 mm	,,	9·36	12·48
Basic opening lights, including weatherstripping			
6V2 600 × 200 mm	,,	8·28	11·04
12V2 1200 × 200 mm	,,	11·37	15·16
18V2 1800 × 200 mm	,,	15·48	20·64
6C9 600 × 900 mm	,,	11·85	15·80
One-piece composites, including weatherstripping			
15FT5 1500 × 500 mm	,,	11·58	15·44
18FT5 1800 × 500 mm	,,	12·66	16·88
6FV9 600 × 900 mm	,,	10·62	14·16
12FV9 1200 × 900 mm	,,	15·33	20·44
6FV15 600 × 1500 mm	,,	12·21	16·28
12FV15 1200 × 1500 mm	,,	17·22	22·96

METALWORK

METAL WINDOWS – *continued*

	Unit	Min. £	Max. £
Steel windows–*continued*			
Reversible windows			
9R7 900 × 700 mm	No.	34·47	45·96
12R9 1200 × 900 mm	„	40·44	53·92
9R11 900 × 1100 mm	„	38·28	51·04
12R13 1200 × 1300 mm	„	44·82	59·76
12RS15 1200 × 1500 mm	„	47·34	63·12
Windows for industrial buildings to B.S. 1787			
SSF23 673 × 1445 mm	„	17·73	23·64
SSF43 1302 × 1445 mm	„	29·25	39·00
SSF54 1616 × 1911 mm	„	43·65	58·20
SS23 673 × 1445 mm	„	48·50	64·68
SS43 1302 × 1445 mm	„	59·20	78·95
SS54 1616 × 1911 mm	„	76·40	101·90

METAL GARAGE DOORS

Galvanized steel 'up and over' type door, spring counterbalanced, for fixing to timber frame

	Unit	
2135 × 1980 mm	No.	57·55
2135 × 2135 mm	„	63·20
2400 × 2125 mm	„	74·55
4270 × 2135 mm	„	244·05

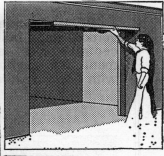

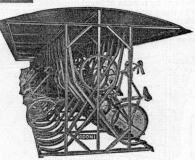
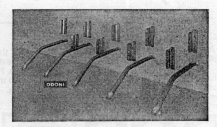

PLUMBING AND ENGINEERING INSTALLATIONS

	Unit £	100 mm £	115 mm £	150 mm £
CAST IRON PIPES, GUTTERS AND FITTINGS				
Half-round gutter to B.S. 460				
1·83 m lengths	No.	5·29	5·54	11·17
Stop ends	„	0·59	0·91	1·31
Stop end outlets	„	1·38	1·83	2·29
Nozzle pieces	„	1·83	2·10	3·59
Angles	„	1·83	2·10	3·59

	Unit £	50 mm £	63 mm £	75 mm £	100 mm £
Rainwater pipe to B.S. 460					
1·83 m lengths	No.	8·23	9·16	9·16	12·48
Extra for:					
Ears per socket	„	0·59	0·59	0·59	0·59
Shoes, eared	„	3·73	4·52	4·52	5·94
Bends	„	1·83	2·74	2·74	4·25
Single branches	„	3·00	3·85	3·85	5·03
Double branches	„	7·17	8·21	8·21	11·77
Offsets, 229 mm projection	„	4·00	5·17	5·17	7·26
„ 305 mm „	„	5·03	5·94	5·94	8·49
Flat heads	„	3·59	3·59	3·59	8·30

	Unit £	75 mm £	100 mm £	150 mm £
Soil, waste and ventilating pipe to B.S. 416				
1·83 m lengths	No.	9·00	12·33	25·46
Extra for:				
Ears per socket	„	0·53	0·53	0·69
Oval access door	„	5·59	5·87	6·63
Bends, short radius	„	4·24	6·34	11·38
Single branches	„	6·39	9·83	18·95
Double branches	„	10·82	12·15	28·60
Offsets, 229 mm projection	„	6·80	9·12	18·78
„ 305 mm „	„	7·73	10·29	24·15
Connecting pipes with 140 mm sockets to take earthenware				
457 mm effective length	„	—	6·46	—
610 mm „	„	—	7·73	—
Long bends with 140 mm sockets to take earthenware				
457 × 457 mm	„	—	17·41	—
610 × 610 mm	„	—	22·36	—
Access pipe, oval door	„	5·59	5·87	6·63

	Unit £	125 mm £	150 mm £	200 mm £
ASBESTOS CEMENT PIPES, GUTTERS AND FITTINGS				
Half-round gutter to B.S. 569				
1·80 m lengths	No.	2·96	4·03	4·72
Stop ends	„	0·48	0·56	0·67
Drop ends	„	1·21	1·74	1·97
Nozzles	„	1·21	1·74	1·97
Angles	„	1·50	1·74	2·18
Fascia brackets	„	0·54	0·62	0·96
Joint pads	„	0·40	0·41	0·56

PLUMBING AND ENGINEERING INSTALLATIONS

ASBESTOS CEMENT PIPES, GUTTERS AND FITTINGS – *continued*

	Unit	406 × 127 × 254 mm £	457 × 127 × 152 mm £	610 × 152 × 229 mm £
Valley gutter to B.S. 569				
1·80 m lengths	No.	10·64	10·64	12·65
Stop ends	,,	2·34	2·49	3·15
Drop ends	,,	6·43	9·88	8·13
Nozzles.	,,	6·43	9·88	8·13
Joint pads	,,	0·85	0·85	0·98

	Unit	279 × 127 × 178 mm £	305 × 152 × 229 mm £	457 × 152 × 305 mm £	559 × 152 × 406 mm £
Boundary wall gutter to B.S. 569					
1·80 m lengths	No.	7·92	10·64	13·03	13·12
Stop ends	,,	1·67	2·46	2·96	3·35
Drop ends	,,	5·24	6·76	7·61	8·46
Nozzles.	,,	5·24	6·76	7·61	8·46
Joint pads	,,	0·66	0·85	0·97	1·04

	Unit	127 × 152 mm £	305 × 203 mm £	381 × 127 mm £
Box gutter to B.S. 569				
1·80 m lengths	No.	9·52	14·31	14·73
Stop ends	,,	2·49	3·42	3·42
Drop ends	,,	6·76	9·28	9·31
Nozzles.	,,	6·76	9·28	9·31
Joint pads	,,	0·66	0·98	0·97

		75 mm £	100 mm £	150 mm £
Rainwater pipe to B.S. 569				
1·80 m lengths	,,	3·91	5·11	10·40
Shoes	,,	1·31	1·74	4·56
Bends	,,	1·53	2·31	4·75
Single branches	,,	2·31	3·07	7·35
Offsets, 225 mm projection	,,	2·31	3·07	5·91
,, 300 mm ,,	,,	2·96	3·62	7·35
Rainwater heads, rectangular	,,	3·80	4·75	9·08

COLOURGLAZE TREATMENT

Colourglaze treatment gloss finish is available on asbestos goods. The extra over the foregoing prices for this finish is

Rainwater pipes and fittings, treated outside only	60%
Gutters and fittings treated on both surfaces	90%

PLUMBING AND ENGINEERING INSTALLATIONS

ALUMINIUM PIPES, GUTTERS AND FITTINGS TO B.S. 2997

	Unit	102 mm £	114 mm £	127 mm £
Half-round gutter				
1·83 m lengths	No.	5·64	5·75	6·30
Stop ends	,,	0·41	0·61	0·92
Stop end outlets	,,	1·38	1·41	1·43
Nozzle pieces	,,	1·61	1·70	1·72
90° angles	,,	1·46	1·60	1·72
Fascia brackets	,,	0·42	0·43	0·68
Side rafter brackets	,,	0·71	0·73	0·74

	Unit	51 mm £	63 mm £	76 mm £	102 mm £
Rainwater pipe					
1·83 m lengths	No.	4·15	5·14	5·88	8·51
Shoes	,,	1·54	1·79	2·43	2·99
Bends	,,	1·50	2·17	2·20	3·71
Single branches	,,	1·70	2·52	3·42	4·85
Offsets, 229 mm projection	,,	3·38	4·60	5·20	5·90
,, 305 mm ,,	,,	3·65	5·00	5·49	6·41
Rainwater heads, rectangular	,,	4·78	4·85	4·95	6·83

COLOURED PVC PIPES, GUTTERS AND FITTINGS

	Unit	76 mm £	112 mm £	152 mm £
Half-round gutter				
2 m lengths	No.	1·86	2·94	6·39
Stop ends	,,	0·35	0·52	1·38
Outlets	,,	0·94	1·17	3·04
Angles	,,	0·96	1·23	4·04
Fascia brackets	,,	0·20	0·32	0·89

	50 mm £	68 mm £	110 mm £
Rainwater pipe			
2 m lengths plain end	2·49	2·98	—
3 m length, plain end	3·74	—	9·44
Pipe connector	0·50	0·53	1·51
Shoes	0·75	0·75	3·00
Bends	1·17	1·40	3·88
Branches	1·63	2·28	5·18
Offsets, 229 mm projection	—	2·23	7·19
,, 305 mm ,,	2·27	2·39	7·38
Fixing brackets	0·34	0·46	0·56

	82 mm £	110 mm £	160 mm £
Soil pipe			
3 m lengths plain end	8·02	9·44	19·69
3 m lengths single socketed	9·03	10·44	23·20
Bends	2·66	3·88	7·68
Single branches	3·93	5·53	15·25
Double branches	6·86	9·32	18·28
Offsets, 229 mm projection	—	7·19	—
,, 305 mm ,,	—	7·38	—

PLUMBING AND ENGINEERING INSTALLATIONS

SUNDRIES	Unit	51 mm £	76 mm £	102 mm £	152 mm £
Fulbora patent rainwater outlets					
Aluminium					
Vertical spigot	No.	17·69	22·58	31·66	40·74
Vertical screw	,,	17·80	22·63	34·81	42·47
45° spigot	,,	20·00	26·62	36·91	—
45° screw	,,	20·84	29·98	42·47	—
90° spigot	,,	20·84	31·55	44·89	—
90° screw	,,	23·78	32·39	46·15	49·61
Two way	,,	24·41	32·97	34·49	46·15
Balloon gratings, g.a.m.	,,	0·46	0·52	0·61	1·82
Straight roof-outlets, cast iron, with lute for asphalt and 305 × 305 mm galvanized flat surface grating	,,	—	—	35·88	42·04
Extra for					
Inner grating	,,	—	—	1·40	1·40
Domical grating	,,	—	—	1·04	1·04
Ditto but with 254 mm diameter grating .	,,	—	23·90	28·08	34·12
Extra for Domical grating	,,	—	3·66	3·66	3·66

LEAD PIPE	Unit	£
Lead pipe in coils to B.S. 602 up to 50 mm diameter	100 kg	109·76
Ditto, 50 mm diameter and over	,,	110·26
Silver–Copper–Lead alloy pipe to B.S. 1085	,,	110·46
Chemical pipe	,,	110·76

TRAPS	Unit	32 mm £	38 mm £
Drawn copper trap to B.S. 1184, prepared for compression type fitting			
'S' trap, 38 mm seal	No.	5·42	6·50
,, 76 mm ,,	,,	5·90	7·10
Polypropylene traps			
Bottle traps			
'P' trap, 38 mm seal	,,	1·45	1·63
,, 75 mm ,, ...	,,	1·58	1·68
'S' trap, 75 mm seal	,,	2·17	2·33
Tubular traps with cleaning eye			
'P' trap, 38 mm seal	,,	1·45	1·63
,, 75 mm ,, ...	,,	1·61	1·71
'S' trap, 75 mm seal	,,	2·01	2·41

PLUMBING AND ENGINEERING INSTALLATIONS

	Unit	½″ £	¾″ £	1″ £	1¼″ £	1½″ £	2″ £
POLYTHENE TUBE							
Polythene tubing, low density (type 32) to B.S. 1972							
Class B . .	100 m	—	39·39	60·35	94·01	123·96	190·64
„ C . .	„	33·70	52·26	81·21	129·08	168·14	262·54
„ D . .	„	41·58	62·44	98·80	159·09	207·72	—
COMPRESSION FITTINGS FOR POLYTHENE TUBE							
Straight coupling .	No.	1·08	1·54	2·73	4·12	6·82	8·78
Tee	„	1·81	2·89	4·24	8·78	10·47	15·70
Elbow . . .	„	1·35	2·04	3·43	6·39	7·97	9·59
Stop end . .	„	1·16	1·58				
Blank cap . .	„	0·27	0·39	0·66	1·04	—	—
Copper liners B.S. 1972							
Class B . .	„	—	0·14	0·16	0·24	0·39	0·48
„ C . .	„	0·12	0·14	0·16	0·24	0·39	0·48
„ D . .	„	0·12	0·14	—	—	—	—
B.S. 3284							
Class B . .	„	—	—	0·16	0·24	0·39	0·48
„ C . .	„	0·12	0·14	0·16	0·24	0·33	0·48
„ D . .	„	0·12	0·14	0·16	0·24	0·33	0·48
PVC TUBE							
Unplasticized PVC tube to B.S. 3505 with plain ends, in 6 metre lengths							
Class D . .	100 m	—	—	—	70·00	88·00	186·00
„ E . .	„	30·00	43·00	59·00	88·00	116·00	211·00
FITTINGS FOR PVC TUBE							
Straight coupling .	No.	0·24	0·27	0·33	0·50	0·72	1·03
Elbow . . .	„	0·29	0·36	0·52	0·98	1·26	1·92
Tee	„	0·41	0·51	0·62	1·17	1·43	2·28
Tank connector .	„	0·59	0·65	—	—	—	—
Cap . . .	„	0·25	0·26	0·33	0·44	0·57	1·03

PLUMBING AND ENGINEERING INSTALLATIONS

	Unit	15 mm £	20 mm £	25 mm £	32 mm £	40 mm £	50 mm £
STEEL TUBE AND FITTINGS							
Galvanized steel tube to B.S. 1387							
Medium . . .	100 m	66·49	80·26	116·19	149·51	177·03	251·40
Heavy . . .	„	77·45	93·49	135·36	174·17	206·23	292·89
Galvanized malleable fittings							
Bend, M & F. .	100	52·87	73·21	105·75	154·55	187·10	284·70
Elbow . . .	„	36·60	48·80	73·20	117·95	162·70	223·70
Tee . . .	„	48·80	65·10	89·50	142·35	183·00	284·70

	Unit	B.S. 2871 Table X £	B.S. 2871 Table Y £	B.S. 2871 Table Z £
COPPER TUBE				
Copper tube				
6 mm diameter	100 m	29·50	36·60	—
8 mm „	„	39·10	49·00	—
12 mm „	„	58·00	73·20	—
15 mm „	„	70·80	107·00	58·30
22 mm „	„	131·00	189·00	99·90
28 mm „	„	166·00	241·00	126·00
35 mm „	„	292·00	381·00	184·00
42 mm „	„	381·00	464·00	274·00
54 mm „	„	493·00	800·00	393·00
67 mm „	„	693·00	1010·00	588·00
76 mm „	„	898·00	1160·00	761·00
108 mm „	„	1270·00	2080·00	1070·00
133 mm „	„	1530·00	—	—
159 mm „	„	2460·00	—	—

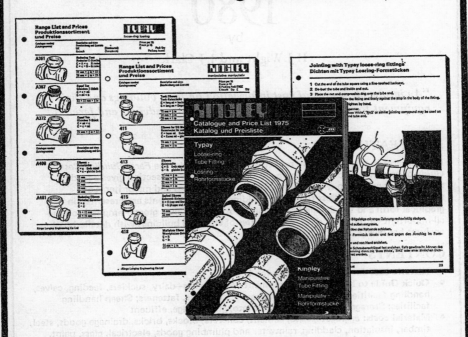

FARM BUILDING COST GUIDE 1980

by

H J Wight and J J Clark

"An invaluable source of cost information on farm buildings and their equipment for designers, consultants, contractors, advisers and farmers."

NEW FEATURES

New features in this completely revised sixth edition make the *Farm Building Cost Guide* of greater use to more people. The new *'Quick Guide'* section gives a ready reference for costs of a wide variety of complete buildings and major items of associated equipment. Cost trends for buildings and components are presented in graphical form and the detailed component cost section gives greater emphasis on the all-important current cost data. A new series of standard buildings with detailed cost analyses has been started.

CONTENTS

- Historic cost data and predictions for 1980
- Quick Guide to building and equipment costs: Cattle—dairy, sucklers, feeding, calves, handling facilities; Pig—dry sows, farrowing, weaners, fatteners; Sheep handling facilities; Storage—general purpose, potato, grain, silage, effluent
- Material costs: aggregate, sand, cement, concrete, blocks, bricks, drainage goods, steel, timber, insulation, cladding, rainwater and plumbing goods, electrical, glass, paint.
- Equipment costs: cattle, pig and general fittings and equipment
- Labour costs
- Plant costs
- Measured rates: excavating, concreting, brickwork, blockwork, roofing, carpenter and joiner work, steelwork, plumbing, electrical, rendering, glazing, painting, drainage work.
- Building cost analyses: calf house, cattle handling pens, flat-deck weaner house, sheep handling pens, below-ground slurry tank.

Farm Building Cost Guide 1980   48 pages   Price £2.00
Sixth edition: ISBN 0 902433 19 9

A full list of SFBIU publications is available on request.

Published by and available from: Scottish Farm Buildings Investigation Unit, Craibstone, Bucksburn, Aberdeen AB2 9TR (cheques etc payable to NOSCA).

PLUMBING AND ENGINEERING INSTALLATIONS

	Unit	15 mm £	22 mm £	28 mm £	35 mm £	42 mm £	54 mm £
FITTINGS FOR COPPER TUBE							
Capillary type to B.S. 864							
Straight couplings							
Copper to copper . .	No.	0·14	0·21	0·31	0·67	0·96	1·94
Copper to iron . .	"	0·41	0·70	0·96	1·70	2·08	3·10
Reducing coupling, copper to copper (large end measured)	"	—	0·33	0·43	0·99	1·63	2·38
Elbow	"	0·23	0·40	0·65	1·47	2·23	5·07
Elbow male iron to copper .	"	0·74	1·05	1·57	1·76	2·56	4·45
Tee	"	0·44	0·67	1·24	2·29	3·40	6·14
Tank connector with two backnuts . . .	"	1·89	3·01	4·12	—	—	—
Compression type to B.S. 864							
Straight couplings							
Copper to copper . .		1·14	1·50	2·25	2·39	4·62	7·35
Copper to iron . .		1·10	1·45	2·11	3·30	4·58	6·38
Copper to lead . .	"	1·28	1·76	2·64	—	—	—
Reducing coupling, copper to copper (large end measured)	"	—	1·50	2·25	3·39	4·62	7·35
Elbows							
Copper to copper . .	"	1·50	1·94	2·95	5·06	7·04	10·56
Ditto, with backplate .	"	2·33	2·95	—	—	—	—
Tee	"	1·98	2·64	4·27	6·60	9·82	14·31
Tank connector . .	"	1·32	1·85	3·08	4·62	5·50	8·80

		15 mm £	20 mm £	25 mm £
PLUMBERS' BRASSWORK				
Stop valves				
For iron, M to F . .	"	3·57	5·35	8·90
With double union for lead	"	4·40	—	—
With compression joints for copper	"	4·82	7·04	11·04
With capillary joints for copper	"	2·13	3·13	5·64
Bibtaps to B.S. 1010, screwed iron				
Crutch head	"	2·37	3·56	6·98
Hose union	"	3·37	5·56	10·25

PLUMBING AND ENGINEERING INSTALLATIONS

	Unit	15 mm £	20 mm £	25 mm £	32 mm £	40 mm £	50 mm £
PLUMBERS' BRASSWORK							
– continued							
Gate valves with wheel heads							
Gunmetal, F to F . .	No.	6·55	8·33	11·30	16·95	21·00	31·50
Ball valves to B.S. 1212, screwed for iron							
High pressure . . .	,,	2·60	5·33	13·72	—	—	—
Equilibrium ball valve . .	,,	9·00	12·30	18·60	—	—	—

		114 mm	125 mm	150 mm			
Copper floats to B.S. 1968 .	,,	1·50	2·00	3·00	—	—	—
Plastic floats . . .	,,	0·34	0·51	1·05	—	—	—

CISTERNS, TANKS AND CYLINDERS

	Unit	£
Galvanized mild steel cistern to B.S. 417, Grade A		
Size No. SCM 270, 191 litres .	No.	24·30
,, SCM 450/1, 327 litres	,,	37·00
,, SCM 1600, 1227 litres	,,	126·56
,, SCM 2720, 2137 litres	,,	182·25
,, SCM 4540, 3364 litres	,,	354·38
Galvanized mild steel tank to B.S. 417, Grade A		
Size No. T 25/1, 95 litres	,,	45·00
,, T 30/1, 114 litres	,,	47·50
,, T 40, 155 litres	,,	57·00
Galvanized mild steel cylinder to B.S. 417, Grade C with one end convex, one end concave, welded construction		
Size No. YM 114, 100 litres	,,	37·00
,, YM 141, 123 litres	,,	39·50
,, YM 218, 195 litres	,,	49·50
,, YM 264, 241 litres	,,	53·50
Copper cylinder to B.S. 699, Grade 3		
Type 1, 74 litres .	,,	30·45
,, 2, 98 litres .	,,	34·15
,, 7, 120 litres .	,,	38·25
,, 8, 144 litres .	,,	42·68
Galvanized welded mild steel indirect cylinder to B.S. 1565, Class B with five tappings for connections		
Size No. BSG 2, 136 litres	,,	65·00
,, BSG 3, 159 litres	,,	73·50
,, BSG 4, 227 litres	,,	140·20
,, BSG 6, 364 litres	,,	193·60
,, BSG 7, 455 litres	,,	233·60
Asbestos cement cistern to B.S. 2777 with lid		
Size No. AC 90 (M), approx. cap. to water line 62 litres .	,,	20·58
,, AC 136 (M), ,, ,, ,, ,, 114 ,, .	,,	23·63
,, AC 227 (M), ,, ,, ,, ,, 188 ,, .	,,	32·19
,, AC 455 (M), ,, ,, ,, ,, 381 ,, .	,,	45·70
Polyester glass fibre cistern with lid		
45 litres .	,,	16·60
182 litres .	,,	34·06
318 litres .	,,	57·38
Polythene cold water cistern with lid		
182 litres .	,,	31·50
318 litres .	,,	53·00

PLUMBING AND ENGINEERING INSTALLATIONS

	Unit	£

SINKS AND WASH TUBS
White glazed inside and out, fireclay ware

	Unit	£
Belfast sinks		
610 × 406 × 203 mm .	No.	24·95
610 × 457 × 254 mm .	,,	31·00
762 × 508 × 254 mm .	,,	51·60
Double sink		
914 × 457 × 254 mm .	,,	78·75
1067 × 533 × 254 mm	,,	106·85
Wash tubs		
610 × 508 × 381 mm .	,,	63·65
Combined stainless steel drainer 1050 × 525 mm and rectangular bowl 462 × 375 × 200 mm deep	,,	64·15
Combined stainless steel double drainer 1575 × 525 mm and rectangular bowl 462 × 375 × 200 mm	,,	84·60

	Unit	*White* £	*Coloured* £
LAVATORY BASINS AND DRINKING FOUNTAINS			
Vitreous china lavatory basin with pair 15 mm chromium-plated easy-clean pillar taps, painted iron towel rail brackets and 32 mm waste plug and chain and union			
559 × 406 mm .	No.	28·40	31·17
635 × 457 mm .	,,	35·55	41·00
Lavatory basin all as last but with pedestal			
559 × 406 mm .	,,	36·17	41·50
635 × 457 mm .	,,	43·32	51·32
Fireclay lavatory basin with pair 15 mm chromium-plated easy-clean pillar taps, waste plug and chain and union, with brackets for building in			
559 × 406 mm .	,,	49·85	—
635 × 457 mm .	,,	59·45	—
Range of four basins 635 × 457 mm all as last with loose overlaps	,,	258·00	—
Extra for each additional basin .	,,	66·25	—
Fireclay circular island wash-fountain, 1067 mm diameter, with chromium-plated fittings comprising central column with circular wave discharge and tray for block soap with waste union, inlet union and mixing valve .	,,	913·00	—
Fireclay drinking fountain, with front jet with chromium-plated fittings controlled by press-action self-closing valve with regulating valve and outlet grating .	,,	69·00	—

BATHS, BATH PANELS AND SHOWERS

		White	*Coloured*
Rectangular pattern, reinforced acrylic bath 1675 mm long with panels, chromium-plated overflow to B.S. 1184, 20 mm chromium-plated pillar taps to B.S. 1010 and 40 mm chromium-plated plug and waste with brass trap .	,,	126·75	131·00

PLUMBING AND ENGINEERING INSTALLATIONS

BATHS, BATH PANELS AND SHOWERS – *continued*

	Unit	£
Chromium-plated shower fitting, comprising 15 mm mixing valve, swan-neck riser, 127 mm shower rose	No.	88·35
762 × 762 × 178 mm white glazed fireclay tray with run through outlet, rubber plug and chain	„	65·55

W.C. SUITES

		White £	Coloured £
Vitreous china W.C. pan, glazed inside and out with 'P' or 'S' trap, to B.S. 1213	„	16·10	20·10
Ditto but with horizontal outlet	„	13·10	16·62

		Black £	
Plastic W.C. seats			
Single	„	4·50	6·00
Double	„	6·85	10·29

Flushing cisterns to B.S. 1125 with fittings, flush pipe, connectors and brackets

Plastic, high level			
9 litres	„	17·40	19·00
14 „	„	19·50	—
Plastic, low level			
9 litres	„	20·70	22·20
14 „	„	22·80	—

		White £	
Vitreous china, high level			
9 litres	„	33·05	40·00
14 „	„	34·80	42·66
Vitreous china, low level			
9 litres	„	20·78	26·85
14 „	„	21·52	28·71

	Unit	£
Closet suite consisting of three white vitreous china pans, 254 mm high, solid plastic open front seats, glass fibre flushing trough with chain pulls, guides, cantilevers and down pipes	Set	233·35

SLOP SINKS

Slop hopper with pedestal, in white glazed fireclay, 432 × 457 mm, with 'P' or 'S' trap comprising hardwood pad on rim, galvanized-iron hinged bucket grating, white vitreous china nine-litre flushing cistern with brackets, chromium-plated chain and china pull, 32 mm galvanized steel flush pipe, for back wall fixing .	No.	169·78

PLUMBING AND ENGINEERING INSTALLATIONS

	Unit	£

URINALS

White glazed fireclay single stall urinal, 1067 mm high to top of stall by 533 mm wide, 152 mm fluted treads, white glazed vitreous china automatic flushing cistern and brackets, polished copper down pipe and domed grating . .   No.   200·85

White glazed fireclay three stall urinal range 1067 mm high, each 533 mm wide, with 76 mm divisions, 152 mm fluted treads, white glazed vitreous china automatic flushing cistern and brackets, copper supply and sparge pipes and brass domed grating . . . .   „   604·00
Extra for each additional stall . . . .   „   202·50

BOILERS

Gas-fired domestic boiler rated at 13 kW providing continuous direct domestic hot-water supply, finished cream or white enamel, with electric controls . .   „   170·00
As above but rated at 22 kW . . . .   „   285·00

RADIATORS

Pressed steel panel type 584 mm high, with air-vent plug
  Heating surface (about)

	Unit	£
1·86 m²	„	16·00
2·23 m²	„	19·20
2·88 m²	„	24·75
3·25 m²	„	28·00

	Unit	Straight £	Angle £
Radiator valves, brass wheel head or lock shield, easy-clean pattern			
15 mm	No.	3·83	3·06
20 mm	„	4·84	3·90

PLUMBING AND ENGINEERING INSTALLATIONS

INSULATING MATERIALS

	Unit	£
Vermiculite loose fill in bags containing 4 cu. ft.	Bag	3·08
Hair felt canvas backed in 7300 × 100 × 6 mm roll	Roll	2·28

Expanded polystyrene lagging sets (without bases) for mild steel cisterns to B.S. 417

	Unit	£
Size No. SCM 270, 191 litres	No.	1·66
„ SCM 450/1, 327 litres	„	2·52
„ SCM 1600, 1227 litres	„	5·90
„ SCM 2720, 2137 litres	„	8·48
„ SCM 4540, 3364 litres	„	11·72

80 mm thick glass-fibre-filled flexible insulating jackets, segmented type, consisting of individual panels, backed internally with stout white cotton canvas, finished externally with pvc sheet with plastic bands and clips for fixing, for

	Unit	£
Copper cylinders to B.S. 699		
Size No. 2	„	5·90
„ 7	„	6·32
„ 8	„	6·93
Mild steel cylinders to B.S. 417		
Size No. Y 25	„	4·50
„ Y 39	„	5·30
„ Y 58	„	5·75

	Unit	Nominal wall thickness		
		19 mm £	25 mm £	32 mm £
Rigid fibreglass pipe insulation in 915 mm lengths, canvas finish complete with metal fixing bands				
For steel tubes				
15 mm nominal bore	m	1·725	1·979	2·503
20 mm „ „	„	1·881	2·146	2·681
25 mm „ „	„	2·062	2·341	2·886
32 mm „ „	„	2·269	2·564	3·123
40 mm „ „	„	2·424	2·731	3·299
50 mm „ „	„	2·802	3·144	3·751
80 mm „ „	„	3·570	3·973	4·627
For copper tubes				
15 mm nominal size	„	1·570	1·811	2·327
22 mm „ „	„	1·805	2·007	2·533
28 mm „ „	„	1·907	2·174	2·710
35 mm „ „	„	2·088	2·368	2·916
42 mm „ „	„	2·269	2·564	3·123
54 mm „ „	„	2·643	2·973	3·569

PLUMBING AND ENGINEERING INSTALLATIONS

	Unit	75 *mm* £	100 *mm* £	150 *mm* £
ASBESTOS-CEMENT FLUE PIPES				
Asbestos-cement, light quality, socketed flue pipe to B.S. 567				
1·80 m lengths	No.	4·20	5·62	9·02
Square and obtuse bends	,,	1·81	2·29	3·51
Square and obtuse equal tees	,,	2·28	2·86	4·42
Split sockets	,,	1·24	1·67	2·07
Cone cap terminal	,,	4·04	5·28	7·20
Extra for cleaning doors	,,	1·44	1·44	1·90

		125 *mm* £	175 *mm* £	225 *mm* £
Asbestos-cement, heavy quality, socketed flue pipe to B.S. 835				
1·80 m lengths	,,	9·15	16·26	21·25
Square and obtuse bends	,,	3·80	6·76	11·57
Square and obtuse equal tees	,,	4·05	7·20	12·30
Loose sockets	,,	2·62	4·00	5·70
Cone cap terminals	,,	8·38	21·33	27·56
Extra for cleaning doors	,,	2·76	3·31	3·31

	Unit	£
SUNDRIES		
Plumbers' solder	kg	5·70
Tinman's solder	,,	7·15
White jointing compound	,,	0·80
Copper tacks		
15 mm	,,	8·21
20 mm	,,	8·21
25 mm	,,	8·21
Linseed oil putty	,,	0·34
Red lead putty	,,	0·37
Philplug jointing composition		
P.C. 3 cord	,,	1·93
P.C. 4 sulphate resisting grade	,,	2·01
Hemp		
Long dressed white jute	,,	4·50
Jute gaskin, untarred in 30 m coils	coil	5·20
,, ,, tarred in 30 m coils	,,	6·10
Evomastic natural in cartridge	No.	0·96
Cast-iron cantilever sink brackets		
406 mm	Pair	3·00
457 mm	,,	3·15
Copper wire, 10–19 s.w.g.	kg	7·40

PLUMBING AND ENGINEERING INSTALLATIONS

SUNDRIES – *continued*	Unit	50 mm £	60 mm £	75 mm £	100 mm £		
Rainwater pipe nails (wire) .	kg	—	—	0·90	0·89		
Gutter screws, galvanized .	100	8·80	10·00	—	—		

		20 mm £	25 mm £	30 mm £	40 mm £		
Gutter bolts and nuts, galvanized . . .	,,	3·66	3·80	4·05	4·50		

		75 mm £	100 mm £	113 mm £	150 mm £		
Cast iron brackets for half-round gutter							
Self-colour . . .	10	3·84	3·84	3·84	4·47		
Galvanized . . .	,,	4·39	4·39	4·39	5·11		

		15 mm £	22 mm £	28 mm £	35 mm £	42 mm £	54 mm £
Pipe brackets, build-in type, brass	,,	7·15	8·25	8·80	11·85	15·75	—
Pipe brackets, screw-on type, brass	,,	2·85	2·95	3·40	4·60	5·80	7·40
Pipe clips Copper	100	2·30	2·85	4·55	7·10	9·85	15·50

		15 mm £	20 mm £	25 mm £	32 mm £	40 mm £	50 mm £
Galvanized steel . .	,,	1·50	1·84	2·35	3·06	4·80	6·20

FINISHINGS

	Unit	£
CEMENT COLOURERS		
Shadecrete 70 powder colours for cement		
Slate Blue	tonne	190·00
Coral Red	,,	200·00
Buff	,,	240·00
Brick Red	,,	250·00
Black	,,	270·00
Deep Red	,,	270·00
Dark Brown	,,	280·00
Tan	,,	310·00
Straw yellow	,,	320·00
Golden brown	,,	340·00
Dark green	,,	680·00
HARDENERS		
Cementone No. 8	litre	0·31
Colemanoid No. 3	,,	0·43
Sealocrete Double Strength Premix	,,	0·46
Silicon carbide grains	50 kg	31·00
GRANITE CHIPPINGS		
Granite chippings 6 mm to dust, to B.S. 1201	tonne	9·92
SAND		
Sand for plastering, to B.S. 1198	,,	5·46
Sand for external rendering, etc., to B.S. 1199	,,	5·46

FINISHINGS

	Unit	£
KEYING LIQUID		
Cemprover	litre	0·98

PLASTER

The following are intended to represent prices for plaster quoted by builders' merchants and are not consistent with the "Prices for Measured Work" which are based upon specialists' prices.

	Unit	£
Plaster to B.S. 1191, Part 1, Class A		
C.B. Stucco	tonne	44·95
Plaster to B.S. 1191, Part 1, Class B		
Browning	,,	50·50
Finish	,,	50·84
Board finish	,,	49·80
Plaster to B.S. 1191, Part 1, Class C		
Sirapite	,,	53·95
Keene's cement to B.S. 1191, Part 1, Class D		
Coarse white	,,	98·15
Premixed lightweight plaster to B.S. 1191, Part 2		
Browning	,,	76·15
Browning HSB	,,	84·50
Bonding	,,	74·80
Metal lathing	,,	73·40
Finishing	,,	60·70
Projection plaster	,,	73·50
Cullamix Tyrolean mixture	,,	81·60

FINISHINGS

	Unit	£
METAL BEADS		
24 B.G. steel beads		
Angle bead	100 lin. ft.	5·28
Screed bead	,,	5·23
Plaster stop	,,	6·93

		Unit	Red £	Brown £
CLAY FLOOR TILES				
Clay floor quarries to B.S. 1286, Class I				
Plain edge				
150 × 150 × 12·5 mm		1000	74·90	101·20
150 × 150 × 19 mm		,,	102·30	117·85
200 × 200 × 19 mm		,,	243·00	299·65
Coved skirting				
150 × 90 × 12·5 mm		,,	106·60	113·00
Angles for above		No.	0·41	0·46
150 × 150 × 12·5 mm		1000	136·80	152·45
Angles for above		No.	0·41	0·46

	Unit	Red £	Black £	Buff £
Vitrified tiles to B.S. 1286				
150 × 75 × 12 mm	m²	5·87	6·83	7·21
150 × 150 × 12 mm	,,	5·83	6·37	7·00
150 × 150 × 18 mm	,,	8·60	10·24	—
Coved skirting				
150 × 150 × 12 mm	100	28·87	30·38	30·38
Angles for above	,,	25·00	25·00	25·00

FINISHINGS

STEEL PAVING	Unit	£
Stelcon anchor steel plates 300 × 300 × 23 mm	No.	1·40
Stelcon steel clad flags 300 × 300 × 28 mm	,,	0·90

	Unit	Thickness	
		18 mm £	25 mm £
VENEERED BLOCKBOARD			
Afrormosia one side, balancer on reverse . . .	10 m²	73·00	—
Figured oak ditto	,,	73·00	90·00
Teak ditto	,,	78·00	95·00
Afrormosia both sides	,,	80·00	—
Figured oak ditto	,,	80·00	97·00
Teak ditto	,,	89·00	106·00

		6 mm £	9 mm £	12 mm £
VENEERED PLYWOOD				
Veneered plywood in standard sizes, ex wharf				
Sapele one side, balancer on reverse . . .	,,	28·00	39·00	48·00
Oak ditto	,,	31·00	42·00	51·00
Afrormosia ditto	,,	29·00	40·00	—
Plain teak ditto	,,	30·00	41·00	48·00

		4 mm £		
DECORATIVE PLYWOOD				
Pre-finished decorative plywood in random V-grooved unmatched panels size 2440 × 1220 mm				
Sapele	,,	39·00	—	—
Afrormosia	,,	56·00	—	—
Knotty pine	,,	42·00	—	—
Cherry	,,	44·00	—	—
Oak	,,	42·00	—	—
Teak	,,	47·00	—	—

FINISHINGS

BUILDING BOARDS	Unit	3·2 mm £	4·8 mm £	6·4 mm £
Hardboard to B.S. 1142				
Standard quality				
Plain	10 m²	6·30	9·60	12·45
Perforated	"	7·30	—	13·15
Tempered	"	8·50	12·30	16·20
Flame retardant to Class 1, B.S. 476	"	12·00	14·56	20·00

		6·4 mm £	9·6 mm £	12·7 mm £
Medium quality				
Interior quality	"	19·00	26·50	—
Exterior quality	"	21·50	31·00	40·00
Flame retardant to Class 1, B.S. 476	"	33·00	49·00	—

		12·5 mm £	19 mm £	
Insulation board to B.S. 1142				
Natural finish	"	10·60	12·25	
Ivory finish	"	10·60	12·25	
Flame retardant to Class 1, B.S. 476	"	18·60	—	
Acoustic tiles and panels				
Treetac tiles, painted finish, bevelled edges	"	—	44·28	
Slotac tiles, painted finish, square edges	"	25·85	33·25	

Market Prices of Materials

FINISHINGS

BUILDING BOARDS *– continued*	Unit	3 mm £	4·5 mm £	6 mm £	9 mm £	12 mm £
Asbestos cement flat sheets						
Semi-compressed . . .	m²	—	1·88	2·55	3·85	6·42
Fully-compressed . . .	,,	—	2·88	3·58	5·41	6·74
Asbestos wood . . .	,,	1·95	3·06	3·56	5·64	7·49
Partition board semi-compressed	,,	1·41	1·86	2·45	—	—
Partition board fully-compressed	,,	2·17	—	—	—	—

	Unit	3·2 mm £		6 mm £		
Stove enamelled asbestos, all colours . . .	,,	7·05	—	12·20		

	Unit	5 mm £			
Surface grained asbestos, grey .	,,	—	4·76	—	—	—

	Unit	6 mm £	9 mm £	12 mm £
Asbestos insulation board .	,,	3·05	4·76	6·28

	Unit	3 mm £	4·5 mm £	6 mm £	9 mm £
Glass reinforced cement flat sheets . . .	,,	1·78	2·53	3·10	4·92

	Unit	1·5 mm £	15 mm £
Decorative laminated sheets, standard and plain colours, Formica, Warerite or similar, price depending on pattern and quality .	m²	5·00	—
Ditto on plywood with balancer on reverse side . . .	,,	—	14·30
Ditto on chipboard ditto	,,	—	8·25

	Unit	9·5 mm £	12·7 mm £
Plasterboard to B.S. 1230			
Plain .			
Insulating . . .	100 m²	65·69	80·30
Plastic faced plasterboard, white	,,	80·56	95·17
Insulating	,,	112·24	126·47

	Unit	57·2 mm £	63·5 mm £
Prefabricated plasterboard partition panels			
Plain . . .	,,	225·87	266·38
Insulating . . .	,,	240·20	280·99

FINISHINGS

GLAZED CERAMIC WALL TILES	Unit	White £	Colours From £	To £
Wall tiles to B.S. 1281				
108 × 108 × 4 mm				
Plain	m²	3·23	3·34	6·25
152 × 152 × 5·5 mm				
Plain	,,	3·82	3·82	9·70
100 × 100 × 8 mm				
Plain	,,	—	7·33	9·59
152 × 152 × 8 mm				
Plain	,,	—	5·55	8·05

LATHING	Unit	9·5 mm £	12·7 mm £
Plasterboard to B.S. 1230			
Gypsum wallboard	100 m²		
Plain		65·69	80·30
Insulating	,,	80·56	95·17
Gypsum lath or baseboard			
Plain	,,	58·29	74·97
Insulating	,,	73·16	89·84

	Unit	£
Metal lathing, to B.S. 1369, type 'a' (plain expanded)		
0·475 mm thick	Sq. yd.	0·87
0·675 mm thick	,,	1·00
Newtonite lathing	m²	2·00

SUNDRIES

Prefabricated 100 mm plasterboard cove	m	0·36
Adhesive for plasterboard cove	kg	0·23
Hessian scrim cloth in rolls 90 mm × 100·00 m	Roll	2·16
Galvanized plasterboard nails	25 kg	19·87
Adhesive cement for acoustic tiles	litre	1·79
Impact adhesive for plastic laminates, plywood, wallboards, etc.	,,	1·57
Ceramic tile adhesive	,,	0·64
Ceramic tile grout	6 kg	3·06

GLAZING

NOTE. *Reductions in certain glass prices are offered by manufacturers for acceptance of specified minimum quantities of one size and substance delivered to one address at one time.*

TRANSPARENT GLASS	Unit	O.Q. £	S.Q. £
Glass to B.S. 952, Section one			
Sheet glass, cut sizes not over 2020 mm one way or 1220 mm both ways			
2 mm	m²	6·58	7·64
3 mm	,,	8·41	10·77
4 mm	,,	9·71	11·38
Thick drawn sheet cut sizes			
5 mm, not over 2100 × 2350 mm or 1750 × 2400 mm	,,	13·74	—
6 mm, not over 9·30 m²	,,	15·34	—
Float glass, cut sizes			£
3 mm, not over 2020 mm one way or 1220 mm both ways	,,		9·57
4 mm, not over 2100 mm one way or 1300 mm both ways	,,		9·71

	Unit	G.G. quality £	S.G. quality £
5 mm not over 4 m² (or 2600 mm long or 1750 mm both ways)	,,	13·74	15·11
6 mm not exceeding 4500 mm one way or 2500 mm both ways	,,	15·34	16·11
10 mm not over 1·41 m²	,,	22·75	—
10 mm not over 4 m² (or 3150 mm long)	,,	25·03	—
10 mm not over 9·30 m² (or 4500 mm long or 2400 mm both ways)	,,	31·09	—
Thick float glass, cut sizes			£
12 mm not over 1·41 m²	,,		30·10
12 mm not over 4 m² (or 3150 mm long)	,,		33·11
12 mm not over 9·30 m² (or 4500 mm long or 2400 mm both ways)	,,		41·13
15 mm over 1·41 m² not over 4 m²	,,		43·40
15 mm over 4 m² not over 9 m²	,,		47·87
19 mm over 1·41 m² not over 4 m²	,,		56·66
19 mm over 4 m² not over 9 m²	,,		62·46
25 mm over 1·41 m² not over 4 m²	,,		88·45
25 mm over 4 m² not over 9 m²	,,		117·95

(Not over 2950 mm long or 2400 mm both ways for 15, 19 and 25 mm substance)

TRANSLUCENT GLASS	Unit	£
Glass to B.S. 952, Section two		
Rough cast glass, cut sizes up to 920 mm wide and up to 3700 mm long		
5 mm	m²	12·44
6 mm	,,	13·30
10 mm	,,	19·27
White patterned glass, cut sizes		
4 mm	,,	9·15
6 mm	,,	16·65
Tinted patterned glass, cut sizes		
3/4 mm	,,	17·73
5/6 mm	,,	22·00

GLAZING

	Unit	£
SPECIAL PURPOSE GLASS		
Glass to B.S. 952, Section four		
Wired glass, cut sizes		
7 mm Georgian wired cast up to 920 mm wide	m²	14·35
Up to 3700 mm long		31·46
6 mm Georgian wired polished plate	„	
Solar control glass, cut sizes		
'Antisun' float	„	19·89
4 mm		28·76
5 mm not over 4 m² (or 2400 mm one way or 1750 mm both ways)	„	28·76
6 mm not over 9·30 m²	„	
10 mm over 4 m² not over 9·30 m²	„	60·39
12 mm over 4 m² not over 9·30 m²	„	78·49
'Spectrafloat'		
6 mm not over 9·30 m²	„	25·15
10 mm over 4 m² not over 9·30 m²	„	52·94
12 mm over 4 m² not over 9·30 m²	„	71·06
Toughened glass, cut sizes, arrised edges		
6 mm rough cast		18·70
over 0·20 m², not over 0·50 m²	„	18·70
over 0·50 m², not over 2·50 m²	„	18·70
over 2·50 m²	„	
10 mm rough cast		27·40
over 0·20 m², not over 0.50 m²	„	24·60
over 0·50 m², not over 2·50 m²	„	23·50
over 2·50 m²	„	
6 mm float		23·70
over 0·20 m², not over 0·50 m²	„	22·25
over 0·50 m², not over 2·50 m²	„	20·75
over 2·50 m²	„	
10 mm float		33·70
over 0·20 m², not over 0·50 m²	„	33·70
over 0·50 m², not over 2·50 m²	„	32·20
over 2·50 m²	„	
12 mm float		40·10
over 0·20 m², not over 0·50 m²	„	40·10
over 0·50 m², not over 2·50 m²	„	38·45
over 2·50 m²	„	
6 mm 'Spectrafloat'		
over 0·50 m², not over 2·50 m²	„	30·80
over 2·50 m²	„	29·45
10 mm 'Spectrafloat'		
over 0·50 m², not over 2·50 m²	„	44·65
over 2·50 m²	„	42·80
12 mm 'Spectrafloat'		
over 0·50 m², not over 2·50 m²	„	53·65
over 2·50 m²	„	51·90
Laminated glass, cut sizes		
Safety glass, clear sheet quality		
4·4 mm not over 2160 × 1200 mm	„	22·83
5·4 mm not over 2160 × 1200 mm	„	23·70
6·8 mm not over 2440 × 1600 mm	„	27·20
Antibandit glass clear sheet quality		
7·5 mm not over 2550 × 1600 mm	„	44·16
Polycarbonate bullet resistant sheet in stock sheet sizes		
33 mm, clear	„	342·32
33 mm, solar bronze	„	375·65

GLAZING

TOUGHENED GLASS DOORS

12 mm toughened clear glass doors, 75 mm aluminium rail finished anodised silver including floor spring, bottom rail with deadlock, socket, door stop, patch fitting top centre housing and up to four drillings.

	Unit	£
762 × 2134 mm	No.	245·75
838 × 2134 mm	,,	253·75
914 × 2134 mm	,,	260·50
914 × 2438 mm	,,	277·00

HOLLOW GLASS BRICKS

Vaculite hollow glass bricks

	Unit	Plain £	Colours £
Type 198, 190 × 190 × 80 mm	No.	1·50	2·25
Type 248, 240 × 240 × 80 mm	,,	2·45	4·25
Type 2411, 240 × 115 × 80 mm	,,	1·35	2·25

INSULIGHT UNITS

Hermetically sealed units composed of two skins of glass separated by a lead spacer and 6 mm or 12 mm cell of dehydrated air

	Unit	Float 3 mm £	Float 4 mm £	Float 5 mm £	Float 6 mm £
0·25 m²	m²	51·15	51·15	52·30	53·55
0·35 m²	,,	44·95	44·95	46·35	48·70
0·50 m²	,,	39·35	39·35	44·80	46·15
0·75 m²	,,	31·35	31·35	41·40	42·85
1·00 m²	,,	26·65	26·65	37·20	38·85
2·00 m²	,,	23·65	23·65	33·55	34·45
4·00 m²	,,	22·95	22·95	32·65	33·80
6·00 m²	,,	—	—	36·20	37·20
8·00 m²	,,	—	—	—	45·40

DOME ROOFLIGHTS

Perspex dome single skin rooflights, clear or diffusing

Circular

	Unit	£
600 mm diameter	No.	15·60
900 mm ,,	,,	27·85
1200 mm ,,	,,	41·60
1800 mm ,,	,,	112·50

Square or rectangular

600 × 600 mm	,,	13·55
900 × 600 mm	,,	23·75
900 × 900 mm	,,	26·35
1200 × 900 mm	,,	31·00
1200 × 1200 mm	,,	37·40
1500 × 1050 mm	,,	58·30
1800 × 1800 mm	,,	171·65

SUNDRIES

General-purpose mastic in cartridge	,,	0·96
Linseed oil putty	kg	0·34
Metal casement putty	,,	0·37
Saferbed plate-glass bedding in 50 yd. coils		
For 6 mm glass and ⅜ in. deep bead	m	0·16
,, 6 mm ,, ½ in. ,,		0·18

PAINTING AND DECORATING

NOTE. *Prices in this section reflect trade discounts allowed by manufacturers to purchasers of bulk quantities.*

	Unit	£
PAINT		
Primers		
Acrylic primer/undercoat	5 litres	5·78
Plaster primer	,,	8·32
Alkali resisting primer	,,	5·47
Lead free wood primer	,,	6·41
Aluminium wood primer	,,	7·56
Calcium plumbate primer	,,	10·28
Zinc chromate primer	,,	8·70
Red lead primer	,,	12·44
Chlorinated rubber zinc primer	,,	15·79
Red oxide chromate primer	,,	7·40
Undercoats and finishings		
Emulsion paint	,,	5·67
Gloss paint		
Undercoat	,,	6·61
Finish	,,	6·79
Aluminium paint	,,	7·50
Anti-condensation paint	,,	7·17
Chlorinated paint		
Primer/undercoat	,,	7·32
Finish	,,	8·72
Road paint		
White	,,	8·04
Yellow	,,	8·04
Bituminous black paint	,,	4·19
Heat resisting paint	,,	6·07
Fire retardant paint	,,	8·00
White Portland cement paint	40 kg	15·40
Sandtex		
Matt	10 litres	23·00
Textured	tonne	1100·00
VARNISHES		
Yacht	5 litres	6·94
Extra pale	,,	6·09
Polyurethane	,,	6·54
Knotting	,,	9·47

PAINTING AND DECORATING

	Unit	£

WALL COVERINGS

	Unit	£
Glass fibre wall covering openweave	m²	1·30
Paper backed woven linen fabric depending on pattern	„	2·01 to 5·33

ADHESIVES

	Unit	£
Adhesives for fibre or paper backed wall coverings and foam polystyrene veneers	5 litres	6·00

STAINS AND PRESERVATIVES

	Unit	£
Creosote	„	2·90
Cuprinol wood preserver for internal or external use, clear	„	5·82
Solignum		
Exterior quality		
Brown	„	4·95
Interior quality		
Oak	„	7·30
Other colours	„	10·20
Colourless	„	7·20
Oil varnish stain	„	6·20
Linseed oil	„	5·00

DRAINAGE

VITRIFIED CLAY PIPES	Unit	100 mm £	150 mm £	225 mm £	300 mm £
Vitrified pipes and fittings:					
Extra strength quality plain end with push-fit flexible couplings					
Straight pipe (including coupling every 1·6 m)	m	1·42	3·00	—	—
Bend	No.	1·12	2·72	—	—
Taper pipe	”	—	2·64	—	—
Single junction	”	2·34	3·99	—	—
Plastic coupling with rubber rings	”	0·57	1·31	—	—
Extra strength quality with flexible socketed joints					
Straight pipe	m	—	3·06	5·87	9·17
Bend	No.	—	5·46	10·78	21·25
Single junction	”	—	6·64	15·00	31·83
Taper pipe	”	—	8·34	12·46	25·49
'Best' or 'Surface Water' quality unjointed					
Straight pipe	m	1·25	2·24	4·52	7·92
Bend	No.	1·16	1·95	6·09	12·03
Single junction	”	2·30	3·91	12·17	24·09
Double junction	”	4·60	7·81	24·31	48·16
Taper pipe	”	3·03	5·10	11·06	21·92
Adjustments to the above for quality					
'B.S. 65'			Plus 10%		
'B.S. Tested'			Plus 50%		
Channel pipe, with or without sockets					
0·9 m lengths	”	0·96	1·57	3·53	6·97
Bend	”	0·98	1·61	5·38	10·67
Taper	”	—	4·07	9·01	17·83
Single junction	”	1·85	3·08	8·82	20·97
Double junction	”	2·78	4·63	14·09	31·50
Branch channel bends					
Half section	”	2·00	3·28	10·82	21·42
Three-quarter section	”	2·21	3·73	13·49	26·75
White glazed channels and fittings					
Channel, half round					
600 mm lengths	”	10·52	14·38	22·90	—
Angle piece	”	14·47	21·45	34·46	—
Stop end	”	9·60	14·25	22·87	—
Channel, block					
300 mm lengths	”	6·31	9·33	—	—
Angle piece	”	19·00	27·97	—	—
Stop end	”	12·66	18·64	—	—

UNPLASTICIZED PVC PIPES AND FITTINGS	Unit	110 mm £	160 mm £
Unplasticized pvc drain pipe and fittings			
6 m lengths single socket pipe	No.	14·35	32·38
swept bend	”	3·26	7·38
knuckle bend	”	3·26	5·92
635 mm radius bend 90°	”	—	11·50
450 mm radius bend 45°	”	5·94	—
single junction	”	4·72	13·04
Taper	”	—	4·20
Adaptor to clayware	”	2·22	4·43
Loose pipe socket	”	1·41	3·24
Coupler double socket	”	1·55	5·12

DRAINAGE

UNPLASTICISED PVC PIPES AND FITTINGS	Unit	£
– *continued*		
Access pit bowl	No.	15·70
Access pit connectors	,,	3·54
Branch 110 × 110 mm	,,	4·72
Branch 160 × 110 mm	,,	9·39
Head of drain bend	,,	3·26

LAND DRAINS	Unit	75 mm £	100 mm £	150 mm £
Clayware pipes to B.S. 1196, 300 mm lengths	100	9·05	16·40	33·01
Single junction	No.	1·38	1·58	2·09

PITCH FIBRE PIPES

Pitch fibre drain and sewer pipes to B.S. 2760

	Unit	75 mm	100 mm	150 mm
2·45 m lengths	No.	2·76	3·40	7·36
Couplings	,,	0·66	0·66	1·54
D rings (rubber)	,,	0·26	0·30	0·65
Knuckle bends, socketed				
45°	,,	2·86	2·86	5·46
85°	,,	3·46	3·46	6·56
Medium radius bends, plain ends				
45°	,,	2·03	2·75	7·60
90°	,,	4·31	4·31	—
Sweep bends, plain ends				
45°	,,	—	5·41	—
90°	,,	—	5·41	—
Equal junctions, plain ends	,,	2·86	—	4·18
Equal junctions, socketed	,,	—	4·59	8·09
Channel, 2·45 m lengths	,,	—	2·32	4·30

CONCRETE PIPES AND FITTINGS	Unit	300 mm £	450 mm £	600 mm £	900 mm £	1200 mm £
Socketed concrete pipes with rubber ring joints to B.S. 556						
Standard	m	10·20	18·50	26·75	35·00	—
Class H.	,,	—	—	—	—	145·90
Extra for bend	,,	39·63	55·90	80·20	105·75	437·00
Extra for single junction	No.	32·83	60·75	—	—	—
Concrete road gully 1·07 m deep	,,	—	14·85	—	—	—

PRECAST CONCRETE MANHOLES		900 mm £	1050 mm £	1200 mm £	1500 mm £	1800 mm £
Precast concrete manholes to B.S. 556						
Chamber or shaft rings	m	28·75	37·30	50·20	90·85	115·50
Taper rings						
0·6 m lengths	No.	21·75	31·60	44·40	85·10	110·00
Precast concrete cover slabs, heavy duty	,,	25·30	33·50	46·00	77·60	122·75
Base section 0·9 m deep with main channel and benching	,,	—	—	—	192·00	277·00
Ditto 0·6 m deep with main channel and one branch channel and benching	,,	53·00	76·60	99·70	—	—
Step irons (fixed)	,,	1·80	1·80	1·80	1·80	1·80

DRAINAGE

	Unit	75 mm £	100 mm £	150 mm £	225 mm £
CAST IRON PIPES AND FITTINGS					
Spigot and socket pipes					
Spun cast iron to B.S. 1211, Class 'B'	m	6·90	8·27	12·59	20·76
Cast iron to B.S. 437	,,	10·17	10·28	15·80	—
Cast iron fittings to B.S. 437					
Large socket connector for stoneware, 305 mm long	No.	—	8·59	12·46	36·00
Bend					
Short radius	,,	6·65	9·14	19·67	75·44
Long radius	,,	9·86	17·11	23·32	—
With socket to receive stoneware pipe	,,	—	11·96	—	—
With small square door shaped to bore of pipe	,,	—	25·98	47·80	—
Branch					
Plain, equal	,,	13·24	14·46	31·40	105·57
With small square door shaped to bore of pipe	,,	—	30·58	61·65	—
Cast-iron gully trap with high invert, 92½°	,,	10·13	13·79	36·17	120·91
Bellmouth and grating, 305 mm high	,,	7·86	8·98	17·39	—
As above but with horizontal inlet	,,	—	14·24	30·29	—
As above but with vertical inlet	,,	13·46	15·07	34·84	—
Rainwater shoe with vertical inlet and rectangular inspection opening	,,	—	24·26	48·79	—
Roof or floor outlet with flat suface grating.	,,	—	37·28	43·44	—
Petrol-trapping bend 381 × 762 mm	,,	—	32·07	50·00	119·12

	Unit	100 × 100 mm £	150 × 100 mm £	150 × 150 mm £	225 × 150 mm £	225 × 225 mm £
Inspection fittings with flat covers as B.S. 1130						
Chamber with no branch as Fig. 010	No.	41·60	—	59·00	—	141·30
Ditto with one branch one side as Fig. 110/011	,,	52·56	75·00	96·15	168·93	199·95
Ditto with one branch on each side as Fig. 111	,,	70·50	91·00	115·59	195·96	242·60
Ditto with three branches each side as Fig. 313	,,	157·35	185·77	266·82	445·65	973·49

DRAINAGE

	Unit	Black £	Galv. £
CAST-IRON CHANNEL GRATINGS			
Light duty plain grating for rabbetted block channels, size 175 × 10 mm	m	13·84	21·05
Medium heavy duty grating and frame			
140 mm wide	,,	14·14	24·41
190 mm ,, ...	,,	15·32	28·90
265 mm ,, ...	,,	18·31	35·90

	Unit	Clear opening		
		450 × 450 mm £	600 × 450 mm £	600 × 600 mm £
MANHOLE COVERS AND FRAMES				
Cast-iron coated manhole cover and frame to B.S. 497, Grade C				
Single seal flat	No.	15·16	11·93	29·76
Double seal flat	,,	22·86	27·49	42·25
Single seal recessed	,,	24·42	30·54	42·11
Double seal recessed	,,	37·99	52·87	69·84

	Unit	£
Cast-iron coated circular manhole cover and frame to B.S. 497, Grade B class 1		
500 mm diameter clear opening	No.	42·50
550 mm ,, ,, ,, ...	,,	51·30
Cast-iron coated rectangular manhole cover and frame to B.S. 497, Grade B		
600 × 450 mm	,,	55·40
600 × 450 mm recessed top	,,	60·85
Cast-iron coated triangular manhole cover and frame to B.S. 497, Grade A		
500 mm clear opening, double	,,	75·00
550 mm ,, ,, ...	,,	77·70

SUNDRIES		
Step iron, galvanized malleable iron, to B.S. 1247		
For half brick wall	,,	2·85
For one brick wall	,,	3·69
Cast-iron hinged grating for road gully to B.S. 497 Grade A, Class 2		
GA2 – 325, 400 mm long grate	,,	32·80
GA2 – 325, 500 mm long grate	,,	40·30
Rectangular light duty stop cock box with hinged lid, clear opening size		
150 × 150 mm	,,	3·40

EXTERNAL WORKS

	Unit	£
GRANITE SETTS		
Granite setts, 100 × 100 × 100 mm	tonne	58·00
COBBLES		
75 mm diameter cobbles	,,	23·00
KERBS, CHANNELS, ETC.		
Granite kerb to B.S. 435 standard 'B' dressing		
127 × 254 mm edge kerb straight	m	14·50
127 × 254 mm ,, ,, radius (up to 3 m)	,,	19·50
152 × 305 mm ,, ,, straight	,,	16·50
152 × 305 mm ,, ,, radius (up to 3 m)	,,	21·50
305 × 152 mm flat kerb straight	,,	18·20
305 × 152 mm ,, ,, radius (up to 3 m)	,,	23·20
Pre-cast concrete kerb and channel to B.S. 340, 254 × 127 mm		
Straight	,,	1·58
Circular on plan, 1·8 m radius and over	,,	2·00
Pre-cast concrete edging with slightly rounded top, 51 × 152 mm	,,	0·67
PAVING SLABS		
Pre-cast concrete paving slabs to B.S. 368, 600 × 750 mm		
50 mm	m²	2·72½
63 mm	,,	3·17½
42 mm Noelite paving in any of five colours	,,	4·52

PAVING BRICKS

Pavers, 215 × 103 × 50 mm			
Red		1000	143·30
Brown		,,	153·30
Dark grey		,,	156·30
Interlocking paving blocks			
Standard and edge blocks natural colour			
225 × 112·5 × 60 mm		m²	5·43
225 × 112·5 × 80 mm		,,	6·09

Extra for colours		60 mm	80 mm
		£	£
Red	m²	0·52	0·68
Bronze	,,	0·65	0·80
Charcoal	,,	0·47	0·56
Yellow	,,	0·73	0·89
Dark brown	,,	0·65	0·80

	Unit	£
GRAVEL		
Gravel hoggin	m³	5·00
COLD BITUMEN EMULSION		
Cold bitumen emulsion	kg	0·22

EXTERNAL WORKS

CABLE CONDUIT	Unit	100 *mm* £	150 *mm* £	225 *mm* £
Vitrified clay cable conduit with polyethylene sleeve bonded to one end				
Straight conduit	m	1·32	2·59	4·74
Bend	No.	1·32	2·59	6·00
Bellmouth	,,	1·32	2·59	6·00
Plastic bellmouth	,,	1·43	2·73	—
Hardwood conduit plugs				
Solid type	,,	0·80	1·27	2·61
Split type	,,	0·99	1·52	2·89

WATER MAINS	Unit	80 *mm* £	100 *mm* £	150 *mm* £	200 *mm* £
Spun grey iron pipes to B.S. 4622 Class 1 with tyton sockets	m	4·40	5·03	7·38	10·58
Grey iron fittings to B.S. 4622					
Bend, double tyton sockets					
90°	No.	10·60	11·59	19·98	39·25
45°	,,	10·50	11·00	18·45	32·75
Tee, all tyton sockets . . .	,,	18·17	19·37	30·54	59·30
45° equal angle branch all tyton sockets .	,,	35·35	44·90	83·40	127·00
Tyton gasket	,,	0·65	0·99	1·10	1·54
Ductile iron fittings to B.S. 4772					
Hydrant duckfoot bend flange and tyton socket	,,	26·15	27·57	51·05	—
Flange and tyton socket	,,	9·40	11·45	17·95	26·70
Flange and tyton spigot	,,	7·90	9·23	15·80	25·10

GAS MAINS					
Spun ductile iron pipes to B.S. 4772 with bolted joints .	m		6·10	8·70	12·75
Ductile iron fittings to B.S. 4772					
90° bend	No.		12·20	21·00	38·15
45° bend	,,		11·55	17·95	31·85
Tee	,,		18·75	29·70	57·15
Flanged socket	,,		11·15	17·45	26·00
Flanged spigot	,,		10·65	18·30	29·00
Bolted joint complete	,,		4·40	6·35	8·60

ASBESTOS-CEMENT PRESSURE PIPES		75 *mm* £	100 *mm* £	150 *mm* £
Asbestos-cement bitumen dipped pressure pipes, to B.S. 486, Class 25	m	2·64	3·34	5·48
Asbestos cement joint	No.	3·71	4·16	6·23
Cast iron special fittings with plain ends				
45° bend	,,	9·57	11·53	29·67
Tee	,,	14·13	18·69	37·59
Flanged branch tee	,,	20·99	30·61	54·93
Taper	,,	—	8·60	14·87

Constants of Labour and Material

'Constants' are given for the major items of work for which prices are given in 'Prices for Measured Work'.

As described later under 'Prices for Measured Work', the 'Constants' have been used to build up the rates for builders' work in that section, that is for work other than that which is normally sub-let. Different organizations will have varying views on 'Constants', which will in any event be affected by the type of job, availability of labour and the extent to which mechanical plant can be used.

The 'Constants' should assist the reader:

(1) *To compare the prices to those used in his own organization.*
(2) *To calculate the effect of changes in wage rates or prices of materials.*
(3) *To calculate analogous prices for work similar to but differing in detail from the examples given.*

COMPUTATION OF LABOUR RATES

From 30th June 1980 the guaranteed minimum weekly earnings in the London area for craft operatives and labourers are £80·60 and £68·80 respectively; to these rates have been added allowances for the items below in accordance with the recommended procedure of the Institute of Building in its 'Code of Estimating Practice'. The resultant hourly rates on which the 'Prices for Measured Work' have generally been based are £2·85 and £2·44 for craft operatives and labourers respectively.

The items referred to above for which allowances have been made are:

Lost time.
Non-productive overtime.
Extra payments under National Working Rule 3.
Construction Industry Training Board Levy.
Holidays with Pay.
Death benefit scheme.
National Insurance.
Severance pay and sundry costs.
Employers liability and third party insurance.

NOTE. *For travelling allowances and site supervision see 'Preliminaries'.*

231

COMPUTATION OF LABOUR RATES

The table which follows shows how the hourly rates referred to above have been calculated. Productive time has been based on a total of 2063 hours worked per year.

		Craft operatives		Labourers	
		£	£	£	£
Wages at standard basic rate, productive time	$51\frac{4}{7}$ wks.	69·20	3568·74	59·00	3042·71
Guaranteed minimum bonus	$46\frac{4}{5}$ wks.	11·40	533·52	9·80	458·64
Lost time allowance	$1\frac{1}{20}$ wk.	80·60	84·63	68·80	72·24
'Non-productive overtime	$3\frac{13}{20}$ wks.	69·20	252·58	59·00	215·35
Extra payments under National Working Rule 3	$46\frac{4}{5}$ wks.	0·96	44·93	0·40	18·72
Sick pay	1 wk.	—	—	—	—
C.I.T.B. levy	1 year	—	33·60	—	6·00
Public holiday pay	$1\frac{3}{5}$ wks.	80·60	128·96	68·80	110·08
Employers' contribution to: –					
Annual holiday pay scheme	48 wks.	8·35	400·80	8·35	400·80
Death benefit scheme	48 wks.	0·10	4·80	0·10	4·80
National Insurance	$47\frac{4}{5}$ wks.	14·35	685·93	12·23	584·59
			5738·49		4913·93
Severance pay and sundry costs	Plus	1%	57·38	1%	49·14
			5795·87		4963·07
Employer's liability and third-party insurance	Plus	1·60%	92·73	1·60%	79·41
Total cost per annum			5888·60		5042·48
Total cost per hour			2·85		2·44

NOTES

1 *Absence due to sickness has been assumed to be for individual days when no payment would be due.*
2 *The annual holiday pay scheme contributions are those which come into force on 4th August 1980 and take account of the contribution made by the management company.*
3 *National Insurance contributions are 'not contracted out' contributions based on assessed gross pay figures of £104·50 for craft operatives and £80·00 for labourers.*

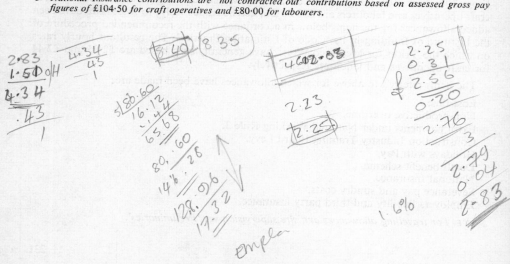

COMPUTATION OF COST OF MATERIALS

To the cost of the following materials (see 'Market Prices' section) should be added the hours of labour shown, to allow for the cost of unloading, stacking, etc. The constants given hereafter *do not* include these labours.

Material	Unit	Labourer Hour
Reinforcing rods	tonne	2·95
Cement	,,	0·67
Lime	,,	0·67
Common bricks	1000	2·00
Light engineering bricks	,,	2·33
Heavy engineering bricks	,,	2·67

COMPUTATION OF MECHANICAL PLANT COSTS

Calculate the weekly hire charge and estimate weekly cost of oil, grease, ropes (where necessary), site servicing and proportion of cartage charges.

Divide the total amount by 30 (assuming 25% idle time) to arrive at cost per hour of plant. To this hourly rate add one hour fuel consumption, one hour for each of the following operators and attendants, the rate to be calculated in accordance with the principles given earlier in this section, i.e. with an allowance for plus rates, etc.

For convenience, the rates per hour used in the calculations of 'Prices for Measured Work' are shown below.

Plant	Labour	Rate per hour £
*Portable compressor (2·86 to 3·70 m³)	—	3·10
Excavator (0·57 m³ dragline or backacter)	Driver and banksman	18·90
Scraper (4·70 m³)	Driver	18·00
Skimmer (caterpillar 951 C 1·34m³ bucket)	Driver	17·70
Bulldozer (65 to 89·9 rated h.p.)	Driver	14·00
Dumper (2·30 m³)	Driver	7·50
Rammer	Operator	3·25
Roller, 0·75 tonnes	Driver	4·45
Roller, 6 to 8 tonnes	Driver	12·90
Concrete Mixer 14/10	Operator	5·50
Mechanical Trencher	Operator and banksman	17·75

* *Operation of compressor assumed by tool operator.*

EXCAVATION AND EARTHWORK

The following constants are applicable to excavation in heavy soil; for different soils, see 'Prices for Measured Work'.

Labourer

SITE PREPARATION per m³
Hour

Excavating vegetable soil average 150 mm deep . . . . 0·30

Plant

EXCAVATION (MECHANICAL) per m³
Hour

Excavating

Surfaces to reduce levels; not exceeding 1 m deep (skimmer deep with 1·34 m³ bucket) . 0·01

    not exceeding 1 m deep (scraper 4·70 m³) . . . . 0·04

Basements; starting from reduced level not exceeding 4 m deep (dragline excavator 0·57 m³ bucket) . . . . 0·03

Pits; starting from reduced level not exceeding 2 m deep (trencher with 0·57 m³ bucket). 0·04

Trenches; starting from reduced level not exceeding 1 m deep (trencher with 0·57 m³ bucket) . . . . 0·04

    not exceeding 2 m deep . . . . 0·04

    not exceeding 4 m deep . . . . 0·04

Labourer

EXCAVATION (HAND) per m³
Hour

Excavating

Surfaces to reduce levels not exceeding 0·25 m deep . . . . . . 2·30

    not exceeding 1 m deep . . . . 2·60

Basements or the like; starting from reduced level not exceeding 2 m deep . . . 3·38

    not exceeding 4 m deep . . . . 4·72

    not exceeding 6 m deep . . . . 6·04

Pits; starting from reduced level not exceeding 1 m deep . . . . . 5·80

    not exceeding 2 m deep . . . . 6·65

    not exceeding 4 m deep . . . . 8·37

Trenches; starting from reduced level not exceeding 0·25 m deep . . . 2·75

    not exceeding 1 m deep . . . . 3·05

    not exceeding 2 m deep . . . . 4·05

    not exceeding 4 m deep . . . . 6·05

BREAKING UP

NOTE. *Where breaking up is measured as extra to excavation an appropriate adjustment must be made.*

	'Cart-away' m³	Two tool compressor Hour	Labourer Hour
Breaking up mass reinforced concrete by mechanical drill per m³			
in open excavation	—	3·90	7·80
in trenches	—	4·43	8·86
Breaking up tarmac and hardcore 150 mm thick and carting to tip, by mechanical drill per m²	0·19	0·05	0·30
Breaking up concrete surfaces (not reinforced) and carting to tip, by mechanical drill per m²			
150 mm thick	0·19	0·20	0·60
300 mm thick	0·38	0·40	1·20

EXCAVATION AND EARTHWORK

EARTHWORK SUPPORT per m²	*Timber* m³	*Labourer* Hour
Sides of excavation; not exceeding 2 m between faces		
not exceeding 1 m deep	0·022	0·20
not exceeding 2 m deep	0·033	0·30
not exceeding 4 m deep	0·044	0·40
over 4 not exceeding 6 m deep	0·055	0·50
over 6 not exceeding 8 m deep	0·070	0·60
not exceeding 2 m deep; next to roadway	0·070	1·50
Sides of excavation; not exceeding 4 m between faces		
not exceeding 2 m deep	0·036	0·33
Sides of excavation; over 4 m between faces		
not exceeding 2 m deep	0·060	0·55
not exceeding 4 m deep	0·070	0·65

NOTE. *Quantities for timber are gross and allowance should be made for the number of uses achieved. Prices for measured work assume 10 uses.*

DISPOSAL (MECHANICAL) per m³

	Labourer Hour
Surplus excavated material	
depositing on site in spoil heaps (2·30 m³ dumper)	0·11
spreading on site average 25 m from excavation (Bulldozer)	0·05
spreading on site average 200 m from excavation (4·70 m³ scraper)	0·04

DISPOSAL (HAND) per m³

	'Cart away' m³	*Labourer* Hour
Surplus excavated material		
depositing on site in spoil heaps	—	1·10
spreading on site average 25 m from excavation	—	1·35
spreading on site average 50 m from excavation	—	1·70
removing from site to tip not exceeding 13 km from site	1·25	1·31

FILLING (MECHANICAL) per m³

Excavated material		
backfilling into excavations (Bulldozer)	—	0·07
filling in making up levels (Bulldozer)	—	0·05

FILLING (HAND) per m³

Excavated material		
backfilling into excavations; depositing and compacting in layers	—	1·15
filling in making up levels; depositing and compacting in layers; multiple handling via spoil heaps average 50 m from excavation	—	3·70

EXCAVATION AND EARTHWORK

	Hardcore m³	Mechanical rammer Hour	Labourer Hour
HARDCORE FILLING OR THE LIKE			
Hardcore			
Filling in making up levels			
over 250 mm thick, per m³	1·25	0·63	0·65
100 mm thick, per m²	0·13	0·10	0·11
150 mm thick, per m²	0·19	0·10	0·13
	Ashes m³		
Coarse ashes			
Filling in making up levels			
over 250 mm thick, per m³	1·25	0·63	0·52
100 mm thick, per m²	0·13	0·10	0·10
150 mm thick, per m²	0·19	0·10	0·12

CONCRETE WORK

	Cement Tonne	Aggregate Tonne	Sand Tonne	Concrete Mixer Hour	Labourer Hour
SITE MIXED BALLAST CONCRETE per m³					
Analysis of basic price for mass concrete					
1:12 nominal mix; 40 mm all in ballast	0·145	2·15	—	0·22	2·94
1:3:6 nominal mix; 40 mm shingle; crushing strength at 28 days 11·50 N/mm².	0·225	1·42	0·75	0·22	2·94
1:2:4 nominal mix; 20 mm shingle; crushing strength at 28 days 21·00 N/mm².	0·311	1·29	0·69	0·22	2·94
1:1½:3 nominal mix, 20 mm shingle; crushing strength at 28 days 25·50 N/mm².	0·379	1·23	0·65	0·22	2·94
Add to the above analysis for:					
Reinforced concrete including working around reinforcement					1·30
Foundations in trenches					
Not exceeding 100 mm thick					2·60
Over 100 not exceeding 150 mm thick					1·96
Over 150 not exceeding 300 mm thick					0·65
Foundations to isolated stanchions					
Over 300 mm thick					1·30
Ground beams					
Not exceeding 0·03 m² sectional area					3·60
Over 0·03 not exceeding 0·10 m² sectional area					3·20
Over 0·10 not exceeding 0·25 m² sectional area					2·80
Over 0·25 m² sectional area					2·40
Beds					
Not exceeding 100 mm thick					2·60
Over 100 not exceeding 150 mm thick					1·96
Over 150 not exceeding 300 mm thick					1·32
Over 300 mm thick					0·65
Suspended slabs					
Not exceeding 100 mm thick					6·20
Over 100 not exceeding 150 mm thick					5·55
Over 150 not exceeding 300 mm thick					4·90
Over 300 mm thick					4·25
Walls					
Not exceeding 100 mm thick					6·55
Over 100 not exceeding 150 mm thick					5·85
Over 150 not exceeding 300 mm thick					5·15
Over 300 mm thick					4·50
Isolated beams					
Not exceeding 0·03 m² sectional area					9·80
Over 0·03 not exceeding 0·10 m² sectional area					7·85
Over 0·10 not exceeding 0·25 m² sectional area					6·55
Over 0·25 m² sectional area					5·90
Deep beams					
Not exceeding 0·03 m² sectional area					7·85
Over 0·03 not exceeding 0·10 m² sectional area					6·55
Over 0·10 not exceeding 0·25 m² sectional area					5·90
Over 0·25 m² sectional area					5·25
Isolated columns					
Not exceeding 0·03 m² sectional area					12·40
Over 0·03 not exceeding 0·10 m² sectional area					9·80
Over 0·10 not exceeding 0·25 m² sectional area					7·65
Over 0·25 m² sectional area					6·55
Steps and staircases					
Generally					5·20
Filling to hollow walls					
Not exceeding 100 mm thick					4·50

Constants of Labour and Material

CONCRETE WORK

ROD REINFORCEMENT per tonne

To the cost of rod reinforcement supplied hooked and bent add for fixing only the hours shown. Allowance must also be made for tying wire and spacers.

25 mm diameter			
20 mm	„		30·00
16 mm	„		33·00
12 mm	„		36·00
10 mm	„		43·00
8 mm	„		54·00
6 mm	„		75·00
			75·00

MESH REINFORCEMENT per m²

Steel wire mesh reinforcement to B.S. 4483, with 150 mm side and 300 mm end laps.

	Fabric m²	*Waste and laps* %	
Weight 2·22 kg/m²	1·00	20	0·11
„ 3·02 kg/m²	1·00	20	0·15
„ 6·16 kg/m²	1·00	20	0·30

BRICKWORK AND BLOCKWORK

MORTAR (MATERIALS ONLY) per m³	Mix	Cement Tonne	Hydrated Lime Tonne	Sand Tonne
Cement lime mortar	1:1:6	0·24	0·11	1·60
ditto	1:2:9	0·16	0·14	1·60
ditto	1:3:9	0·16	0·21	1·60
Cement mortar	1:3	0·47	—	1·60
ditto	1:4	0·37	—	1·60

BRICKWORK IN COMMON BRICKS per m²

	Bricks No.	Waste %	Mortar m³	Bricklayer Hour	Labourer Hour
Half brick thick walls . . .	60	5	0·023	1·44	0·96
Half brick thick walls; built curved mean radius 6 m	64	5	0·024	1·65	1·10
Half brick thick walls; built curved mean radius 1·50 m	68	5	0·026	1·87	1·25
One brick thick walls . . .	120	5	0·061	2·60	1·73
One brick thick walls; filling old openings .	120	5	0·061	3·00	2·00
One and a half brick thick walls . .	180	5	0·093	3·60	2·40

BRICKWORK IN ENGINEERING BRICKS per m²

	Bricks No.	Waste %	Mortar m³	Bricklayer Hour	Labourer Hour
Half brick thick walls	60	2½	0·023	1·58	1·05
One brick thick walls	120	2½	0·061	2·87	1·91
One and a half brick thick walls . . .	180	2½	0·093	3·96	2·64

NOTE. *Constants for brickwork based on brick size of 215 × 102·5 × 65 mm*

BRICKWORK AND BLOCKWORK

	Bricks No.	Waste %	Mortar m³	Bricklayer Hour	Labourer Hour
BRICK FACEWORK per m²					
Extra over common bricks for fair face and flush pointing	—	—	—	0·30	0·20
Extra over common bricks for facing bricks, Flemish bond, flush pointing . . .	75	5	—	0·90	0·60
BRICKWORK ENTIRELY OF FACING BRICKS per m²					
Ordinary Facings					
Half brick wall, stretcher bond pointed both sides	60	5	0·023	2·28	1·52
One brick wall pointed both sides Flemish bond	120	5	0·061	3·63	2·42

NOTE. *Constants for brickwork based on brick size of 215 × 102·5 × 65 mm.*

WOODWORK

	Nails kg	Carpenter Hour

CARCASSING TIMBERS per m³

Softwood
To the cost of timber per m³ add 7½% waste and the value of nails and labour
hours given

	Nails kg	Carpenter Hour
Floors	1·60	3·20 26·50
Partitions	4·80	3·20 39·90
Flat roofs	1·60	3·20 26·50
Pitched roofs including ceiling joists	1·60	3·20 31·50
Kerbs, bearers and the like	2·40	3·20 11·65
Kerbs, bearers and the like fixed by bolting (bolts and holes measured separately)		35·30

(handwritten annotations: .14% ✓ 8"x2" 1325/100m = 0.1325 13.25 133% 143%; 200% 19.95 200% 200%; 17½% 13.25 133% 155%; 18½% 1575 158% 175%; 7½% 5·25 60% 130%; 95% 17.65 178% 178%)

FIRST FIXINGS per 100 m²

	Boarding m²	Waste %	Nails kg	Carpenter Hour	Labourer Hour
Softwood					
Board flooring; butt joints					
19 mm thick	103·00	5	24	53·80	10·75
22 mm thick	103·00	5	24 *35*	53·80	10·75
25 mm thick	103·00	5	24	53·80	10·75
Board flooring; butt joints laid diagonally					
19 mm thick	103·00	5	24 *42*	67·80	10·75
Board flooring; tongued and grooved joints					
19 mm thick	109·00	5	26	70·00	10·75
22 mm thick	109·00	5	26 *66*	70·00	10·75
25 mm thick	109·00	5	26	70·00	10·75
Matchboarding; tongued and grooved joints; veed one side					
12 mm thick; to walls	113·00	5	34	91·50	10·75
19 mm thick; to walls	113·00	5	34	91·50	10·75
19 mm thick; to ceilings	113·00	5	34	136·00	10·75
Roof boarding; butt joints laid to slope					
19 mm thick	103·00	5	24	69·94	10·75
25 mm thick	103·00	5	24	69·94	10·75
Roof boarding; butt joints laid diagonally to slope					
19 mm thick	103·00	5	24	83·40	10·75
Roof boarding; butt joints laid diagonally to tops and cheeks of dormers					
19 mm thick	103·00	5	24	107·60	21·50
Roof boarding; tongued and grooved joints laid to falls					
19 mm thick	109·00	5	26 *44*	70·00	10·75
25 mm thick	109·00	5	26	70·00	10·75
Roof boarding; tongued and grooved joints laid to slope					
19 mm thick	109·00	5	26 *55*	91·00	10·75
25 mm thick	109·00	5	26	91·00	10·75
Chipboard					
Board flooring; butt joints					
18 mm thick	100·00	5	10 *25*	30·00	10·75
22 mm thick	100·00	5	10	30·00	10·75
Board flooring, tongued and grooved joints					
18 mm thick	100·00	5	10	35·00	10·75
22 mm thick	100·00	5	10 *30*	35·00	10·75

(handwritten: +150%)

SECOND FIXINGS – Softwood

	Skirting m	Waste %	Nails kg	Carpenter Hour	Labourer Hour
SKIRTINGS per 100 m					
19 × 45 mm	100·00	10	3·70	11·50	0·80
19 × 75 mm	100·00	10	3·70	11·50	0·80
25 × 75 mm	100·00	10	4·40	14·75	0·80
25 × 100 mm	100·00	10	4·40	14·75	0·80

207% (handwritten)

	Architrave m				
ARCHITRAVES per 100 m					
19 × 38 mm	100·00	10	3·70	11·50	0·80
19 × 75 mm	100·00	10	3·70	14·75	0·80
25 × 75 mm	100·00	10	4·40	14·75	0·80

	Shelves m				
SHELVES; 230 mm wide per 100 m					
19 mm thick	100·00	5	5	16·00	—
22 mm thick	100·00	5	5	16·00	—

248% (handwritten)

	Window boards m				
WINDOW BOARDS; 150 mm wide per 100 m					
25 mm thick	100·00	5	9	40·00	—

COMPOSITE ITEMS – Softwood

	Doors No.				
PURPOSE MADE DOORS each					
Matchboarded or panelled doors					
38 mm thick	1	—	—	1·50	—
50 mm thick	1	—	—	2·00	—

× twice and half for making (handwritten)

	Jambs heads m				
JAMBS AND HEADS per 100 m	315%				
32 × 63 mm	100·00	5	4	9·00	0·80
32 × 100 mm	100·00	5	4	9·53	0·80
32 × 140 mm	100·00	5	4	11·95	0·80

JAMBS AND HEADS: rebated and grooved per 100 m					
38 × 115 mm	100·00	5	4	11·65	0·80
38 × 140 mm	100·00	5	4	14·20	0·80
50 × 100 mm	100·00	5	4	13·33	0·80
63 × 88 mm	100·00	5	4	14·78	0·80
63 × 100 mm	100·00	5	4	16·80	0·80

245% (handwritten)

	Sills m				
SILLS per 100 m					
63 × 150	100·00	5	4	16·80	1·60
75 × 150	100·00	5	4	20·00	1·60

IRONMONGERY

All the constants given in this trade are for **fixing only** to softwood and hardwood as shown.

Surface fixing	Unit	Carpenter Softwood Hour	Hardwood Hour
door furniture	Set	0·25	0·33
numerals 50 mm	No.	0·10	0·13
hat and coat hooks	,,	0·10	0·13
shelf brackets	,,	0·20	0·27
cabin hooks and eyes	,,	0·25	0·33
casement stay and pins	,,	0·30	0·40
finger plates	,,	0·20	0·27
postal plates and knockers	,,	1·50	2·00
sash lifts	,,	0·20	0·27
bolts, barrel			
100 mm	,,	0·30	0·40
150 mm	,,	0·36	0·48
300 mm	,,	0·50	0·67
bolts, necked			
100 mm	,,	0·33	0·44
150 mm	,,	0·40	0·53
bolts, monkey tail			
375 mm	,,	1·00	1·33
600 mm	,,	1·13	1·50
750 mm	,,	1·20	1·60
Norfolk latches	,,	1·00	1·33
Flush fixing			
overhead door closers	,,	2·50	3·33
butt hinges, skew, 50 to 100 mm	Pair	0·20	0·20
floor spring door closers double action	No.	4·00	5·33
floor spring door closers double action with check	,,	5·00	6·67
spring hinges			
single action			
75 and 100 mm long	Pair	2·00	2·67
125 and 150 mm long	,,	2·50	3·33
double action			
75 and 100 mm long	,,	2·50	3·33
125 and 150 mm long	,,	3·00	4·00
bolts lever action 200 mm	No.	1·00	1·33
bolts mortice	,,	0·75	1·00
bolts panic, for single door	,,	2·50	3·33
bolts panic, for pair of doors	,,	3·00	4·00
cupboard lock	,,	0·75	1·00
cylinder rim latches	,,	1·25	1·67
fanlight catches	,,	0·30	0·40

FLOOR, WALL AND CEILING FINISHINGS

PLAIN SHEET FINISHINGS per m²

To the cost of sheet material per m² add the value of nails etc. and labour hours given

	Waste %	Nails kg	Carpenter Hour
Fibre insulating board sheet, butt joints to walls			
13 mm thick	5	0·08	0·36
19 mm thick	5	0·08	0·42
Laminated plastic sheet, fixing with adhesive to timber		*Adhesive* litre	
1·5 mm thick	10	0·68	1·20

	Waste %	Scrim m	Nails kg	
Plasterboard butt joints to walls				
9·5 mm thick	5	1·65	0·08	0·50
12·7 mm thick	5	1·65	0·08	0·56

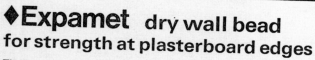

PAINTING AND DECORATING

	Paint kg	Painter Hour
CEMENT PAINT per 100 m²		
Two coats cement paint external		
Walls		
stone, brickwork or blockwork	44·00	27·00
cement render	41·00	24·00
concrete	30·00	25·00

	litre	
EMULSION PAINT per 100 m²		
Two coats emulsion paint internal		
Walls or ceilings		
stone, brickwork or blockwork	21·00	27·00
cement render	14·00	24·00
concrete	16·00	25·00
plaster	14·00	24·00

	Primer litre	Paint kg	Painter Hour
OIL PAINT per 100 m²			
Iron or steel, internal			
General surfaces			
One coat primer one undercoat	8·00	8·00	42·00
One additional undercoat	—	8·00	17·00
One finishing coat	—	8·00	19·00
Plaster surfaces, internal			
Walls or ceilings			
One coat primer, one undercoat	9·50	8·00	42·00
One additional undercoat	—	8·00	17·00
One finishing coat	—	8·00	19·00

	Knotting litre	Stopping kg	Primer litre	Paint kg	Painter Hour
Wood surfaces, internal					
General surfaces					
One coat primer, one undercoat	0·70	2·00	8·00	8·00	48·00
One additional undercoat	—	—	—	8·00	17·00
One finishing coat	—	—	—	8·00	19·00

NOTE. *Add 5% to value of materials and labour for sundries and brushes.*

Girders up to 150mm + 50%.

" " " 300mm + 33%

(67 hrs)

DRAINAGE

EXCAVATING TRENCHES BY MECHANICAL TRENCHER per m

To receive pipes, grading bottoms, planking and strutting, filling with excavated material and compacting disposal of surplus soil by spreading on site average 25 m from excavation.
Refer to following table for quantity of excavation and allow:

0·11 hours of machine and operator per m³
1·26 hours labour for attendance on machine and disposal of excavated material per m³
0·39 hours labour for trimming bottom of trench per m²
Earthwork support as given in 'Excavation'.
Prices for Measured Work assume that earthwork support will only be required for trenches 1 m or over in depth.

Diameter of pipe mm	Width of trench m		Depth of trench m					
		0·75	1·50	2·00	2·50	3·00	3·50	
				Quantity of excavation m³				
Up to 225	0·70	0·53	1·05	1·40	1·75	2·10	2·45
„ 300	0·75	0·56	1·13	1·50	1·88	2·25	2·63
„ 375	0·85	0·64	1·28	1·70	2·13	2·55	2·98
„ 450	1·00	0·75	1·50	2·00	2·50	3·00	3·50
„ 600	1·20	—	1·80	2·40	3·00	3·60	4·20
„ 750	1·45	—	2·18	2·90	3·63	4·35	5·08
„ 900	1·60	—	2·40	3·20	4·00	4·80	5·60
„ 1050	1·75	—	2·63	3·50	4·38	5·25	6·13
„ 1200	1·90	—	2·85	3·80	4·75	5·70	6·65

EXCAVATING TRENCHES BY HAND per m

To receive pipes, grading bottoms, earthwork support, filling with excavated material and compacting disposal of surplus soil by spreading on site average 25 m from excavation.
Assume the widths of trenches given below. Calculate the quantity of excavation and allow labour hours as follows:

3·05 hours per m³ up to 1 m deep
4·05 „ „ 1 to 2 m deep
6·05 „ „ 2 to 4 m deep
Allow 1·15 hours per m³ for filling, compacting or disposal on site.
Allow for earthwork support as given in 'Excavation'. Prices for Measured Work assume that earthwork support will only be required for trenches 1 m or over in depth.

Diameter of pipe mm	0·75 m to 1·00 m	Depth of trench 1·25 m to 2·00 m Width of trench	2·25 m to 3·50 m
	m	m	m
Up to 225	0·61	0·69	0·84
„ 300	0·69	0·69	0·91
„ 375	0·76	0·84	0·99
„ 450	0·91	0·99	1·14
„ 600	—	1·22	1·37
„ 750	—	1·45	1·60
„ 900	—	1·60	1·75
„ 1050	—	1·75	1·91
„ 1200	—	1·91	2·06

DRAINAGE

IN SITU CONCRETE per m

Normal mix 11·50 N/mm²
40 mm aggregate in beds and benchings
For basic price of concrete see 'Concrete Work' and add 2·60 hours labour per m³
Concrete 300 mm wider than external diameter of pipe.

Diameter of pipes mm	Bed only		Bed and benching		Square section surround	
	100 mm thick m³	150 mm thick m³	100 mm thick m³	150 mm thick m³	100 mm thick m³	150 mm thick m³
100	0·044	0·067	0·084	0·100	0·121	0·163
150	0·049	0·074	0·092	0·117	0·151	0·196
230	0·059	0·088	0·117	0·142	0·198	0·248
300	0·067	0·100	0·142	0·176	0·243	0·305
375	0·076	0·115	0·176	0·209	0·289	0·360
450	0·085	0·128	0·209	0·251	0·338	0·416
600	0·102	0·153	0·268	0·318	0·435	0·531
750	0·109	0·162	0·360	0·418	0·548	0·657
900	0·124	0·186	0·443	0·510	0·665	0·788
1050	0·140	0·209	0·535	0·610	0·789	0·929
1200	0·155	0·232	0·652	0·719	0·921	1·077

PIPEWORK per m

	Pipe m	Waste %	Labourer Hour
Clay pipes butt jointed			
75 mm internal diameter	1·00	5	0·16
100 mm ,, ,, ...	1·00	5	0·16
150 mm ,, ,, ...	1·00	5	0·23
Vitrified clay pipes, mortar joints			
100 mm internal diameter	1·00	5	0·35
150 mm ,, ,, ...	1·00	5	0·45
230 mm ,, ,, ...	1·00	5	0·55
300 mm ,, ,, ...	1·00	5	0·80

NOTE. *Add for jointing material.*

EXTERNAL WORKS

	Hardcore m³	5 tonne roller Hour	Labourer Hour
HARDCORE OR THE LIKE			
Hardcore			
Filling in making up levels			
over 250 mm thick per m³	1·33	0·13	0·65
75 mm thick per m²	0·10	0·01	0·10
150 mm thick per m²	0·20	0·02	0·14

	Ashes m³	2 tonne roller Hour	
Coarse ashes			
Filling in making up levels			
over 250 mm thick per m³	1·33	0·13	0·52
75 mm thick per m²	0·10	0·01	0·10
150 mm thick per m²	0·20	0·02	0·14

	Concrete m³	Kerb m	Waste %	Bricklayer Hour	
PRECAST CONCRETE					
125 × 250 mm kerb, haunching with in-situ concrete mix 11·50 N/mm², 40 mm aggregate one side per m					
straight	0·02	1·00	5	0·15	0·30
curved	0·02	1·00	5	0·23	0·46
NOTE. *Allow small amount of mortar for jointing.*					

	Flags m²	Waste %	Mortar m³		
50 mm concrete flags, bedding and pointing in mortar per m²	1·00	5	0·02	0·30	0·30

	Gravel tonne	Roller Hour	
IN SITU FINISHINGS			
Gravel paving in two layers to pavements to hardcore base per m²			
50 mm thick	0·10	0·02	0·80
63 mm thick	0·13	0·03	0·10

Prices for Measured Work

The 'Prices for Measured Work' are intended to apply to a project costing about £855,000 in the London area and assume that reasonable quantities of all types of work are required. Similarly it has been necessary to assume that the size of the job warrants the sub-letting of all types of work normally sub-let.

The prices for builders' work include 10% for overhead charges and profit and are based on labour rates per hour of £2·85 for craft operatives and £2·44 for labourers. The prices for specialist work are based upon sub-contractors' prices with the addition of 5% for the general contractors' attendance, overhead charges and profit. This does not apply to the section 'Landscaping and Planting' which has been compiled in a different manner from the remainder of the book and the reader is advised to study the notes which preface that section.

The distinction between builders' work and work normally sub-let is stressed, because prices for work which can be sub-let may well be quite inadequate for the contractor who is called upon to carry out relatively small quantities of such work himself.

The prices for measured work are set out as in a Bill of Quantities and exclude normal preliminary items which are dealt with under the heading of 'Preliminaries'.

Reference to the section dealing with 'Preliminaries' will show that, in the absence of detailed calculations, at least 15% must be added to all prices if one is concerned with the total cost (including 'Preliminaries') as opposed to the estimated price for a particular item of measured work.

The letters 'P.C.', where they occur, mean the prime cost of the materials or goods delivered to site. Their use indicates that there exists a range of prices for such products and that, to avoid repetition, the prices for measured work have been based upon typical market prices.

No allowance has been made in the prices for Value Added Tax.

Reference to the section 'Constants of Labour and Material' will assist the reader to make adjustments to prices for measured work because of different labour or materials costs or to calculate analogous rates for work similar to but differing in detail from the examples given.

PRELIMINARIES

The number of items priced in the 'Preliminaries' section of Bills of Quantities and the manner in which they are priced vary considerably between Contractors. Some Contractors, by modifying their percentage factor for overheads and profit, attempt to cover the costs of 'Preliminary' items in their 'Prices for Measured Work'. However, the cost of 'Preliminaries' will vary widely according to job size and complexity, site location, accessibility, degree of mechanization practicable, position of the Contractor's head office and relationships with local labour/domestic sub-contractors. It is therefore usually far safer to price 'Preliminary' items separately on their merits according to the job.

The normal clause descriptions from the 'Preliminaries' section are given below together with further details against those items which are usually priced in tenders. At the end of these priced items an example is given to demonstrate how the details are applied. In every case the examples assume the same project in which the form of contract used is the Standard Form of Building Contract 1980 Edition and the value, excluding 'Preliminaries' is £855,000. The contract is estimated to take 80 weeks to complete and the value is built up as follows:

	£
Labour value .	306,000
Material value	281,000
Provisional sums and all sub-contractors .	268,000
	£855,000

At the end of the section the examples are summarized to give a total value of 'Preliminaries' for the project.

General conditions and preliminaries

Preliminary particulars

Item
1. **Project, parties and consultants** – *Not priced*
2. **Description of site** – *Not priced*
3. **Drawings and other documents** – *Not priced*

Contract
4. **Form, type and conditions of contract**
(The Standard Form of Building Contract 1980 Edition is assumed)

Clause No.
1. **Interpretation, definitions etc.** – *Not priced*
2. **Contractor's obligations** – *Not priced*
3. **Contract Sum – adjustment – Interim Certificates** – *Not priced*
4. **Architect's Instructions** – *Not priced*
5. **Contract documents – other documents – issue of certificates**

The contract conditions may require a master programme to be prepared. This will normally form part of head office overheads and therefore is *Not priced* separately here.

NOTE. *The term 'Not Priced' where used throughout this section means either that the cost implication is negligible or that it is usually included elsewhere in the tender.*

PRELIMINARIES

6. **Statutory obligations, notices, fees and charges** – *Not priced unless the Contractor is specifically instructed to allow for these items.*
7. **Levels and setting out of the works** – *Not priced*
8. **Materials, goods and workmanship to conform to description, testing and inspection** – *Not priced*
9. **Royalties and patent rights** – *Not priced*
10. **Person-in-charge**

Under this heading are usually priced any staff that will be required on site. The staff required will vary considerably according to the size, layout and complexity of the scheme, from one foreman-in-charge to a site agent, general foreman, assistants, checkers and storemen as well as a percentage of total time for production controllers, structural engineers, etc.

The costs included for such people should include not only their wages, but their total cost to the site including statutory payments, pension, expenses, holiday relief, overtime, etc.

Part of the foreman's time, together with that of an assistant will be spent on setting out the site. Allow say £1 per day for levels, staff, pegs and strings, plus the assistant's time if not part of the general management team. Most sites usually include for one operative to clean up generally and do odd jobs around the site; if no allowance has been made for an operator against odd items of plant used only intermittently, then it may be prudent to cost this person under this heading as well.

Cost of other staff, such as Buyers, and Quantity Surveyors, are usually part of head office overhead costs.

	£
A typical build-up of a foreman's costs might be:	
Annual salary	7,000·00
Expenses	800·00
Bonus	1,000·00
Employer's National Insurance contribution on salary and bonus (13·7% on £8,000·00)	1,096·00
Training levy	26·40
Pension scheme (say 5% of salary and bonus)	400·00
Sundries, including Employer's Liability and Third Party (say 2·62% of salary and bonus)	209·60
	10,532·00
÷ 47 to allow for holidays: Per week	224·08
Say	£225·00

Corresponding costs for other site staff should be calculated in a similar manner.

PRELIMINARIES

Example

Site administration

	£
General foreman 80 weeks @ £225	18,000·00
Holiday relief 4 weeks @ £225	900·00
Assistant foreman 60 weeks @ £140	8,400·00
Storeman/checker 60 weeks @ £90	5,400·00
Trades foreman 120 weeks @ £125	15,000·00
Ganger 80 weeks @ £100	8,000·00
	55,700·00
Add 10% for overheads and profit	5,570·00
	61,270·00

11. Access for Architect to the Works – *Not priced*
12. Clerk of Works – *Not priced*
13. Variations and provisional sums – *Not priced*
14. Contract Sum – *Not priced*
15. Value added tax – supplemental provisions

The majority of new and alteration works are zero-rated and on such work, although the Contractor will recover VAT payments from the Inland Revenue, he should allow for all incidental costs and expenses which he may incur thereby. No guidance can be given for this item as costs will vary depending on whether the work is taxable or zero-rated.

16. Materials and goods unfixed or off-site – *Not priced*
17. Practical completion and Defects Liability

Inevitably some defects will arise after practical completion and an allowance will often be made to cover this. An allowance of say 0·25 to 0·5% should be sufficient.

Example

Defects after completion

	£
Based on £0·25% of the contract sum	
£855,000 @ £0·25%	2,137·50
Add 10% for overheads and profit	213·75
	2,351·25

18. Partial possession by Employer – *Not priced*
19. Assignment and Sub-Contracts – *Not priced*
19A. Fair Wages – *Not priced*
20. Injury to persons and property and Employers indemnity
(See Clause No. 21)
21. Insurance against injury to persons and property

The Contractor's Employer's Liability and Public Liability policies (which would both be involved under this heading) are often in the region of 0·5 to 0·6% on the value of his

PRELIMINARIES

own contract work (excluding provisional sums and work by sub-contractors whose prices should allow for these insurances). However, this allowance is usually included in the all-in hourly rate used in the calculation of 'Prices for Measured Work' (see page 232).

Under Clause 21.2 no requirement is made upon the contractor to insure as stated by the clause unless a provisional sum is allowed in the Contract Bills.

22. Insurance of the works against Clause 22 Perils

If at the contractor's risk the insurance cover must be sufficient to include the full cost of reinstatement, all increases in cost, professional fees and any consequential costs such as demolition. The average provision for fire risk is £0·10% of the value of the work after adding for increased costs and professional fees.

Example

	£
Contractor's Liability – Insurance of the works against fire etc.	
Contract value (including 'Preliminaries')	983,250·00
Estimated increased costs during contract period say 10% . .	98,325·00
	1,081,575·00
Estimated increased costs incurred during period of reinstatement say an average of 13%	140,604·75
	1,222,179·75
Fees @ 16%	195,548·76
	1,417,728·51
Allow £0·10% on say £1,417,800 plus 10% overheads and profit	£1,560

NOTE : *Insurance premiums are liable to considerable variation, depending on the contractor, the nature of the work and the market in which the insurance is placed.*

23. **Date of possession, completion and postponement** – *Not priced*
24. **Damages for non-completion** – *Not priced*
25. **Extension of Time** – *Not priced*
26. **Loss and expense caused by matters materially affecting regular progress of the works** – *Not priced*
27. **Determination by Employer** – *Not priced*
28. **Determination by Contractor** – *Not priced*
29. **Works by Employer or persons employed or engaged by Employer** – *Not priced*
30. **Certificates and payments** – *Not priced*
31. **Finance (No. 2) Act 1975 – Statutory tax deduction scheme** – *Not priced*
32. **Outbreak of hostilities** – *Not priced*
33. **War damage** – *Not priced*
34. **Antiquities** – *Not priced*
35. **Nominated Sub-Contractors**
    Not priced here. An amount should be added to the relevant P.C. sums, if required, for profit and a further sum for attendance.

PRELIMINARIES

36. Nominated Suppliers
Not priced here. An amount should be added to the relevant P.C. Sums, if required, for profit.

37. Choice of fluctuation provisions – entry in Appendix
The amount which the Contractor may recover under the fluctuations clauses (Clause Nos. 38, 39 and 40) will vary depending on whether the Contract is 'firm', i.e. Clause No. 38 is included, or 'fluctuating' and if 'fluctuating', whether the traditional method of assessment is used, i.e. Clause No. 39 or the formula method, i.e. Clause No. 40.

An allowance should be made for any shortfall in reimbursement under fluctuating contracts.

38. Contribution Levy and Tax Fluctuations (*see Clause No. 37*)
39. Labour and Materials Cost and Tax Fluctuations (*see Clause No. 37*)
40. Use of Price Adjustment Formulae (*see Clause No. 37*)

5. Contractor's Liability – *Not priced*
(If the Standard Form of Building Contract 1980 Edition is assumed this item is covered under Item 4. If a different Form of Contract is used it may be necessary to include for Contractor's insurances, etc. under this item.)

6. Employer's liability – *Not priced*
(If additional insurances are to be allowed for in the contract sum such cost shall be given as a provisional sum.)

7. Local Authorities' fees and charges – *Not priced*
(If the Standard Form of Building Contract 1980 Edition is assumed this Item is covered under Item 4 otherwise the fees and charges which the Contractor is required to pay under this Item shall be given as provisional sums.)

8. Obligations and restrictions imposed by the Employer
These include the following items and costs can only be assessed in the light of circumstances on a particular job.

(*a*) Access to and possession of use of the site.
(*b*) Limitations of working space.
(*c*) Limitations of working hours.
(*d*) The use or disposal of any materials found on site.
(*e*) Hoardings, fences, screens, temporary roofs, temporary name boards and advertising rights.
(*f*) The maintenance of existing live drainage, water, gas and other main or power services on or over the site.
(*g*) The execution or completion of the work in any specific order or in sections or phases.
(*h*) Maintenance of specific temperature and humidity levels.
(*j*) Temporary accommodation and facilities for the use of the Employer including heating, lighting, furnishing and attendance.

This will include an office for the Clerk of Works if there is to be one on the site. Against this item the following should be priced:
Hire of Clerk of Works office
Transport to and from the site
Erecting on a suitable base and later dismantling
Lighting, heating and attendance on office
Local Authority rates

PRELIMINARIES

Example

Mobile office for Clerk of Works	£
15m² × 76 weeks @ £1.80 m²	2,052·00
Haulage to and from site, say	40·00
Lighting, heating and attendance on office, say 76 weeks @ £50·00 .	3,800·00
Rates on temporary buildings based on £2·50 m² per annum . .	55·00
	5,947·00
Add 10% for overheads and profit	594·70
	£6,541·70

) The installation of telephones for the use of the Employer and the cost of his telephone
calls shall be given as a provisional sum.
) Any other obligations or restriction.

Additional obligations may include the provision of a performance bond. If the Con-
actor is required to provide sureties for the fulfilment of the work the usual method of
oviding this is by a bond provided by one or more insurance companies. The cost of a
rformance bond depends largely on the financial standing of the applying contractor.
gures tend to range from £0.25 to £0·50% of the contract sum.

Works by nominated sub-contractors, goods and materials from nominated suppliers and works by public bodies

Works by nominated sub-contractors – *Not priced*
Work to be carried out by a nominated sub-contractor shall be given as a prime cost
m to which an amount should be added, if required, for profit and a further sum for
tendance.

. Goods and materials from nominated suppliers – *Not priced*
Goods and materials which are required to be obtained from a nominated supplier
all be given as a prime cost sum to which an amount should be added, if required, for
ofit.

. Works by Public Bodies – *Not priced*
Works which are to be carried out by a Local Authority or public undertaking shall
 given as a provisional sum.

. Works by others directly engaged by the Employer – *Not priced*
A description shall be given of works by others directly engaged by the Employer and
y attendance that is required shall be priced in the same way as works by nominated
b-contractors.

General facilities and obligations

. Pricing
For convenience in pricing the following items will be listed. The Contractor should
clude maintaining any temporary works in connection with the items, adapting, clearing
vay and making good and all notices and fees to local authorities and public under-

PRELIMINARIES

takings. On fluctuating contracts, i.e. where Clause No. 39 or No. 40 is incorporated ther
is no allowance for fluctuations in respect of plant and temporary works and in sucl
instances allowances must be made for any increases likely to occur over the contrac
period.

(a) Plant, tools and vehicles

The sixth edition of the 'Standard Method of Measurement for Building Works
provides for these items to be priced at the beginning of those trades, involving their use
However, for some items their use tends to be 'across' two or three trades and to date ver
few Contractors have adopted this method of pricing in tendering documents received b
the Editors. Prices for Measured Work include the cost of mechanical plant, except fo
hoists, and the hourly rates used are given on page 233.

Small plant and hand tools are usually assessed as between $\frac{1}{2}$ % and $1\frac{1}{2}$ % of total labou
value.

Example

	£
Plant, tools and vehicles	
Hoist, on hire say 30 weeks @ £40·00 per week	1,200·00
Haulage to and from site, say	40·00
Fuel and oil	10·50
Operator 40 hours @ £2·51	100·40
Scaffold tower, two lifts @ £100 per lift	200·00
Small plant and hand tools @ 1 % of labour value . . .	3,060·00
	4,610·90
Add 10 % for overheads and profit	461·09
	£5,071·99

(b) Scaffolding

Scaffolding is usually undertaken by specialist sub-contractors who will submit quota
tions based on the specific requirements of the works. It is not possible to give rates her
for the various types of scaffolding that may be required but for the purposes of this sectio
it is assumed that the cost of supplying, erecting, maintaining and subsequently dis
mantling the scaffolding required would amount to £8,500·00 inclusive of overheads an
profit.

(c) Site administration and security

NOTE: *The cost of administrative staff is normally included against Clause No. 10 of th
Conditions of Contract.*

When required allow for the provision of a watchman or inspection by a securit
organization.

(d) Transport for workpeople

The labour rates per hour on which 'Prices for Measured Work' have been based d
not cover travel and lodging allowances which must be assessed according to the appro
priate working rule agreement.

PRELIMINARIES

Example

Transport for workpeople

Assuming all labour can be found within the London region, the labour value of £306,000 represents approximately 2400 man weeks. Assume each man receives an allowance of £0·965 per day or £4·825 per week of five days.

	£	£
2400 man weeks at £4·825	11,580·00	
Add 10% for overheads and profit	1,158·00	12,738·00

) Protecting the works from inclement weather

In areas likely to suffer particularly inclement weather, some nominal allowance ɪould be included for tarpaulins, polythene sheeting, battening, etc. and the effect of any ɘlays in concreting or brickwork by such weather.

) Water for the works

Charges should properly be ascertained from the local Water Authority. If these are ɔt readily available, an allowance of £0·25% of the value of the contract is probably ɑequate, providing water can be obtained directly from the mains. Failing this, each case ɪust be dealt with on its merits. In all cases an allowance should also be made for temp-ɾary plumbing including site storage of water if required.

Useful rates for temporary plumbing include:

Piping	£4·00 per metre
Connection	£100·00
Standpipe	£40·00

Plus an allowance for barrels and hoses.

Example

Water for the works

	£	£
£0·25% on £855,000·00	2,137·50	
Temporary plumbing	250·00	
	2,387·50	
Add 10% for overheads and profit	238·75	2,626·25

ɟ) Lighting and power for the Works

The Contractor is usually responsible for providing all temporary lighting and power ɔr the works and all charges involved. On large sites this could be expensive and involve ɪub-stations and the like, but on smaller sites it is often limited to general lighting (depend-ɪg upon time of year), power for power operated tools, a small diesel generator and some ɾansformers.

Typical costs are:

Low voltage diesel generator	£30·00–£40·00 per week
1½ to 3 VA transformer	£ 5·00–£ 7·00 per week

A typical allowance, including charges, issuing and fitting costs could be 1% of contract ɑlue.

PRELIMINARIES

Example

	£	£
Lighting and power for the works		
Dependent on the nature of the work, time of year and incidence of power operated tools, allow say, 1% on £855,000·00	8,550·00	
Add 10% for overheads and profit	855·00	9,405·00

(*h*) Temporary roads, hardstandings, crossings and similar items

Quite often consolidated bases of eventual site roads are used throughout a contract to facilitate movement of materials around the site. However, during the initial setting up of a site, with drainage works outstanding, this is not always possible and occasionally temporary roadways have to be formed and ground levels later reinstated.

Typical costs are:

Removal of topsoil and provision of 225 mm stone/hardcore base blinded with ashes as a temporary roadway 3·50 mm wide and subsequent reinstatement:

 – on level ground . . . . £20·00 per metre
 – on sloping ground including 1 m of cut or fill . . £25·00 per metre

Removal of topsoil and provision of 225 mm stone/hardcore base blinded with ashes as a temporary hardstanding . . £6·00 per metre^2

NOTE: *Any allowance for special hardcore hardstandings for piling sub-contractors is usually priced against the 'special attendance' clause after the relevant Prime Cost Sum.*

(*j*) Temporary accommodation for the use of the Contractor

This includes all temporary offices and sheds for the Contractors and his domestic sub contractors' use (temporary office for a Clerk of Works is covered under obligations and restrictions imposed by the Employer).

Typical costs for mobile offices are as follows, based upon a twelve months minimum hire period they exclude furniture which could add a further £8·50 per week.

Size	Rate per week £
12′ × 7′6″ (8·36 m²)	12·60 (£1·51 m²)
16′ × 7′6″ (11·15 m²)	14·90 (£1·30 m²)
22′ × 7′6″ (15·33 m²)	19·10 (£1·25 m²)
32′ × 10′0″ (29·72 m²)	26·70 (£0·90 m²)

Typical rates for timber huts are as follows:

Size	Rate per week £
6′ × 12′ (6·69 m²)	4·95 (£0·74 m²)
18′ × 12′ (20·00 m²)	9·20 (£0·46 m²)
24′ × 12′ (26·75 m²)	11·80 (£0·44 m²)
30′ × 12′ (33·44 m²)	14·50 (£0·62 m²)

Allowance must also be made for transport, erection, dismantling, etc. as previously shown under the item for Clerk of Work's office.

PRELIMINARIES

Example

	£
Temporary accommodation	
Foreman's office 1 No.	
15m² × 76 weeks @ £1·80m²	2,052·00
Haulage to and from site say	40·00
Storage sheds 2 No.	
30 m² × 70 weeks @ £0·44 m²	924·00
Erection and dismantle say	170·00
Haulage to and from site say	60·00
	3,246·00
Lighting, heating and attendance on offices say, 76 weeks @ £50 . .	3,800·00
Rates on temporary buildings based on £2·50 per m² per annum . .	155·76
	7,201·76
Add 10% for overheads and profit	720·18
	£7,921·94

(*k*) Temporary telephones for the use of the Contractor

Against this item should be included the cost of installation, rental and an assessment of the cost of calls made during the contract.

Installation costs	£55·00 (for more than 12 months rental)
	£60·00 (for less than 12 months rental)
Rental	£11·25 per quarter
Cost of calls	For sites with one telephone allow about £10·00 per week.

Example

	£
Temporary telephones	
Connection charges for telephone	55·00
for outside bell	10·00
Rental	
for telephone 6 quarters @ £11·25	67·50
for outside bell 6 quarters @ 60p	3·60
Calls – 76 weeks @ £10	760·00
	896·10
Add 10% for overheads and profit	89·61
	£985·71

(*l*) Traffic regulations

Waiting and unloading restrictions can occasionally add considerably to costs, resulting in forced overtime or additional weekend working. Any such restrictions must be carefully assessed for the job in hand.

PRELIMINARIES

(*m*) Safety, health and welfare of workpeople

The Contractor is required to comply with the Code of Welfare Conditions for the Building Industry which sets out welfare requirements as follows:

1. Shelter from inclement weather
2. Accommodation for clothing
3. Accommodation and provision for meals
4. Provision of drinking water
5. Sanitary conveniences
6. Washing facilities
7. First Aid
8. Site conditions

A variety of self-contained mobile or jack-type units are available for hire and a selection of rates is given below:

Kitchen with cooker, fridge, sink unit, water heater and basin

12′ × 7′6″	£19·00 per week
24′ × 9′0″	£27·00 per week

Mess room with water heater, wash basin and seating

12′ × 7′6″	£16·00 per week
16′ × 9′0″	£18·00 per week

Welfare unit with drying rack, lockers, tables, seating, cooker, heater, sink and basin

22′ × 7′6″	£28·00 per week

Toilets (mains type)

Single unit	£9·00 per week
Three unit	£21·00 per week
Four unit	£28·00 per week

Allowance must be made in addition for transport costs to and from site, setting up costs, connections to mains, fuel supplies and attendance.

Site first aid kit	£3·00 per week

A general provision to comply with the above code is often £0·50 to £0·75% of the contract value.

The costs of safety supervisors (required for firms employing more than 20 people) are usually part of head office overhead costs.

Example

	£
Safety, health and welfare	
Based on £0·75% of £855·000	6,412·50

(*n*) Disbursements arising from the employment of workpeople

Travelling and lodging allowances have been dealt with under 'Transport for Workpeople' and usually all other on-costs and disbursements are included in the all-in hourly rate used in the calculation of 'Prices for Measured Work' (see page 232). However, it is as well to check that all such disbursements have been included elsewhere.

PRELIMINARIES

(p) Maintenance of public and private roads

Some additional insurance or value may be required against this item to insure against damage to entrance gates, kerbs or bridges caused by extraordinary traffic in the execution of the works.

(q) Removing rubbish, protective casings and coverings and cleaning the works on completion

This includes removing surplus materials and final cleaning of the site prior to handover. Allow for sufficient 'bins' for the site throughout contract duration and for some operative time at the end of the contract for final clearing and cleaning ready for handover.
Cost of 'bins' – approx. £20 each.
A general allowance of 0·20% of contract value is probably sufficient.

Example

	£
Removing rubbish, etc. and cleaning	
Say £0·2% of contract value of £855,000·00	1,710·00

(r) Drying the works

Use or otherwise of an installed heating system will probably determine the value to be placed against this item.
Dependent upon the time of year, say allow 0·1% to 0·2% of the contract value to cover this item.

	£
Say £0·2% of contract value of £855,000·00	1,710·00

(s) Temporary fencing, hoardings, screens, fans, planked footways, guardrails, gantries and similar items

This item must be considered in some detail as it is dependent on site perimeter, phasing of the work, work within existing buildings, etc.
Useful rates include

Hoarding 2·3 m high of 18 mm plywood with 50 × 100 mm sawn softwood studding, rails and posts including later dismantling	
– undecorated	£23·00 (£10·00 m²)
– decorated one side	£29·00 (£12·60 m²)
Pair of gates in above hoarding	extra £100·00 per pair
Cleft chestnut fencing 1·2 m high including dismantling . .	£5·00 m

Example

Temporary hoarding

	£	£
Plywood decorated hoarding		
100 metres @ £29·00	2,900·00	
Extra for one pair of gates	100·00	3,000·00

PRELIMINARIES

(*t*) Control of noise, pollution and all other statutory obligations

The Local Authority may impose restrictions on the timing of certain operations, particularly noisy or dust-producing operations, and may necessitate the carrying out of these works outside normal working hours or using special tools and equipment.

The situation is most likely to occur in built-up areas such as city centres, etc. where the site is likely to be in close proximity to offices or residential property.

Contingencies

14. Generally

Any provision for contingencies shall be given as a provisional sum.

Example Summary

Item		£
4.10	Site administration	61,270·00
4.17	Defects after completion	2,351·25
4.22	Insurance of the works against fire, etc.	1,560·00
8.*j*	Clerk of Works' office	6,541·70
13.*a*	Plant, tools and vehicles	5,071·99
13.*b*	Scaffolding	8,500·00
13.*d*	Transport for workpeople	12,738·00
13.*f*	Water for the works	2,626·25
13.*g*	Lighting and power for the works	9,405·00
13.*j*	Temporary accommodation	7,921·94
13.*k*	Temporary telephones	985·71
13.*m*	Safety, health and welfare	6,412·50
13.*q*	Removing rubbish, etc. and cleaning	1,710·00
13.*r*	Drying the works	1,710·00
13.*s*	Temporary hoarding	3,000·00

TOTAL £131,804·34

It is emphasized that the above is an example only of the way in which 'Preliminaries' may be priced and it is essential that for any particular contract or project the items set out in 'Preliminaries' should be assessed on their respective values.

It will be seen, however, that the value of 'Preliminaries' for a typical contract can add about 15 % to the value of the work; this fact should not be forgotten when using the rates given in the trades following this section.

15.9% nearer the figure [handwritten annotation]

EXCAVATION AND EARTHWORK

Prices are applicable to excavation in heavy soil. Multiplying factors for other soils are as follows:

Firm soil	× 0·83
Compact gravel	× 1·40
Soft chalk	× 2·50
Hard rock	× 5·00

SITE PREPARATION

	Unit	£
Preserving turf		
Lifting 50 mm thick		
laying out on site for future use	m²	1·13
Clearing undergrowth		
Clearing site of bushes, shrub, undergrowth; grubbing up roots	m²	0·13
Removing trees or hedges		
Cutting down trees; grubbing up roots		
small	No.	48·30
over 600 mm not exceeding 900 mm girth	,,	64·50
Cutting down hedges; grubbing up roots		
hawthorn 1500 mm high	m	4·40
,, 2000 ,,	,,	5·90

EXCAVATION (MECHANICAL)

Assumes 80% of earth moved by mechanical plant and 20% by hand (trimming sides and bottom)

	Unit	£
Preserving vegetable soil		
Excavating average 150 mm deep	m²	0·28

Excavating	Unit	Excavating only £	Excavating and hand filling working space £
Surfaces to reduce levels (using skimmer with 1·34 m³ bucket)			
not exceeding 0·25 m deep	m³	1·39	4·48
,, ,, 1 m deep	,,	1·55	4·64
Surfaces to reduce levels (using 4·70 m³ scraper)			
not exceeding 0·25 m deep	,,	1·86	4·95
,, ,, 1 m deep	,,	2·02	5·11
Basements or the like; starting from reduced level (using a dragline excavator with 0·57 m³ bucket)			
not exceeding 2 m deep	,,	2·32	5·41
,, ,, 4 m deep	,,	3·02	6·11
over 4 m not exceeding 6 m deep	,,	3·80	6·89
,, 6 m ,, ,, 8 m deep	,,	4·61	7·70
Pits; to receive bases; starting from reduced level (using mechanical trencher with 0·57 m³ bucket)			
not exceeding 1 m deep	,,	3·74	6·83
,, ,, 2 m deep	,,	4·19	7·28
,, ,, 4 m deep	,,	5·20	8·29
Trenches; to receive foundations; starting from reduced level (using mechanical trencher with 0·57 m³ bucket)			
not exceeding 0·25 m deep	,,	2·18	5·27
,, ,, 1 m deep	,,	2·26	5·35
,, ,, 2 m deep	,,	2·80	5·89
,, ,, 4 m deep	,,	3·87	6·96

EXCAVATION AND EARTHWORK

EXCAVATION (MECHANICAL) – *continued*
Excavating – *continued*

	Unit	Excavating only £	Excavating and filling working space £
Extra over all kinds of excavation			
for excavating below ground water level	m³	1·56	
for excavating alongside services or groups of services . .	,,	8·05	
for excavating across services	,,	10·75	

EXCAVATION (HAND)
Preserving vegetable soil
	Unit	Excavating only £	Excavating and filling working space £
Excavating average 150 mm deep	m²	0·81	—

Excavating
	Unit	Excavating only £	Excavating and filling working space £
Surfaces to reduce levels			
not exceeding 0·25 m deep	m³	6·17	9·26
,, ,, 1 m deep	,,	6·98	10·07
Basements or the like; starting from reduced level			
not exceeding 2 m deep	,,	9·07	12·16
,, ,, 4 m deep	,,	12·67	15·76
over 4 m not exceeding 6 m deep	,,	16·21	19·30
,, 6 m ,, ,, 8 m deep	,,	19·75	22·84
Pits; to receive bases; starting from reduced level			
not exceeding 1 m deep	,,	15·57	18·66
,, ,, 2 m deep	,,	17·85	20·94
,, ,, 4 m deep	,,	22·47	25·56
Trenches; to receive foundations; starting from reduced level			
not exceeding 0·25 m deep	,,	7·38	10·47
,, ,, 1 m deep	,,	8·19	11·28
,, ,, 2 m deep	,,	10·87	13·96
,, ,, 4 m deep	,,	16·24	19·33
,, ,, 1 m deep, not exceeding 0·30 m wide . .	m	3·22	
Extra over all kinds of excavation			
for excavating below ground water level	m³	1·23	

BREAKING UP
	Unit	£
Breaking up by mechanical drill		
reinforced concrete		
In open excavation	m³	34·65
In trenches	,,	39·35
brickwork		
In open excavation	,,	17·33
In trenches	,,	18·75
150 mm surface concrete	m²	3·34
300 mm surface concrete	,,	6·68
150 mm tarmac and hardcore	,,	2·00
Taking up 50 mm concrete flags; storing flags for re-use . .	,,	1·07

EXCAVATION AND EARTHWORK

EARTHWORK SUPPORT
Sides of excavation

	Unit	Distance between opposing faces		
		not exceeding 2m £	2m-4m £	over 4m £
not exceeding 1 m deep	m²	0·86	0·95	—
„ „ 2 m deep	„	1·29	1·42	2·37
„ „ 4 m deep	„	1·72	1·89	2·78
over 4 m not exceeding 6 m deep . . .	„	2·16	2·37	3·20
„ 6 m „ „ 8 m deep . . .	„	2·64	2·90	3·62
not exceeding 2 m deep; next to roadways . .	„	5·21	—	—
„ „ 4 m deep; „ „ „ . .	„	—	—	7·47
over 4 m not exceeding 6 m deep; . . .	„	—	—	8·61
not exceeding 1 m deep; next to existing buildings .	„	2·72	—	—
„ „ 2 m „ ; „ „ „ „ . .	„	3·44	—	—
„ „ 4 m „ ; „ „ „ „ . .	„	4·16	—	—

	Unit	£
Extra over all kinds of earthwork support		
for support below ground water level	m²	3·45
for support in running silt/sand or the like	„	3·03
Earth support left in	„	12·10

DISPOSAL (MECHANICAL)
Surplus excavated material

		£
depositing on site in spoil heaps (2·30 m³ dumper)	m³	0·91
spreading on site; average 25 m from excavation (D.6. bulldozer) . .	„	0·77
„ „ „ ; „ 200 m „ „ (4.70 m³ scraper) . .	„	0·73
removing from site; to tip not exceeding 13 km from site (using lorries) .	„	3·96

DISPOSAL (HAND)
Surplus excavated material

		£
depositing on site in spoil heaps	„	2·95
spreading on site; average 25 m from excavation	„	3·62
„ „ „ ; „ 50 m „ „	„	4·56
removing from site; to tip not exceeding 13 km from site (using lorries) .	„	10·25

FILLING (MECHANICAL)
Excavated material

		£
backfilling into excavations (D.6. bulldozer)	„	1·27
filling in making up levels (D.6. bulldozer)	„	0·90

FILLING (HAND)
Excavated material

		£
back filling into excavations; depositing and compacting in layers . .	„	3·09
filling in making up levels; depositing and compacting in layers; multiple handling via spoil heaps average 25 m from excavation	„	9·93

EXCAVATION AND EARTHWORK

FILLING (HAND) – *continued*

											Unit	£
Hardcore												
filling in making up levels over 250 mm thick; depositing and compacting in layers											m³	7·42
filling in making up levels average 100 mm thick											m²	1·00
”	”	”	150	”	.	.	.	.	.		”	1·23
”	”	”	200	”	.	.	.	.	.		”	1·54
”	”	”	225	”	.	.	.	.	.		”	1·93
”	”	”	250	”	.	.	.	.	.		”	2·00
Coarse ashes												
filling in making up levels over 250 mm thick; depositing and compacting in layers											m³	12·64
filling in making up levels average 75 mm thick											m²	1·29
”	”	”	100	”	.	.	.	.	.		”	1·56
”	”	”	125	”	.	.	.	.	.		”	1·80
”	”	”	150	”	.	.	.	.	.		”	2·05

SURFACE TREATMENTS

		£
Bottoms of excavations		
levelling; compacting	”	0·19
grading to cambers; compacting	”	0·32
Sides of cuttings		
trimming to slope	”	0·64
Surfaces of hardcore		
levelling; blinding; compacting	”	0·48
hand-packing hardcore to form vertical or battering face over 300 mm wide	”	1·61
Surfaces of coarse ashes		
levelling; compacting	”	0·19
Sand		
blinding with sand 50 mm thick	”	0·91

PILING

BORED CONCRETE PILES

The following approximate prices for the quantities of piling quoted, are for work on clear open sites with reasonable access. They are based on **475** mm nominal diameter pneumatic caisson-type piles normal concrete mix 21·00 N/mm² reinforced for loading up to 40,000 kg dependent on ground conditions and include any necessary bulb at the base of the pile and up to 0·16 m of projecting reinforcement at top of pile.

	Unit	£
Provision of all plant, including transport to and from site, erection and dismantling at each pile position for 100 No. piles	item	4000·00
Boring, concrete and reinforcement to a depth of 10 metres . .	m	25·00
Add for additional depth of pile up to a maximum depth of 15 metres .	,,	22·00
Deduct for reduction in the length of pile	,,	16·00
Cutting of heads of piles	No.	7·00

STEEL INTERLOCKING PILING

Approximate rates for providing, pitching and driving of rolled steel sheet sections of the type and weight (kg/m²) described.

	Unit	£ to £
Type 1U, weight 106 kg/m²	m²	41·00 ,, 43·00
,, 2 , ,, 122 kg/m²	,,	48·00 ,, 51·00
,, 3 , ,, 155 kg/m²	,,	62·00 ,, 65·00
,, 4B, ,, 200 kg/m²	,,	77·00 ,, 80·00
,, 5 , ,, 238 kg/m²	,,	92·00 ,, 97·00

CONCRETE WORK

PLAIN IN SITU CONCRETE	Unit	Mix 40 *mm* aggregate (1 : 12) £	Mix 11·50 N/mm² 40 *mm* aggregate (1 : 3 : 6) £	Mix 21·00 N/mm² 20 *mm* aggregate (1 : 2 : 4) £	Mix 25·50 N/mm² 20 *mm* aggregate (1 : 1½ : 3) £	Mix 31·00 N/mm² 20 *mm* aggregate (1 : 1 : 2) £
Foundations in trenches over 300 mm thick						
Normal Portland cement . .	m³	28·07	31·93	34·90	37·52	42·25
Sulphate-resistant cement . .	„	29·06	33·47	37·00	40·10	45·65
Normal Portland cement with water-repellent additive . .	„	—	—	37·05	39·70	43·75
Aluminous cement . . .	„	36·40	44·84	52·75	59·25	70·90
Rapid-hardening cement . .	„	28·42	32·47	35·63	38·42	43·42

	Unit	£
A D D to the above for		
Reinforced in situ concrete	m³	3·50
Foundations in trenches		
not exceeding 100 mm thick	„	6·98
over 100 mm not exceeding 150 mm thick	„	5·26
„ 150 mm „ „ 300 mm thick	„	1·74
Foundations to isolated stanchions		
over 300 mm thick	„	3·49
Ground beams		
Sectional area not exceeding 0·03m².	„	9·66
sectional area over 0·03 m² not exceeding 0·10 m²	„	8·59
„ „ „ 0·10 m² „ „ 0·25 m²	„	7·52
„ „ „ 0·25 m³	„	6·44
Beds		
not exceeding 100 mm thick	„	6·98
over 100 mm not exceeding 150 mm thick	„	5·26
„ 150 mm „ „ 300 mm thick	„	3·54
„ 300 mm thick	„	1·74
Suspended slabs		
not exceeding 100 mm thick	„	16·64
over 100 mm not exceeding 150 mm thick	„	14·90
„ 150 mm „ „ 300 mm thick	„	13·15
„ 300 mm thick	„	11·40
Coffered or troughed slabs		
over 150 mm not exceeding 300 mm thick	„	15·03
„ 300 mm thick	„	13·15
Walls		
not exceeding 100 mm thick	„	17·58
over 100 mm not exceeding 150 mm thick	„	15·70
„ 150 mm „ „ 300 mm thick	„	13·82
„ 300 mm thick	„	12·08
Isolated beams		
sectional area not exceeding 0·03 m²	„	26·30
„ „ over 0·03 m² not exceeding 0·10 m²	„	21·07
„ „ „ 0·10 m² „ „ 0·25 m²	„	17·58
„ „ „ 0·25 m²	„	15·84
Deep beams		
sectional area not exceeding 0·03 m²	„	21·07
„ „ over 0·03 m² not exceeding 0·10 m²	„	17·58
„ „ „ 0·10 m² „ „ 0·25 m²	„	15·84
„ „ „ 0·25 m²	„	14·09

CONCRETE WORK

PLAIN IN SITU CONCRETE – *continued*

	Unit	£
Isolated columns		
sectional area not exceeding 0·03 m²	m³	33·28
„ „ over 0·03 m² not exceeding 0·10 m²	„	26·30
„ „ „ 0·10 m² „ „ 0·25 m²	„	20·53
„ „ „ 0·25 m²	„	17·58
Steps and staircases		
generally	„	13·95
Filling to hollow walls		
not exceeding 100 mm thick	„	12·08

LABOURS ON CONCRETE OF ANY DESCRIPTION

	Unit	£
Treating surfaces of unset concrete		
grading to cross falls	m²	0·56
„ „ spade finishing	„	0·37
with powered float to receive floor coverings	„	0·84

REINFORCEMENT

(The following are average prices for reinforcement in structures)

	Unit	Mild steel B.S. 4449 £	High-tensile steel B.S. 4461 £
Rolled steel bars; bends, hooks, tying wire, distance blocks and ordinary spacers			
25 mm	t	365·75	374·55
20 „	„	375·05	383·74
16 „	„	389·41	398·21
12 „	„	442·18	450·98
10 „	„	488·21	497·00
8 „	„	564·00	572·80
6 „	„	573·15	581·95

	Unit	£
Fabric; B.S. 4483; 150 mm side laps, 300 mm end laps		
Mesh 200 × 200 mm × 1·54 kg/m²	m²	0·87
„ 200 × 200 „ × 2·22 „	„	1·24
„ 200 × 200 „ × 3·02 „	„	1·66
„ 200 × 200 „ × 3·95 „	„	2·18
„ 200 × 200 „ × 6·16 „	„	3·36
„ 100 × 200 „ × 3·05 „	„	1·67
„ 100 × 200 „ × 5·93 „	„	3·31
„ 100 × 200 „ × 8·14 „	„	4·49
„ 100 × 200 „ × 10·90 „	„	6·17

FORMWORK TO PLAIN (OR REINFORCED) IN SITU CONCRETE

Formwork generally

		£
Edges and faces		
foundations; exceeding 1 m high	„	9·05
Soffits; horizontal		
slabs	„	10·65
slabs; strutting over 3·50 m not exceeding 5 m high	„	11·90
„ ; „ „ 5 m „ „ 6·50 m high	„	13·75
„ ; 400–500 mm thick	„	11·85

CONCRETE WORK

FORMWORK TO PLAIN (OR REINFORCED) IN SITU
CONCRETE – *continued*

	Unit	£
Formwork generally – *continued*		
Soffits; horizontal; less than 1 m high		
slabs	m²	12·80
Soffits; sloping over 15° from horizontal		
slabs	,,	13·07
Soffits of coffered slab; including 325 mm deep pans		
full boarded soffit	,,	12·72
patent steel shuttering	,,	8·00
Sides; vertical or battering		
walls	,,	11·15
projections/pilasters (average superficial rate)	,,	13·57
,, ,, ; 500 mm girth	m	7·53
,, ,, ; 1000 mm girth	,,	14·15
,, ,, ; 1500 mm girth	,,	18·80
,, ,, ; 2000 mm girth	,,	25·05
isolated columns (average superficial rate)	m²	12·00
,, ,, ; 750 mm girth	m	9·88
,, ,, ; 1500 mm girth	,,	17·60
,, ,, ; 2000 mm girth	,,	23·35
,, ,, ; 3000 mm girth	,,	34·00
Sides and soffits		
wall openings, recesses (average superficial rate)	m²	16·86
,, ,, ,, ; 100 mm girth	m	1·87
,, ,, ,, ; 200 mm girth	,,	2·71
,, ,, ,, ; 300 mm girth	,,	4·14
Sides and soffits		
attached beams (average superficial rate)	m²	13·00
,, ,, ; 750 mm girth	m	10·48
,, ,, ; 1500 mm girth	,,	19·22
,, ,, ; 3000 mm girth	,,	38·00
,, ,, ; 5000 mm girth	,,	62·38
deep beams (average superficial rate)	m²	11·65
,, ,, ; 3000 mm girth	m	35·23
,, ,, ; 5000 mm girth	,,	57·80
isolated beams (average superficial rate)	m²	13·92
,, ,, ; 750 mm girth	m	11·25
,, ,, ; 1500 mm girth	,,	20·58
,, ,, ; 3000 mm girth	,,	40·76
,, ,, ; 5000 mm girth	,,	66·80
Edges, risers or faces		
not exceeding 250 mm	,,	2·52
over 250 mm not exceeding 500 mm	,,	5·03
,, 500 mm ,, ,, 1 m	,,	10·05
Grooves		
12 × 12 mm	,,	0·19
25 × 25 ,,	,,	0·28
32 × 32 ,,	,,	0·31
Labours		
raking cutting	,,	1·10
curved cutting	,,	2·11
ADD to prices for formwork generally for		
curved radius 6 m	%	37½
,, ,, 2 ,,	,,	50
coating with retarding agent generally	m²	0·21
formwork finish using plywood sheeting	,,	1·88

CONCRETE WORK

PRECAST CONCRETE
Normal; mix 21·00 N/mm² 19 mm aggregate (1 : 2 : 4)

	Unit	£
Copings		
175 × 75 mm; weathered once; grooved twice; surface finish 400 mm girth	m	5·24
325 × 75 mm; „ „ ; „ „ ; „ „ 550 mm girth	„	7·18
Duct covers; reinforced with 200 × 200 mm × 2·22 kg/m² mesh		
50 mm thick; 300 mm wide	„	5·40
75 mm thick; „ „	„	6·00
100 mm thick; „ „	„	7·18
Lintels; plate; suitably reinforced		
100 × 65 mm × 1200 mm long	No.	3·25
150 × 65 mm × 1200 mm long	„	3·92
220 × 65 mm × 1200 mm long	„	5·32
265 × 65 mm × 1800 mm long	„	9·15
Lintels; traditional		
100 × 150 × 1200 mm long; reinforced with one 12 mm mild steel bar	„	6·46
150 × 150 × 1800 mm long; „ „ two 12 mm „ „ bars	„	10·05
225 × 225 × 1800 mm long; „ „ two 16 mm „ „ „	„	23·15
Lintels; traditional boot		
150 × 215 × 1500 mm long; reinforced with two 12 mm mild steel bars; surface finish 200 mm girth	„	13·85
260 × 290 × 2100 mm long; reinforced with two 20 mm mild steel bars; surface finish 300 mm girth	„	36·60
Padstones; surface finish to one exposed edge		
300 × 100 × 75 mm deep	„	1·45
225 × 225 × 150 mm deep	„	2·40
450 × 450 × 150 mm deep	„	6·42

HOLLOW BLOCK SUSPENDED CONSTRUCTION
Ribs and topping concrete mix 21·00 N/mm²; hollow clay blocks size 300 × 300 mm at 380 mm centres; keyed soffits

	Unit	£
Suspended slabs; horizontal		
140 mm thick (100 mm thick tiles)	m²	8·84
190 „ (150 „ „)	„	10·35
241 „ (200 „ „)	„	12·00
Extra; slip tiles 300 × 75 × 15 mm	„	0·23

Reinforcement
For prices of reinforcement see page 269

Formwork
For prices of formwork see page 269

CONCRETE WORK

SUNDRIES	Unit	£
Hacking by mechanical means to form key for		
asphalt or in situ finishings	m²	1·07
Building paper; 150 mm lapped joints		
horizontal on foundations	„	0·23
0·13 mm polythene sheet; 150 mm lapped joints		
horizontal on slabs	„	0·28
Cold bitumen solution; two coats		
horizontal on slabs	„	0·97
'Synthaprufe'; two coats; blinding with sand		
horizontal on slabs	„	1·74
Expansion joints in concrete; formwork		
10 × 100 mm; filling with bitumen impregrated joint filler 75 mm wide; sealing		
with polyurethene sealant backer and polyurethene sealant	m	2·80
P.V.C. water stop; cast into concrete		
190 mm wide	„	3·80
Formed channel in concrete; formwork		
300 × 150 mm wide	„	5·95
Formed mortice in concrete; formwork; filling with cement		
for ragbolt; per 25 mm depth	No.	0·45
Formed hole in concrete; formwork		
not exceeding 1 m girth; 100 mm deep	„	2·54
over 1 m not exceeding 2 m girth; 100 mm deep	„	4·40
„ 2 m „ „ 4 m girth; 100 mm deep	„	7·35
Add for each additional 25 mm thickness up to 300 mm thick		
not exceeding 1 m girth	„	0·50
over 1 m not exceeding 2 m girth	„	0·75
„ 2 m „ „ 4 m girth	„	1·30
Cut mortice in concrete; filling with cement		
for ragbolt; per 25 mm depth	„	0·54
Cut hole in concrete; 100 mm deep		
150 × 150 mm	„	4·25
250 × 250 mm	„	6·08
500 × 500 mm	„	13·35
Add for each additional 25 mm thickness up to 300 mm thick		
150 × 150 mm	„	0·85
250 × 250 mm	„	1·22
500 × 500 mm	„	2·68
Add for making good fair finish per side		
150 × 150 mm	„	0·45
250 × 250 mm	„	0·50
500 × 500 mm	„	0·70
18 Gauge (1.22 mm) galvanized steel anchor slots; canting into concrete		
150 mm long	„	0·50

BUILDER'S WORK IN CONNECTION WITH PLUMBING AND ENGINEERING INSTALLATIONS		Small pipe	Large pipe
Cut holes for pipes or the like	Unit	£	£
100 mm concrete	No.	1·63	2·72
Add for each additional 25 mm thickness up to 300 mm thick	„	0·36	0·61
Add to above for making good fair finish per side	„	0·25	0·53

BRICKWORK AND BLOCKWORK

BRICKWORK

	Unit	Common bricks p.c. £45·00 1000	
		in gauged mortar (1 : 1 : 6)	in cement mortar (1 : 3)
		£	£
Half brick thick			
walls	m²	11·10	11·23
walls; built curved, mean radius 6 m	,,	12·40	12·52
,, ; ,, ,, ,, ,, 1·50 m	,,	13·75	13·90
honeycomb walls	,,	7·60	7·70
skins of hollow walls	,,	11·10	11·23
,, ,, ,, ,, ; in narrow trenches	,,	11·64	11·76
One brick thick			
walls	,,	21·20	21·50
skins of hollow walls	,,	21·20	21·50
walls; filling old openings	,,	23·17	23·50
One and a half brick thick			
walls	,,	30·35	30·85
Two brick thick			
walls	,,	37·60	38·25
battering walls	,,	42·65	43·50
projections of footings	,,	38·95	39·60
Extra: grooved bricks	,,	0·02	0·02

Labours on common bricks in any mortar	Unit	£
Rough cutting		
to chamfered angle	m	1·46
Rough chases		
50 mm wide, 50 mm deep; vertical	,,	2·57
100 mm wide, 100 mm deep; vertical	,,	4·50
Bonding ends to old common brickwork		
one brick walls	,,	4·15

Engineering bricks in cement mortar (1 : 3)		Engineering bricks p.c. £122·00 1000
Half brick thick		
walls	m²	17·10
One brick thick		
walls	,,	33·20
walls; built curved; mean radius 6 m	,,	36·30
,, ; ,, ,, ; ,, ,, 1·50 m	,,	39·60
walls; in narrow trenches	,,	34·20
One and a half brick thick		
walls	,,	48·15

Labours on engineering bricks in any mortar		
Bonding ends to old common brickwork		
one brick walls	m	4·40

NOTE. *Prices based on bricks size 215 × 102·5 × 65 mm.*

BRICKWORK AND BLOCKWORK

BRICK FACEWORK

Extra over common bricks in any mortar for

	Unit	£
Fair face; flush pointing as work proceeds; Flemish bond walls or the like .	m²	1·48

	Unit	Machine made facings p.c. £122·00 1000 £	Hand made facings p.c. £187·00 1000 £
Extra over common bricks *p.c.* £45·00 1000 in any mortar for			
Facing bricks; flush pointing as work proceeds; Flemish bond			
walls or the like .	m²	11·10	16·94
Add or deduct for variation of £1·00 1000 in p.c. of facing bricks .	„	0·09	0·09

BRICKWORK FAIR BOTH SIDES OR ENTIRELY OF FACING BRICKS

Common bricks in cement–lime mortar (1 : 2 : 9)

	Unit	Common bricks p.c. £45·00 1000 £
Half brick thick faced both sides; flush pointing both sides as work proceeds; stretcher bond walls .	m²	14·00
One brick thick faced both sides; flush pointing both sides as work proceeds; Flemish bond walls .	„	24·00

Engineering bricks in cement mortar (1 : 3)

		Engineering bricks p.c. £122·00 1000 £
Half brick thick faced both sides; flush pointing both sides as work proceeds; stretcher bond walls .	„	20·05
One brick thick faced both sides; flush pointing both sides as work proceeds; Flemish bond walls .	„	36·10

Refractory bricks in cement fireclay mortar (1 : 4)

		£
Half brick thick faced one side; flush pointing one side as work proceeds; stretcher bond		
linings to flues; built 50 mm clear of flues; one header per m² abutting flue; keeping cavity clear .	„	27·75
paving to flues .	„	27·15

BRICKWORK AND BLOCKWORK

	Machine made facings p.c. £122·00 1000	Hand made facings p.c. £187·00 1000

BRICKWORK FAIR BOTH SIDES OR ENTIRELY OF FACING BRICKS – *continued*

	Unit	£	£
Facing bricks in cement–lime mortar (1: 2: 9)			
Half brick thick faced one side; flush pointing one side as work proceeds; stretcher bond			
walls	m²	19·05	23·65
Half brick thick faced one side; flush pointing one side as work proceeds; Flemish bond			
walls	„	20·80	25·80
Half brick thick faced both sides; flush pointing both sides as work proceeds; stretcher bond			
walls	„	20·55	25·15
One brick thick, flush pointing both sides as work proceeds; Flemish bond			
walls	„	36·80	46·05
Facework			
to margins	m	0·37	0·37
Flat arches half brick thick; flush pointing one side as work proceeds			
half brick wide face; flush pointing half brick wide exposed soffit	„	5·19	6·23
one brick wide face; flush pointing half brick wide exposed soffit	„	5·19	6·23
Sills; all headers-on-edge; flush pointing top and one side as work proceeds			
150 mm × half brick; horizontal; set weathering	„	5·19	6·23
Copings; all headers-on-edge; flush pointing top and both sides as work proceeds			
one brick × half brick; horizontal	„	5·19	6·23

		Machine made special p.c. £25·50 100 £	Hand made special p.c. £40·70 100 £
Sills; all headers-on-edge; one angle rounded 53 mm radius; flush pointing top and one side as work proceeds			
150 mm × half brick horizontal; set weathering	„	7·20	9·56

		Machine made special p.c. £30·50 100 £	Hand made special p.c. £48·70 100 £
Copings; all headers-on-edge; two angles rounded 53 mm radius; flush pointing top and both sides as work proceeds			
one brick × half brick horizontal	„	8·00	10·80
Extra; tile creasing; flush pointing as work proceeds			
two courses; projecting 50 mm from face of wall; horizontal	„	5·82	5·82

		Engineering bricks p.c. £32·60 100 £
Engineering bricks in cement mortar (1:3)	Unit	
Steps; all headers-on-edge; one angle rounded 53 mm radius; flush pointing top and one side as work proceeds		
one brick × half brick; horizontal; set weathering	m	8·15
Returned ends; flush pointing	No.	1·63
one and a half brick × half brick; horizontal; set weathering	m	10·90
Returned ends; flush pointing	No.	2·45

BRICKWORK AND BLOCKWORK

LABOURS ON BRICK FACEWORK

Facing bricks in any mortar		Unit	£
Fair cutting			
to curve		m	2·95
Fair cut angles			
squint		,,	5·17
birdsmouth		,,	1·70

BLOCKWORK

	Unit	75 mm thick £	90 mm thick £	100 mm thick £	140 mm thick £	190 mm thick £	215 mm thick £
Hollow clay blocks to B.S. 3921; keyed both sides; in cement–lime mortar (1 : 2 : 9)							
Walls or partitions . . .	m²	6·84	—	8·09	—	—	
Thermalite blocks; keyed both sides; in cement–lime mortar (1 : 2 : 9)							
Walls or partitions . . .	,,	6·85	7·77	8·26	10·85	13·85	
Thermalite blocks; smooth-faced both sides; in cement–lime mortar (1 : 2 : 9)							
Walls or partitions . . .	,,	7·90	8·90	9·40	12·20	16·95	19·00
Solid dense concrete masonry blocks in cement–lime mortar (1 : 2 : 9)							
Walls or partitions . . .	,,	—	12·50	—	17·15	22·95	
Labours on blockwork in any mortar							
Bonding ends to brickwork							
blockwork . . .	m	1·52	1·77	1·85	2·26	2·72	2·94
		£					
Rough cutting							
to chamfered angle . . .	,,	1·00					
Fair cutting							
to chamfered angle . . .	,,	2·17					

Extra over blockwork in any mortar for fair face; flush pointing as work proceeds	Unit	£
Walls or partitions	m²	0·74

DAMP-PROOF COURSES

Bitumen; hessian base; 200 mm laps		
Horizontal		
over 225 mm wide	,,	2·80
,, 225 ,, with cavity gutters in hollow walls	,,	5·30
Bitumen; hessian base and lead; 150 mm laps		
Horizontal		
over 225 mm wide	,,	5·08
,, 225 ,, with cavity gutters in hollow walls	,,	7·60
Pitch polymer; 150 mm laps		
Horizontal		
over 225 mm wide	,,	3·46
,, 225 ,. with cavity gutters in hollow walls	,,	6·00
Two courses slates		
Horizontal		
over 225 mm wide	,,	25·70
Vertical		
over 225 mm wide	,,	31·25

BRICKWORK AND BLOCKWORK

SUNDRIES	Unit	£
Forming cavities in hollow walls; two wall ties per m²		
50 mm wide	m²	0·54
Closing at jambs with half brick common brickwork	m	2·90
„ sills with one course common brickwork	„	2·90
24 s.w.g. (0·56 mm) asphaltum dipped expanded metal reinforcement in walls; 150 mm laps		
64 mm wide	„	0·24
Preparing to receive new walls		
tops of old one brick walls	„	1·45
Weather fillets in cement mortar (1 : 3)		
100 mm face width	„	1·43
Bedding in cement mortar (1 : 3)		
plates not exceeding 125 mm wide	„	0·31
wood frames or sills	„	0·31
wood frames or sills; pointing one side with cement mortar (1 : 3)	„	0·41
wood frames or sills; pointing both sides with cement mortar (1 : 3)	„	0·50
Wedging and pinning up to underside of old construction with slates in cement mortar		
one brick walls	„	10·90
Cutting grooves in brickwork or blockwork		
for water bars or the like	„	2·07
Raking out joint in brickwork or blockwork for turn-in edge of flashing		
horizontal; pointing with cement mortar (1 : 3)	„	1·03
stepped; pointing with cement mortar (1 : 3)	„	1·54
Raking out and enlarging joint in brickwork or blockwork for nib of asphalt		
horizontal; pointing with cement mortar (1 : 3)	„	1·54
Building in Module 100 range metal windows; cutting and pinning lugs; bedding in cement mortar; pointing one side with mastic (*see 'Metalwork' for supply of windows*)		
6F9, 600 mm wide × 900 mm high; lugs to brick jambs, concrete head, brick sill	No.	6·65
12F9, 1200 mm wide × 900 mm high; lugs to brick jambs, concrete head, brick sill	„	11·35
18F9, 1800 mm wide × 900 mm high; lugs to brick jambs, concrete head, brick sill	„	17·25
Building in 'Industrial type' metal windows to B.S. 1787; cutting and pinning lugs; bedding in cement mortar; pointing one side with mastic (*see 'Metalwork' for supply of windows*)		
SSF 23, 673 × 1445 mm; lugs to brick jambs, concrete head, brick sill	„	7·00
SSF 43, 1302 × 1445 mm; lugs to brick jambs, concrete head, brick sill	„	14·00
SSF 54, 1616 × 1911 mm; lugs to brick jambs, concrete head, brick sill	„	19·40
Building in metal door frames to B.S. 1245, profile A; building in lugs; bedding in cement mortar		
610 × 1981 mm; lugs to brick jambs	„	6·40
Building in metal door frames to B.S. 1245, profile A; building in lugs; bedding in cement mortar; pointing one side with mastic		
762 × 1981 mm; lugs to brick jambs	„	7·90
Mortices in brickwork; running with cement mortar (1 : 1)		
for 20 mm bolt; 75 mm deep	„	0·47
for 20 mm bolt; 150 mm deep	„	0·94
75 × 75 mm; 225 mm deep	„	0·61
75 × 75 mm; 300 mm deep	„	1·22
Forming openings; one ring arch over		
225 × 225 mm; one brick facing brickwork; making good facings both sides	„	8·95

BRICKWORK AND BLOCKWORK

SUNDRIES – *continued*

	Unit	230 × 75 mm £	230 × 150 mm £	230 × 230 mm £
Forming openings; slate lintel over; providing and building in red terracotta air brick; pointing with cement mortar one side	Unit			
one brick wall; rendering all round with cement mortar	No.	1·02	1·80	3·50
two half brick skins of hollow wall; sealing 50 mm cavity with slates in cement mortar . . .	„	1·80	3·10	5·25
Forming openings; slate lintel over; providing and building in galvanized cast-iron air brick B.S. 493 type A; pointing with cement mortar one side				
one brick wall; rendering all round with cement mortar	„	1·90	3·55	5·10
two half brick skins of hollow wall; sealing 50 mm cavity with slates in cement mortar . . .	„	2·65	4·80	6·85

	Unit	£
Parging and coring flues with lime mortar		
sectional area not exceeding 0·25 m²	m	3·00
Galvanized steel brick anchors; setting into anchor slot and building into joints of brickwork		
standard No. 14 gauge (2 mm) 102 mm projection	No.	0·26
Tile surround, preslabbed; 406 mm wide firebrick back, cast-iron stool bottom, black vitreous enamelled fret, *p.c.* £120·00		
1372 × 864 × 152 mm; assembling, setting and pointing firebrick back in fireclay mortar and backing with fine concrete finished to splay at top, laying hearth tiles in cement mortar and pointing in white cement . . .	„	161·00
Closed stove, vitreous enamelled finish, *p.c.* £130·00		
571 × 606 × 267 mm; setting in position	„	152·00
Closed stove, vitreous enamelled finish, fitted with mild steel barffed boiler *p.c.* £170·00		
571 × 606 × 267 mm; setting in position	„	196·00

BRICKWORK AND BLOCKWORK

BUILDER'S WORK IN CONNECTION WITH PLUMBING AND ENGINEERING INSTALLATIONS

	Unit	£
Cutting chases in brickwork		
for one pipe; small; vertical	m	2·57
for one pipe; large; vertical	,,	4·55
Cutting and pinning to brickwork or blockwork; ends of supports		
for pipes not exceeding 55 mm bore	,,	2·00
for 100 mm diameter cast iron pipes	No.	3·15
radiator stays or brackets	,,	2·35

	Unit	Holes £	Holes; making good facings one side £
Holes for pipes or the like; small	No.	1·57	1·88
half brick wall	,,	2·76	3·07
one brick wall	,,	3·92	4·23
one and a half brick wall	,,	0·69	—
75 mm blockwork	,,	0·85	—
100 mm blockwork			
Holes for pipes or the like; large	,,	2·60	3·07
half brick wall	,,	4·55	5·02
one brick wall	,,	6·49	6·96
one and a half brick wall	,,	1·03	—
75 mm blockwork	,,	1·29	—
100 mm blockwork			

	Unit	£
Pits for underground valves, meters or the like; half brick thick walls in common bricks in cement mortar (1 : 3); bedding on normal concrete mix 21·00N/mm² 20 mm aggregate 100 mm thick		
100 × 100 mm × 750 mm high internally; holes for one small pipe; cast iron hinged box cover; bedding in cement mortar (1 : 3)	No.	18·85

CENTERING

To brickwork

	Unit	£
Flat soffits		
not exceeding 2 m span; over 300 mm wide	m²	11·40
,, 2 ,, 115 mm wide	m	1·63
,, 2 ,, 225 ,,	,,	2·64
Semi-circular arches		
1·50 m span, 115 mm wide	No.	10·70

UNDERPINNING

The items which follow are for work carried out in short lengths. Prices are not given for temporary supports which are dependent on the particular circumstances in which the work is being carried out.

	Unit	Excavating only £	Excavating and filling working space £
EXCAVATION			
Excavating			
Preliminary trenches to level of base of existing foundations; starting from surface level			
not exceeding 1 m deep	m³	8·20	11·29
,, ,, 2 m deep	,,	10·90	14·00
,, ,, 4 m deep	,,	13·55	16·65
Trenches to receive foundations below level of base of existing foundations; starting 2 m below surface level			
not exceeding 0·25 m deep	,,	13·40	16·50
,, ,, 1 m deep	,,	15·60	18·70
,, ,, 2 m deep	,,	17·85	20·95

	Unit	£
Cutting away projecting plain concrete foundations		
150 mm wide, 150 mm thick	m	0·70
150 ,, 225 ,,	,,	1·05
150 ,, 300 ,,	,,	1·40
Cutting away projecting common brick footings in any mortar		
one course high	,,	0·29
two courses high	,,	0·58
three courses high	,,	0·87
Prepare underside of existing construction; to receive pinning up of new		
750 mm wide	,,	3·37
1·50 m wide	,,	6·73
Disposal		
Surplus excavated material		
removing from site; to tip not exceeding 13 km from site	m³	10·25
Filling		
Excavated material		
backfilling into excavations; depositing and compacting in layers	,,	3·09

	Unit	£
EARTHWORK SUPPORT		
Sides of preliminary excavation; next to existing building		
not exceeding 1 m deep	m²	2·58
,, ,, 2 m deep	,,	3·25
,, ,, 4 m deep	,,	3·95
over 4 m not exceeding 6 m deep	,,	4·65
Sides of excavation below level of base of existing foundations		
Starting 2 m below surface level; not exceeding 2 m between opposing faces		
not exceeding 1 m deep	,,	3·55
,, ,, 2 m deep	,,	4·48
,, ,, 4 m deep	,,	5·30
over 4 m not exceeding 6 m deep	,,	6·25

UNDERPINNING

	Unit	£
IN SITU CONCRETE; PLAIN		
Normal; mix 11·50 N/mm² 40 mm aggregate		
Foundations in trenches		
over 300 mm thick	m³	46·00

FORMWORK TO PLAIN IN SITU CONCRETE		
Formwork generally		
Edges and faces		
foundations, exceeding 1 m high	m²	12·27

BRICKWORK		
Common bricks *p.c.* £45·00 per 1000 in cement mortar (1 : 3)		
One brick thick		
walls	"	24·65
Two brick thick		
walls	"	43·40

SUNDRIES		
Wedging and pinning up to underside of old construction with slates in		
cement mortar (1 : 3)		
one brick walls	m	11·70
one and a half brick walls	"	17·55
two brick walls	"	23·40

MASONRY

The following prices are for Portland Limestone. Approximate prices for other kinds of stone can be obtained by using the prices for Portland Limestone varied in accordance with the following table:

Bath	*Doulting*
Deduct 5%	As Portland Limestone

NATURAL STONEWORK

Portland Limestone; bedding and jointing in masons mortar; flush pointing as work proceeds; slurrying with weak lime mortar and cleaning down on completion (wall ties, anchors, dowels, cramps, mortices, cutting and pinning measured separately)

	Unit	£
Facework built against brickwork backing		
50 mm thick stones; one face plain and rubbed	m²	73·40
75 mm thick stones; one face plain and rubbed	„	86·00
100 mm thick stones; one face plain and rubbed	„	99·30
Slab surrounds to openings built against brickwork backing		
175 mm wide on face, 75 mm thick; one face splayed; rubbed	m	24·30
200 mm wide on face, 100 mm thick; one face splayed; rubbed	„	29·00
225 mm wide on face, 125 mm thick; one face splayed; rubbed	„	32·00
200 mm wide on face, 75 mm thick; one face sunk splayed; rubbed	„	26·25
250 mm wide on face, 75 mm thick; one face sunk splayed; rubbed	„	29·80
300 mm wide on face, 75 mm thick; one face sunk splayed; rubbed	„	33·50
Grooving rubbed face	„	3·20
Throating rubbed face	„	2·00
Rebating rubbed face; 50 mm girth		
Extra; stoolings plain and rubbed	No.	5·00
Moulded surrounds to openings built against brickwork backing		
225 mm wide on face, 125 mm thick; moulded on face; rubbed	m	46·75
250 mm wide on face, 150 mm thick; moulded on face; rubbed	„	61·60
300 mm wide on face, 150 mm thick; moulded on face; rubbed	„	63·30
Extra; internal angle	No.	12·40
Extra; external angle	„	10·00
Copings; horizontal or raking		
300 × 50 mm; one face weathered; once throated; rubbed	m	22·60
300 × 75 mm; one face weathered; once throated; rubbed	„	27·45
375 × 100 mm; one face weathered; once throated; rubbed	„	40·30
Extra; internal angle	No.	8·85
Extra; external angle	„	8·40
Mortices in limestone masonry for		
Metal dowel	„	0·24
Metal cramp	„	0·70

ASPHALT WORK

		B.S. 1097 £	B.S. 1418 £
DAMP-PROOFING AND TANKING	*Unit*		
Mastic asphalt			
13 mm one coat coverings to brickwork (or concrete) base; flat or to falls or slopes not exceeding 10 degrees from horizontal			
over 300 mm wide; subsequently covered	m²	4·00	5·67
over 150 mm not exceeding 300 mm wide; subsequently covered	m	2·00	2·64
13 mm two coat coverings to brickwork (or concrete) base; vertical or sloping over 45 degrees from horizontal			
over 300 mm wide; subsequently covered	m²	12·35	14·06
over 150 mm not exceeding 300 mm wide; subsequently covered	m	4·12	4·75
20 mm two coat coverings to brickwork (or concrete) base; flat or to falls or slopes not exceeding 10 degrees from horizontal			
over 300 mm wide; subsequently covered	m²	5·00	7·19
over 150 mm not exceeding 300 mm wide; subsequently covered	m	6·88	3·52
20 mm three coat coverings to brickwork (or concrete) base; vertical or sloping over 45 degrees from horizontal			
over 300 mm wide; subsequently covered	m²	15·80	18·00
over 150 mm not exceeding 300 mm wide; subsequently covered	m	5·16	5·90
30 mm three coat coverings to brickwork (or concrete) base; flat or to falls or slopes not exceeding 10 degrees from horizontal			
over 300 mm wide; subsequently covered	m²	7·00	10·20
over 150 mm not exceeding 300 mm wide; subsequently covered	m	3·34	4·40
Rounded edges	"	0·72	0·72
Internal angle fillets in two coats; subsequently covered	"	1·36	1·76

		B.S. 1076 £	B.S. 1410 £
PAVING AND SUB-FLOORS			
Mastic asphalt; black			
20 mm one coat coverings to concrete base; flat or to falls or slopes not exceeding 10 degrees from horizontal			
over 300 mm wide	m²	4·60	7·05
" 150 " not exceeding 300 mm wide	m	2·68	3·52
25 mm one coat coverings to concrete base; flat or to falls or slopes not exceeding 10 degrees from horizontal			
over 300 mm wide	m²	5·90	8·75
" 150 " not exceeding 300 mm wide	m	3·43	4·40
20 mm one coat skirtings to brickwork base			
150 mm wide on face; one fair edge; one coved internal angle	"	3·60	4·15
Angles	No.	0·10	0·10
Mastic asphalt; acid resisting			
20 mm one-coat coverings to concrete base; flat or to falls or slopes not exceeding 10 degrees from horizontal			
over 300 mm wide	m²	5·60	—
" 150 " not exceeding 300 mm wide	m	3·20	—
35 mm one coat coverings to concrete base; flat or to falls or slopes not exceeding 10 degrees from horizontal			
over 300 mm wide	m²	8·43	—
" 150 " wide not exceeding 300 mm wide	m	4·90	—
Extra; felt underlay	m²	0·64	—

ASPHALT WORK

	Unit	B.S. 1451 Brown £	B.S. 1451 Red £
PAVING AND SUB-FLOORS – *continued*			
Mastic asphalt; coloured			
15 mm one-coat coverings to concrete base; flat or to falls or slopes not exceeding 10 degrees from horizontal			
over 300 mm wide	m²	5·00	5·36
„ 150 „ not exceeding 300 mm wide	m	2·44	2·67
15 mm one-coat skirtings to brickwork base			
75 mm wide on face; one fair edge; one coved internal angle	„	3·53	3·72
Angles	No.	0·22	0·22

	Unit	B.S. 988 £	B.S. 1162 £
ROOFING			
Mastic asphalt			
20 mm two-coat coverings; felt isolating membrane; to concrete (or timber) base; flat or to falls or slopes not exceeding 10 degrees from horizontal			
over 300 mm wide	m²	5·00	6·94
over 150 mm not exceeding 300 mm wide	m	2·88	4·00
Extra; covering with 10 mm limestone chippings in hot bitumen	m²	1·00	1·00
Extra; covering with solar reflective paint . . .	„	1·10	1·10
Extra; covering with 300 × 300 × 8 mm asbestos cement tiles in hot bitumen	„	11·15	11·15
Extra; working into shallow channels; 150 mm girth on face; two arrises; to falls	m	2·45	2·45
Extra; working into shallow channels; 250 mm girth on face; two arrises; to falls	„	3·06	3·06
Cutting to line; jointing to old asphalt	„	1·01	1·01
Working to flashings	„	1·01	1·01
Working into outlet pipes, dishing to gullies or the like .	No.	6·05	6·05
13 mm two-coat fascias to concrete base			
150 mm wide on face; one drip; one water check roll . .	m	4·15	4·40
13 mm two-coat skirtings to brickwork base			
average 150 mm wide on face; one internal angle fillet, one turning into groove	„	2·88	3·30
13 mm two-coat skirtings; expanded metal lathing reinforcement nailed to timber base			
average 150 mm wide on face; one internal angle fillet, one turning into groove	„	3·90	4·50
Angles	No.	0·11	0·11
13 mm two-coat linings to gutter of slated roof; expanded metal lathing reinforcement; nailing lathing to timber base			
average 600 mm girth on face; two arrises, one internal angle fillet; one dressing over tilting fillet; to falls . . .	m	11·40	13·50
13 mm two-coat linings to bottom and sides of opening; to brickwork base			
250 × 150 mm through one brick wall; fair edges, internal angle fillets, working to lead apron	No.	7·20	7·44
13 mm two-coat collars to metal base			
100 mm high around small pipes or the like; internal angle fillet at junction with covering; fair edge and arris . .	„	3·50	3·50
100 mm high around large pipes or the like; internal angle fillet at junction with covering; fair edge and arris . .	„	4·12	4·12

ROOFING

		500 × 250 mm × 75 mm lap £	600 × 300 mm × 100 mm lap £	600 × 350 mm × 100 mm lap £
SLATE ROOFING	*Unit*			
Asbestos-cement slates; Turnall blue or similar; uniform size; 19 × 50 mm softwood battens fixed with galvanized nails				
Coverings; fixing with two copper nails and one copper disc-rivet per slate				
sloping not exceeding 50° from horizontal . .	m²	11·42	10·32	12·08
vertical or sloping over 50° from horizontal . .	,,	12·08	10·80	12·47
Square cutting to large openings . . .	m	1·71	1·55	1·81
Raking cutting	,,	2·57	2·32	2·72
Eaves				
double course	,,	3·43	3·10	3·62
Verges				
generally; extra single undercloak course; bedding and pointing in cement mortar (1 : 4) . .	,,	4·28	3·87	4·53
Valleys				
cutting	,,	2·57	2·32	2·72
Hips				
cutting prior to covering	,,	2·57	2·32	2·72
Sundries				
holes for pipes or the like; small . . .	No.	0·82	0·82	0·82
,, ,, ,, large . . .	,,	1·65	1·65	1·65

		305 × 255mm £	405 × 255 mm £	510 × 255 mm £	610 × 305 mm £
Natural slates; B.S. 680; Welsh blue; uniform size; 19 × 38 mm softwood battens fixed with galvanized nails	*Unit*				
Coverings; centre fixing with two copper nails per slate; 76 mm lap					
sloping not exceeding 50° from horizontal	m²	17·60	16·05	18·50	19·10
vertical or sloping over 50° from horizontal . .	,,	18·85	16·90	19·16	19·50
Square cutting to large openings .	m	2·64	2·41	2·78	2·86
Raking cutting	,,	3·96	3·61	4·16	4·29
Eaves					
double course	,,	5·28	4·82	5·55	5·72
Verges					
generally; extra single undercloak course; bedding and pointing in cement mortar (1 : 4)	,,	6·60	6·02	6·94	7·15
Valleys					
cutting	,,	3·96	3·61	4·16	4·29
Hips					
cutting prior to covering . . .	,,	3·96	3·61	4·16	3·29
Sundries					
holes for pipes or the like: small . .	No.	0·82	0·82	0·82	0·82
,, ,, ,, large . .	,,	1·65	1·65	1·65	1·65

ROOFING

SLATE ROOFING – *continued*		305 × 255 *mm*	405 × 255 *mm*	510 × 255 *mm*	610 × 305 *mm*
Fittings to slated roofs	Unit	£	£	£	£
Cappings to hips (or ridges) red terracotta; 152 × 152 mm plain angle; butt jointed; bedding and pointing in cement mortar (1 : 4)	m	5·00	5·00	5·00	5·00
Mitred angles	No.	1·65	1·65	1·65	1·65
„ intersections . . .	„	3·29	3·29	3·29	3·29
Fixing					
lead soakers	„	0·14	0·14	0·14	0·14

	Unit	£
Natural slates; Westmorland green; random lengths 457–229 mm proportionate widths; 50 × 25 mm softwood battens fixed with galvanized nails		
Coverings; centre fixing with two copper nails per slate; 75 mm lap; in diminishing courses		
sloping not exceeding 50° from horizontal	m²	31·60
vertical or sloping over 50° from horizontal	„	33·20
Square cutting to large openings	m	4·75
Raking cutting	„	7·10
Eaves		
double course	„	9·50
Verges		
generally; extra single undercloak course, slates 152 mm wide; bedding and pointing in cement mortar (1 : 4)	„	11·85
Valleys		
cutting	„	7·10
Hips		
cutting prior to covering	„	7·10
mitreing	„	10·70
Sundries		
holes for pipes or the like; small	No.	1·00
„ „ „ large	„	2·00
Fittings to slated roofs		
Cappings to ridges		
152 × 152 mm matching colour; plain angle; butt jointed; bedding and pointing in cement mortar (1 : 4)	m	9·80
mitred angles	No.	1·65
Fixing		
lead soakers	„	0·14

ROOFING

TILE ROOFING Plain tiles; 267 × 165 mm; 19 × 38 mm softwood battens fixed with galvanized nails	Unit	Concrete tiles; B.S. 473 £	Broseley or Staffordshire machine-made tiles; B.S. 402 £	Best hand-made sand-faced tiles; B.S. 402 £
Coverings; hanging by nibs; fixing every fourth course with two aluminium nails per tile; to 64 mm lap				
sloping not exceeding 50°; from horizontal	m²	13·25	13·95	21·15
Coverings; hanging by nibs; fixing every tile with two aluminium nails; to 64 mm lap				
sloping over 50°; from horizontal	,,	15·50	16·20	23·40
Coverings; hanging by nibs; fixing every tile with two aluminium nails; to 38 mm lap				
vertical	,,	13·70	14·30	20·65
Square cutting to large openings	m	2·00	2·10	3·17
Raking cutting	,,	3·00	3·14	4·75
Eaves				
double course	,,	3·97	4·18	6·34
Verges				
generally; extra single undercloak course; bedding and pointing in cement mortar (1 : 4)	,,	4·96	5·23	7·93
Valleys				
cutting	,,	3·00	3·14	4·75
extra; valley tiles	,,	8·16	12·72	13·93
Hips				
cutting prior to covering	,,	2·98	3·14	4·75
extra; bonnet hip tiles	,,	8·39	12·95	14·15
Vertical angles; external; supplementary fixing with aluminium nails				
extra; 90° angle tiles	,,	3·93	6·27	7·55
Sundries				
holes for pipes or the like; small	No.	0·82	0·82	0·82
,, ,, ,, large	,,	1·65	1·65	1·65
Fittings to tiled roofs				
Cappings to ridges				
half round; butt jointed; bedding and pointing in cement mortar (1 : 4)	m	5·80	6·13	7·00
Mitred angles	No.	1·65	1·65	1·65
,, intersections	,,	3·29	3·29	3·29
Fixing				
lead soakers	,,	0·14	0·14	0·14

ROOFING

	Red plain Lincolnshire pantiles 337 × 241 mm; to 76 mm head and 38 mm side laps

TILE ROOFING – *continued*
Pantiles; 19 × 38 mm softwood battens fixed with galvanized nails

	Unit	£
Coverings; hanging by nibs; fixing every second course with two aluminium nails per tile		
sloping not exceeding 50° from horizontal	m²	9·85
Square cutting to large openings	m	1·48
Raking cutting	„	2·22
Eaves; supplementary fixing with aluminium nails		
extra undercourse plain tiles 267 mm wide	„	2·95
Verges		
generally; extra single undercloak course, plain tiles; bedding and pointing in cement mortar (1 : 4)	„	3·70
Valleys		
cutting; extra single undercloak course; bedding and pointing in cement mortar (1 : 4)	„	5·10
Hips		
cutting prior to covering	„	2·22
extra; hip tiles	„	7·00
Sundries		
holes for pipes or the like; small	No.	0·82
„ „ „ large	„	1·65

Fittings to tiled roofs

	Unit	£
Cappings to ridges		
half round; butt jointed; bedding and pointing in cement mortar (1 : 4)	m	7·00
Mitred angles	No.	1·65
„ intersections	„	3·29

ROOFING

	Concrete tiles; B.S. 550 type A; 413 × 330 mm; fixing every second course with two aluminium nails per tile	Concrete Modern tiles; 413 × 330 mm; fixing every course with one aluminium clip per tile
Unit	£	£

TILE ROOFING – *continued*
Interlocking tiles; 19 × 38 mm softwood battens fixed with galvanized nails

	Unit	£	£
Coverings; hanging by nibs			
sloping not exceeding 50° from horizontal	m²	4·75	5·00
Square cutting to large openings	m	0·71	0·75
Raking cutting	„	1·07	1·12
Verges			
generally; extra single undercloak course plain tiles; bedding and pointing in cement mortar (1:4)	„	0·79	1·04
Valleys			
cutting	„	1·07	1·12
extra; valley tiles	„	5·39	5·39
Hips			
cutting prior to covering	„	1·07	1·12
extra; hip tiles	„	3·67	3·67
Sundries			
holes for pipes or the like; small	No.	0·82	0·82
„ „ „ large	„	1·65	1·65
Fittings to tiled roofs			
Cappings to ridges			
half round; butt jointed; bedding and pointing in cement mortar (1:4)	m	4·45	6·00
Mitred angles	No.	1·65	1·65
„ intersections	„	3·30	3·30

	Unit	£
SLATE AND TILE ROOFING		
Underfelting; B.S. 747 type 1B bitumen felt weighing 14 kg/10 m²		
Fixing with galvanized clout nails; 75 mm laps		
to sloping or vertical surfaces	m²	1·40
Underfelting: B.S. 747 type 1F reinforced felt weighing 22·5 kg/15 m²		
Fixing with galvanized clout nails; 75 mm laps		
to sloping or vertical surfaces	„	1·85

ROOFING

		Big Six sheeting £	Big Six insulated sheeting incorporating lining panels and 25 mm glass fibre infill £
CORRUGATED OR TROUGHED SHEET ROOFING	Unit		
Asbestos cement; natural both sides			
Coverings; fixing to timber purlins at 1375 mm general spacing with galvanized steel drive screws with waterproof covers and washers; 150 mm end laps and one-corrugation side laps bedded in bitumen mastic			
sloping not exceeding 50° from horizontal	m²	7·05	13·80
Coverings; fixing to steel purlins at 1375 mm general spacing with galvanized hook bolts with waterproof covers and washers; 150 mm end laps and one corrugation side laps bedded in bitumen mastic			
sloping not exceeding 50° from horizontal	„	7·65	14·40
Coverings; fixing to timber rails at 1825 mm general spacing with galvanized drive screws with waterproof covers and washers; 150 mm end laps and one-corrugation side laps bedded in bitumen mastic			
vertical	„	7·50	14·55
Square cutting around large openings	m	1·25	2·50
Raking cutting	„	1·87	3·75
Curved „	„	2·28	4·65
		£	
Accessories; fixing with galvanized steel seam bolts with nuts and washers; bedding and pointing in bitumen mastic			
filler pieces; at eaves	„	3·86	
closure pieces; at jambs	„	4·75	
flashing pieces	„	3·60	
flashing pieces; below glazing	„	4·27	
cappings, two piece; to ridges	„	6·80	
barge boards; to verges	„	4·50	
corner pieces; to vertical angles	„	4·40	
Sundries			
holes for pipes or the like; small	No.	0·82	
„ „ „ large	„	1·65	

ROOFING

	Unit	£

ROOF DECKING

Galvanized steel; 0·71 mm; natural soffit; fibre board lined top surface; felt roofing; stone chippings

Coverings; fixing to timber, steel or concrete; flat or sloping not exceeding 50° from horizontal; 13 mm fibre board; three layers bitumen bonded 17 kg/10m roll felt; hot bitumen bonding compound; 10 mm stone chippings in hot bitumen

supports at 2 m general spacing	m²	19.00
„ 3 m „ „	„	20·00
„ 3·50 m „ „	„	20·70
„ 4·25 m „ „	„	22·00

Aluminium; 0·90 mm; natural soffit; fibre board lined top surface; felt roofing; stone chippings

Coverings; fixing to timber, steel or concrete; flat or sloping not exceeding 50° from horizontal; 13 mm fibre board; three layers bitumen bonded 17 kg/10 m roll felt; hot bitumen bonding compound; 10 mm stone chippings in hot bitumen

supports at 1·50 m general spacing	„	23·00
„ 2·50 m „ „	„	25·00
„ 3 m „ „	„	26·00
„ 3·50 m „ „	„	28·00
Accessories in aluminium: fixing with rivets		
water check curb and fascia	m	5·66
fascia	„	6·00

Wood wool; B.S. 1105 type B; natural soffit; natural top surface

Coverings; fixing to timber or steel with galvanized nails or clips; flat or sloping not exceeding 50° from horizontal

nominal thickness 50 mm; supports at 910 mm general spacing	m²	5·85
„ „ 75 mm; „ 1220 „ „	„	7·20
Extra; pre-screeded and proofed deck	„	1·10
Extra; pre-felted deck	„	1·66

Wood wool; interlocking reinforced high density; natural soffit; natural top surface

Coverings; fixing to steel with clips; flat or sloping not exceeding 50° from horizontal

nominal thickness 50 mm; supports at 2030 mm general spacing	„	11·70
„ „ 75 mm; „ „ „	„	17·65 ~ 17·65
Extra; pre-screeded deck and soffit.	„	3·15
Extra; pre-screeded deck and proofed deck; pre-screeded soffit.	„	4·10

ROOFING

	Unit	£

BITUMEN FELT ROOFING

Felt; B.S. 747; type 2B fine granule surface asbestos base 18 kg/10 m²; hot bitumen bonding compound

Three-layer coverings; 75 mm laps; overall bonding between layers and first layer to concrete base; flat or to falls or cross-falls or sloping not exceeding 10° from horizontal

    over 300 mm wide .   .   .   .   m²   5·32

Felt; B.S. 747; first layer type 3G venting base layer glass fibre base 32 kg/10 m²; second and third layers type 3B fine granule surface glass fibre base 18 kg/10 m²; hot bitumen bonding compound

Three-layer coverings; 75 mm laps; overall bonding between layers and first layer to concrete base; flat or to falls or cross-falls or sloping not exceeding 10° from horizontal

    over 300 mm wide . . . .   ,,   5·00

Three-layer skirtings; overall bonding between layers and first layer to brickwork base

    150 mm width in contact with base; one dressing over tilting fillet; one turning into groove . . . .   m   2·40

    225 mm width in contact with base; one dressing over tilting fillet; one turning into groove . . . .   ,,   2·52

Three-layer collars; overall bonding between layers and first layer to metal base

    150 mm high around small pipes or the like; hole in three-layer roofing .   No.   2·75

    ,,   ,,   large ,,   ,,   ,,   ,,   ,,   ,,   3·41

Felt; B.S. 747; first and second layers type 1B fine granule surface fibre base 18 kg/10 m²; third layer type 1B fine granule surface fibre base 25 kg/10 m²; hot bitumen bonding compound

Three-layer coverings; 75 mm laps; overall bonding between layers and first layer to dry insulation base; flat or to falls or cross-falls or sloping not exceeding 10° from horizontal

    over 300 mm wide . . . .   m²   4·60

    Extra covering with:

        13 mm granite chippings in hot bitumen . . .   ,,   1·68

        300 × 300 × 25 mm concrete tiles in hot bitumen . .   ,,   9·25

        300 × 300 × 8 mm asbestos-cement tiles in hot bitumen .   ,,   13·50

Felt; B.S. 747; first layer type 2B fine granule surfaced asbestos base 18 kg/10 m²; second layer type 2E mineral surfaced asbestos base; hot bitumen bonding compound; galvanized clout nails

Two-layer coverings; 75 mm laps; overall bonding between layers and nailing first layer to timber base; sloping over 10° not exceeding 50° from horizontal

    over 300 mm wide . . . .   ,,   4·60

Felt; B.S. 747; first and second layer type 2B fine granule surfaced asbestos base 16 kg/10 m²; third layer type 2E mineral surfaced asbestos base; hot bitumen bonding compound; galvanized clout nails

Three-layer coverings; 75 mm laps; overall bonding between layers and nailing first layer to timber base; sloping over 10° not exceeding 50° from horizontal

    over 300 mm wide . . . .   ,,   6·35

ROOFING

SHEET METAL ROOFING

	Unit	£
Milled lead; B.S. 1178		
No. 6 roof coverings; fixing with No. 5 milled lead tacks nailed; copper nails		
flat or to falls or cross-falls or slopes not exceeding 10° from horizontal .	m²	64·50
Dressing over glass and glazing bars	m	2·10
Welting edges	,,	1·30
Soldered angles	,,	12·60
,, seams	,,	12·60
Copper nailing	,,	0·90
Bossed ends to rolls	No.	1·95
No. 5 roof coverings; fixing with No. 4 milled lead tacks nailed; copper nails		
sloping over 10° not exceeding 50° from horizontal	m²	55·50
No. 4 roof coverings; fixing with No. 4 milled lead tacks nailed; copper nails		
vertical or sloping over 50° from horizontal	,,	46·20
No. 5 gutter coverings; fixing with No. 4 milled lead tacks nailed; copper nails		
flat	,,	55·50
Zinc; B.S. 849		
0·81 mm roof coverings; fixing with 0·81 mm zinc clips nailed; galvanized nails		
flat or to falls or cross-falls or slopes not exceeding 10° from horizontal .	,,	17·40
sloping over 10° not exceeding 50° from horizontal	,,	17·40
vertical or sloping over 50° from horizontal	,,	17·40
Dressing over glass and glazing bars	m	1·30
Welting edges	,,	0·40
Beading edges	,,	0·40
0·81 mm gutter coverings; fixing with 0·81 mm zinc clips nailed; galvanized nails		
flat	m²	18·60
Aluminium		
0·90 mm commercial grade roof coverings; fixing with 0·90 mm clips nailed; aluminium nails		
flat or to falls or cross-falls or slopes not exceeding 10° from horizontal .	,,	17·20
sloping over 10° not exceeding 50° from horizontal	,,	17·20
vertical or sloping over 50° from horizontal	,,	17·20
Welting edges	m	0·40

		0·56 mm	0·61 mm
Copper; B.S. 2870	Unit	£	£
Roof coverings; fixing with clips nailed; copper nails			
flat or to falls or cross-falls or slopes not exceeding 10° from horizontal	m²	25·65	27·75
sloping over 10° not exceeding 50° from horizontal . .	,,	25·65	27·75
vertical or sloping over 50° from horizontal . . .	,,	25·65	27·75
Welting edges	m	0·40	0·40

ROOFING

SHEET METAL FLASHINGS   *Unit*   £
Milled lead; B.S. 1178

No. 4 flashings; fixing with No. 4 milled lead tacks
  150 mm girth; 100 mm laps not allowed in girth; once wedging into groove;
  lead wedges .   .   .   .   m   6·55
No. 4 stepped flashings; fixing with No. 4 milled lead tacks
  150 mm girth; 100 mm laps not allowed in girth; once wedging into groove;
  lead wedges .   .   .   .   „   7·55

Zinc; B.S. 849

0·81 mm flashings: fixing with 0·81 mm zinc clips
  150 mm girth; 100 mm laps not allowed in girth; once wedging into groove;
  zinc wedges .   .   .   .   „   3·80
0·81 mm stepped flashings; fixing with 0·81 mm zinc clips
  225 mm girth; 100 mm laps not allowed in girth; once wedging into groove;
  zinc wedges .   .   .   .   „   5·25

Aluminium

0·90 mm commercial quality stepped flashings; fixing with 0·90 mm clips
  300 mm girth; 100 mm laps not allowed in girth; once wedging into groove;
  oak wedges .   .   .   .   „   4·85

		0·56 mm	0·61 mm
Copper; B.S. 2870	*Unit*	£	£
Flashings; fixing with clips			
150 mm girth; 100 mm laps not allowed in girth; once wedging			
into groove; copper wedges	m	4·40	4·70
Stepped flashings; fixing with clips			
225 mm girth; 100 mm laps not allowed in girth; once wedging			
into groove; copper wedges	„	6·15	6·60

WOODWORK

+15% *30%*
+15%
+15%.

	Unit	Softwood £
ARCASSING 145%		+. Bin
oors		
50 × 100 mm	m	1·14 (0·60) 0·79
50 × 125 mm	„	1·43
50 × 150 mm	„	1·71 — 145%
50 × 175 mm	„	2·00 +
50 × 200 mm	„	2·28 (+·20)
50 × 225 mm	„	2·57
rtitions 200%		
50 × 50 mm	„	0·68
50 × 75 mm	„	1·01
50 × 100 mm	„	1·35
75 × 100 mm	„	2·03
100 × 100 mm	„	2·70
at roofs 155%		
50 × 100 mm	„	1·14
50 × 150 mm	„	1·71
50 × 200 mm	„	2·28
50 × 225 mm	„	2·57
tched roofs 175%		
50 × 50 mm	„	0·61
50 × 75 mm	„	0·91
50 × 100 mm	„	1·22
50 × 125 mm	„	1·52
50 × 150 mm	„	1·83
50 × 175 mm	„	2·13
50 × 200 mm	„	2·44 +·20·
50 × 225 mm	„	2·74
erbs, bearers, plates and the like 130%		
50 × 50 mm	„	0·45
50 × 75 mm	„	0·68
50 × 100 mm	„	0·91 0·60
75 × 100 mm	„	1·36
erbs, bearers, plates and the like; fixing by bolting 178%		
50 × 50 mm	„	0·63
50 × 75 mm	„	0·95
50 × 100 mm	„	1·26
75 × 100 mm	„	1·89
erringbone strutting		
50 × 38 mm between 225 mm deep joists	„	2·65
lid strutting		
50 × 150	„	2·28 103%
50 × 200	„	2·66
rming openings		
1·50 × 0·50 m; trimming 200 × 50 mm joists to two short sides	No.	6·27
2·50 × 1 m; trimming 300 × 100 mm joists to two short sides	„	18·80
eats		
25 × 100 × 75 mm	„	1·32
rought faces		
generally	m²	2·04
otching and fitting timber to metal	No.	0·53
e-treatment of timber by vacuum/pressure impregnation with Celcure pre- vative against fungal decay and insect attack, prices are for impregnation ly and exclude cost of transport to and from plant, any necessary seasoning d treatment subsequent to impregnation		
timbers for interior work, minimum nett dry salt retention 4 kg/m³	m³	12·50
„ exterior „ „ „ „ 5·30 kg/m³	„	15·12
e-treatment as above but Rentokil flame proof		
timbers for all purposes, minimum nett dry salt retention 36 kg/m³	„	47·15

Handwritten: 64·66, 113·44

WOODWORK

	Unit	Butt joints £	Tongued and grooved joints £
FIRST FIXINGS			
Boardings and flooring – softwood			
Board flooring; wrought			
19mm thick; 150 mm widths	m²	5·55	6·27
15mm „ ; ×175 m wredths laid diagonally	„	6·00	6·70
22 mm thick; 150 mm widths	„	6·14	6·90
25 mm „ ; „ „	„	6·74	7·53
Matchboarding; tongued and grooved joints; veed one side; wrought			
12 mm thick; 150 mm widths; to walls; vertical	„	—	6·11
19 mm „ ; „ „ ; „ ; „	„	—	7·15
19 mm „ ; „ „ ; to ceilings; horizontal	„	—	8·55
Roof-boarding (butt-jointed-sawn; tongued and grooved jointed-wrought)			
19 mm thick; flat to falls	„	5·55	6·27
25 mm „ ; „ „ „	„	6·14	6·90
19 mm thick; sloping not exceeding 50° from horizontal	„	6·05	6·95
25 mm „ ; „ „ „ „ „ „	„	6·65	7·55
Roof boarding; laid diagonally			
19 mm thick; flat to falls	„	6·00	6·70
19 mm „ ; sloping not exceeding 50° from horizontal	„	6·47	7·48
19 mm „ ; to tops and cheeks of dormers	„	7·52	8·75
Gutter boarding and sides			
19 mm thick; sloping to falls	„	9·07	9·81
25 mm „ ; „ „	„	9·67	10·44

	Unit	£
Extra 38 × 50 mm bearers at 300 mm centres	„	3·30
Eaves and verge boarding, fascias and barge boards; wrought		
19 mm thick; over 300 mm wide	„	10·95
19 × 150 mm wide; once grooved	m	1·22
25 × 150 mm wide; „ „	„	1·32
32 × 225 mm wide; „ „ ; moulded not exceeding 100 mm girth	„	4·00

	Unit	19 mm thick £	25 mm thick £
Raking cutting	„	0·64	0·69
Curved cutting	„	0·85	0·89
Tongued edges	„	0·92	0·92
Mitred angles	„	1·57	1·61

	Unit	Butt joints £	Tongued and grooved joints £
Boardings and flooring – chipboard			
Board flooring			
18 mm thick	m²	3·58	3·97
22 mm thick	„	4·26	4·72

	Unit	£
Firrings, drips and bearers – softwood		
Firring pieces		
50 × 38 mm (average) deep	m	1·35
50 × 75 mm („) „	„	1·73
Bearers		
38 × 50 mm	„	0·71
50 × 50 mm	„	0·93
50 × 75 mm	„	1·06
Nosing; wrought		
19 × 75 mm; once rounded	„	2·28
25 × 75 mm; „ „	„	2·34

WOODWORK

Fillets, rolls, grounds, battens and framework – softwood	Unit		£
Tilting fillet			
50 × 75 mm	m		0·90
125 × 150 mm	,,		2·14
Rolls			
50 × 75 mm	,,		1·29

	Unit	Nailed £	Plugged and screwed £
Grounds and battens			
25 × 50 mm; 300 mm centres one way	m²	1·89	4·10
13 × 19 mm	m	0·41	1·28
13 × 32 mm	,,	0·43	1·31
25 × 50 mm	,,	0·57	1·44
Framing			£
25 × 50 mm; 300 mm centres both ways	m²		4·05
25 × 50 mm	m		0·61
38 × 50 mm	,,		0·80
50 × 50 mm	,,		0·94
50 × 75 mm	,,		1·19
75 × 75 mm	,,		1·53

Softwood – treated			£
Floor fillets; in or on concrete			
50 × 50 mm	,,		0·64
Floor battens; resilient buffers; lapped and staggered on concrete to receive strip flooring			
38 × 50 mm	,,		3·27

+ 30 for conjecture
+ 33% ordinary

SECOND FIXINGS

Unframed second fixings – wrought softwood	Unit		£
Skirtings, picture rails, dado rails and the like			
19 × 45 mm; splayed once or moulded	,,		0·88
19 × 75 mm; ,, ,, ,, ,,	,,		1·18
25 × 75 mm; ,, ,, ,, ,,	,,		1·52
25 × 100 mm; ,, ,, ,, ,,	,,		1·74
Architraves, cover fillets and the like			
19 × 38 mm; splayed once or moulded	,,		0·89
19 × 75 mm; ,, ,, ,, ,,	,,		1·57
25 × 44 mm; ,, ,, ,, ,,	,,		1·13
25 × 75 mm; ,, ,, ,, ,,	,,		1·57
Stops; screwed on *+ 50p/m for screwing.*			
19 × 38 mm	,,		0·91
25 × 38 mm	,,		0·96
25 × 50 mm	,,		1·12
Glazing beads and the like *150%*			
13 × 19 mm	,,		0·52
13 × 25 mm	,,		0·59
13 × 25 mm; fixing with brass cups and screws	,,		0·76
19 × 36 mm	,,		1·42

	Unit	19 mm thick £	25 mm thick £	32 mm thick £
Isolated shelves, worktops, seats and the like				
150 mm wide	m	2·05	2·31	2·86
230 mm ,,	,,	3·06	3·46	4·30
300 mm wide; cross-tongued joints	,,	4·88	5·45	6·42
450 mm ,,	,,	6·53	7·39	8·84
600 mm ,, ; cross-tongued joints	,,	8·86	9·96	11·90
Shelves; slatted				
50 mm longitudinal slats at 75 mm centres	m²	14·05	15·75	17·75

WOODWORK

SECOND FIXINGS – *continued*
Unframed second fixings – wrought softwood – *continued*

	Unit	25 mm thick £	32 mm thick £	38 mm thick £
Window boards, nosings, bed moulds and the like				
75 mm wide; rebated once; rounded once . . .	m	2·20	2·33	2·60
150 mm „ ; „ „ ; „ „ . . .	„	3·39	3·71	4·27
225 mm „ ; „ „ ; „ „ . . .	„	4·58	5·42	5·98
300 mm „ ; „ „ ; „ „ . . .	„	6·08	7·02	7·87
Ends; returned fitted	No.	0·91	1·05	1·18

	Unit	Nailed £	Plugged and screwed £
Battens; beneath window board			
38 × 19 mm	m	1·51	3·17

	Unit	£
Handrails; rounded		
37 × 87 mm	„	3·26
50 × 75 mm	„	3·69
63 × 87 mm	„	4·95
75 × 100 mm	„	5·86
Handrails; moulded		
37 × 87 mm	„	3·57
50 × 75 mm	„	4·03
63 × 87 mm	„	5·46
75 × 100 mm	„	6·08

Sheet linings and casings

	Unit	6 mm thick £	9 mm thick £	12 mm thick £	18 mm thick £
Plywood					
Internal quality wall lining; plain finish for painting; butt jointed		*2·07 -*		*O.K. at Jan 81.*	
Over 300 mm wide	m²	5·41	6·37	(7·55)	9·90
Internal quality wall lining; West African Mahogany veneered; butt jointed; W.A.M. cover strips		*+19%*			
Over 300 mm wide	„	7·57	8·82	10·10	—

	Unit	12 mm thick £	15 mm thick £	18 mm thick £
Chipboard				
Internal quality wall lining; plain finish for painting; butt jointed				
Over 300 mm wide	m²	4·81	5·20	5·59
Internal quality wall lining; white matt melamine faced one side; balancer the other side; butt jointed; masking strips				
Over 300 mm wide	„	10·45	11·00	11·65
Internal quality lining to sides of isolated columns; white matt melamine faced one side; balancer the other side; butt jointed; aluminium cover strips				
Over 300 mm wide	„	15·63	16·35	18·00

	Unit	6 mm plywood £	18 mm plywood £	18 mm chipboard £	18 mm laminated chipboard £
Raking cutting	m	0·42	0·78	0·47	0·96
Curved cutting	„	0·55	0·94	0·62	1·28
Holes for pipes and the like; small . .	No.	0·19	0·25	0·25	0·38
„ „ „ „ „ „ ; large . .	„	0·28	0·38	0·38	0·56

WOODWORK

COMPOSITE ITEMS

	Unit	762 × 1981 *mm* £	838 × 1981 *mm* £
Doors – wrought softwood			
Matchboarded, ledged and braced doors; 25 mm ledges and braces; 19 mm tongued, grooved and vee jointed one side vertical boarding one side	No.	32·30	35·00
Matchboarded, framed, ledged and braced doors; 44 mm framing; 25 mm intermediate and bottom rails; 19 mm tongued, grooved and vee jointed one side vertical boarding as filling			
44 mm thick	,,	36·00	38·75
Panelled doors; one open panel for glass; including glazing beads			
44 mm thick	,,	29·00	29·60
Panelled doors; two open panels for glass; including glazing beads			
44 mm thick	,,	28·25	28·75
Panelled doors; six panels raised and fielded; mouldings worked on solid both sides			
44 mm thick	,,	—	122·00

	Unit	44 *mm* thick doors £	50 *mm* thick doors £
Rebates; rounded edges or heels	m	1·08	1·08

	Unit	762 *mm* wide door £	838 *mm* wide door £
Weatherboard; fixed to bottom rail	No.	2·50	2·55

	Unit	626 × 2040 *mm* £	726 × 2040 *mm* £	826 × 2040 *mm* £
Doors – softwood/composition				
Flush doors; skeleton or cellular core; hardboard facing both sides				
40 mm (f) thick	No.	14·60	14·70	14·80
Flush doors; skeleton or cellular core; plywood facing both sides; lipping on two long edges				
40 mm (f) thick	,,	20·15	20·25	20·35
Flush doors; solid core; Sapele veneered plywood facing both sides; matching hardwood lipping				
44 mm (f) thick	,,	38·60	38·75	38·85
Flush doors; half hour fire resistance; veneered plywood facing both sides; matching hardwood lipping				
44 mm (f) thick	,,	37·30	37·40	37·50
Flush doors; one hour fire resistance; plywood facing both sides; lipping on two long edges				
54 mm (f) thick	,,	68·50	68·65	68·75

WOODWORK

COMPOSITE ITEMS – *continued*
Door frames and lining sets – wrought softwood

	Unit	£
Jambs and heads		
32 × 63 mm	m	1·74
32 × 100 mm	,,	2·58
32 × 140 mm	,,	3·54
38 × 115 mm; rebated once; rounded once	,,	3·23
38 × 140 mm; ,, ,, ; ,, ,,	,,	3·92
50 × 100 mm; ,, ,, ; ,, ,,	,,	3·68
63 × 88 mm; ,, ,, ; ,, ,,	,,	4·09
63 × 100 mm; ,, ,, ; ,, ,,	,,	4·64
Mullions and transoms		
32 × 63 mm	,,	1·63
32 × 100 mm	,,	2·58
32 × 140 mm	,,	3·59
38 × 115 mm; rebated twice; rounded twice	,,	3·50
38 × 140 mm; ,, ,, ; ,, ,,	,,	4·26
50 × 100 mm; ,, ,, ; ,, ,,	,,	4·00
Sills		
63 × 100 mm; sunk weathered once; rebated once; grooved three times	,,	6·70
75 × 150 mm; ,, ,, ,, ; ,, ,, ; ,, ,, ,,	,,	7·30

Casements and frames, window surrounds, etc – 'treated' wrought softwood
(references refer to nearest equivalent Imperial code)

Casement windows without glazing bars; with 75 × 150 mm softwood sills; opening casements and ventilators hung on rust-proof hinges; fitted with aluminized laquered finish casement stays and fasteners; knotting and pining by manufacturer before delivery

			Unit	£
600 × 750 mm reference 126P			No.	18·70
600 × 900 mm	,,	136V	,,	19·00
1200 × 1050 mm	,,	236T	,,	36·90
1200 × 1200 mm	,,	240T	,,	38·65
1800 × 1050 mm	,,	336V	,,	52·75
1800 × 1200 mm	,,	340V	,,	55·35
1800 × 1350 mm	,,	346V	,,	59·30

Top hung windows; with 75 × 150 mm softwood sills; opening casements and ventilators hung on rust-proof hinges; fitted with aluminized laquered finish casement stays; knotting and priming by manufacturer before delivery

			Unit	£
600 × 750 mm reference TH126			,,	17·35
600 × 900 mm	,,	TH130	,,	18·65
1200 × 1050 mm	,,	TH236	,,	32·20
1200 × 1200 mm	,,	TH240	,,	34·00
1200 × 1350 mm	,,	TH246S	,,	36·00
1800 × 1050 mm	,,	TH336	,,	39·10
1800 × 1200 mm	,,	TH340	,,	41·00
1800 × 1350 mm	,,	TH346S	,,	44·75
2400 × 1500 mm	,,	TH450S	,,	53·70

WOODWORK

COMPOSITE ITEMS – *continued*

Casements and frames – *continued*

High performance pivot windows; with 75 × 150 mm softwood sills; adjustable ventilators; weather stripping; opening sashes and fanlights hung on rust-proof hinges; fitted with aluminized laquered espagnolette bolts; knotting and priming by manufacturer before delivery

	Unit	£
900 × 900 mm	No.	55·00
1200 × 1050 mm	,,	60·70
1200 × 1200 mm	,,	62·65
1800 × 1050 mm	,,	74·35
1800 × 1200 mm	,,	76·85
1800 × 1350 mm	,,	91·85
2400 × 1500 mm	,,	102·30

Modified B.S.S. double hung sash windows with glazing bars; 63 × 175 mm softwood sills; standard flush external linings; spiral spring balances and sash catch; knotting and priming by manufacturer before delivery

593 × 1096 mm	,,	80·65
821 × 1096 mm	,,	89·40
821 × 1401 mm	,,	98·40
1050 × 1401 mm	,,	110·10
1050 × 1706 mm	,,	120·50
1617 × 1401 mm	,,	209·75

Staircases – wrought softwood

Stairs; 25 mm treads with rounded nosings; 12 mm plywood risers; 32 mm strings once rounded; bullnose bottom tread; 50 × 75 mm hardwood handrail; two 32 × 140 mm balustrade knee rails; 32 × 50 mm stiffeners and 100 × 100 mm newel posts with hardwood newel caps on top

Straight flight; 838 mm wide; 2688 mm going; 2600 mm rise; with two newel posts	,,	149·00
Straight flight; 838 mm wide; 2600 mm rise; with two newel posts and three top treads winding	,,	208·00
Dogleg staircase; 838 mm wide; 2600 mm rise; quarter space landing third riser from top; with three newel posts	,,	169·00
As last but with half space landing; one 100 × 200 mm newel post; two 100 × 100 mm newel posts	,,	197·50

Stairs; 25 mm treads with rounded nosings; 12 mm plywood risers; 32 mm strings once rounded with string cappings, bullnose bottom tread, 50 × 75 mm hardwood handrail; two 32 × 32 mm balusters per tread and 100 × 100 mm newel post with hardwood newel caps on top

Straight flight; 838 mm wide 2688 mm going, 2600 mm rise with two newel posts	,,	158·00

Balustrades – wrought softwood

Landing balustrade; 50 × 75 mm hardwood handrail; three 32 × 140 mm balustrade knee rails; two 32 × 50 mm stiffeners; one end jointed to newel post; other end built into wall (newel post and mortices both measured separately)

3 m long	,,	64·85
Landing balustrade; 50 × 75 mm hardwood handrail; 32 × 32 mm balusters; one end of handrail jointed to newel post; other end built into wall; balusters housed in at bottom (newel post and mortices both measured separately)	,,	53·28

WOODWORK

COMPOSITE ITEMS – *continued*

Fittings – wrought softwood – supplied and fixed

	Unit	£
Backs, fronts or sides; cross-tongued joints		
25 mm thick; over 300 mm wide	m²	19·10
Divisions; cross-tongued joints		
25 mm thick; over 300 mm wide	,,	19·10
Work-tops; cross-tongued joints		
25 mm thick; over 300 mm wide	,,	19·10
Flush doors; softwood skeleton core; 4 mm plywood facing both sides, fixed with adhesive under hydraulic pressure; 13 mm softwood lippings all edges		
35 mm thick, size 450 × 750 mm	No.	13·50
Bearers		
38 × 19 mm	m	1·00
50 × 25 ,,	,,	1·31
50 × 50 ,,	,,	2·19
75 × 50 ,,	,,	2·70
Bearers; framed		
38 × 19 mm	,,	1·02
50 × 26 ,,	,,	1·42
50 × 50 ,,	,,	2·65
75 × 50 ,,	,,	3·56
Framing to backs, fronts or sides		
38 × 19 mm	,,	1·02
50 × 25 ,,	,,	1·42
50 × 50 ,,	,,	2·65
75 × 50 ,,	,,	3·58

Plywood; birch, grade 2	Unit	6 mm thick £	9 mm thick £	12 mm thick £
Backs, fronts or sides				
over 300 mm wide	m²	13·20	15·08	17·28
Divisions				
over 300 mm wide	,,	13·20	15·08	17·28

Blockboard; beech or gaboon	Unit	19 mm thick £	25 mm thick £
Shelves			
over 300 mm wide	m²	20·08	23·22
Backs, fronts or sides			
over 300 mm wide	,,	20·75	24·00
Divisions			
over 300 mm wide	,,	20·75	24·00
Work-tops			
over 300 mm wide	,,	20·75	24·00
Flush doors; 13 mm hardwood lipping all edges			
450 × 750 mm	No.	8·30	10·10
600 × 900 mm	,,	12·25	13·85

WOODWORK

Formica Standard Range +10⁵/₀ (handwritten)

Fixing fittings	Unit	£
Kitchen fittings; fixing to backgrounds requiring plugging		
Wall units	No.	3·80
500 × 600 × 300 mm	No.	3·80
500 × 900 × 300 mm	,,	4·12
1000 × 600 × 300 mm	,,	5·00
1000 × 900 × 300 mm	,,	5·45
Floor units		
500 × 900 × 500 mm	,,	3·18
1000 × 900 × 500 mm	,,	4·85
1500 × 900 × 500 mm	,,	6·00
Store units		
600 × 500 × 1950 mm	,,	5·19
Sink units		
1000 × 900 × 500 mm	,,	6·05
1200 × 900 × 600 mm	,,	7·25

FIRST FIXINGS

Boardings and flooring – wrought hardwood

Strip flooring, 25 mm thick; tongued and grooved joints; tongued and grooved heading joints; secret nailed; surface sanded after laying

	Unit	£
American oak	m²	16·50
Canadian maple	,,	13·80
Gurjun	,,	17·50
Iroko	,,	19·20
Sapele	,,	23·00
raking cutting (all types)	m	1·75
curved cutting (all types)	,,	2·75

SECOND FIXINGS

	Unit	West African Mahogany £	Afrormosia £
Unframed second fixings – wrought hardwood; selected			
Skirtings, picture rails, dado rails and the like			
19 × 45 mm; splayed once or moulded	m	1·51	2·17
19 × 75 mm; ,, ,, ,, ,,	,,	2·10	3·20
25 × 75 mm; ,, ,, ,, ,,	,,	2·85	3·62
25 × 100 mm; ,, ,, ,, ,,	,,	3·07	4·57
Architraves, cover fillets and the like			
19 × 38 mm; splayed once or moulded	,,	1·59	2·47
19 × 75 mm; ,, ,, ,, ,,	,,	2·71	3·79
25 × 44 mm; ,, ,, ,, ,,	,,	2·26	3·12
25 × 75 mm; ,, ,, ,, ,,	,,	3·00	4·33
Stops; screwed on			
19 × 38 mm	,,	1·94	2·19
25 × 38 mm	,,	2·09	2·44
25 × 50 mm	,,	2·36	2·63
Glazing beads and the like			
13 × 19 mm	,,	0·79	0·88
13 × 25 mm	,,	0·98	1·03
13 × 25 mm; fixing with brass cups and screws	,,	1·24	1·57
19 × 36 mm ,, ,, ,, ,, ,, ,,	,,	2·32	2·73

WOODWORK

SECOND FIXINGS – *continued*

| | | West African Mahogany | | | Afrormosia | | |
	Unit	19 mm thick	25 mm thick	32 mm thick	19 mm thick	25 mm thick	32 mm thick
Isolated shelves, work-tops, seats and the like		£	£	£	£	£	£
150 mm wide . .	m	3·47	4·25	5·05	4·58	5·86	7·00
230 mm ,, . .	,,	5·19	6·41	7·62	6·87	8·84	10·55
300 mm wide; cross-tongued joints . .	,,	7·95	9·05	10·30	10·45	12·25	14·45
450 mm wide; cross-tongued joints . .	,,	10·75	12·40	14·25	14·40	17·10	20·40
600 mm wide; cross-tongued joints . .	,,	14·55	16·70	19·15	19·37	22·95	27·35
Shelves; slatted 50 mm longitudinal slats at 75 mm centres	m²	20·80	23·80	27·90	25·37	29·80	35·55

| | | West African Mahogany | | | Afrormosia | | |
	Unit	25 mm thick	32 mm thick	38 mm thick	25 mm thick	32 mm thick	38 mm thick
Window boards, nosings, bed moulds and the like		£	£	£	£	£	£
75 mm wide; rebated once; rounded once .	m	3·51	4·00	4·45	4·60	5·00	5·75
150 mm wide; rebated once; rounded once .	,,	5·63	6·55	7·47	7·68	8·50	10·00
225 mm wide; rebated once; rounded once .	,,	7·75	9·13	10·50	10·77	12·00	14·25
300 mm wide; rebated once; rounded once .	,,	10·44	12·22	13·78	14·00	15·90	17·80
Ends; returned fitted	No.	1·57	1·83	2·07	2·10	2·39	2·67

	Unit	West African Mahogany £	Afrormosia £
Handrails; rounded			
37 × 87 mm .	m	5·62	10·18
50 × 75 mm .	,,	6·37	11·68
63 × 87 mm .	,,	8·64	16·36
75 × 100 mm .	,,	11·76	20·20
Handrails; moulded			
37 × 87 mm .	,,	6·36	10·56
50 × 75 mm .	,,	7·24	12·12
63 × 87 mm .	,,	9·92	17·00
75 × 100 mm .	,,	12·86	21·75

WOODWORK

		West African Mahogany		Afrormosia	
	Unit	726 × 2040 mm £	826 × 2040 mm £	726 × 2040 mm £	826 × 2040 mm £

COMPOSITE ITEMS

Doors – wrought hardwood; selected

Panelled doors; four panels; mouldings worked on the solid both sides

	Unit	726 × 2040 £	826 × 2040 £	726 × 2040 £	826 × 2040 £
50 mm thick; 19 mm panels	No.	83·00	90·95	122·75	135·65
63 mm thick; 25 mm panels	,,	95·60	105·25	142·70	158·40

Panelled doors; one open panel for glass; mouldings worked on the solid one side; 19 × 13 mm beads one side; fixing with brass screws and cups

50 mm thick	,,	64·85	69·50	82·60	86·15
63 mm thick	,,	71·50	75·55	90·30	97·15

Panelled doors; 250 mm wide cross-tongued intermediate rail; two open panels for glass; mouldings worked on the solid one side; 19 × 13 mm beads one side; fixing with brass screws and cups

50 mm thick	,,	68·70	73·90	87·00	100·00
63 mm thick	,,	74·75	82·15	103·50	111·45

Panelled doors; 150 mm wide stiles in one width; 430 mm wide cross-tongued bottom rail; five panels, raised and fielded one side; mouldings worked on the solid both sides

50 mm thick; 25 mm panels	,,	152·00	165·00	198·00	331·00
63 mm thick; 25 mm panels	,,	175·00	186·00	217·00	236·00

	Unit	50 or 63 mm thick doors £
rebates; beaded	m	0·63
rounded edges or rounded heels	,,	0·63

		West African Mahogany		Afrormosia	
	Unit	38 mm thick £	50 mm thick £	38 mm thick £	50 mm thick £
Casements					
rebated once, moulded	m²	15·40	20·05	18·80	23·10
rebated once, moulded, divided into medium panes	,,	18·70	25·00	24·35	29·25
fitting and hanging	No.	2·35	2·35	2·51	2·51

WOODWORK

COMPOSITE ITEMS – *continued*

	Unit	West African Mahogany £	Afrormosia £
Door frames and lining sets – wrought hardwood; selected			
Jambs and heads			
32 × 63 mm	m	2·90	4·19
32 × 100 mm	,,	4·66	6·24
32 × 140 mm	,,	5·34	8·30
38 × 115 mm; rebated once; rounded once	,,	5·49	7·36
50 × 100 mm; ,, ,, ; ,, ,,	,,	6·28	12·64
63 × 88 mm; ,, ,, ; ,, ,,	,,	6·36	9·80
63 × 100 mm; ,, ,, ; ,, ,,	,,	7·22	11·13
Mullions and transomes			
32 × 63 mm	,,		
32 × 100 mm	,,	2·95	4·07
32 × 140 mm	,,	4·53	6·34
38 × 115 mm; rebated twice; rounded twice	,,	5·78	8·17
38 × 140 mm; ,, ,, ; ,, ,,	,,	5·63	7·98
50 × 100 mm; ,, ,, ; ,, ,,	,,	6·87	9·70
Sills		6·44	9·12
63 × 100 mm; sunk weathered once; rebated once; grooved three times	,,	11·80	16·48
75 × 150 mm; sunk weathered once; rebated once; grooved three times	,,	12·30	18·84
Weather-fillets			
50 × 25 mm; rebated once, moulded; fixing with screws, countersinking and flush pellating; bedding in white lead	,,	2·66	3·48
Ends; moulded	No.	0·90	0·90

	Unit	Mahogany £
Casements and frames, window surrounds etc – wrought hardwood; selected (Phillipine Mahogany)		
Side hung windows; with 45 × 140 mm hardwood sills; weather stripping; opening sashes on canopy hinges; including fasteners; aluminized laquered finish ironmongery; preservative stain finished wood		
600 × 600 mm reference SS0606/L	No.	35·32
600 × 900 mm ,, SS0609/L	,,	40·32
1200 × 1050 mm ,, SS1210/L	,,	70·60
1200 × 1200 mm ,, SS1212/L	,,	74·85
1800 × 1050 mm ,, SS180D/O	,,	115·45
1800 × 1200 mm ,, SS182D/O	,,	122·35
2400 × 1200 mm ,, SS242D/O	,,	137·45
Top hung windows; with 45 × 140 mm hardwood sills; weather stripping; opening sashes on canopy hinges; including fasteners; aluminized laquered finish ironmongery; preservative stain finished wood		
600 × 600 mm reference ST0606/O		
900 × 600 mm ,, ST0906/O	,,	35·70
1200 × 1050 mm ,, ST1210/O	,,	46·20
1200 × 1200 mm ,, ST1212/O	,,	67·80
1200 × 1500 mm ,, ST1215/O	,,	74·00
1800 × 1050 mm ,, ST1810/R	,,	81·70
1800 × 1200 mm ,, ST1812/R	,,	90·15
1800 × 1350 mm ,, ST1813/R	,,	98·00
2400 × 1500 mm ,, ST2415/R	,,	103·60
	,,	130·15

WOODWORK

COMPOSITE ITEMS – *continued*

Screen components – wrought hardwood; selected	Unit	West African Mahogany 32 mm frame 12 mm panels £	38 mm frame 12 mm panels £	50 mm frame 19 mm panels £	Afrormosia 32 mm frame 12 mm panels £	38 mm frame 12 mm panels £	50 mm frame 19 mm panels £
Panelled partitions over 300 mm wide	m²	40·70	43·80	53·70	50·00	54·00	60·00
Panelled partitions: mouldings worked on the solid both sides Over 300 mm wide	,,	45·80	50·65	55·55	55·20	59·00	65·50
Panelled partitions: mouldings planted on both sides over 300 mm wide	,,	52·55	57·00	61·60	62·00	65·60	71·30
Panelled partitions: diminishing stiles over 300 mm wide; upper portion open panels for glass divided into medium panes	,,	71·80	76·80	81·15	84·25	87·50	93·00

Staircase components – wrought hardwood; selected	Unit	West African Mahogany 25 mm thick £	32 mm thick £	European Oak 25 mm thick £	32 mm thick £
Board landings; cross-tongued joints; 100 × 50 mm sawn softwood bearers	m²	32·30	34·40	65·65	75·50

	Unit	25 mm thick treads, 19 mm thick risers £	32 mm thick treads, 25 mm thick risers £	25 mm thick treads, 19 mm thick risers £	32 mm thick treads, 25 mm thick risers £
Treads, cross-tongued joints and risers; rounded nosings; tongued, grooved, glued and blocked together; one 75 × 50 mm sawn softwood carriage	,,	43·55	50·45	84·00	92·55
Ends; quadrant	No.	22·70	24·60	36·70	42·15
Ends; housed to hardwood	,,	1·82	1·82	1·82	1·82
Winders, cross-tongued joints and risers in one width; rounded nosings; tongued, grooved, glued and blocked together; one 75 × 50 mm sawn softwood carriage	m²	51·65	59·90	92·90	101·70
Wide ends; housed to hardwood	No.	2·22	2·22	2·22	2·22
Narrow ends; housed to hardwood	,,	1·82	1·82	1·82	1·82

WOODWORK

COMPOSITE ITEMS – *continued*

Staircase components – wrought hardwood; selected – *continued*	Unit	West African Mahogany			European Oak		
		32 mm thick £	38 mm thick £	50 mm thick £	32 mm thick £	38 mm thick £	50 mm thick £
Closed strings; 280 mm wide; rounded once; fixing with screws; plugging 450 mm centres	m	14·10	16·00	20·00	27·50	31·40	36·75
Ends; fitted	No.	2·90	2·90	2·90	2·87	2·87	2·87
Ends; framed to hardwood	,,	3·37	3·37	3·37	3·33	3·33	3·33
Extra; short ramps	,,	6·30	7·45	7·70	12·00	12·35	14·10
Extra; tongued heading joint	,,	2·76	2·76	3·34	2·73	2·73	3·30
Closed strings; ramped; cross-tongued joints 280 mm wide; rounded once; fixing with screws; plugging 450 mm centres	m	14·05	16·15	20·00	27·50	31·40	36·75
Closed strings; in one width 230 mm wide; rounded twice	,,	11·67	13·45	16·35	21·45	24·60	31·75

	Unit	West African Mahogany		European Oak	
		19 mm thick £	26 mm thick £	19 mm thick £	25 mm thick £
Apron-linings; in one width 230 mm wide	m	6·60	8·30	12·50	16·18

	Unit	West African Mahogany			European Oak		
		44 × 50 mm £	63 × 75 mm £	75 × 100 mm £	44 × 50 mm £	63 × 75 mm £	75 × 100 mm £
Handrails							
rounded	m	4·74	7·52	11·76	7·94	13·95	19·95
moulded	,,	5·62	8·62	12·86	8·71	14·50	21·50
Handrails; ramped							
rounded	,,	10·57	17·42	23·12	14·05	25·00	35·00
moulded	,,	11·84	19·62	24·90	16·30	27·70	37·00
ADD to above for							
grooved once	,,	0·43	0·43	0·43	0·43	0·43	0·43
Ends; framed	No.	2·56	2·78	2·78	2·53	2·75	2·75
Ends; framed on rake	,,	3·08	3·24	3·35	3·03	3·19	3·30
Mitres; handrail screws	,,	1·47	1·62	1·91	1·45	1·62	1·89
Heading joints; on rake; handrail screws	,,	3·21	3·49	3·76	3·17	3·44	3·72

WOODWORK

COMPOSITE ITEMS – *continued*

Staircase components – wrought hardwood; selected – *continued*	Unit	West African Mahogany £	European Oak £
Balusters			
914 × 32 × 32 mm; both ends housed	No.	4·95	5·26
914 × 32 × 32 mm; both ends housed on rake	,,	5·95	6·23
Newels			
75 × 75 mm	m	8·54	15·25
Ends; rounded	No.	1·67	1·67
100 × 100 mm	m	13·78	26·50
Ends; rounded	No.	1·95	1·95
Newel caps			
100 × 100 × 50 mm; rounded; housed to hardwood; fixing with adhesive	,,	7·47	7·31
125 × 125 × 50 mm; rounded; housed to hardwood; fixing with adhesive	,,	7·85	7·69

Fitting components – wrought hardwood; selected – supplied and fixed		Iroko £	Teak £
Backs, fronts or sides; cross-tongued joints			
25 mm thick; over 300 mm wide	m²	34·74	50·24
Divisions; cross-tongued joints			
25 mm thick; over 300 mm wide	,,	34·74	50·24
Work-tops; cross-tongued joints			
25 mm thick; over 300 mm wide	,,	34·74	50·24
Draining boards; cross-tongued joints			
25 mm thick; over 300 mm wide	,,	38·42	53·55
grooves; cross grain	m	0·56	0·68
flutes; stopped	,,	0·68	0·68

		West African Mahogany £	
Bearers			
38 × 19 mm	,,	1·95	1·92
50 × 25 ,,	,,	2·63	2·59
50 × 50 ,,	,,	4·55	4·50
75 × 50 ,,	,,	5·67	5·62
Bearers; framed			
38 × 19 mm	,,	2·41	2·39
50 × 25 ,,	,,	2·97	2·94
50 × 50 ,,	,,	4·86	4·82
75 × 50 ,,	,,	6·63	6·59
Framing; to backs, sides or fronts			
38 × 19 mm	,,	2·41	2·39
50 × 25 ,,	,,	2·97	2·94
50 × 50 ,,	,,	4·86	4·82
75 × 50 ,,	,,	6·63	6·59

WOODWORK

SUNDRIES	Unit	£
Plugging brickwork		
300 mm centres; both ways	m²	2·07
300 mm centres	m	0·69
Plugging concrete		
300 mm centres; both ways	m²	2·82
300 mm centres	m	0·94
Holes for bolts and the like		
12 mm softwood	No.	0·09
25 mm ,,	,,	0·16
50 mm ,,	,,	0·31
100 mm ,,	,,	0·63
Holes for pipes, bars, cables and the like		—
12 mm softwood	,,	0·19
25 mm ,,	,,	0·25
50 mm ,,	,,	0·47
100 mm ,,	,,	0·85
Building paper; single sided reflective; 150 mm laps		
over 300 mm wide; pinned	m²	1·00
Expanded polystyrene; butt joints		
38 mm thick; over 300 mm wide; fixing with adhesive	,,	1·97
Glass fibre quilt; butt joints		
80 mm thick; over 300 mm wide; placing in position	,,	2·30
100 mm ,, ; ,, ,, ,, ,, ; ,, ,, ,,	,,	2·90
Metalwork – mild steel; sherardized		
Clips for floor fillets		
setting into concrete; fixing to softwood	No.	0·33
Straps; tie down		
30 × 5 × 600 mm long; fixing to softwood with screws	,,	1·56
30 × 5 × 1000 mm ,, ; ,, ,, ,, ,, ,,	,,	3·03
30 × 5 × 1500 mm ,, ; ,, ,, ,, ,, ,,	,,	2·51
Shoes		
for 50 × 150 mm joists; fixing to softwood with screws	,,	1·53
,, 50 × 225 mm ,, ; ,, ,, ,, ,, ,,	,,	2·00
,, 50 × 150 mm ,, ; building into brickwork	,,	1·83
,, 50 × 225 mm ,, ; ,, ,, ,,	,,	2·30
Timber connectors		
51 mm diameter; double sided; for 12 mm bolts	,,	0·14
75 mm ,, ; ,, ,, ,, ,,	,,	0·20
Framing anchors; Trip-L-Grip		
anchor	,,	0·53

10/2/81

IRONMONGERY

s ironmongery is largely a matter of selection, the prices given are for fixing only. The prices for doors
'Joinery' include for fixing ordinary butt hinges and only the extra cost of fixing other hinges are given.

	Unit	To softwood £	To hardwood or the like £
IXING			
urface fixing			
door furniture	Set.	0·78	1·03
numerals; 50 mm high	No.	0·31	0·41
hat and coat hooks	"	0·31	0·41
shelf brackets	"	0·63	0·85
bow handles; 150 mm	"	0·63	0·85
cabin hooks and eyes	"	0·78	1·03
casement stays and pins; 305 mm	"	0·94	1·25
drawer pulls	"	0·31	0·41
finger plates	"	0·63	0·85
postal plates and knockers; apertures through 50 mm	"	4·70	6·27
sash lifts	"	0·63	0·85
bolts; barrel			
102 mm	"	0·94	1·25
152 "	"	1·13	1·50
305 "	"	1·57	2·10
bolts; necked			
102 mm	"	1·03	1·38
152 "	"	1·25	1·66
bolts; monkey tail			
381 mm	"	3·14	4·17
610 "	"	3·54	4·70
762 "	"	3·76	5·02
rod door closers			
406 mm	"	2·35	3·14
457 "	"	3·14	4·17
Norfolk latches	"	3·14	4·17
Flush fixing			
overhead door closers	"	7·85	10·45
butt hinges; skew			
51 mm	Pair	0·63	0·63
64 "	"	0·63	0·63
76 "	"	0·63	0·63
102 "	"	0·63	0·63
floor spring door closers; double action; filling box with lubricating oil	No.	12·55	16·70
floor spring door closers; double action with check, filling box with lubricating oil	"	15·70	20·90
spring hinges; single action			
76 mm	Pair	6·27	8·37
102 "	"	6·27	8·37
127 "	"	7·84	10·44
152 "	"	7·84	10·44
178 "	"	9·41	12·55
spring hinges; double action			
76 mm	"	7·84	10·45
102 "	"	7·84	10·45
127 "	"	9·41	12·55
152 "	"	9·41	12·55
178 "	"	11·00	14·65

(handwritten) X3 to be really competitive.

(handwritten, circled) 2·20   X4   X4

IRONMONGERY

+26%

FIXING – *continued*	Unit	To softwood £	To hardwood or the like £
Flush fixing – *continued*			
bolts; lever action; 203 mm	No.	3·14	4·17
„ mortice	„	2·35	3·14
„ panic; automatic: for single doors	„	7·85	10·45
„ „ „ „ pairs of doors	„	9·40	12·55
„ „ „ locking; for single doors	„	9·40	12·55
„ „ „ „ pairs of doors	„	11·00	14·65
cupboard locks	„	2·35	3·14
cylinder rim latches	„	3·92	5·24
fanlight catches	„	0·94	1·25
handles	„	1·57	2·10
mortice deadlocks	„	3·92	5·24
„ latches	„	3·14	4·17
„ locks	„	3·92	5·24
„ „ ; rebated	„	6·27	8·37
rim latches	„	2·35	3·14
„ locks	„	2·35	3·14
ring catches	„	1·57	2·10
sash fasteners; mortice plates	„	1·25	1·66
„ screws	„	1·57	2·10
quadrant stays	„	1·57	2·10

Fixing; sliding door gear, for top hung timber doors, weight not exceeding 363 kg	Unit	£
To softwood; flush fixing		
bottom guides	No.	0·78
hangers	„	3·14
To concrete or the like		
floor channel; to groove; bedding and jointing in cement mortar (1 : 2)	m	1·57
track brackets; fixing by bolting	No.	1·19
To brackets		
track	m	1·03
To softwood; surface fixing		
detachable locking bar and padlock	No.	1·57

STRUCTURAL STEELWORK

The prices for columns and beams are based on sections costing £231·50 per tonne. Prices for other sections vary roughly in proportion to the price of the steel.

MILD STEEL

	Unit	Delivered to site £
Fabricated steelwork		
Single beams; joists and channels	t	390·00
Single columns; joists and channels	,,	470·00
Built-up beams; joists, channels and plates	,,	480·00
Built-up columns; joists, channels and plates	,,	526·00
Single hollow rectangular columns	,,	572·00
Single hollow circular columns	,,	572·00
Latticed beams; angle sections	,,	390·00
Latticed beams; hollow circular tubes	,,	572·00
Roof trusses; angle sections	,,	765·00
Roof trusses; hollow circular tubes	,,	572·00
Rivetted fittings; consisting of cleats, brackets etc for site bolted connections	,,	1170·00
Rivetted fittings; consisting of cleats, brackets etc for and including site welded connections	,,	1170·00
Galvanizing	,,	100·00
Shot blasting	m²	3·00
Painting at works		
one coat zinc chromate primer	,,	1·72
Erection of fabricated steelwork		£
on site	t	165·00
Wedging		
stanchion bases	No.	1·40
Holes for other trades		
13 mm diameter; 10 mm metal; made on site	,,	0·70

	Unit	Delivered to site £
Unfabricated steelwork		
Single beams; joists and channels	t	295·00
		£
Erection (hoisting and positioning) on site	,,	165·00

METALWORK

ROLLED PLATES, BARS, SECTIONS AND TUBES	*Unit*	£
Mild steel; B.S. 4360		
Handrails		
44 × 13 mm half oval bar; brackets at 1 m centres, welded, fixing with steel wood screws; plugging	m	13·65
Core-rails		
38 × 10 mm bar; brackets at 1 m centres, welded; fixing with steel wood screws; plugging	„	11·00
Counter-sunk holes for wood screws; 10 mm metal . . .	No.	0·21
Ends fanged	„	0·40
Ends scrolled	„	1·80
Ramps in thickness	„	1·80
Wreaths	„	2·16
Balusters		
19 × 19 mm × 914 mm long bar; one end ragged; one 76 × 25 × 6 mm flange plate welded on, ground to smooth finish, counter-sunk drilled and tap screwed to handrail	„	4·45
Bearers		
51 × 10 mm bar	m	2·50
64 × 13 mm bar	„	3·64
Ends fanged	No.	0·40
50 × 50 × 6 mm angle section	m	2·87
Ends fanged	No.	0·39
Mat-frames; welded construction; galvanized after manufacture; 40 × 40 × 5 mm angle; six fanged lugs; wedging in position, temporary wedges		
overall size 1000 × 500 mm	„	27·40
„ 1150 × 600 „	„	29·00
„ 1200 × 700 „	„	30·50
Balustrades; welded construction; galvanized after manufacture		
1070 mm high; 51 × 51 × 3 mm r.h.s. top rail; 38 × 13 mm bottom rail; 51 × 51 × 3 mm r.h.s. standards at 1830 mm centres with base plate drilled and bolted to concrete; 13 × 13 mm balusters at 102 mm centres . .	m	31·00
Cat ladders; welded construction; 64 × 13 mm bar strings; 19 mm rungs at 250 mm centres; fixing by bolting		
0·46 m wide, 3·05 m high	No.	71·35
Extra; ends of strings, bent once; holed once for 13 mm diameter bolt .	„	0·80
Extra; ends of strings, fanged	„	0·35
Extra; ends of strings, bent in plane	„	4·00
Aluminium polished		
Mat frames; brazed construction; 38 × 5 mm bar; six fanged lugs; wedging in position; temporary wedges		
overall size 1000 × 500 mm	„	15·70
„ 1150 × 600 „	„	19·35
„ 1200 × 700 „	„	20·00

STANDARD UNITS

Windows

Refer to 'Market Prices of Materials' for costs of these items and to 'Brick-work and Blockwork' of 'Prices for Measured Work' for fixing.

METALWORK

SUNDRIES	*Unit*	**£**
Rag-bolts; mild steel		
10 mm diameter × 120 mm long; one nut; one washer	No.	0·80
13 „ „ × 160 „ „ „ „	„	0·95
22 „ „ × 200 „ „ „ „	„	1·10

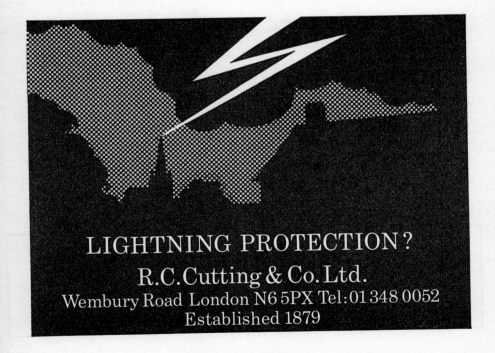

PLUMBING AND ENGINEERING INSTALLATIONS

RAINWATER INSTALLATIONS

Gutterwork aluminium half-round gutters and fittings, B.S. 2997	Unit	102 *mm* £	114 *mm* £	127 *mm* £
Gutters; bolted red-lead joints in the running length; brackets at 914 mm centres; fixing with galvanized screws to timber	m	6·25	6·45	7·22
Extra; angles	No.	3·18	3·33	3·87
Extra; stopped ends	,,	2·64	2·67	2·82
Extra; nozzle outlets	,,	2·91	3·00	3·16

Rainwater pipework; aluminium pipes and fittings; B.S. 2997, ears cast on	Unit	51 *mm* £	63 *mm* £	76 *mm* £	102 *mm* £
Pipes; red-lead joints in the running length; fixing with galvanized nails to backgrounds requiring plugging	m	4·44	5·41	6·40	8·49
Extra; shoes	No.	3·84	4·40	5·37	6·51
Extra; bends	,,	3·39	4·31	4·54	6·46
Extra; 229 mm projection offsets	,,	5·45	6·99	7·84	8·87
Extra; 305 mm projection offsets	,,	5·75	7·43	8·15	9·43
Connection to vitrified clay pipes; cement and sand (1:2) joint	,,	0·42	0·47	0·53	0·61
Galvanized wire balloon gratings; B.S. 416 for pipes or outlets	,,	0·86	0·95	1·08	2·47

Gutterwork; asbestos cement box gutters and fittings, B.S. 569	Unit	127 × 152 *mm* £	305 × 203 *mm* £	381 × 127 *mm* £
Gutters; bolted expanded butyl joints in the running length; fixing in supports measured elsewhere	m	10·59	15·22	15·82
Extra; stopped ends	No.	5·40	7·68	7·61
Extra; nozzle outlets	,,	10·12	14·07	13·89

Gutterwork; asbestos cement half-round gutters and fittings, B.S. 569		125 *mm* £	150 *mm* £	200 *mm* £
Gutters; bolted expanded butyl joints in the running length; brackets at 914 mm centres; fixing with galvanized screws to timber	m	5·22	6·40	7·80
Extra; angles	No.	4·27	4·90	6·14
Extra; stopped ends	,,	2·06	2·39	2·85
Extra; nozzle outlets	,,	3·40	4·26	4·90

PLUMBING AND ENGINEERING INSTALLATIONS

RAINWATER INSTALLATIONS – *continued*

Gutterwork; asbestos cement valley gutters and fittings, B.S. 569	Unit	406 × 127 × 254 *mm* £	457 × 127 × 152 *mm* £	610 × 152 × 229 *mm* £
Gutters; bolted expanded butyl joints in the running lengths; fixing in supports measured elsewhere . .	m	12·49	13·26	15·24
Extra; stopped ends	No.	5·95	6·47	7·67
Extra; nozzle outlets	„	10·73	15·23	13·69

Gutterwork; asbestos cement boundary wall gutters and fittings, B.S. 569	Unit	279 × 127 × 178 *mm* £	305 × 152 × 229 *mm* £	457 × 152 × 305 *mm* £	559 × 152 × 406 *mm* £
Gutters; bolted expanded butyl joints in the lengths; fixing in supports measured elsewhere	m	9·59	11·46	14·89	15·38
Extra; stopped ends	No.	4·60	5·68	7·16	7·84
Extra; nozzle outlets	„	8·91	10·28	12·46	13·80

Rainwater pipework; asbestos cement pipes and fittings, B.S. 569	Unit	75 *mm* £	100 *mm* £	150 *mm* £
Pipes; mastic joints in the running length; fixing in supports measured elsewhere	m	5·32	6·57	11·83
Extra; shoes	No.	4·29	5·40	10·19
Extra; bends	„	4·15	5·52	9·37
Extra; 225 mm projection offsets	„	5·01	6·36	10·64
Extra; 300 mm projection offsets	„	5·72	6·97	12·23
Connection to vitrified clay pipes; cement and sand (1:2) joint	„	0·53	0·61	0·78
Standard pipe clips; galvanized by manufacturer; fixing with galvanized screws to backgrounds requiring plugging	„	1·84	2·50	2·95
Galvanized wire balloon gratings; B.S. 416 for pipes or outlets	„	0·98	1·14	2·58

PLUMBING AND ENGINEERING INSTALLATIONS

RAINWATER INSTALLATIONS –
continued

Gutterwork; cast iron half-round gutters and fittings, B.S. 460	Unit	100 *mm* £	115 *mm* £	150 *mm* £
Gutters; one priming coat inside and outside by manufacturer, bolted red-lead joints in the running length; brackets at 914 mm centres; fixing with galvanized screws to timber	m	6·06	6·34	10·37
Extra; angles	No.	3·61	3·86	5·00
Extra; stopped ends	„	1·77	2·12	—
Extra; nozzle outlets	„	3·22	3·48	4·57

Rainwater pipework; cast iron pipes and fittings, B.S. 460, ears cast on		50 *mm* £	63 *mm* £	75 *mm* £	100 *mm* £
Pipes; one priming coat by manufacturer, red-lead joints in the running length; fixing with galvanized nails and galvanized pipe distance pieces to backgrounds requiring plugging	m	7·08	8·03	8·56	11·13
Extra; shoes	No.	5·89	7·05	7·38	9·40
Extra; bends	„	2·99	4·19	4·52	6·31
Extra; 229 mm projection offsets	„	5·37	6·86	7·19	9·62
Extra; 305 mm projection offsets	„	6·51	7·71	8·04	10·97
Connection to vitrified clay pipes; cement and sand (1:2) joint	„	0·42	0·47	0·53	0·61
Galvanized wire balloon gratings; B.S. 416 for pipes or outlets	„	0·86	0·95	1·08	2·47
Cast iron roof outlets; Fulbora vertical spigot type with grating, hook bolt and clamping device; caulked lead joint to pipe	„	22·75	—	30·06	43·61

Gutterwork; coloured PVC half-round gutters and fittings	Unit	76 *mm* £	112 *mm* £	152 *mm* £
Gutters; combined fascia brackets and union clips in the running length and intermediate brackets at 914 mm centres; fixing with galvanized screws to timber	m	2·96	4·15	7·66
Extra; angles	No.	3·16	3·90	8·40
Extra; stopped ends	„	1·26	1·44	2·39
Extra; nozzle outlets	„	2·52	3·19	6·37

Rainwater pipework; PVC pipes and fittings		50 *mm* £	68 *mm* £	110 *mm* £
Pipes; push fit joints in the running length; fixing in supports measured elsewhere	m	2·88	3·18	5·79
Extra; shoes	No.	2·57	2·57	5·63
Extra; bends	„	3·03	3·29	6·60
Extra; 229 mm projection offsets	„	—	3·64	9·06
Extra; 305 mm projection offsets	„	3·78	3·82	9·27
Connection to vitrified clay pipes; cement and sand (1:2) joint	„	0·42	0·53	0·61
Standard PVC pipe brackets; fixing with brass screws to backgrounds requiring plugging	„	1·34	1·49	1·61

PLUMBING AND ENGINEERING INSTALLATIONS

SANITARY INSTALLATIONS	Unit	22 mm £	28 mm £	35 mm £	42 mm £	54 mm £
Overflow pipework; copper pipes, B.S. 2871, Table X; brass capillary fittings						
Pipes; couplers in the running length; fixing with standard brass two-piece spacing clips to backgrounds requiring plugging	m	3·39	3·84	5·54	6·99	8·79
Extra; made bends . .	No.	1·92	2·91	4·36	4·83	5·82
Extra; couplings with union to iron	,,	2·52	3·09	4·20	4·91	6·32
Extra; reducing couplings with union to copper . .	,,	1·53	1·93	2·83	3·83	4·95
Extra; sweep tees with union to copper (equal or reducing, largest branch measured) .	,,	2·60	3·52	4·96	6·47	9·78
Overflow pipework; copper pipes, B.S. 2871, Table X; brass compression fittings						
Pipes; couplers in the running length; fixing with standard brass two-piece spacing clips to backgrounds requiring plugging	m	3·39	3·84	5·54	6·99	8·79
Extra; made bends . .	No.	1·92	2·91	4·36	4·83	5·82
Extra; couplings with union to iron	,,	3·52	4·59	6·19	7·95	10·22
Extra; reducing couplings with union to copper . .	,,	2·93	4·10	5·65	7·35	10·65
Extra; sweep tees with union to copper (equal or reducing, largest branch measured) .	,,	4·94	7·08	9·94	13·83	19·06

	Unit	20 mm £	25 mm £	32 mm £	40 mm £	50 mm £
Overflow pipework; lead pipes, B.S. 602						
Pipes; soldered joints in the running length; fixing with lead tacks to timber . .	m	4·64	5·78	6·49	6·91	9·79
Extra; made bends . .	No.	0·87	0·93	1·05	1·46	2·91
Extra; soldered branch joints	,,	7·19	8·64	10·09	11·54	14·78
Connections to brass pipe fittings; soldered joint . .	,,	6·21	7·48	8·81	10·08	12·98

Waste pipework; copper pipes. B.S. 2871, Table X; brass capillary fittings	Unit	35 mm £	42 mm £	54 mm £
Pipes; couplers in the running length; fixing with standard brass two-piece spacing clips to backgrounds requiring plugging	m	5·96	7·51	9·53
Extra; made bends	No.	4·36	4·83	5·82
Extra; couplings with union to iron . .	,,	4·20	4·91	6·32
Extra; reducing couplings with union to copper .	,,	2·83	3·83	4·95
Extra; sweep tees with union to copper (equal or reducing largest branch measured)	,,	4·96	6·47	9·78

PLUMBING AND ENGINEERING INSTALLATIONS

SANITARY INSTALLATIONS – *continued*

Waste pipework; copper pipes, B.S. 2871, Table X; brass compression fittings

	Unit	35 mm £	42 mm £	54 mm £
Pipes; couplers in the running length; fixing with standard brass two-piece spacing clips to backgrounds requiring plugging	m	6·30	8·22	10·57
Extra; made bends	No.	4·36	4·83	5·82
Extra; couplings with union to iron . . .	,,	6·19	7·95	10·22
Extra; reducing couplings with union to copper .	,,	5·65	7·35	10·65
Extra; sweep tees with unions to copper (equal or reducing, largest branch measured)	,,	9·94	13·83	19·06

Waste pipework; lead pipes, B.S. 602

		32 mm £	40 mm £	50 mm £
Pipes; soldered joints in the running length; fixing with lead tacks to timber	m	7·96	8·59	11·95
Extra; made bends	No.	1·05	1·45	2·91
Extra; soldered branch joints	,,	10·08	11·54	14·79
Connections to brass pipe fittings; soldered joint .	,,	8·80	10·08	12·98

Soil and vent pipework; cast iron pipes and fittings, B.S. 416 type A sockets, ears cast on

		75 mm £	100 mm £	150 mm £
Pipes; caulked lead joints in the running length; fixing with galvanized nails and galvanized pipe distance pieces to backgrounds requiring plugging . . .	m	9·72	12·80	25·72
Extra; bends	No.	8·49	11·87	19·46
Extra; long bends 457 × 457 mm with large socket .	,,	—	25·15	—
Extra; single branches	,,	11·94	17·17	29·94
Extra; double branches	,,	17·90	21·19	42·71
Extra; 229 mm projection offsets	,,	11·30	14·93	27·60
Extra; 305 mm projection offsets	,,	12·33	16·21	33·50
Extra; connecting pipes 457 mm long with large socket	,,	—	13·40	—
Extra; oval door bolted and sealed to side of any fitting	,,	6·15	6·46	7·29
Isolated caulked lead joints	,,	3·76	5·10	8·79
Connections to vitrified clay pipes; cement and sand (1:2) joint	,,	0·53	0·62	0·78

Soil and vent pipework; lead pipes, B.S. 602

	Unit	65 mm £	75 mm £	90 mm £	100 mm £
Pipes; soldered joints in the running length; plain lead tacks; fixing with brass screws to backgrounds requiring plugging . .	m	26·70	32·21	36·93	41·07
Extra; made bends . . .	No.	5·64	5·99	6·40	6·98
Extra; soldered branch joints . . .	,,	17·97	21·27	23·59	25·90
Connections to brass pipe fitting; soldered joint		15·81	18·71	20·85	23·00

PLUMBING AND ENGINEERING INSTALLATIONS

SANITARY INSTALLATIONS – *continued*

			32 *mm*	38 *mm*
Solid drawn copper traps; joint to copper pipe; red-lead joint to waste fitting		Unit	£	£
P; 38 mm seal; inlet with coupling nut and collar; outlet with compression fitting		No.	11·66	14·57
P; 76 mm seal; inlet with coupling nut and collar; outlet with compression fitting		,,	12·20	15·19
S; 38 mm seal; inlet with coupling nut and collar; outlet with compression fitting		,,	12·15	15·10
S; 76 mm seal; inlet with coupling nut and collar; outlet with compression fitting		,,	12·68	15·76

		Unit	£
Appliances		Unit	£
Sinks, white glazed fireclay; B.S. 1206; cast iron cantilever brackets			
610 × 457 × 254 mm *p.c.* £31·00 each		No.	54·60
762 × 508 × 254 mm *p.c.* £51·60 each		,,	77·40
Wash tubs, white glazed fireclay; B.S. 1229; cast iron cantilever brackets to B.S. 1225			
610 × 508 × 381 mm *p.c.* £63·65 each		,,	90·65
Sinks, stainless steel combined bowl and draining board; chain and self colour plug to B.S. 3380			
1050 × 525 mm with bowl 457 × 343 × 178 mm *p.c.* £64·15 each . .		,,	82·20
Sinks, stainless-steel combined bowl and double draining board; chain and self colour plug to B.S. 3380			
1575 × 525 mm with bowl 559 × 381 × 178 mm *p.c.* £84·60 each . .		,,	110·50
Lavatory basins, white glazed vitreous china; B.S. 1188; 32 mm chromium-plated waste, chain, stay and plug; painted cantilever brackets; pair 13 mm chromium-plated easy clean pillar taps to B.S. 1010			
559 × 406 mm *p.c.* £28·50 each; fixing brackets with screws to backgrounds requiring plugging		,,	44·00
635 × 457 mm *p.c.* £35·55 each; fixing brackets with screws to backgrounds requiring plugging		,,	53·10
Lavatory basins and pedestals, white vitreous china; B.S. 1188; 32 mm chromium-plated waste, chain, stay and plug; pair 13 mm chromium-plated easy clean pillar taps to B.S. 1010			
559 × 406 mm *p.c.* £36·15 each; fixing basin and pedestal with screws to backgrounds requiring plugging		,,	52·60
635 × 457 mm *p.c.* £43·30 each; fixing basin and pedestal with screws to backgrounds requiring plugging		,,	61·60
Lavatory basin ranges, overlap joints, white glazed fireclay; 32 mm chromium-plated waste, chain, stay and plug; painted cast iron cantilever brackets; pair 13 mm chromium-plated easy clean pillar taps to B.S. 1010			
Range of four, 635 × 457 mm *p.c.* £258·00 per range; fixing brackets to backgrounds requiring plugging		,,	334·25
ADD for each additional basin in the range		,,	85·50
Washing fountains and pedestals, white glazed fireclay; chromium-plated supply column with umbrella spray fitting and soap tray; 38 mm chromium-plated waste; 19 mm water inlet			
1067 mm diameter, 813 mm high; *p.c.* £913·00 each; bedding and pointing pedestal in cement mortar (1:2)		,,	1039·20
Drinking fountains, white glazed fireclay; 19 mm chromium-plated waste; 13 mm selfclosing non-concussive tap			
381 × 305 × 140 mm; *p.c.* £69·00 each; fixing with chromium-plated screws to backgrounds requiring plugging		,,	93·36

PLUMBING AND ENGINEERING INSTALLATIONS

SANITARY INSTALLATIONS – *continued*
Appliances – *continued*

	Unit	£
Baths, coloured reinforced Acrylic, rectangular pattern, 40 mm chromium-plated overflow chain and plug; 40 mm chromium-plated waste; cast brass P trap with plain outlet and overflow connection to B.S. 1184; pair 20 mm chromium-plated easy clean pillar taps to B.S. 1010		
1675 mm long; *p.c.* £126·75 each	No.	156·88
Shower trays, white glazed fireclay with hooded outlet; chain and plug; 762 × 762 × 178 mm *p.c.* £65·55 each; bedding and pointing in cement mortar (1:2)	"	89·60
Shower fittings, riser pipe with mixing valve and shower rose; chromium-plated 15 mm diameter riser pipe; 127 mm diameter shower rose; *p.c.* £88·35 each; plugging and screwing mixing valve and pipe bracket to backgrounds requiring plugging	"	132·00
WC suites, high level; white vitreous china pan, black plastic seat, 9-litre white vitreous china cistern and brackets; low-pressure ball valve; galvanized steel flush pipe and clip		
WC suite *p.c.* £60·00 each; fixing with screws to backgrounds requiring plugging; mastic joint to drain	"	82·30
WC suites, low level; white vitreous china pan, black plastic seat, 9-litre white vitreous china cistern and brackets, low pressure ball valve, and plastic flush pipe		
WC suite *p.c.* £46·00 each; fixing with screws to backgrounds requiring plugging; mastic joint to drain	"	67·00
Slop sinks, white glazed fireclay with hardwood pad; aluminium bucket grating; vitreous china cistern and porcelain-enamelled brackets; galvanized steel flush pipe and clip		
Slop sink; *p.c.* £170·00 each; fixing with screws to backgrounds requiring plugging; mastic joint to drain	"	206·50
Urinal ranges, white glazed fireclay, integral back, divisions, ends and floor channels; fluted treads; outlets and connectors; white vitreous china automatic flushing cistern; mild-steel brackets; copper flush pipes and spreaders		
Range of three stalls 1905 × 1067 mm; *p.c.* £604·00 per range; bedding, jointing and pointing slabs, channels and treads in cement mortar (1:2); fixing cistern brackets, flush pipes and spreaders with screws to backgrounds requiring plugging	"	729·00
ADD for each additional stall in the range	"	243·00

PLUMBING AND ENGINEERING INSTALLATIONS

**COLD-WATER AND
HOT-WATER
INSTALLATIONS**

Service pipework; copper pipes, B.S. 2871, Table X; capillary fittings, B.S. 864	Unit	15 *mm* £	22 *mm* £	28 *mm* £	35 *mm* £	42 *mm* £	54 *mm* £
Pipes; couplers in the running length; fixing with standard two-piece-spacing clips to backgrounds requiring plugging	m	2·92	3·74	4·27	6·10	7·67	9·72
Extra; made bends . . .	No.	1·45	1·92	2·91	4·36	4·83	5·82
Extra; couplings with union to iron	,,	1·91	2·52	3·09	4·20	4·91	6·32
Extra; reducing couplings with union to copper	,,	—	1·53	1·93	2·83	3·83	4·95
Extra; elbows with union to copper	,,	1·13	1·60	2·17	3·36	4·49	7·90
Extra; tees with union to copper (equal or reducing, largest branch measured)	,,	2·06	3·70	3·52	4·96	6·47	9·78
Extra; tank connectors; to tank with backnut	,,	3·65	5·17	6·69	—	—	—

Service pipework; copper pipes, B.S. 2871, Table X; compression fittings, B.S. 864							
Pipes; couplers in the running length; fixing with standard two-piece spacing clips to backgrounds requiring plugging	m	3·21	4·12	4·84	6·63	8·73	11·26
Extra; made bends . . .	No.	1·45	1·92	2·91	4·36	4·83	5·82
Extra; couplings with union to iron	,,	2·84	3·52	4·59	6·19	7·95	10·22
Extra; reducing couplings with union to copper	,,	—	2·93	4·10	4·55	7·35	10·65
Extra; elbows with union to copper	,,	2·64	3·41	4·87	7·48	10·60	14·18
Extra; tees with union to copper (equal or reducing, largest branch measured)	,,	3·93	4·94	7·08	10·92	13·83	19·06
Extra; tank connectors; to tank with backnut	,,	3·20	4·07	5·78	7·75	9·07	13·00

Service pipework; lead pipes, B.S. 602		12 *mm* £	20 *mm* £	25 *mm* £	32 *mm* £	40 *mm* £	50 *mm* £
Pipes; soldered joints in the running length; fixing with clips to timber .	m	5·73	6·88	8·05	10·02	12·51	15·32
Extra; made bends . . .	No.	0·81	0·87	0·93	1·05	1·45	2·91
Extra; soldered branch joints .	,,	5·81	7·19	8·64	10·09	11·54	14·78
Connections to brass pipe fitting; soldered joint	,,	4·93	6·21	7·48	8·81	10·08	13·00

PLUMBING AND ENGINEERING INSTALLATIONS

**COLD-WATER AND
HOT-WATER
INSTALLATIONS** – *continued*

Service pipework; silver–copper–lead alloy, B.S. 1085	Unit	12 mm £	20 mm £	25 mm £	32 mm £	40 mm £	50 mm £
Pipes; soldered joints in the running length; fixing with clips to timber .	m	6·18	7·55	8·94	11·15	13·86	17·12

Service pipework; polythene pipes, B.S. 1972, Class C; compression fittings, B.S. 864		13 mm £	19 mm £	25 mm £	32 mm £	38 mm £	51 mm £
Pipes; couplers in the running length; fixing with standard two-piece spacing clips to backgrounds requiring plugging	,,	2·63	2·86	3·21	3·56	4·14	5·25

Service pipework; steel pipes and screwed fittings, B.S. 1387, medium weight		15 mm £	20 mm £	25 mm £	32 mm £	40 mm £	50 mm £
Pipes; galvanized; red-lead joints in the running length; fixing with tinned clips to backgrounds requiring plugging	,,	3·50	4·17	5·16	6·58	7·43	8·80
Extra; made bends . . .	No.	1·16	2·04	2·79	3·49	4·36	6·69
Extra; malleable iron bends .	,,	2·33	3·01	4·08	5·55	7·00	10·12
Extra; malleable iron elbows .	,,	2·15	2·75	3·71	5·14	6·74	9·45
Extra; malleable iron tees (equal or reducing, largest branch measured)	,,	2·28	2·93	3·90	5·40	6·96	10·12
Isolated screwed red-lead joints .	,,	2·23	2·52	2·93	3·81	4·68	6·43
Connections to steel tank; red-lead joint, backnuts and lead washers .	,,	3·88	4·44	5·59	7·13	8·76	12·61

Mains pipework; copper pipes, B.S. 2871, Table Y		15 mm £	22 mm £	28 mm £	35 mm £	42 mm £	54 mm £
Pipes; no joints in the running length; in trenches	m	1·62	2·60	3·23	4·92	5·93	9·87

Mains pipework; lead pipes, B.S. 602		Unit	12 mm £	20 mm £	25 mm £
Pipes; soldered joints in the running length; in trenches		m	6·48	10·28	15·07
Pipes; soldered joints in the running length; fixing with clips to backgrounds requiring plugging . . .		,,	8·96	11·73	16·71

PLUMBING AND ENGINEERING INSTALLATIONS

COLD-WATER AND HOT-WATER
INSTALLATIONS – *continued*

	Unit	12 *mm* £	20 *mm* £	25 *mm* £
Mains pipework; silver–copper–lead alloy, B.S. 1085	*Unit*			
Pipes; soldered joints in the running length; in trenches	m	6·52	10·35	15·17
Pipes; soldered joints in the running length; plain lead tacks; fixing with brass screws to backgrounds requiring plugging	,,	9·00	11·80	16·80

	Unit	13 *mm* £	19 *mm* £	25 *mm* £	32 *mm* £	38 *mm* £	51 *mm* £
Mains pipework; polythene pipes, B.S. 1972, Class C	*Unit*						
Pipes; no joints in the running length, in trenches	m	1·03	1·24	1·60	2·19	2·71	3·90

		15 *mm* £	20 *mm* £	25 *mm* £	32 *mm* £	40 *mm* £	50 *mm* £
Mains pipework; steel pipes and screwed fittings, B.S. 1387, heavy weight							
Pipes; galvanized; red-lead joints in the running length; in trenches .	,,	3·28	3·54	4·11	5·92	7·92	9·11

	Unit	15 *mm* £	20 *mm* £	25 *mm* £
Bib taps; B.S. 1010; joints to pipes	*Unit*			
crutch head; one end screwed male iron . .	No.	4·84	6·44	—
crutch head; screwed; screwed boss, for lead . .	,,	7·81	10·40	—
cross top, easy clean; one end screwed male iron .	,,	6·07	8·75	—
cross top, easy clean; screwed boss, for lead . .	,,	8·64	12·18	—
crutch head; screwed for hose union; one end screwed male iron	,,	5·94	8·64	—
Stop valves, B.S. 1010; joints to pipes				
crutch head; both ends screwed male iron . .	,,	8·39	10·90	15·66
crutch head; both ends with union for lead .	,,	9·77	—	—
crutch head; both ends compression joints for copper	,,	7·05	9·60	14·40
crutch head; both ends compression joints for polythene.	,,	8·34	11·20	—
Ball valves; B.S. 1212 pattern II; copper float for high pressure; joints to pipes				
inlet screwed male iron; fixing to steel cistern .	,,	7·44	11·40	22·25

PLUMBING AND ENGINEERING INSTALLATIONS

COLD-WATER AND HOT-WATER INSTALLATIONS – *continued*

Equipment	Unit	£
Mild-steel cisterns; B.S. 417, grade A; galvanized by manufacturer		
191 litres actual capacity; reference SCM270	No.	38·40
327 ,, ,, ,, ,, SCM450/1	,,	60·00
1227 ,, ,, ,, ,, SCM1600	,,	194·50
2137 ,, ,, ,, ,, SCM2720	,,	285·00
3364 ,, ,, ,, ,, SCM4540	,,	518·00
Mild-steel tanks; B.S. 417, grade A; galvanized by manufacturer		
95 litres capacity; reference T25/1	,,	59·20
114 ,, ,, ,, T30/1	,,	67·80
155 ,, ,, ,, T40	,,	86·00
Mild-steel cylinders; B.S. 417, grade C; welded construction; galvanized by manufacturer		
100 litres capacity; reference YM114	,,	56·25
123 ,, ,, ,, YM141	,,	62·85
195 ,, ,, ,, YM218	,,	81·60
241 ,, ,, ,, YM264	,,	90·00
Mild-steel indirect cylinders; B.S. 1565, annular heaters; for vertical fixing; galvanized by manufacturer		
136 litres nominal capacity; reference BSG2	,,	94·75
159 ,, ,, ,, ,, BSG3	,,	108·00
227 ,, ,, ,, ,, BSG4	,,	185·25
364 ,, ,, ,, ,, BSG6	,,	251·75
455 ,, ,, ,, ,, BSG7	,,	303·50
Copper cylinders; B.S. 699, grade 3		
74 litres capacity; reference 1	,,	45·15
98 ,, ,, ,, 2	,,	51·10
120 ,, ,, ,, 7	,,	59·50
144 ,, ,, ,, 8	,,	68·30

HEATING INSTALLATION

NOTE. *Large central-heating and hot-water installations are normally carried out by specialist firms, the price varying considerably in accordance with the type and size of the installations. The following prices are intended to apply to the small-scale work undertaken by the General Contractor. (For prices of steel and copper pipes and brasswork see 'Cold-water and Hot-water Installations'.)*

Equipment	Unit	£
Gas fired domestic boilers, cream or white enamelled casing		
13 kW output, 32 mm Bsp female flow and return tappings, 102 mm flue socket, 13 mm Bsp male draw-off outlet; electric controls.	No.	197·00
As above but 22 kW output	,,	324·00
Radiators, pressed steel panel type 610 mm high, 3 mm chromium-plated air valve, 15 mm chromium-plated easy clean straight valve with union, 13 mm straight union		
1·86 m² heating surface	,,	29·30
2·23 ,, ,, ,,	,,	32·80
2·88 ,, ,, ,,	,,	41·80
3·25 ,, ,, ,,	,,	45·40

PLUMBING AND ENGINEERING INSTALLATIONS

HEATING INSTALLATION – *continued*

Smoke-flue pipework; asbestos cement pipes and fittings B.S. 567 (Light quality)	Unit	75 mm £	100 mm £	150 mm £
Pipes; asbestos yarn and composition joints in the running length; fixing in supports measured elsewhere . .	m	5·12	6·41	10·14
Extra; bends	No.	5·47	6·68	10·53
Extra; loose sockets	,,	3·10	3·90	6·15
Asbestos cement terminal cone caps; asbestos yarn and composition joint to pipe	,,	6·18	7·88	11·80

Smoke-flue pipework; asbestos cement pipes and fittings, B.S. 835 (Heavy quality)		125 mm £	175 mm £	225 mm £
Pipes; asbestos yarn and composition joints in the running length; fixing in supports measured elsewhere . .	m	9·73	15·06	19·17
Extra; bends	No.	8·44	13·25	20·24
Extra; loose sockets	,,	5·86	8·47	11·00
Asbestos cement terminal cone caps; asbestos yarn and composition joint to pipe	,,	12·20	27·50	35·00

Thermal insulation	Unit	£
½ in. (13 mm) thick hessian backed needle felt flexible wrapping secured with copper wire; to copper or steel pipework; working over pipe fittings		
around 13 mm pipes	m	0·56
,, 19 ,, ,, ...	,,	0·71
,, 25 ,, ,, ...	,,	0·79
,, 32 ,, ,, ...	,,	0·95
,, 38 ,, ,, ...	,,	1·03
,, 51 ,, ,, ...	,,	1·25
19 mm thick rigid fibreglass sectional lagging secured with metal bands, white, pvc finish; to copper or steel pipework; working over pipe fittings		
around 13 mm pipes	,,	2·21
,, 19 ,, ,, ...	,,	2·48
,, 25 ,, ,, ...	,,	2·72
,, 32 ,, ,, ...	,,	3·00
,, 38 ,, ,, ...	,,	3·20
,, 51 ,, ,, ...	,,	3·70

ELECTRICAL INSTALLATIONS

ELECTRICAL INSTALLATION

Approximate prices for wiring of lighting and power points complete, including accessories and socket outlets with plugs but excluding lighting fittings, consumer control units shown separately.

	Unit	£
Consumer control units		
8-way 60 amp S.P. & N. surface mounted insulated consumer control units fitted with miniature circuit breakers including 2 m long 32 mm screwed welded conduit with three runs of 16 mm² PVC cables ready for final connections by the supply authority	No.	82·00
As above but 100 amp metal cased consumer control unit and 25 mm² PVC cables	,,	94·50
Lighting circuits		
Wired in PVC insulated and PVC sheathed cable in flats and houses installed in cavities and roof space protected, where buried, by light gauge conduit	,,	20·45
As above but in commercial property	,,	24·70
Wired in PVC insulated cable in screwed welded conduit in flats and houses	,,	35·20
As above but in commercial property	,,	44·10
As above but in industrial property	,,	50·40
Wired in M.I.C.C. cable in flats and houses	,,	33·60
As above but in commercial property	,,	38·30
As above but in industrial property	,,	50·00
Single 13 amp switched socket outlets		
Wired in PVC insulated and PVC sheathed cable in flats and houses on a ring main circuit protected, where buried, by light gauge conduit	,,	24·15
As above but in commercial property	,,	28·60
Wired in PVC insulated cable in screwed welded conduit throughout on a ring main circuit in flats and houses	,,	37·30
As above but in commercial property	,,	38·60
As above but in industrial property	,,	47·75
Wired in M.I.C.C. cable on a ring main circuit in flats and houses	,,	38·00
As above but in commercial property	,,	43·60
As above but in industrial property	,,	56·00
Cooker control units		
30 amp circuit including unit, wired in PVC insulated and PVC sheathed cable, protected where buried, by conduit	,,	47·25
As above but wired in PVC insulated cable in screwed welded conduit	,,	58·50
As above but wired in M.I.C.C. cable	,,	79·80

	Unit	£
LIGHTNING CONDUCTORS		
Flag staff terminal	No.	9·45
Copper strip roof or down conductors fixed with bracket or saddle clips		
20 × 3 mm	m	7·15
25 × 3 mm	,,	8·60
Aluminium strip roof or down conductors fixed with bracket or saddle clips		
20 × 3 mm	,,	6·30
25 × 3 mm	,,	7·90
Joints in tapes	No.	4·00
Bonding connections to roof and structural metalwork	,,	6·00
Testing points	,,	7·80
Earth electrodes		
16 mm driven copper electrodes in 1220 mm sectional lengths (2440 mm minimum)		
First 2440 mm driven and tested	,,	33·60
25 × 3 mm copper strip electrode in 457 mm deep prepared trench	m	5·50

FLOOR, WALL AND CEILING FINISHINGS

IN SITU FINISHINGS

Mortar; cement and sand (1:3); steel trowelled	Unit	32 mm work £	50 mm work £	75 mm work £
One-coat work to floors; level; to concrete base over 300 mm wide	m²	2·75	3·70	5·25
ADD to above for				
to falls, crossfalls or slopes not exceeding 15° from horizontal	„	0·11	0·11	0·11
two coats sodium silicate solution surface hardener brushed on	„	0·40	0·40	0·40
water-repellent additive incorporated in the mix (average price)	„	0·18	0·29	0·40
oil-repellent additive incorporated in the mix	„	0·62	0·97	1·45
Labour finishing around pipes and the like:				
not exceeding 0·30 m girth	No.	0·47	0·47	0·47
over 0·30 m not exceeding 1 m girth	„	1·03	1·03	1·03

Granolithic; cement and granite chippings 6 mm down (1 : 2½); steel trowelled	Unit	20 mm work £	25 mm work £	38 mm work £	50 mm work £
One-coat work to floors; level; laid in one operation with concrete base over 300 mm wide	m²	2·90	3·50	4·60	5·55
ADD to above for					
two coats sodium silicate solution surface hardener brushed on	„	0·40	0·40	0·40	0·40
liquid hardening additive incorporated in the mix (average price)	„	0·24	0·30	0·45	0·59
oil-repellent additive incorporated in the mix	„	0·39	0·49	0·73	0·97
1 kg/m² carborundum grains trowelled in	„	0·71	0·71	0·71	0·71
fair joints to flush fair edges of other finishings	m	0·47	0·47	0·47	0·47
Labour finishing around pipes and the like:					
not exceeding 0·30 m girth	No.	0·47	0·47	0·47	0·47
over 0·30 m not exceeding 1 m girth	„	1·03	1·03	1·03	1·03

13 mm one-coat skirtings to brickwork or blockwork base 75 mm wide on face; rounded edge, coved junction with paving not exceeding 30 mm radius	Unit m	£ 3·05
Ends; fair	No.	1·02
Internal angles	„	1·02
External angles	„	1·02

FLOOR, WALL AND CEILING FINISHINGS

	Unit	£
IN SITU FINISHINGS – *continued*		
Terrazzo; cement and white Sicilian marble aggregate; polished		
16 mm one-coat work to floors; level; to cement and sand base		
over 300 mm wide	m²	12·60
ADD to above for		
coloured marble aggregate	„	0·63
white or coloured cement	„	0·63
carborundum grains incorporated at the rate of 0·27 kg/m²	„	3·15
16 mm one-coat work to steps; 16 mm cement and sand (1 : 3) screed; to concrete base		
460 mm girth; one rounded nosing	m	31·50
ADD to above for		
white or coloured cement	„	0·32
projecting nosing	„	3·15
38 mm carborundum inlay	„	4·20
return end to step	No.	4·20
6 mm one-coat skirtings; to and including 13 mm cement and sand base 150 mm wide on face; fair edge with wall finishing; coved junction with paving not exceeding 13 mm radius	m	7·90
Accessories		
dividing strips; 25 × 3 mm ebonite; setting in finishing	„	1·90
„ 25 × 3 „ brass; setting in finishing	„	4·20

	Unit	B.S. 1450 Black £	B.S. 3672 Brown £	B.S. 3672 Red £
Pitchmastic; wood floated				
15 mm one-coat work to floors; level; to cement and sand base				
over 300 mm wide	m²	4·70	4·90	5·12
„ 150 „ not exceeding 300 mm wide	m	2·37	2·43	2·53
20 mm one-coat work to floors; level; to cement and sand base				
over 300 mm wide	m²	5·37	5·50	5·90
„ 150 „ not exceeding 300 mm wide	m	2·64	2·72	2·89
25 mm one-coat work to floors; level; to cement and sand base				
over 300 mm wide	m²	6·05	6·25	6·60
„ 150 „ not exceeding 300 mm wide	m	3·00	3·05	3·25
15 mm skirtings; to cement and sand base 75 mm wide on face; rounded edge; coved junction with paving not exceeding 13 mm radius	„	3·43	3·52	3·62
Internal angles	No.	0·20	0·20	0·20
External angles	„	0·20	0·20	0·20

FLOOR, WALL AND CEILING FINISHINGS

IN SITU FINISHINGS – *continued*

	Unit	£

P.V. Acetate based composition; steel trowelled

6 mm one-coat work to floors; level; to cement and sand base

	Unit	£
over 300 mm wide	m²	4·90

6 mm one-coat skirtings; to cement and sand base

75 mm wide on face; rounded edge junction with wall finishing; coved

	Unit	£
junction with paving not exceeding 13 mm radius	m	3·20
Ends	No.	0·68
Internal or external angles	,,	0·79

150 mm wide on face; rounded edge junction with wall finishing; coved

	Unit	£
junction with paving not exceeding 13 mm radius	m	6·46
Ends	No.	0·79
Internal or external angles	,,	0·95

Polyester flooring; steel trowelled

6 mm one-coat work to floors; level; to cement and sand base

	Unit	£
over 300 mm wide	m²	21·00

Polyester flooring; polished

3 mm one-coat work to floors; level; to cement and sand base

	Unit	£
over 300 mm wide	,,	16·20

Mortar; first and finishing coat of cement and sand (1 : 3); steel trowelled

19 mm work to walls; to brickwork or blockwork base

	Unit	£
over 300 mm wide	,,	3·70
not exceeding 150 mm wide	m	1·30
over 150 mm not exceeding 300 mm wide	,,	1·84
fair joints to surrounds to openings	,,	0·21
edges; fair	,,	1·16
quirks	,,	0·95
arrises	,,	0·84

ADD to above over 300 mm wide for

	Unit	£
first coat in water-repellent cement	m²	0·79
finishing coat in white cement	,,	0·53
oil-repellent additive incorporated in the mix	,,	1·58

Tyrolean decorative rendering; 13 mm first coat of cement, lime and sand (1 : 1 : 6); finishing, three coats of Cullamix applied with approved hand-operated machine; external

To walls; to brickwork or blockwork base

	Unit	£
over 300 mm wide	,,	8·95
not exceeding 150 mm wide	m	2·63
over 150 mm not exceeding 300 mm wide	,,	4·46
arrises	,,	1·05

FLOOR, WALL AND CEILING FINISHINGS

IN SITU FINISHINGS – *continued*

Plaster; first and finishing coat of carlite plaster; steel trowelled   *Unit*   £

13 mm work to walls; to brickwork or blockwork base

over 300 mm wide	m²	2·63
,, ,, ,, ,, ; in staircase area	,,	3·31
,, ,, ,, ,, ; in compartment not exceeding 4m²	,,	3·31
not exceeding 300 mm wide	m	1·31

13 mm work to ceilings; to concrete base

over 300 mm wide	m²	2·85
fair joints to surrounds to openings	m	0·21
edges; fair	,,	1·05
quirks	,,	0·84
arrises	,,	0·74

Plaster; first coat of cement and sand (1:3); finishing coat of gypsum plaster; steel trowelled

13 mm work to walls; to brickwork or blockwork base

over 300 mm wide	,,	3·68

Plaster; first coat of gypsum plaster and sand (1:2) and hair; second coat of gypsum plaster and sand (1:1½) and hair; finishing coat of gypsum plaster; steel trowelled

19 mm work to ceilings; to metal lathing base

over 300 mm wide	,,	5·25

Plasterboard B.S. 1230 type b; 3 mm joints; filled with plaster

10 mm thick wall or ceiling finishings; fixing with nails

over 300 mm wide	,,	2·42
cutting and fitting around pipes and the like		
not exceeding 0·30 m girth	No.	0·53

Gypsum plaster; steel trowelled

5 mm one-coat work to walls or ceilings; to plasterboard base

over 300 mm wide	m²	1·58
Labour finishing around pipes and the like		
not exceeding 0·30 m girth	No.	0·53

FLOOR, WALL AND CEILING FINISHINGS

TILE, SLAB OR BLOCK FINISHINGS

	Unit	150 × 150 × 12·5 mm red floor tiles £	150 × 150 × 19 mm brown floor tiles £	200 × 200 × 19 mm brown floor tiles £
Clay floor quarries, B.S. 1286 Class 1; bedding in 10 mm cement mortar (1:3); jointing and flush pointing with cement mortar (1:3)				
Joints straight both ways; to floors; level; to cement and sand base				
over 300 mm wide	m²	8·00	9·50	11·25
not exceeding 150 mm wide	m	2·00	2·40	2·81
over 150 mm not exceeding 300 mm wide	„	3·15	3·75	4·50
fair square cutting against flush fair edges of other finishings	„	1·15	1·15	1·15
raking cutting	„	1·65	1·65	1·65
curved cutting	„	2·47	2·47	2·47

	Unit	150 × 90 × 12·5 mm £	150 × 150 × 12·5 mm £	
Skirtings; to cement and sand base 150 mm extreme width; rounded edge; coved junction with paving	„	0·90	1·08	—
Ends	No.	0·26	0·26	—
Internal angles; coved	„	0·51	0·56	—
External angles; rounded	„	0·51	0·56	—

	Unit	£
Wood blocks, 25 mm thick; tongued and grooved joints; herring-bone pattern; two block border; fixing with adhesive; surface prepared to receive finishings after laying; to floors; level; to cement and sand base; over 300 mm wide	m²	
Gurjun	m²	14·00
Iroko	„	17·00
Merbau	„	17·00
Oak	„	18·25
Redwood	„	17·00
Raking cutting (all types)	m	1·76
Curved cutting (all types)	„	2·77
Extra; cutting and fitting into recessed duct covers or the like 450 mm wide; jointing to line up with adjoining work	„	8·50
Extra; cutting and fitting into recessed manhole covers 600 × 450 mm; jointing to line up with adjoining work	No.	6·95
Extra; cutting and fitting into recessed manhole covers 600 × 600 mm; jointing to line up with adjoining work	„	7·70
ADD to wood block flooring over 300 mm wide for sanding; one coat sealer, one coat wax polish	m²	0·44
„ two coats sealer; buffing with steel wool	„	0·44
„ three coats polyurethane lacquer; buffing down between coats	„	1·80

FLOOR, WALL AND CEILING FINISHINGS

	Unit	£
TILE, SLAB OR BLOCK FINISHINGS – *continued*		

Cork tiles; heavy density; butt joints straight both ways; fixing with adhesive; three coats of polyurethane; to floors; level; to cement and sand base; over 300 mm wide

	Unit	£
305 × 305 × 5 mm thick	m²	8·60
305 × 305 × 6 ,, ,,	,,	9·70
305 × 305 × 8 ,, ,,	,,	11·20

Linoleum tiles, B.S. 810; butt joints straight both ways; fixing with adhesive; to floors; level; to cement and sand base; over 300 mm wide

Marbled patterns

	Unit	£
3·20 mm thick	,,	7·25

Rubber tiles, plain colours or marbled patterns; butt joints straight both ways; fixing with adhesive; to floors; level; to cement and sand base; over 300 mm wide

	Unit	£
3·75 mm thick	,,	13·00
5 mm thick	,,	15·50

Terrazzo tiles, cement and white Sicilian marble aggregate; bedding in 13 mm cement mortar (1 : 3); margins laid in situ; polished

305 × 305 × 25 mm; butt joints straight both ways; to floors; level; to concrete base

	Unit	£
over 300 mm wide	,,	17·85

305 × 305 × 25 mm; butt joints straight both ways; to quarter-space or half-space landings; level; to concrete base

	Unit	£
over 300 mm wide	,,	21·00

Thermoplastic tiles 3 mm thick; butt joints straight both ways; fixing with adhesive; to floors; level; to cement and sand base; over 300 mm wide

	Unit	£
'B' range	,,	4·70

Vinyl tiles; butt joints straight both ways; fixing with adhesive to floors; level; to cement and sand base; over 300 mm wide

	Unit	£
2 mm thick	,,	4·78
2·50 mm thick	,,	5·00

Vinyl asbestos tiles; butt joints straight both ways; fixing with adhesive to floors; level; to cement and sand base; over 300 mm wide

	Unit	£
2 mm thick	,,	4·65
2·50 mm thick	,,	5·15

Vinyl tiles; anti-static; butt joints straight both ways; fixing with adhesive to floors; level; to cement and sand base; over 300 mm wide

	Unit	£
2 mm thick; insulating membrane	,,	20·15

FLOOR, WALL AND CEILING FINISHINGS

TILE, SLAB OR BLOCK FINISHINGS – *continued*

Ceramic tiles, B.S. 1281; fixing with approved adhesive; grouted with white cement	Unit	White standard quality tiles £	Coloured tiles from £	to £
108 × 108 × 4 mm; butt joints straight both ways; to walls; to plaster base				
over 300 mm wide	m²	8·45	8·60	12·45
fair square cutting to surround to openings	m	0·41	0·41	0·41
raking cutting	,,	0·56	0·56	0·56
curved cutting	,,	0·69	0·69	0·69
cutting and fitting around pipes and the like; not exceeding 0·30 m girth	No.	0·69	0·69	0·69

Fibreboard acoustic tiles; butt joints straight both ways; fixing with staples; to walls or ceilings; over 300 mm wide	Unit	£
305 × 305 × 13 mm, *p.c.* £2·40 per m²	m²	5·60
305 × 305 × 19 ,, *p.c.* £3·20 ,,	,,	6·55
305 × 305 × 25 ,, *p.c.* £4·00 ,,	,,	7·50

Fibreboard acoustic tiles; butt joints straight both ways; fixing with adhesive; to walls or ceilings; to cement and sand base; over 300 mm wide		
305 × 305 × 13 mm, *p.c.* £2·40 per m²	,,	9·15
305 × 305 × 19 ,, *p.c.* £3·20 ,,	,,	10·05
305 × 305 × 25 ,, *p.c.* £4·00 ,,	,,	11·00

FLOOR, WALL AND CEILING FINISHINGS

PLAIN SHEET FINISHINGS

Linoleum sheet, B.S. 810; fixing with adhesive; butt joints; to floors; level; to cement and sand base; over 300 mm wide

	Unit	£
Plain colours		
3·20 mm thick	m²	5·00
Marbled patterns		
3·20 mm thick	„	5·40

Rubber sheet, plain colours or marbled patterns; fixing with adhesive

To floors; level; to cement and sand base; over 300 mm wide

3.75 mm thick	„	12·00
5 „ „	„	14·25
To treads; to cement and sand base		
5 mm thick; 300 mm wide	m	6·48
To risers; to cement and sand base		
5 mm thick; 150 mm wide	„	3·28

Plasterboard sheet, B.S. 1230; fixing with nails; butt joints; to walls; over 300 mm wide	Unit	9·5 mm thick £	12·7 mm thick £
wallboard	m²	2·44	2·79
insulating wallboard, aluminium veneer one side	„	2·61	2·96
White plastic faced plasterboard sheet; fixing with nails; butt joints; to walls; over 300 mm wide			
insulating wallboard, aluminium veneer one side	„	2·93	3·28

FLOOR, WALL AND CEILING FINISHINGS

BEDS OR BACKINGS

	Unit	32 mm thick £	50 mm thick £	75 mm thick £
Mortar; cement and sand (1:3)				
One-coat beds, screeded; to floors; level; to concrete base over 300 mm wide	m²	2·60	3·40	5·00
One-coat beds, trowelled with a plain surface; to floors; level; to concrete base over 300 mm wide	,,	2·86	3·68	5·25

	Unit	£
13 mm one-coat backings, screeded; to walls; to brickwork or blockwork base over 300 mm wide	m²	3·31
Latex-cement		
3 mm one-coat beds, trowelled with a plain surface; to floors; level; to cement and sand base over 300 mm wide	,,	2·42
5 mm two-coat beds, trowelled with a plain surface; to floors; level; to cement and sand base over 300 mm wide	,,	4·20

	Unit	25 mm thick £	50 mm thick £	75 mm thick £
Synthanite				
One-coat beds, trowelled with a plain surface; to floors; level; paper felt underlay; to concrete base over 300 mm wide	m²	3.90	6·10	8·30

LATHING OR BASEBOARDING

	Unit	£
Metal lathing, B.S. 1369 figure 1; 50 mm laps		
For ceiling finishings; fixing with staples		
over 300 mm wide	m²	3·12
not exceeding 150 mm wide	m	0·89
raking cutting	,,	1·75
curved cutting	,,	2·56
For ceiling finishings; fixing with wire		
over 300 mm wide	m²	3·63
Building paper; 75 mm laps		
For floor finishings		
over 300 mm wide	,,	0·23
Linovent underlay; 75 mm laps		
For floor finishings		
over 300 mm wide	,,	1·65
Hardboard; 3·2 mm thick; butt joints		
For floor finishings; fixing with nails		
over 300 mm wide	,,	2·08
Plywood; 6 mm thick; butt joints		
For floor finishings; fixing with nails		
over 300 mm wide		5·70

FLOOR, WALL AND CEILING FINISHINGS

FIBROUS PLASTER

	Unit	£
Fibrous plaster; fixing with screws, countersinking, stopping, filling and pointing joints with plaster		
16 mm plain slab coverings to ceilings		
over 300 mm wide	m²	23·40
Fibrous plaster; fixing with plaster wadding; filling and pointing joints with plaster		
16 mm plain slab coverings to ceilings; to steel base		
over 300 mm wide	„	26·36
16 mm plain casings to stanchions; to steel base		
per 25 mm girth	m	0·62
16 mm plain casings to beams; to steel base		
per 25 mm girth	„	0·62
Fibrous plaster coves; fixing with screws, countersinking, stopping, filling and pointing joints with plaster		
not exceeding 150 mm girth; per 25 mm girth	„	0·62
over 150 mm girth; per 25 mm girth	„	0·68
Fibrous plaster cornices; fixing with screws, countersinking, stopping, filling and pointing joints with plaster		
per 25 mm girth	„	1·00
Fibrous plaster cornices; enrichments; fixing with screws, stopping, filling and pointing joints with plaster		
per 25 mm girth; dependent on degree of enrichments	„	1·35
Plasterboard coves; fixing with adhesive; filling and pointing joints with plaster		
125 mm girth; to plaster base	„	1·46
internal angles	No.	0·44
external angles	„	0·44

GLAZING

The following are based upon **sub-contractor's** estimates and assume that full quantity discounts are obtained.

GLASS TO METAL OR WOOD IN OPENINGS Sheet; OQ	Unit	To wood or metal with putty £	To wood with putty and beads £	To wood or metal with screwed or clipped beads £	To wood or metal with non-setting compound and screwed or clipped beads £
3 mm thick					
over 0·10 m² not exceeding 0·50 m² •	m²	8·68	8·74	12·13	12·33
Sheet; SQ					
3 mm thick					
over 0·10 m² not exceeding 0·50 m² •	,,	8·85	8·90	12·30	12·48
Patterned					
4 mm thick					
over 0·10 m² not exceeding 0·50 m² •	,,	9·25	9·30	12·69	12·89
6 mm thick					
over 0·10 m² not exceeding 0·50 m² •	,,	13·23	13·28	16·67	16·86
Rolled; rough cast					
5 mm thick					
over 0·10 m² not exceeding 0·50 m² •	,,	11·30	11·35	14·74	14·94
6 mm thick					
over 0·10 m² not exceeding 0·50 m² •	,,	11·75	11·80	15·18	15·38
Rolled; rough cast Georgian wired					
7 mm thick					
over 0·10 m² not exceeding 0·50 m² •	,,	12·67	12·73	16·12	16·31
Rolled; polished Georgian wired					
6 mm thick					
all sizes • • • • • •	,,	24·96	25·27	26·68	27·33
Clear float; GG					
4 mm thick					
over 0·10 m² not exceeding 0·50 m² •	,,	9·31	9·37	12·76	12·95
5 mm thick					
not exceeding 4 m² • • • •	,,	13.99	14·30	15·71	16·36
6 mm thick					
not exceeding 4 m² • • • •	,,	14·93	15·25	16·65	17·30
Toughened clear float					
6 mm thick					
over 0·20 m² not exceeding 0·50 m² •	,,	29·83	30·15	31·55	32·20
over 0·50 m² not exceeding 2·50 m² •	,,	28·36	28·68	30·08	30·73
over 2·50 m² • • • • •	,,	26·84	27·15	28·56	29·21
10 mm thick					
over 0·20 m² not exceeding 0·50 m² •	,,	40·07	40·38	41·79	42·44
over 0·50 m² not exceeding 2·50 m² •	,,	40·07	40·38	41·79	42·44
over 2·50 m² • • • • •	,,	38·55	38·86	40·27	40·92

GLAZING

GLASS TO METAL OR WOOD IN OPENINGS – *continued*

	Unit	To wood or metal with putty £	To wood with putty and beads £	To wood metal with screwed or clipped beads £	To wood or metal with non-setting compound and screwed or clipped beads £
Laminated clear sheet					
4·4 mm thick, safety over 0·10 m²	m²	29·55	29·86	31·27	31·92
5·4 mm thick, safety over 0·10 m²	„	30·46	30·78	32·18	32·83
7·5 mm thick, antibandit over 0·10 m²	„	51·94	52·20	53·67	54·32

		Unit	With non-setting compound £
Solar control float			
4 mm thick, 'Antisun'		m²	30·42
10 mm thick, 'Antisun' over 4 m² not exceeding 9·30 m²		„	72·94
12 mm thick 'Antisun' over 4 m² not exceeding 9·30 m²		„	91·95
Hermetically sealed units; Insulight			
Double; two panes 4 mm clear float GG; one 13 mm air space		„	36·30
Double; two panes 6 mm clear float GG; one 13 mm air space		„	47·63
			£
Sundries			
Imitation wash leather strips 32 mm wide; bedding to edges of glass . .		m	0·38
Black ribbon velvet strips 32 mm wide; bedding to edges of glass . .		„	0·38
Patent glazing			
6 mm rolled rough cast Georgian wired glass; aluminium alloy bars to roof; bars 2438 mm long at 622 mm spacing; metal bearings		m²	29·75

PAINTING AND DECORATING

The following prices include for preparing surfaces and those for painting woodwork also include for knotting prior to applying the priming coat and stopping of nail holes, etc.

TWO COATS CEMENT PAINT

Stonework, brickwork or blockwork surfaces; external	Unit	£
Walls over 300 mm girth	m²	1·04

Cement render surfaces

Walls over 300 mm girth	„	0·93

Concrete surfaces

Walls over 300 mm girth	„	0·93

ONE COAT PLASTIC COMPOUND

Concrete surfaces

Walls or ceilings over 300 mm girth	„	1·02

Plasterboard surfaces

Walls or ceilings over 300 mm girth	„	0·76

EMULSION PAINT

Stonework, brickwork or blockwork surfaces; internal	Unit	Two coats £	One mist and two coats £
Walls over 300 mm girth	m²	1·16	1·48

Cement render surfaces

Walls or ceilings over 300 mm girth	„	0·97	1·27

Concrete surfaces

Walls over 300 mm girth	„	1·03	1·32

Plaster surfaces

Walls or ceilings over 300 mm girth	„	0·97	1·27

PAINTING AND DECORATING

OIL PAINT	Unit	One undercoat ready primed Int. £	Ext. £	One coat primer one undercoat Int. £	Ext. £	One additional undercoat Int. £	Ext. £	One coat full gloss finish Int. £	Ext. £
Iron or steel surfaces									
General surfaces									
over 300 mm girth . . .	m²	0·80	0·86	1·69	1·83	0·68	0·73	0·75	0·81
not exceeding 150 mm girth . .	m	0·18	0·20	0·39	0·43	0·16	0·17	0·18	0·19
over 150 mm not exceeding 300 mm girth	,,	0·29	0·31	0·61	0·67	0·25	0·27	0·27	0·30
Windows or the like									
large panes	m²	1·04	1·14	2·19	2·40	0·88	0·97	0·98	1·08
medium panes	,,	1·31	1·43	2·76	3·02	1·11	1·22	1·24	1·36
small panes	,,	1·38	1·51	2·91	3·18	1·18	1·29	1·31	1·43
opening edges . . .	m	0·25	0·27	0·52	0·57	0·21	0·23	0·23	0·25
Members of roof trusses or the like									
generally	m²	1·12	1·22	2·38	2·58	0·96	1·04	1·06	1·15
Ornamental railings or the like; each side measured									
generally	,,	1·45	1·58	3·06	3·34	1·24	1·35	1·37	1·50
Eaves, gutters or the like									
not exceeding 150 mm girth .	m	—	0·60	—	1·28	—	0·52	—	0·57
over 150 mm not exceeding 300 mm girth	,,	—	0·94	—	2·00	—	0·80	—	0·89
Pipes and conduit									
over 300 mm girth . . .	m²	1·12	1·22	2·38	2·58	0·96	1·04	1·06	1·16
not exceeding 150 mm girth . .	m	0·27	0·29	0·57	0·62	0·23	0·25	0·25	0·28
over 150 mm not exceeding 300 mm girth	,,	0·41	0·45	0·87	0·95	0·35	0·38	0·39	0·43
Plaster surfaces									
Walls or ceilings									
over 300 mm girth . . .	m²	—	—	1·68	—	0·68	—	0·75	—
Wood surfaces									
General surfaces									
over 300 mm girth . . .	,,	0·79	0·85	1·84	2·00	0·68	0·73	0·75	0·81
not exceeding 150 mm girth . .	m	0·18	0·20	0·44	0·47	0·16	0·17	0·18	0·19
over 150 mm not exceeding 300 mm girth	,,	0·29	0·31	0·67	0·73	0·25	0·27	0·27	0·30
Windows or the like									
large panes	m²	1·04	1·13	2·47	2·70	0·88	0·97	0·98	1·08
medium panes	,,	1·31	1·43	3·11	3·41	1·11	1·22	1·24	1·36
small panes	,,	1·38	1·51	3·28	3·60	1·18	1·29	1·31	1·43
opening edges . . .	m	0·25	0·27	0·58	0·63	0·21	0·23	0·23	0·25
Windows or the like (casements painted alternative colour)									
large panes	m²	1·23	1·35	2·94	3·22	1·05	1·15	1·17	1·28
medium panes	,,	1·55	1·70	3·71	4·07	1·33	1·45	1·48	1·62
small panes	,,	1·64	1·80	3·91	4·29	1·40	1·53	1·56	1·71
opening edges	m	0·29	0·32	0·68	0·75	0·25	0·27	0·28	0·30

PAINTING AND DECORATING

HEAT-RESISTING PAINT

	Unit	One coat; ready primed £	One coat primer; one coat £	One additional coat £
Iron or steel surfaces; internal				
Radiators or the like				
generally	m²	0·98	2·06	0·90
Pipes and conduit				
over 300 mm girth	„	1·40	2·98	1·29
not exceeding 150 mm girth	m	0·34	0·71	0·32
over 150 mm not exceeding 300 mm girth	„	0·52	1·08	0·49

STAIN; TWO COATS VARNISH; FULL GLOSS FINISH

	Unit	£
Wood surfaces; internal		
General surfaces		
over 300 mm girth	m²	2·12
not exceeding 150 mm girth	m	0·50
over 150 mm not exceeding 300 mm girth	„	0·77

ONE COAT PRIMER; ONE UNDERCOAT; TWO COATS VARNISH; FULL GLOSS FINISH

	Unit	Brush and comb grain £	Figure grain £
Wood surfaces; internal			
General surfaces			
over 300 mm girth	m²	4·92	6·00
not exceeding 150 mm girth	m	1·18	1·45
over 150 mm not exceeding 300 mm girth	„	1·81	2·22

TWO COATS WOOD PRESERVATIVE

	Unit	Creosote £	Proprietary branded wood preservative £
Wood surfaces; external			
General surfaces			
over 300 mm girth	m²	1·13	1·24
not exceeding 150 mm girth	m	0·27	0·29
over 150 mm not exceeding 300 mm girth	„	0·41	0·45

TWO COATS BOILED LINSEED OIL

	Unit	£
Hardwood surfaces; internal		
General surfaces		
over 300 mm girth	m²	1·36
not exceeding 150 mm girth	m	0·31
over 150 mm not exceeding 300 mm girth	„	0·49

SEAL; TWO COATS OF SYNTHETIC RESIN LACQUER; WAX PASTE DECORATIVE FLATTED FINISH

	Unit	£
Wood surfaces; internal		
General surfaces		
over 300 mm girth	m²	5·09
not exceeding 150 mm girth	m	1·38
over 150 mm not exceeding 300 mm girth	„	1·73

PAINTING AND DECORATING

BODYING IN WITH FRENCH POLISH; ONE COAT OF LACQUER OR VARNISH

Wood surfaces; external   *Unit*   £

General surfaces
    over 300 mm girth . . . . m²   5·46
    not exceeding 150 mm girth . . . . m   1·23
    over 150 mm not exceeding 300 mm girth . . . . „   1·66

STAINING, BODYING IN AND FULLY FRENCH POLISHING; FULL GLOSS FINISH

Hardwood surfaces; internal

General surfaces
    over 300 mm girth . . . . m²   10·70
    not exceeding 150 mm girth . . . . m   2·40
    over 150 mm not exceeding 300 mm girth . . . . „   3·27

STAINING, BODYING IN AND FULLY FRENCH POLISHING; EBONIZED FINISH

Hardwood surfaces; internal

General surfaces
    over 300 mm girth . . . . m²   12·60
    not exceeding 150 mm girth . . . . m   2·84
    over 150 mm not exceeding 300 mm girth . . . . „   3·84

SEALING, BODYING IN AND WAX POLISHING; DULL GLOSS FINISH

Hardwood surfaces; internal

General surfaces
    over 300 mm girth . . . . m²   3·57
    not exceeding 150 mm girth . . . . m   0·81
    over 150 mm not exceeding 300 mm girth . . . . „   1·09

SIGNWRITING

Painted or varnished surfaces; internal

Plain capital or lower-case letters or numerals
    per coat; per 25 mm in height . . . . No.   0·33
Stops; per coat . . . . „   0·16

LINING PAPER

Plaster surfaces

Papering; *p.c.* £0·70 per piece
    Walls . . . . m²   0·84

DECORATIVE PAPER

Plaster surfaces

Papering; *p.c.* £4·00 per piece
    Walls . . . . „   1·82

DRAINAGE – DRAINS

Prices for drain trenches are for excavation in heavy soil, and it has been assumed that earthwork support will only be required for trenches 1 m or more in depth.

Attention is drawn to the notes on excavation in other soils under 'Excavation and Earthwork'. The necessity for and the type of earthwork support required do, of course, vary with the type of soil, and this factor should be borne in mind if adjustments are being made.

EXCAVATING TRENCHES
To receive pipes; grading bottoms; earthwork support; filling with excavated material and compacting; disposal of surplus spoil by spreading on site average 50 m from excavation

Excavating and filling trenches by hand
Internal diameter of pipes

Starting from natural ground level; average depth	Unit	Not exceeding 200 mm £	225 mm £	300 mm £	375 mm £	450 mm £	600 mm £	750 mm £	900 mm £	1050 mm £	1200 mm £
0·75 m	m	5·86	5·86	6·63	7·27	8·67	—	—	—	—	—
1·00 „	„	10·93	10·93	11·83	12·84	14·76	17·69	20·62	—	—	—
1·25 „	„	13·63	13·63	14·78	16·05	18·48	22·17	25·74	28·17	—	—
1·50 „	„	16·46	16·46	17·73	19·26	22·20	26·53	31·00	33·80	36·73	39·80
1·75 „	„	22·07	22·07	23·67	25·52	28·78	34·40	39·70	43·14	46·72	50·43
2·00 „	„	29·80	29·80	31·85	34·03	37·97	44·38	50·50	54·75	59·09	63·04
2·25 „	„	34·07	34·07	36·29	38·80	43·54	50·77	57·87	62·58	67·60	72·35
2·50 „	„	38·35	38·35	40·90	43·72	48·96	57·16	65·23	70·60	76·26	81·50
2·75 „	„	42·62	42·62	45·50	48·64	54·53	63·54	72·59	77·82	84·92	90·81
3·00 „	„	46·90	46·90	50·09	53·57	59·95	70·10	79·95	86·62	93·58	100·00
3·25 „	„	53·40	53·40	57·00	60·86	68·03	79·17	90·20	97·67	105·40	112·60
3·50 „	„	58·53	58·53	62·50	66·80	74·54	87·05	99·10	107·33	115·90	123·63

Excavating by mechanical trencher, filling by hand
Internal diameter of pipes

Starting from natural ground level: average depth	Unit	Not exceeding 200 mm £	225 mm £	300 mm £	375 mm £	450 mm £	600 mm £	750 mm £	900 mm £	1050 mm £	1200 mm £
0·75 m	m	3·51	3·51	3·72	4·24	4·97	—	—	—	—	—
1·00 „	„	6·53	6·53	6·84	7·47	8·42	9·67	11·24	—	—	—
1·25 „	„	8·00	8·00	8·37	9·11	10·26	11·78	13·66	14·80	—	—
1·50 „	„	9·43	9·43	9·90	10·79	12·10	13·88	16·13	17·44	18·80	20·11
1·75 „	„	13·22	13·22	13·69	14·73	16·25	18·30	20·86	22·38	23·90	25·31
2·00 „	„	14·97	14·97	15·55	16·70	18·43	20·73	23·60	25·34	27·06	28·80
2·25 „	„	16·78	16·78	17·40	18·66	20·60	23·16	26·36	28·29	30·23	32·17
2·50 „	„	18·53	18·53	19·26	20·68	22·77	25·60	29·16	31·25	33·40	35·50
2·75 „	„	20·34	20·34	21·07	22·64	24·94	28·03	31·90	34·21	36·50	38·87
3·00 „	„	22·09	22·09	22·93	24·60	27·12	30·47	34·65	37·17	39·68	42·20
3·25 „	„	25·40	25·40	26·29	28·07	30·80	34·40	38·90	41·63	44·35	47·07
3·50 „	„	27·27	27·27	28·26	30·20	33·08	36·95	41·82	44·70	47·63	50·50

DRAINAGE – DRAINS

EXCAVATING TRENCHES – *continued*

To receive land drain pipes; grading bottoms; earthwork support; filling 300 mm deep with hardcore, remainder with excavated material; disposal of surplus spoil by spreading on site average 50 m from excavation

Starting from natural ground level; average depth

	Unit	£
0·75 m	m	7·78
1·00 ,,	,,	11·82
1·25 ,,	,,	14·27

FILLING

Sand

Beds to receive pitch fibre pipes

			Unit	£
50 mm thick; 600 mm wide			,,	0·63
50 ,,	700 ,,		,,	0·74
50 ,,	800 ,,		,,	0·93

		Beds		Beds and benchings		Beds and coverings	
		100 *mm* thick	150 *mm* thick	100 *mm* thick	150 *mm* thick	100 *mm* thick	150 *mm* thick
Granular	*Unit*	£	£	£	£	£	£
Internal diameter of pipe							
100 mm	m	0·38	0·58	0·76	0·90	1·09	1·46
150 ,,	,,	0·42	0·65	0·83	1·05	1·36	1·77
225 ,,	,,	0·51	0·76	1·05	1·28	1·79	2·24
300 ,,	,,	0·58	0·87	1·28	1·59	2·19	2·75

IN SITU CONCRETE; PLAIN

Normal mix 11·50 N/mm²
40 mm aggregate
Internal diameter of pipe

		Beds 100 mm	Beds 150 mm	Benchings 100 mm	Benchings 150 mm	Coverings 100 mm	Coverings 150 mm
100 mm	,,	1·78	2·72	3·45	4·11	5·03	6·78
150 ,,	,,	1·99	3·00	3·78	4·80	6·28	8·15
225 ,,	,,	2·39	3·57	4·80	5·83	8·24	10·31
300 ,,	,,	2·72	4·05	5·83	7·23	10·11	12·69
375 ,,	,,	3·08	4·66	7·23	8·58	12·02	14·97
450 ,,	,,	3·44	5·19	8·58	10·31	14·06	17·30
610 ,,	,,	4·13	6·20	11·00	13·06	18·09	22·08
750 ,,	,,	4·42	6·56	14·78	17·16	22·79	27·33
900 ,,	,,	5·02	7·54	18·19	20·94	27·66	32·77
1050 ,,	,,	5·67	8·47	22·00	25·05	32·82	38·64
1200 ,,	,,	6·28	9·40	26·77	29·52	38·31	44·80

PIPEWORK

Clay pipes, B.S. 1196

Pipes; no joints in the running length

			Unit	£
75 mm internal diameter			m	0·80
100 mm	,,	,,	,,	1·10
150 mm	,,	,,	,,	1·93

DRAINAGE – DRAINS

PIPEWORK – *continued*

Vitrified clay pipes and fittings; joints in the running length	Unit	100 £	150 £	225 £	300 £	400 £	450 £
		Diameter of pipes in mm					
Extra strength quality; plains ends with plastic sleeve couplings							
Straight pipe	m	2·35	4·31	—	—	—	—
" " ; in branches not exceeding 3 m long	"	2·59	4·63	—	—	—	—
Extra; bends	No.	1·37	2·73	—	—	—	—
Extra; single branches . .	"	3·04	4·09	—	—	—	—
Extra; socket adapter . .	"	1·73	3·37	—	—	—	—
Extra; taper	"	—	2·61	—	—	—	—
Extra strength quality, socketted; with flexible joints							
Straight pipe	m	—	4·53	7·72	12·51	22·44	29·23
" " ; in branches not exceeding 3 m long	No.	—	4·89	8·17	13·29	23·53	30·80
Extra; bends	No.	—	6·21	11·70	23·00	51·86	68·53
Extra; single branches . .	"	—	7·22	15·50	33·30	65·10	77·58
Extra; double branches . .	"	—	14·86	32·38	69·16	—	—
Extra; taper	"	—	9·22	13·07	27·04	—	—
Best quality; cement mortar joints							
Straight pipe	m	2·52	4·01	6·96	11·71	19·91	25·25
" " ; in branches not exceeding 3 m long	"	2·85	4·48	9·53	15·80	26·65	33·77
Extra; bends	No.	1·82	2·64	7·03	13·25	48·35	59·54
Extra; single branches . .	"	3·42	5·05	13·71	26·15	46·06	56·73
Extra; double branches . .	"	6·49	9·93	27·88	53·74	—	—
Extra; taper	"	—	6·05	12·32	23·92	—	—
B.S.65 Tested quality; cement mortar joints							
Straight pipe	m	3·23	5·31	9·57	16·29	28·04	35·61
Extra; bends	No.	2·31	3·45	9·90	19·06	71·38	87·83
Extra; single branches . .	"	4·39	6·66	19·44	37·78	67·06	59·33
Extra; double branches . .	"	8·67	13·59	40·20	78·54	—	—
Extra; taper	"	—	8·58	17·85	35·07	—	—

Concrete pipes and fittings, B.S.556; rubber ring joints in the running length	Unit	300 *mm* £	450 *mm* £	600 *mm* £	900 *mm* £	1200 *mm* £
		Diameter of pipes				
Standard	m	14·75	26·34	36·95	55·68	—
Extra; bends . . .	No.	24·70	31·00	44·00	61·25	—
Extra; single branches . .	"	39·40	73·20	—	—	188·65
Class H	m	—	—	—	—	234·00
Extra; bends	No.	—	—	—	—	

	Unit	£
Concrete road gullies, B.S. 556; cement mortar (1:2) joint to concrete pipe; bedding and surrounding in normal concrete mix 11·50 N/mm² 40 mm aggregate trapped; 450 mm diameter; 1·07 m deep	No.	41·56

Vitrified clay gullies; B.S. 539; flexible joint to vitrified clay pipe; bedding and surrounding in normal concrete mix 11·50 N/mm² 40 mm aggregate	Unit	150 × 150 mm top 100 mm outlet £	225 × 225 mm top 150 mm outlet £
trapped; cast iron grating; coated	No.	11·80	20·00
ADD to above for			
100 mm horizontal inlets	"	3·09	7·77
100 mm vertical inlets	"	3·09	7·77
raising pieces	"	4·25	9·56

DRAINAGE – DRAINS

PIPEWORK – *continued*

		225 mm diameter 100 mm outlet	300 mm diameter 100 mm outlet
Vitrified clay gullies; B.S. 539 Table 24; flexible joint to vitrified clay pipe; bedding and surrounding in normal concrete mix 11·50 N/mm² 40 mm aggregate	*Unit*	£	£
trapped; perforated bucket; galvanized; Stanford stopper; cast iron grating; coated	No.	41·10	58·25
Vitrified clay mud gullies; B.S. 539 Table 25; flexible joint to vitrified clay pipe; bedding and surrounding in normal concrete mix 11·50 N/mm² 40 mm aggregate			
trapped; cast iron grating; coated; perforated bucket; coated .	,,	26·95	44·25

Vitrified clay street gullies; B.S. 539 Table 26; flexible joint to vitrified clay pipe; bedding and surrounding in normal concrete mix 11·50 N/mm² 40 mm aggregate

	Unit	£
trapped; 375 mm diameter; 750 mm deep 100 mm outlet	No.	31·70
,, 375 mm ,, 900 mm ,, 150 mm ,, . . .	,,	34·85
,, 450 mm ,, 900 mm ,, 150mm ,, . . .	,,	43·60

Cast iron road gratings and frames; B.S. 497 Table 12: coated; bedding and pointing in cement mortar (1:3); one-course half-brick-thick wall in common bricks in cement mortar (1:3)

		£
400 × 350 mm weight 99 kg	,,	43·40
500 × 350 ,, ,, 124 ,, ...	,,	53·80

		Internal diameter		
		75 mm	100 mm	150 mm
Pitch fibre pipes and fittings, B.S. 2760	*Unit*	£	£	£
Pipes; couplers in the running length . . .	m	1·93	2·36	4·91
Extra; 90; bends	No.	2·94	2·41	—
Extra; single branch junctions .	,,	3·26	5·08	4·19

		Internal diameter			
Spun cast iron pipes, B.S. 1211, Class B; Cast iron fittings, B.S. 1130; coated; joints in the running length		75 mm	100 mm	150 mm	225 mm
	Unit	£	£	£	£
Pipes; caulked lead joints; in trenches . .	m	9·17	11·31	16·97	29·20
Extra; fig. No. 1 bends	No.	10·12	13·78	28·18	94·52
Extra; fig. No. 4 branches .	,,	17·53	19·82	41·41	127·41
Pipes; caulked lead joints; in ducts; supported on piers (m.s.)	m	9·47	11·75	17·52	30·43
Pipes; caulked lead joints; bracketted off walls (brackets m.s.)	,,	9·77	12·18	18·07	31·69
Pipes; caulked lead joints; suspended from soffits (hangers m.s.)	,,	10·18	12·77	18·80	33·32

DRAINAGE – MANHOLES

PIPEWORK – *continued*

Cast iron traps; B.S. 1130; coated; caulked lead joint to cast iron pipe; bedding and surrounding in normal concrete mix 11·50 N/mm² 40 mm aggregate

	Unit	*Internal diameter*			
		75 *mm* £	100 *mm* £	150 *mm* £	225 *mm* £
Gully trap; high invert	No.	16·56	21·53	49·60	147·20
ADD to above for					
Bellmouth 305 mm high and grating; caulked lead joint to cast iron . .	„	11·28	13·38	25·33	—
Bellmouth 305 mm high and grating; one horizontal inlet; caulked lead joint to cast iron	„	—	19·16	39·50	—
Bellmouth 305 mm high and grating; one vertical inlet; caulked lead joint to cast iron	„	17·44	20·10	44·55	—

Cast iron rainwater shoes; B.S. 1130: coated; caulked lead joint to cast iron pipe; bedding and surrounding in normal concrete mix 11·50 N/mm² 40 mm aggregate

	Unit	*Internal diameter*	
		100 *mm* £	150 *mm* £
Rainwater shoe; vertical inlet; inspection cover fitted with No. 4 gunmetal screws and felt washer	No.	30·40	60·20

Cast iron gratings; light duty for block channels

		Coated £	Galvanized £
175 × 10 mm	m	17·42	25·36

Cast iron medium-duty gratings and frames with lugs; bedding and pointing to frames in cement mortar (1:3); cutting and pinning lugs to concrete

		£	£
140 mm wide	„	24·36	35·65
190 „ „	„	25·66	40·60
265 „ „	„	28·95	48·30

Composite cleaning eyes; vitrified clay pipe 600 mm long and short radius bend, best quality; cement mortar (1:2) joints; bedding and surrounding in concrete mix 11·50 N/mm² 40 mm aggregate; 152 × 127 mm cast iron and hinged cover and frame; coated; 457 × 457 × 102 mm concrete mix 21·00 N/mm² 20 mm aggregate surround; trowelling; formwork

	Unit	£
100 mm internal diameter	No.	21·95

Kerbs to gullies; common bricks on edge to three sides in cement mortar (1:3); rendering in cement mortar (1:3) to top and two sides and skirting to brickwork 230 mm high, dishing in cement mortar (1:3) to gully; steel trowelled . . .

230 × 230 mm internally	„	5·05

Kerbs to gullies: common bricks on edge to three sides in cement mortar (1:3); rendering in cement mortar (1:3) to top and two sides and skirting to brickwork 230 mm high; steel trowelled 100 mm straight vitrified clay best-quality half-section channel; 450 mm long; bedded in cement mortar (1:3); 102 mm concrete mix 11·50 N/mm² 40 mm aggregate base; dishing in cement mortar (1:3) to channel and gully; steel trowelled

690 × 230 mm internally	„	6·16

DRAINAGE – MANHOLES

	Unit	£
EXCAVATION		
Excavating		
Pits; starting from natural ground level		
not exceeding 1 m deep	m³	11·50
,, ,, 2 m deep	,,	12·00
,, ,, 4 m deep	,,	14·25
Disposal		
Surplus excavated material		
depositing on site in permanent spoil heaps	,,	1·45
Filling		
Excavated material		
backfilling into excavation	,,	1·27
Surface treatments		
Bottoms of excavations		
levelling, compacting	m²	0·19
EARTHWORK SUPPORT		
Sides of excavations; not exceeding 2 m between opposing faces		
not exceeding 1 m deep	,,	1·58
,, ,, 2 m ,,	,,	2·09
,, ,, 4 m ,,	,,	2·59
IN SITU CONCRETE; PLAIN		
Normal; mix 11·50 N/mm² 40 mm aggregate		
Beds		
over 100 mm not exceeding 150 mm thick	m³	37·20
,, 150 mm ,, ,, 300 mm ,,	,,	35·50
Normal; mix 25·50 N/mm² 20 mm aggregate		
Benching in bottoms		
over 150 mm not exceeding 300 mm thick	,,	77·00
IN SITU CONCRETE; REINFORCED		
Normal; mix 21·00 N/mm² 20 mm aggregate		
Isolated cover slabs		
Not exceeding 100 mm thick	,,	60·25
REINFORCEMENT		
Fabric; B.S. 4483		
Mesh 200 × 200 mm × 3·02 kg/m²		
in cover slabs	m²	1·66
FORMWORK TO REINFORCED IN SITU CONCRETE		
Formwork generally		
Isolated cover slabs		
soffits	,,	17·95
edges; not exceeding 250 mm high	m	2·52

DRAINAGE – MANHOLES

PRECAST CONCRETE		Diameter of chamber or shaft sections internally				
		900 mm	1050 mm	1200 mm	1500 mm	1800 mm
Inspection chamber units; B.S. 556; bedding, jointing and pointing in cement mortar (1:3)	Unit	£	£	£	£	£
Chamber or shaft sections . .	m	61·29	76·15	96·70	152·45	190·65
Sections 600 mm high tapering from 675 mm diameter to .	No.	42·00	56·18	74·12	135·95	172·15
Cover slabs for chambers or shaft sections						
Heavy duty	„	38·38	48·38	64·87	103·27	156·50

BRICKWORK		Common bricks p.c. £45·00 1000	Engineering bricks p.c. £122·00 1000
Bricks in cement mortar (1 : 3)	Unit	£	£
One brick thick walls	m²	22·54	37·25
One and a half brick thick walls	„	30·60	51·35
Two brick thick projection of footings or the like	„	38·20	64·85
Extra over common or engineering bricks in any mortar for			
Fair face; flush pointing as work proceeds; English bond walls or the like	„	1·48	1·48

IN SITU FINISHINGS	Unit	£
Mortar; cement and sand (1:3); steel trowelled		
13 mm one coat work to manhole walls; to brickwork or blockwork base over 300 mm wide	m²	4·10

SUNDRIES

Building into brickwork; ends of pipes; making good facings or rendering		
100 mm diameter	No.	0·77
150 „ „	„	0·98
225 „ „	„	1·49
225 „ „ half brick ring arch	„	2·00
300 „ „ „ „	„	2·65

DRAINAGE – MANHOLES

STANDARD UNITS

Cast iron inspection chambers; B.S. 1130; coated; access covers fitted with 16 mm gunmetal bolts and nuts and felt washers; bedding in cement mortar (1:3)	Unit	*Internal diameter*		
		100 *mm* £	150 *mm* £	225 *mm* £
Fig. No. 13; one branch	No.	66·57	115·40	236·10
„ 14; „ each side	„	84·55	140·00	288·00
„ 18; three branches each side	„	187·00	318·00	—
Cast iron petrol interceptor bends; coated 381 × 762 mm .	„	36·00	56·00	132·50

Step irons; B.S. 1247 malleable cast iron; galvanized; building into joints	Unit	£
general-purpose pattern for one brick walls	No.	4·74

Access covers and frames, B.S. 497 Grade 'C'; bedding frame in cement mortar (1:3), cover in grease and sand		*Clear opening sizes* *Coated*		
	Unit	450 × 450 *mm* £	600 × 450 *mm* £	600 × 600 *mm* £
single seal flat	No.	18·75	15·20	35·20
double seal flat	„	27·22	32·31	48·95
single seal recessed	„	28·94	35·67	48·80
double seal recessed	„	43·86	60·23	79·30

Access covers and frames, B.S. 497 Grade 'B'; coated; bedding frame in cement mortar (1:2), cover in grease and sand	Unit	£
clear opening 500 mm diameter	No.	50·60
„ 550 „ „	„	60·25
„ 600 × 450 mm, solid top	„	64·75
„ 600 × 450 „ recessed top	„	70·75

Access covers and frames, B.S. 497 Grade 'A' double triangular; coated; bedding frame in cement mortar (1:3), cover in grease and sand		
clear opening 500 mm diameter	„	86·30
„ 550 „ „	„	89·30

CHANNELS

Vitrified clay, best quality; bedding and jointing in cement mortar (1 : 2)	Unit	100 *mm* £	150 *mm* £	225 *mm* £	300 *mm* £
Half section					
straight	m	2·56	4·02	7·00	11·70
bends	No.	2·47	3·54	8·39	15·10
straight, tapered; 600 mm long . .	„	—	6·04	12·19	22·85

Vitrified clay, best quality; ceramic glazed; bedding and jointing in cement mortar (1 : 2)					
Half section					
straight	m	21·70	29·90	47·00	—
angle pieces	No.	18·00	26·45	42·00	—

DRAINAGE – MANHOLES

INTERCEPTING TRAPS

Vitrified clay; inspection arm; brass stopper; iron lever, chain and staple; galvanized; staple cut and pinned to brickwork; cement mortar (1 : 2) joints to vitrified clay pipe and channel; bedding and surrounding in normal concrete mix 11·50 N/mm² 40 mm aggregate; cutting and fitting brickwork; maknig good facings

	Unit	£
100 mm inlet, 100 mm outlet	No.	15·75
150 ,, ,, 150 ,, ,, ...	,,	22·35
225 ,, ,, 225 ,, ,, ...	,,	52·00

FENCING

STANDARD FENCING

Strained wire; 3·25 mm galvanized mild steel line wire; galvanized steel components; concrete posts and struts	Unit	Height of fencing			
		915 mm £	1070 mm £	1220 mm £	1375 mm £
fencing with three lines plain wire threaded through posts and strained with eye bolts; setting posts at 2750 mm centres, 610 mm below ground; excavating holes; filling with excavated material; replacing top soil; disposing of surplus spoil by removing from site . .	m	2·68	3·00	3·10	3·70
Extra; straining posts with one strut; post and strut set 610 mm below ground; excavating holes; filling to within 150 mm of ground level with concrete mix 1 : 12 40 mm aggregate; replacing top soil; disposal of surplus spoil by removing from site .	No.	14·33	15·25	15·85	17·00
Extra; straining posts with two struts; post and struts set 610 mm below ground; excavating holes; filling to within 150 mm of ground level with concrete mix 1 : 12 40 mm aggregate; replacing top soil; disposal of surplus spoil by removing from site . .	,,	21·15	22·90	24·00	25·85

Chain link, 3 mm, 51 mm mesh; galvanized mild-steel mesh, line wires and tying wires; galvanized steel components; mild-steel angle posts and struts	Unit	Height of fencing		
		915 mm £	1220 mm £	1830 mm £
fencing with line wires threaded through posts and strained with eye bolts; setting posts at 3050 mm centres, 610 mm below ground (760 mm below ground for 1830 mm high fencing); excavating holes; filling with excavated material; replacing top soil; disposal of surplus spoil by removing from site .	m	5·75	6·90	9·10
Extra; straining posts with one strut; post and strut set 610 mm below ground (760 mm below ground for 1830 mm high fencing); excavating holes; filling to within 150 mm of ground level with concrete 1 : 12 40 mm aggregate; replacing top soil; disposal of surplus spoil by removing from site	No.	13·10	14·25	19·50
Extra; straining posts with two struts; post and struts set 610 mm below ground; excavating holes; filling to within 150 mm of ground level with concrete 1 : 12 40 mm aggregate; replacing top soil; disposal of surplus spoil by removing from site	,,	19·00	21·60	30·80

FENCING

STANDARD FENCING – *continued*

			Height of fencing	
	Unit	915 mm £	1220 mm £	1830 mm £
Chain link, 3 mm, 51 mm mesh; galvanized mild-steel mesh, line wires and tying wires; galvanized steel components; concrete posts and struts				
fencing with line wires threaded through posts and strained with eye bolts; setting posts at 3050 mm centres, 610 mm below ground (760 mm below ground for 1830 mm high fencing); excavating holes; filling with excavated material; replacing top soil; disposal of surplus spoil by removing from site	m	5·90	7·30	10·60
Extra; straining posts with one strut; post and strut set 610 mm below ground (760 mm below ground for 1830 mm high fencing); excavating holes; filling to within 150 mm of ground level with concrete 1 : 12 40 mm aggregate; replacing top soil; disposal of surplus spoil by removing from site	No.	14·33	15·85	21·65
Extra; straining posts with two struts; post and struts set 610 mm below ground (760 mm below ground for 1830 mm high fencing); excavating holes; filling to within 150 mm of ground level with concrete 1 : 12 40 mm aggregate; replacing top soil; disposal of surplus spoil by removing from site		21·15	24·00	32·35

	Unit	Fencing 2745 mm high with standards at 3660 mm centres £	Fencing 3660 mm high with standards at 2745 mm centres £
Chain link; 2 mm, 44 mm mesh; galvanized mild-steel mesh, line wires and tying wires; galvanized steel components; 44 × 44 × 5 mm mild-steel angle standards, straining posts and struts			
fencing surrounding tennis court size 36 × 18 m; gate size 1070 × 1980 mm complete with hinges, lock and ironmongery	No.	1050·00	1400·00

		Height of fencing		
	Unit	915 mm £	1070 mm £	1220 mm £	1375 mm £
Cleft chestnut pale; pales spaced 51 mm apart; two lines of galvanized wire; galvanized tying wire; 64 mm diameter posts; 76 × 51 mm struts					
fencing; driving posts at 2740 mm centres 610 mm below ground	m	2·10	3·40	3·90	4·10
Extra; straining posts with one strut; driving post and strut 610 mm below ground	No.	6·70	7·10	8·15	9·10

FENCING

STANDARD FENCING – *continued*

Mild steel unclimbable; in panels 2440 mm long; 44 × 13 mm flat section top and bottom rails; two 44 × 19 mm flat section standards, one with foot plate and 38 × 13 mm raking stay with foot plate; 19 mm diameter pointed verticals at 120 mm centres; two 44 × 19 mm supports 760 mm long with ragged ends to bottom rail; coating with red oxide primer

		Height of fencing	
	Unit	1675 *mm* £	2130 *mm* £
riveted type; bolting panels together; setting standards and stays at 2440 mm centres and supports at 815 mm centres 610 mm below ground	m	44·50	46·80
pairs of gates to match fencing; two 102 × 102 mm hollow section posts with cap and foot plates; hinges; lock and handles; drop bolt; gate stop; holding back catches			
2440 mm wide	No.	400·00	465·00
4880 „	„	550·00	610·00

Close boarded; concrete posts, rails and gravel boards; oak pales

		Height of fencing			
	Unit	915 *mm* £	1070 *mm* £	1220 *mm* £	1375 *mm* £
fencing with two rails; 89 × 19 mm pales lapped 13 mm; setting posts at 2740 mm centres 610 mm below ground . .	m	15·60	16·15	16·70	17·70

		Height of fencing	
	Unit	1675 *mm* £	1830 *mm* £
Fencing with three rails; 89 × 19 mm pales lapped 13 mm; setting posts at 2740 mm centres 760 mm below ground	m	23·00	23·20

Close boarded; oak posts, softwood rails, pales, gravel boards and stumps

		Height of fencing		
	Unit	1070 *mm* £	1375 *mm* £	1830 *mm* £
fencing with two 76 × 38 mm rectangular rails; 152 × 25 mm gravel boards; 89 × 19 mm pales lapped 13 mm; setting posts at 2740 mm centres 610 mm below ground	m	10·85	12·50	—
fencing with three 76 × 38 mm rectangular rails; 152 × 25 mm gravel boards; 89 × 19 mm pales lapped 13 mm; setting posts at 2740 mm centres 610 mm below ground	„	—	—	17·65

Slab fencing; concrete posts; grooved twice; concrete slabs

	Height of fencing		
	1220 *mm* £	1520 *mm* £	1830 *mm* £
fencing with 305 × 38 mm slabs 1753 mm long; setting posts at 1830 mm centres 610 mm below ground (760 mm below ground for 1830 mm high fencing)	21·00	24·90	31·10

EXTERNAL WORKS

EXCAVATION

NOTE. *The prices in 'Excavation and Earthwork' will apply for such works in this section.*

	Unit	£
SOILING, SEEDING OR TURFING		
Vegetable soil selected from spoil heaps		
150 mm thick on general surfaces; grading; preparing for turfing	m²	1·42
150 „ on cuttings or embankments; grading; preparing for turfing	„	1·61
Lightly raking; grass seed, 57 g per m²		
general surfaces	„	0·87
cuttings or embankments	„	0·87
Turf; selected from stacks on site		
general surfaces	„	1·32
cuttings or embankments	„	1·98
Planting only; hedge or shrub plants		
not exceeding 600 mm high	No.	0·63
„ 1200 „	„	1·10
„ 1800 „	„	2·35
Planting only; saplings		
not exceeding 3000 mm high	„	6·27

ROADS AND FOOTPATHS
HARDCORE OR THE LIKE

	Unit	£
Hardcore		
filling in making up levels over 250 mm thick; depositing and compacting in layers by 5 tonne roller	m³	7·23
filling in making up levels average 75 mm thick	m²	0·68
„ „ „ 150 „	„	1·21
Surfaces of hardcore		
levelling; blinding; compacting	„	0·61
Coarse ashes		
filling in making up levels over 250 mm thick; depositing and compacting in layers by 2 tonne roller	m³	12·81
filling in making up levels average 75 mm thick	m²	1·13
„ „ „ 150 „	„	2·10
Surfaces of coarse ashes		
grading to cambers; compacting	„	0·19

IN SITU CONCRETE; PLAIN

	Unit	£
Normal; mix 1 : 12 40 mm aggregate		
Beds		
not exceeding 100 mm thick	m³	35·05
Normal mix 11·50 N/mm² 40 mm aggregate		
Foundations in trenches		
over 150 mm not exceeding 300 mm thick	„	37·20

IN SITU CONCRETE; REINFORCED

	Unit	£
Normal; mix 21·00 N/mm² 20 mm aggregate		
Roads		
not exceeding 100 mm thick	„	41·85
over 100 mm not exceeding 150 mm thick	„	40·15
over 150 mm not exceeding 300 mm thick	„	38·43

LABOURS ON CONCRETE OF ANY DESCRIPTION

	Unit	£
Treating surfaces of unset concrete		
grading to cambers; tamping with a 75 mm thick steel shod tamper	m²	0·81

EXTERNAL WORKS

REINFORCEMENT	Unit	£
Fabric; B.S. 4483; 150 mm side laps, 300 mm end laps		
Mesh 200 × 200 mm × 2·22 kg/m²		
in roads, footpaths or pavings	m²	1·24
Mesh 200 × 200 mm × 3·02 kg/m²		
in roads, footpaths or pavings	,,	1·66

FORMWORK TO PLAIN (OR REINFORCED) IN SITU CONCRETE

Formwork generally

Edges, risers or faces		
not exceeding 250 mm high	m	2·13
over 250 mm not exceeding 500 mm high	,,	4·27
ADD to above for		
curved, radius 6 m	,,	0·32

Steel road forms

Edges, sides or risers		
beds or the like; 150 mm wide	,,	0·91

PRECAST CONCRETE

Units to B.S. 340

Kerbs		
127 × 254 mm fig. 7; haunching with in situ concrete mix 11·50 N/mm²		
40 mm aggregate one side; formwork	,,	3·91
Kerbs; curved, mean radius 5 m		
127 × 254 mm fig. 7; haunching with in situ concrete mix 11·50 N/mm²		
40 mm aggregate one side; formwork	,,	5·07
Edgings		
51 × 152 mm fig. 10; haunching with in situ concrete mix 11·50 N/mm²		
40 mm aggregate both sides; formwork	,,	2·11
Channels		
127 × 254 mm fig. 8; bedding and jointing in cement mortar (1 : 3)	,,	3·25
Channels; curved, mean radius 5 m		
127 × 254 mm fig. 8; bedding and jointing in cement mortar (1 : 3)	,,	4·41

SUNDRIES

Expansion joints in concrete; formwork		
butt joints in 100 mm thick beds or the like; one coat bitumen, one edge	,,	0·92
ADD to above for		
each additional 25 mm width	,,	0·28
Expansion joints in concrete; formwork		
10 × 100 mm; filling with bitumen impregnated joint filler 75 mm wide; sealing with polyethylene sealant backer and polysulphide sealant	,,	2·80
Building paper; 150 mm lapped joints		
horizontal on foundations or the like	m²	0·23

IN SITU FINISHINGS

Fine cold asphalt (category A traffic), B.S. 4897 section 2.3.7 rolled with a 0·25 tonne roller; covering with bitumen coated 14 mm granite chippings rolled in

19 mm work to pavements; to falls, crossfalls or slopes not exceeding 15° from horizontal; to tarmacadam base		
over 300 mm wide	,,	2·75

EXTERNAL WORKS

	Unit	£
IN SITU FINISHINGS – *continued*		

Tarmacadam, B.S. 4987 section 2.2.3 rolled with an 8 tonne roller

75 mm one-coat work to pavements; to falls, crossfalls or slopes not exceeding 15° from horizontal; to hardcore base

over 300 mm wide	m²	4·76

Tarmacadam; B.S. 4987 section 2.2.1 50 mm base coat of 40 mm graded material section 2.3.1; 25 mm wearing coat of 14 mm graded material; rolled with an 8 tonne roller

75 mm work to pavements; to falls, crossfalls or slopes not exceeding 15° from horizontal; to hardcore base

over 300 mm wide	”	5·17

Tarmacadam, B.S. 4987 section 2.2.2 38 mm base coat; section 2.3.3 12 mm wearing coat; rolled with a 0·75 tonne roller; covering with bitumen coated chippings rolled in

50 mm work to pavements; to falls, crossfalls or slopes not exceeding 15° from horizontal; to hardcore base

over 300 mm wide	”	4·76

Gravel paving; first layer coarse aggregate; wearing layer fine aggregate; watered; rolled with a 0·75 tonne roller

50 mm work to pavements; to falls, crossfalls or slopes not exceeding 15° from horizontal; to hardcore base

over 300 mm wide	”	0·86

63 mm work to pavements; to falls, crossfalls or slopes not exceeding 15° from horizontal

over 300 mm wide	”	1·13

TILE, SLAB OR BLOCK FINISHINGS

Concrete flags; B.S. 368 natural finish; bedding in lime mortar; jointing and pointing in cement mortar

750 × 600 × 50 mm; 6 mm joints straight joints both ways; to pavements; to falls, crossfalls or slopes not exceeding 15° from horizontal; to ash base

over 300 mm wide	”	5·33
raking cutting	m	0·82
curved cutting	”	1·63

Noelite slabs; mixed colours; bedding in lime mortar; joints filled with sifted earth

Random sizes, 42 mm thick; 13 mm joints; to pavements; to falls, crossfalls or slopes not exceeding 15° from horizontal; to ash base

over 300 mm wide	m²	8·35

Stock bricks, *p.c.* £110·00 1000; bedding in 10 mm cement mortar (1 : 3); jointing and flush pointing with cement mortar (1 : 3) as work proceeds

To pavements; to falls, crossfalls or slopes not exceeding 15° from horizontal; to concrete base

over 300 mm wide; laid flat; straight joints one way	”	10·12
raking cutting	m	0·50
curved cutting	”	0·78
over 300 mm wide; laid flat; herringbone pattern	m²	11·40
raking cutting	m	0·50
curved cutting	”	0·78
over 300 mm wide; laid on edge; straight joints one way . .	m²	14·36
raking cutting	m	0·63
curved cutting	”	0·94
over 300 mm wide; laid on edge; herringbone pattern . . .	m²	16·00
raking cutting	m	0·63
curved cutting	”	0·94

EXTERNAL WORKS

TILE, SLAB OR BLOCK FINISHINGS – *continued*

215 × 103 × 50 mm paving bricks, *p.c.* £153·00 1000; bedding in 10 mm cement mortar (1 : 3); jointing and flush pointing with cement mortar (1 : 3) as work proceeds

To pavements; to falls, crossfalls or slopes not exceeding 15° from horizontal; to concrete base

	Unit	£
over 300 mm wide; straight joints one way	m²	13·11
raking cutting	m	1·32
curved cutting	,,	1·98
over 300 mm wide; herringbone pattern	m²	14·77
raking cutting	m	1·32
curved cutting	,,	1·98

WATER AND GAS INSTALLATIONS

NOTE. *The prices in 'Drainage' for trenches will apply for such works in this section.*

Mains pipework; asbestos cement pressure pipes, B.S. 486, Class C	Unit	Internal diameter 75 mm £	100 mm £	150 mm £
Pipes; coated inside and outside by manufacturer; asbestos cement joints in the running length; in trenches .	m	5·44	6·52	10·70
Extra; cast-iron 45° bends; caulked lead joints . .	No.	15·50	19·15	43·30
Extra; cast-iron tees; caulked lead joints . . .	,,	26·35	34·65	63·90
Extra; cast-iron tapers; caulked lead joints . . .	,,	—	15·90	26·90
Extra; cast-iron flanged branches; caulked lead joints .	,,	33·90	47·75	82·95

Mains pipework; spun grey iron pipes, B.S. 4622, Class 1		Internal diameter 100 mm £	150 mm £	200 mm £
Pipes; coated inside and outside by manufacturer; flexible joints in the running length; in trenches . . .	m	8·50	12·20	16·40
Extra; 45° bends	No.	16·55	26·50	45·25
Extra; 45° branches	,,	56·10	101·00	153·60
Extra; tees	,,	30·25	46·00	83·65

Mains pipework; spun ductile iron pipes, B.S. 4772, Class K9				
Pipes; coated inside and outside by manufacturer; bolted gland joints in the running length; in trenches . .	m	10·63	15·12	20·80
Extra; 45° bends	No.	22·40	33·70	55·00
Extra; tees	,,	40·00	60·60	102·75
Extra; flanged socket	,,	21·95	33·15	48·50
Extra; flanged spigot	,,	21·40	34·10	51·85

EXTERNAL WORKS – LANDSCAPING AND PLANTING

This section has been prepared in collaboration with, and the prices provided by, the following body:

British Association of Landscape Industries

The Institute of Landscape Architects has indicated its approval to the contents.

Landscaping is often carried out independently of a building contract, therefore this section does vary from the remainder of the book, in that no allowance has been made in the prices for discount, profit or attendance for or by a general building contractor. The prices are intended as average for a contract with a minimum of 5000 m² of soft surface work and 500 m² of planting. Allowance has been made for all overheads, insurances and profit for the landscape contractor, but not, as mentioned earlier, for a general building contractor.

For these reasons prices given here are not consistent with those given elsewhere for work of a similar character carried out as part of a building contract.

The prices for earth works are applicable for 'medium soil' as defined in B.S. 3882.

The prices for cutting down and disposal of trees assume reasonable access and positions for felling and removal and that blasting is permissible and possible. The prices for cutting down and disposal of trees do not allow for any credit value of timber.

The prices for planting include for working in accordance with B.S. 4428, but do not include the cost of trees, shrubs or hedges.

Particular attention is drawn to the units of measurement which have been used in certain instances in order that a realistic rate may be shown.

	Unit	£
SITE PREPARATION		
Cutting and raking rough grass and weeds and burning on site	100 m²	7·75
Clearing scrub, grubbing up roots and burning on site	„	15·60
Cutting down hedges not exceeding 2000 mm high; grubbing up roots; burning on site	m	5·26
Removing trees or hedges		
Cutting down trees; grubbing up or blasting roots; removing from site		
small	No.	20·90
over 600 mm not exceeding 900 mm girth	„	26·00
„ 900 „ „ 1200 „	„	34·50
„ 1200 „ „ 1500 „	„	49·60
„ 1500 „ „ 1800 „	„	73·00
„ 1800 „ „ 2100 „	„	95·00
„ 2100 „ „ 2400 „	„	112·00
„ 2400 „ „ 2700 „	„	138·00
„ 2700 „ „ 3000 „	„	167·00
„ 3000 „ „ 3300 „	„	198·00
Extra; every additional 300 mm girth	„	34·00
HAND EXCAVATION		
Preserving vegetable soil		
Excavating; depositing in spoil heaps not exceeding 100 m from excavation		
average 100 mm deep	m²	1·39
„ 150 „	„	1·96
„ 375 „	„	4·56
Excavating		
Surfaces to reduce levels; over 300 mm deep; depositing on site in spoil heaps not exceeding 100 m from excavation	m²	12·30

EXTERNAL WORKS – LANDSCAPING AND PLANTING

	Unit	£
HAND EXCAVATION – *continued*		
Disposal		
Surplus excavated material		
removing from site; to tip not exceeding 13 km from site . . .	m³	14·90

Soiling, seeding or turfing

Vegetable soil selected from spoil heap; removing not exceeding 100 m; spreading and lightly consolidating on general surfaces

100 mm thick	m²	1·58
150 ,, ...	,,	2·14
375 ,, ...	,,	4·68

Vegetable soil; imported *p.c.* £6·50 per m³ delivered site; spreading and lightly consolidating on general surfaces

100 mm thick	,,	3·07
150 ,, ...	,,	4·00
375 ,, ...	,,	9·78

SURFACE TREATMENTS
Sides of embankments; trimming

exceeding 15°, not exceeding 30° slope	,,	0·47
,, 30° slope	,,	0·94

MECHANICAL EXCAVATION
Preserving vegetable soil

Excavating; depositing in temporary spoil heaps not exceeding 100 m from excavation

average 100 mm deep	,,	0·55
,, 150 ,, ...	,,	0·65
,, 375 ,, ...	,,	1·58

Excavating

Surfaces to reduce levels; over 300 mm deep; depositing on site in spoil heaps not exceeding 100 m from excavation	m³	4·56

Disposal

Surplus excavated material		
removing from site; to tip not exceeding 13 km from site . . .	,,	7·08

Soiling, seeding or turfing

Vegetable soil selected from spoil heap; remove not exceeding 100 m; spreading and lightly consolidating

100 mm thick on general surfaces	m²	0·65
150 ,, ,, ,, ...	,,	0·84
375 ,, ,, ,, ...	,,	1·76
100 mm thick on cuttings or embankments not exceeding 30° slope .	,,	0·84
150 ,, ,, ,, ,, ,, ,, ,, . .	,,	1·02
375 ,, ,, ,, ,, ,, ,, ,, . .	,,	2·14

Vegetable soil; imported *p.c.* £6·50 per m³ delivered site; spreading and lightly consolidating

100 mm thick on general surfaces	,,	1·76
150 ,, ,, ,, ...	,,	2·51
375 ,, ,, ,, ...	,,	6·00
100 mm thick on cuttings or embankments not exceeding 30° slope .	,,	2·22
150 ,, ,, ,, ,, ,, ,, ,, . .	,,	3·48
375 ,, ,, ,, ,, ,, ,, ,, . .	,,	8·58

EXTERNAL WORKS – LANDSCAPING AND PLANTING

	Unit	£

LAND DRAINAGE

The average depths of pipes refer to the inverts when laid
The prices for land drainage include for excavating by mechanical means and filling by hand
Excavate trenches; provide and lay land drains; the clay pipes B.S. 1196 and concrete pipes B.S. 1194; fill trench to within 100 mm of surface with coarse rubble, clinker or similar material; relay vegetable soil; spread surplus excavated material on surrounding ground

					Unit	£
75 mm diameter clay pipes; average 0·50 m deep					m	5·48
,,	,,	,,	,,	0·60 ,,	,,	6·06
,,	,,	,,	,,	0·75 ,,	,,	6·84
100 mm	,,	,,	,,	0·50 ,,	,,	5·70
,,	,,	,,	,,	0·60 ,,	,,	6·38
,,	,,	,,	,,	0·75 ,,	,,	7·08
150 mm	,,	,,	,,	0·50 ,,	,,	6·38
,,	,,	,,	,,	0·60 ,,	,,	7·08
,,	,,	,,	,,	0·75 ,,	,,	7·76
225 mm	,,	concrete pipes; average 0·60 m deep			,,	10·06
,,	,,	,,	,,	0·75 ,,	,,	12·60
,,	,,	,,	,,	1·00 ,,	,,	14·05

Extra over for laying French drain with pipes but trenches filled to surface with ballast rejects. and spreading all excavated material on surrounding ground

		Unit	£
75 mm, 100 mm or 150 mm diameter clay pipes		,,	1·03
225 mm diameter concrete pipes		,,	1·61

Strip turf; excavate trenches in existing playing fields or lawns; provide and lay land drains; clay pipes B.S. 1196 concrete pipes B.S. 1194; fill trench to within 150 mm of surface with coarse rubble, clinker or similar material; relay vegetable soil; relay turf; remove surplus excavated material from site

					Unit	£
75 mm diameter clay pipes; average 0·50 m deep					,,	8·92
,,	,,	,,	,,	0·60 ,,	,,	9·60
,,	,,	,,	,,	0·75 ,,	,,	10·20
100 mm	,,	,,	,,	0·50 ,,	,,	9·24
,,	,,	,,	,,	0·60 ,,	,,	9·90
,,	,,	,,	,,	0·75 ,,	,,	10·50
150 mm	,,	,,	,,	0·50 ,,	,,	9·90
,,	,,	,,	,,	0·60 ,,	,,	10·50
,,	,,	,,	,,	0·75 ,,	,,	11·28
225 mm	,,	concrete pipes, average 0·60 m deep			,,	14·60
,,	,,	,,	,,	,, 0·75 ,,	,,	16·70
,,	,,	,,	,,	,, 1·00 ,,	,,	18·85
Mole drainage with wheeled tractor; 375 mm deep at 2000 mm centres					5000 m²	117·60

CULTIVATION BY MACHINERY

	Unit	£
Plough to depth of 150 mm; break up furrows; cultivate to fine tilth	100 m²	15·60
Extra for sub-soiling at 1000 mm centres	,,	4·70
Evenly grading general surfaces to finish levels	,,	7·80
Fertilizer *p.c.* £0·17 per kg spreading at rate of 57 g per m²; harrowing in	,,	5·10
Selected grass seed *p.c.* £1·80 per kg sowing at rate of 28 g per m² in two applications in transverse directions; lightly harrowing in and rolling		
general surfaces	,,	23·40
cuttings or embankments not exceeding 30° slope	,,	34·20

EXTERNAL WORKS – LANDSCAPING AND PLANTING

	Unit	£
CULTIVATION BY HAND		
Cultivation by pedestrian-operated machine to general surfaces . . .	100 m²	17·16
Evenly grading and raking to finish levels		
general surfaces		34·40
cuttings or embankments not exceeding 30° slope	,,	42·85
Clear stones 50 mm and over from general surfaces and remove from site .	,,	8·45
Fertilizer *p.c.* £0·17 per kg spreading at 57 g per m²; raking in . .	,,	6·30
Selected grass seed *p.c.* £1·80 per kg delivered site; sowing at rate of 42 g per m² in two applications in transverse directions; raking; rolling		
general surfaces	,,	48·00
cuttings or embankments not exceeding 30° slope	,,	51·60
Selected meadow turf *p.c.* £0·48 per m² delivered site and laying with broken joints		
general surfaces	,,	185·00
cuttings or embankments exceeding 15°, not exceeding 30° slope .	,,	192·00
cuttings or embankments exceeding 30° slope; pegging . . .	,,	226·00
MAINTENANCE OF GRASSED AREAS		
Watering turf till established; per occasion (excluding charges for water used)	,,	5·22
Picking all stones over 38 mm from newly sown grassed areas; removing from site .	,,	7·80
First cut of newly sown grass when 75 mm high down to 25 mm; remove cuttings from site	,,	7·20
First cut of newly sown grass; leaving cuttings.	,,	4·80
Mowing with 914 mm wide mower; leaving cuttings. . . .	,,	2·34
Mowing with 914 mm wide mower with box; stacking mowings on site .	,,	3·36
Tractor gang mowing; leaving cuttings	,,	1·12
Rotary or flail mowing; leaving cuttings	,,	1·39
PLANTING STANDARD NURSERY STOCK TREES		
Excavate pit size 1000 × 1000 × 600 mm deep; set aside material for re-use	No.	4·56
Provide manure and fork into lower strata	,,	1·20
Plant tree; providing and fixing one stake and two ties; refilling excavated material	,,	5·34
Provide tree guard *p.c.* £9·00 and fix in position	,,	14·40
Keeping ground reasonably clear of weeds for one season and attention to trees, stakes and ties as necessary	m²	1·76
PLANTING SHRUBS		
Hand dig ground to depth of 225 mm; incorporate suitable fertilizer .	,,	1·39
Provide peat and mulch ground to depth of 25 mm . . .	,,	1·20
Plant small shrubs, ground cover and herbaceous plants . . .	No.	0·28
Plant rose trees	,,	0·55
Plant shrubs not exceeding 750 mm high	,,	0·94
Plant shrubs exceeding 750 not exceeding 1500 mm high . .	,,	1·38
Keeping ground reasonably clear of weeds for one season and attention to shrubs and plants as necessary	m²	2·14
PLANTING HEDGES		
Excavate trench 450 mm wide and 300 mm deep, provide manure and fork into bottom and plant three plants per metre	m	2·14
Dig ground 450 mm wide and 225 mm deep; provide and incorporate suitable fertilizer and plant hedge	,,	1·58
Provide peat and mulch ground to depth of 25 mm . . .	,,	0·74
Keeping ground reasonably clear of weeds for one season and attention to hedge as necessary	,,	0·94

Alterations and Additions

Apart from the nature and quantities of work involved, the prices for alterations and additions are governed by circumstances peculiar to each project. The ease or otherwise of site access, amount of space available for plant and storage of materials, whether the existing building will be wholly vacated at the date for possession or vacated in portions for sectional completion are but a few of the factors which have a bearing on prices. Lack of storage space could mean the extra cost of materials being delivered in less than full lorry loads; executing the works in sections means the return of trades after the completion of the initial section.

The prices have been based on alterations, additions and redecorations to a hypothetical two-storey structure, with reasonable site access and storage space wholly vacated at the date for possession and with a reasonable amount of the items included in the works. Should there be abnormal circumstances, they can be covered by lump-sum additions or by suitably modifying the percentage factor covering overheads and profits in the prices. No allowance has been made for Value Added Tax which would be payable in respect of certain items of work of a repair and maintenance nature included in this section.

PRELIMINARIES

When pricing 'Preliminaries' all factors affecting the execution of the works must be considered; some of the more obvious have already been mentioned above.

The reader is referred to 'Preliminaries' (page 250) for guidance in pricing the specific items given therein, but care must be exercised that all adverse factors are covered.

Where the Standard Form of Contract applies two clauses which will affect the pricing of Preliminaries should be noted.

(a) Insurance of the works against Clause 22 Perils
 Clause 22 C will apply whereby the employer and not the contractor effects the insurance.

(b) Fluctuations
 An allowance for any shortfall in recovery of increased costs under whichever clause is contained in the Contract may be covered by the inclusion of a lump sum in the Preliminaries or by increasing the prices by a suitable percentage.

ADDITIONS AND NEW WORKS WITHIN EXISTING BUILDING

It has been assumed the additions and new works within the existing building include reasonable amounts of the items and that the prices in 'Prices for Measured Work' will therefore apply.

It is likely, however, that the excavations for foundations preclude the use of mechanical plant, and the prices for hand excavation will apply.

The prices include a percentage addition of 10% for overhead charges and profit which can, of course, be adjusted as necessary.

365

As less than what might be termed 'normal quantities' are likely to be involved, it is stressed that actual quotations should be invited for specialist sub-contractors works.

DRAINAGE

The works involved might comprise only short runs of drains, a few manholes and a connection to the existing system. Each scheme must be considered on its merits and a suitable percentage addition applied to the prices in 'Prices for Measured Work' if deemed necessary.

EXTERNAL WORKS

If only small quantities in extension of existing works a suitable percentage addition must be applied to the prices in 'Prices for Measured Work' as considered necessary.

ALTERATIONS

The items of alteration are not exhaustive, but comprise the usual items encountered in works of the nature envisaged in this example.

As the example includes both alterations and additions, the prices for alterations include the same percentage, i.e. 10% for overhead charges and profit. Here again it is a simple calculation to adjust the prices for any other percentage required.

JOBBING WORK

Jobbing work is outside the scope of this example, and no attempt has been made to include prices for such work.

DEMOLITIONS

The cost of demolitions of complete structures (normally executed by Specialists) is so dependent on the construction, for example whether in reinforced concrete, steel framed, the brickwork in lime mortar or in cement mortar and also on the value of the materials salvaged that no reliable prices can be given.

Prices for the demolition of individual walls and small buildings such as out-houses can, however, be assessed from the information given.

CONCRETE WORK

DEMOLITIONS AND FORMING OPENINGS INCLUDING REMOVING DEBRIS FROM SITE

	Unit	£
Break up plain concrete bed		
100 mm thick	m²	2·70
150 mm „	„	3·80
200 mm „	„	5·10
300 mm „	„	7·55
Break up reinforced concrete bed		
100 mm thick	„	4·00
150 mm „	„	5·40
200 mm „	„	7·20
300 mm „	„	10·50
Take down concrete column or cut away casing to steel column	m³	74·00
Take down concrete beam or cut away casing to steel beam	„	80·00
Demolish reinforced concrete wall		
100 mm thick	m²	7·40
150 mm „	„	11·05
225 mm „	„	16·70
300 mm „	„	22·10
Demolish reinforced concrete structural slabs		
100 mm thick	„	6·75
150 mm „	„	10·15
225 mm „	„	15·20
300 mm „	„	20·30
Cut openings in reinforced concrete walls		
150 mm thick	„	25·30
225 mm „	„	36·80
300 mm „	„	48·00
Cut openings in reinforced concrete structural slabs		
150 mm thick	„	31·00
225 mm „	„	40·50
300 mm „	„	50·00
Diamond cutting reinforced concrete structural slabs		
150 mm thick	„	33·00
225 mm „	„	49·50
300 mm „	„	66·00
Break up plain concrete plinths and make good floor beneath	m³	34·40
Break up precast concrete kerb	m	2·00
Take out precast concrete window sill and set aside for re-use	„	5·35
Break out concrete hearth	No.	6·20

	Unit	Small pipe £	Large pipe £	Extra large pipe £
Cut hole in plain concrete for pipes or the like 100 mm thick concrete	No.	1·55	2·80	4·00
Add for each additional 25 mm thickness up to 300 mm thick	„	0·30	0·53	0·72
Add to above for making good fair finish per side	„	0·30	0·50	0·70
Cut hole in reinforced concrete for pipes or the like 100 mm thick concrete	„	2·50	4·05	5·65
Add for each additional 25 mm thickness up to 300 mm thick	„	0·47	0·72	0·95
Add to above for making good fair finish per side	„	0·30	0·50	0·70

	Unit	£
Drill mortices in reinforced concrete structure to accommodate starter bars	No.	1·15
20 mm high tensile steel starter bars 450 mm long grouted in	„	2·35

CONCRETE WORK

FILLING OPENINGS	Unit	£
Reinstate plain ground floor slab		
100 mm thick	m²	8·15
150 mm „	„	11·20
Reinstate mesh reinforced ground floor slab		
100 mm thick	„	11·15
150 mm „	„	14·45
Reinforced concrete (25·50 N/mm² 20 mm aggregate) infilling to floor openings including mesh reinforcement and formwork to soffit		
150 mm thick	m²	37·00
225 mm „	„	45·00
300 mm „	„	54·00
Plain concrete (25·50 N/mm² 20mm aggregate) infilling to openings		
150 × 150 × 150 mm deep including formwork to soffit	No.	4·15

REPAIRS AND RENEWALS		
Clean surfaces of existing concrete to receive new damp proof membrane (measured elsewhere)	m²	0·50
Clean out and fill existing minor cracks with cement mortar mixed with a bonding agent	m	1·30
Cut out old crack to form 20 × 20 mm groove, treat with bonding agent and fill with fine concrete mixed with a bonding agent	„	3·50
Cut small mortice in concrete and make good	No.	0·85

INCIDENTAL WORK		
Hack concrete as key	m²	1·30
Precast concrete (1:2:4) cover stones reinforced with steel fabric weighing 2·22 kg/m² bedded in cement mortar (1:3) to steel beams and wedged and pinned up brickwork over with slates in cement mortar		
300 × 50 mm	m	20·90
300 × 75 mm	„	22·20
300 × 100 mm	„	24·20
Precast concrete (1:2:4) plate lintel, suitably reinforced including pinning up to brickwork over		
100 × 65 mm × 1200 mm long	No.	4·95
150 × 65 mm × 1200 mm „	„	5·85
220 × 65 mm × 1200 mm „	„	7·70
265 × 65 mm × 1800 mm „	„	13·10
Precast or in situ concrete (1:2:4) lintels reinforced with mild steel rods, including formwork and pinning up to brickwork over		
100 × 150 mm × 1200 mm long with one 12 mm rod	„	9·25
150 × 150 mm × 1200 mm „ with two 16 mm rods	„	14·00
225 × 225 mm × 1800 mm „ with three 16 mm rods	„	27·35
Precast concrete (1:2:4) padstone and bedding and pointing in cement mortar (1:3)		
225 × 225 × 150 mm thick	„	5·70
280 × 225 × 150 mm „	„	7·00
340 × 340 × 225 mm „	„	11·60

BRICKWORK AND BLOCKWORK

DEMOLITIONS AND FORMING OPENINGS INCLUDING REMOVING DEBRIS FROM SITE

	Unit	Lime Mortar £	Cement Mortar £
Demolish external walls	m²		
Half brick		3·25	4·55
One brick or two half brick skins	,,	5·95	7·85
One and a half brick	,,	8·65	11·25
Two brick	,,	10·65	13·35

	Unit	£
Add for plaster, render or pebbledash per side	m²	0·50
Pull down internal partitions in lime mortar		
Half brick	,,	4·70
One brick	,,	8·40
One and a half brick	,,	11·65
75 mm block	,,	3·15
100 mm ,,	,,	4·05
150 mm ,,	,,	5·35
215 mm ,,	,,	7·90
Add for plaster, per side	,,	0·60
Pull down internal partitions in cement mortar		
Half brick	,,	6·05
One brick	,,	10·30
One and a half brick	,,	14·05
Add for plaster, per side	,,	0·60
Demolish brick plinths	m³	21·00
Demolish one brick bund walls or piers in cement mortar	m²	6·00
Demolish one brick walls to brick ventilator housing	,,	7·20
Clean off lime mortar and stack old bricks for re-use	1000	40·00
Pull down brick chimney stack to 300 mm below roof level, seal flues with slates		
680 × 680 × 900 mm high above roof	No.	59·60
Add for each additional 300 mm of height	,,	13·10
680 × 1030 × 900 mm high above roof	,,	88·20
Add for each additional 300 mm of height	,,	19·40
1030 × 1030 × 900 mm high above roof	,,	137·00
Add for each additional 300 mm of height	,,	30·10
Cut back chimney breast flush with adjacent wall		
One brick thick	m²	11·50
One and a half brick thick	,,	17·00
Remove brick on edge copings and prepare walls for raising		
One brick thick	,,	2·70
One and a half brick thick	,,	3·90
Cut back brick pier flush with adjacent wall		
225 mm wide × 102·5 mm thick	,,	3·55
450 mm wide × 225 mm thick	,,	5·40
Make good adjacent brickwork where wall removed		
Half brick	,,	2·90
One brick	,,	4·05
100 mm block	,,	2·35
215 mm ,,	,,	3·20

BRICKWORK AND BLOCKWORK

DEMOLITIONS AND FORMING OPENINGS INCLUDING REMOVING DEBRIS FROM SITE – *continued*

	Unit	Lime Mortar £	Cement Mortar £
Cut openings through walls or partitions for doors or windows			
Half brick	m²	10·00	13·00
One brick	,,	17·80	21·90
One and a half brick	,,	24·40	29·70
Two brick	,,	31·00	36·80
75 mm block	,,	6·90	—
100 mm ,,	,,	8·80	—
150 mm ,,	,,	12·10	—
215 mm ,,	,,	15·40	—.
Cut through walls or partitions for lintels or beams			
Half brick	,,	19·40	24·60
One brick	,,	34·10	42·30
One and a half brick	,,	46·50	57·00
Two brick	,,	58·90	70·50
75 mm ,,	,,	12·15	—
100 mm ,,	,,	15·20	—
150 mm ,,	,,	20·80	—
215 mm ,,	,,	26·45	—

	Unit	£
Quoin up jambs in cement–lime mortar as the work proceeds		
Half brick	m	7·45
One brick	,,	11·80
One and a half brick	,,	16·15
Two brick	,,	21·40
75 mm block	,,	6·05
100 mm ,,	,,	7·60
150 mm ,,	,,	10·25
215 mm ,,	,,	13·00

Quoin up jambs in facing bricks to match existing *p.c.* £122·00 per 1000 in cement–lime mortar and rake out joints and point with a neat struck weathered joint

Half brick wall or skin of hollow wall	,,	10·40
One brick wall	,,	16·45
One and a half brick wall	,,	22·15
Two brick wall	,,	27·10

Return half brick skin of hollow wall to close cavity, including lead-lined hessian based bitumen vertical damp-proof course 113 mm wide   . .   ,,

FILLING OPENINGS

Fill openings flush with common bricks or partition blocks in cement–lime mortar (any cutting, toothing and bonding required measured separately)

Half brick	m²	15·20
One brick	,,	29·50
One and a half brick	,,	42·90
Two brick	,,	57·40
75 mm block	,,	9·80
100 mm ,,	,,	12·35
150 mm ,,	,,	16·50
215 mm ,,	,,	21·60

Extra over common brickwork for fair face and pointing with a neat struck joint as the work proceeds	,,	2·20
Extra over common brickwork for facings to match existing *p.c.* £122.00 per 1000 and point with a neat struck joint as the work proceeds . .	,,	13·80
Add or deduct for variation of £1·00 per 1000 in *p.c.* of facings . .	,,	0·11

BRICKWORK AND BLOCKWORK

	Unit	£
FILLING OPENINGS – *continued*		
Hollow wall consisting of inner half brick skin in common and outer skin in facing bricks to match existing *p.c.* £122·00 per 1000 and point with a neat struck joint as the work proceeds	m²	39·30
Cut tooth and bond new to existing in cement–lime mortar	m	
Half brick		2·40
One brick	,,	4·80
One and a half brick	,,	7·20
75 mm blocks	,,	1·90
100 mm ,,	,,	2·25
150 mm ,,	,,	2·90
215 mm ,,	,,	3·65
Half brick in facings to match existing *p.c.* £122·00 per 1000	,,	4·65
Lead-lined hessian-base bitumen horizontal damp-proof course		
75 mm wide	,,	0·50
100 mm ,,	,,	0·60
113 mm ,,	,,	0·75
225 mm ,,	,,	1·25
325 mm ,,	,,	1·90
Cut out ends of joists and plates from walls and make good in common bricks in cement mortar		
175 mm joists	,,	5·70
225 mm ,,	,,	7·20
INCIDENTAL WORK		
Rake out joints and hack face of brickwork	m²	1·05
Thicken brick walls, including extra labour and materials block bonding to existing		
Half brick in common bricks in cement mortar (1:3)	,,	17·30
One brick ,, ,, ,, ,, ,, ,, ,,	,,	30·60
Extra for fair face and point with a neat flush joint as work proceeds	,,	1·50
Attached piers in common bricks in cement mortar (1:3) including extra labour and materials block bonding to existing wall	m	
225 mm wide × 113 mm projection		4·50
225 mm wide × 225 mm projection	,,	6·10
338 mm wide × 225 mm projection	,,	9·20
50 mm cavity wall insulation injected through drilled outer skin afterwards made good		
Foamed in situ urea formaldehyde	m²	2·60
Blown in situ polystyrene beads	,,	3·10
Blown in situ rock or glass fibre	,,	4·50
REPAIRS AND RENEWALS		
Rake out decayed joints and re-point brickwork with a neat struck weather joint in cement mortar (1:1)		
Walls	m²	4·80
Chimney stacks	,,	7·00
Clean off moss and lichen from brick walls	,,	2·40
Rake out joints, re-wedge and repoint horizontal lead flashings	m	1·60
,, ,, ,, ,, ,, ,, stepped ,, ,, ,,	,,	2·35
Cut out defective brickwork and replace with new in cement–lime mortar and point to match existing in small areas		
Half brick in commons	m²	38·30
One brick ,, ,,	,,	72·00
Half brick in facings *p.c.* £122·00 per 1000	,,	57·20

BRICKWORK AND BLOCKWORK

REPAIRS AND RENEWALS – *continued*

	Unit	£
Cut out defective brickwork and replace with new in cement–lime mortar and point to match existing, individual bricks		
Half brick	No.	2·00
Half brick in facings *p.c.* £122·00 per 1000	”	2·85
Cut out staggered cracks along brick joints and repoint to match existing in cement mortar (1:1)	m	2·50
Cut out crack in brickwork and make good with new bricks in cement–lime mortar and point to match existing		
Half brick in commons	”	22·00
One brick ,, ,,	”	42·00
One and a half brick in ditto	”	60·00
Half brick in facings *p.c.* £122·00 per 1000	”	33·05
Hollow wall consisting of inner half brick skin in commons and outer skin in facing bricks to match existing *p.c.* £122·00 per 1000	”	55·80
Cut out defective brick soldier arch and renew in facing bricks to match existing *p.c.* £122·00 per 1000	”	14·80
Take down defective parapet wall one brick thick 600 mm high with two courses of tiles and brick coping over and rebuild in new facing bricks in cement–lime mortar and point to match existing	m	54·80
Cut away and renew 50 mm cement mortar (1:1) angle fillets	”	2·30
Cut and pin ends of joists to existing brickwork	No.	3·15

	Unit	Holes £	Holes: making good facings one side £
Make good where small pipe removed			
Half brick	No.	1·10	1·55
One brick	”	1·55	2·05
One and a half brick	”	2·05	2·50
75 mm blockwork	”	0·80	—·
100 mm ,,	”	0·95	—
150 mm ,,	”	1·10	—
Make good where large pipe removed			
Half brick	”	2·20	3·45
One brick	”	3·15	4·40
One and a half brick	”	4·10	5·30
75 mm blockwork	”	1·60	—
100 mm ,,	”	1·80	—
150 mm ,,	”	2·15	—
Make good where extra large pipe removed			
Half brick	”	3·80	5·60
One brick	”	5·00	6·90
One and a half brick	”	6·30	8·15
75 mm blockwork	,,	2·50	—
100 mm ,,	”	2·80	—
150 mm ,,	”	3·45	—

	Unit	£
Damp proof one brick wall by silicone injection and make good brickwork	m	10·00

RUBBLE WALLING

DEMOLITION INCLUDING REMOVING DEBRIS FROM SITE

	Unit	£
Demolish external walls in lime mortar		
300 mm stonework	m²	6·30
400 mm „	„	8·05
600 mm „	„	11·30
Carefully pull down stone walls in lime-mortar, clean off and set aside for re-use		
300 mm stonework	„	10·60
400 mm „	„	13·40
600 mm „	„	18·50

REPAIRS AND RENEWALS

	Unit	£
Rake out decayed joints and re-point in cement–lime mortar	„	7·85
Carefully cut out defective stonework and rebuild in cement–lime mortar to match existing		
300 mm stonework	„	38·60
400 mm „	„	46·70
600 mm „	„	60·10
Carefully remove defective stone capping from top of wall, remove haunching, replace stones and rehaunch in cement–lime mortar		
300 mm stonework	m	10·40
400 mm „	„	12·20
600 mm „	„	15·15

ASPHALT WORK

REPAIRS AND RENEWALS INCLUDING REMOVING DEBRIS FROM SITE

	Unit	£
Hack up flooring	m²	2·00
Hack up roofing	,,	3·35
Hack off skirtings	m	0·55
Cut out crack in flooring and make good to match existing		
13 mm one coat	,,	4·70
20 mm two coat	,,	6·25
30 mm three coat	,,	7·85
Cut out crack in roofing and make good to match existing		
20 mm two coat	,,	7·50

ROOFING

PREPARATORY WORK INCLUDING REMOVING DEBRIS FROM SITE

	Unit	£
Strip off roof coverings		
Slating, and clear away	m²	1·55
Slating, and set aside for re-use	,,	2·05
Nibbed tiling, and clear away	,,	1·20
Nibbed tiling, and set aside for re-use	,,	2·60
Bitumen felt (and prepare base to receive new)	,,	1·10
Remove chippings from bitumen felt roof	,,	2·30
Corrugated asbestos	,,	1·40
Corrugated metal	,,	1·55
Sheet metal	,,	2·00
Remove tiling battens, withdraw nails from rafters	,,	0·75
Remove underfelt and nails	,,	0·60
Remove rafters, purlins, ceiling joists and plates etc., complete (measured flat on plan)	,,	1·75
Remove defective flashings	m	0·70
,, ,, stepped flashings	,,	0·95
Remove bitumen felt roof covering and boarding to allow access for work to tops of walls or beams beneath	,,	3·65
Turn back bitumen felt and later dress up face of new brickwork as skirting	,,	4·35

REPAIRS AND RENEWALS

	Unit	Asbestos 508 × 254 mm £	Welsh 510 × 225 mm £	Welsh 610 × 305 mm £
Carefully remove damaged slates and replace with new				
Individual slates	No.	3·00	4·10	4·90
Small areas	m²	29·00	43·00	43·80
Carefully take off slates and refix including 25% new and all labours	,,	10·40	14·40	14·00

		Plain £	Concrete Interlocking £
	Unit		
Carefully remove damaged nibbed tiles 267 × 165 mm and replace with new			
Individual tiles	No.	2·65	2·90
Small areas	m²	30·00	15·70
Carefully take off nibbed tiles 267 × 165 mm and refix including 25% new and all labours	,,	13·85	8·70

	Unit	£
Renail loose and replace 25% of existing 19 × 50 mm battens with new .	m²	1·20
,, ,, ,, ,, ,, ,, 19 × 38 mm ,, ,, .	,,	2·00
Remove existing 19 × 50 mm battens and replace with new . . .	,,	2·60
,, ,, 19 × 38 mm ,, ,, ,, ,, ,, . . .	,,	4·45

	Unit	14 kg/10 m² £	22·5 kg/10 m² £
Prepare, strip old nails and fix new underfelt to existing roof .	m²	1·55	2·00

Labour rate £3.35

WOODWORK +30%.

DEMOLITION INCLUDING REMOVING DEBRIS FROM SITE

	Unit	£
Remove softwood floor construction		
100 mm joists at ground-floor level	m²	1·05
175 mm joists at first-floor level	,,	2·05
125 mm joists at roof level	,,	2·85
Cut out and remove individual floor or roof joists	m	1·05
Cut out and remove infected or decayed floor plates	,,	1·20
Remove boarding, withdraw nails from joists		
25 mm softwood flooring	m²	1·60
25 mm softwood roof boarding	,,	2·95
25 mm softwood gutter boarding	,,	4·50
22 mm chipboard flooring	,,	2·00
Remove tilting fillet or roll	m	0·65
Remove fascias or barge boards	,,	2·90
Take down softwood stud partition including plaster on lath or plasterboard both sides	m²	3·00
Take down half glazed softwood stud partition including glass and plasterboard both sides	,,	2·80
Take down plain sheeting, including battening behind		
from walls	,,	1·80
from ceilings	,,	2·50
Take down matchboarding including battening behind		
from walls	,,	2·50
from ceilings	,,	3·20
Carefully remove oak dado panelling and stack for re-use	,,	4·70
Remove skirtings, picture rails, dado rails, architraves and the like	m	0·45
Remove shelves, window boards and the like	,,	1·25
Carefully remove existing softwood skirting and put aside for re-use in making good	,,	0·90
Remove		
Single door	No.	4·80
Single door and frame	,,	8·35
Pair of doors	,,	7·30
Pair of doors and frame	,,	11·55
Extra for taking out floor spring box	,,	6·15
Window and frame	,,	9·00
Sash window and surround	,,	12·75
Carefully dismantle and remove sash window and store for re-use elsewhere	,,	17·50
Pair of french windows	,,	30·45
Remove handrail and brackets	m	1·00
Remove straight flight of stairs including handrails	No.	53·00
Remove dogleg flight ,, ,, ,, ,,	,,	96·00
Remove sloping timber ramp in corridor at change of levels	,,	15·90
Take off bath panel and bearers	,,	3·15
Remove kitchen fittings		
Wall units	,,	4·10
Floor units	,,	3·30
Larder units	,,	5·15
Remove built in wall cupboards	,,	16·15
Remove pipe casings	m	2·50
Remove door or window furniture in preparation for redecoration and refix including providing new screws as necessary	No.	1·85
Remove nameplates or numerals from face of door and make good	,,	2·80
Remove flyscreen and frame from face of window and make good	,,	5·30
Remove curtain track and rail from head of window and make good	m	0·90
Remove carpet fixing strips from floors	,,	0·15

WOODWORK

DEMOLITION INCLUDING REMOVING DEBRIS
FROM SITE – *continued*

	Unit	£
Take down small notice boards and make good	No.	3·15
Take down fire extinguisher and bracket from wall and make good . .	„	4·70

INCIDENTAL WORK

25 mm softwood plain edge flooring and bearers in making good where partitions removed or openings formed (boards running in direction of removed partition)

150 mm wide	m	4·00
225 mm „	„	5·00
300 mm „	„	6·50

25 mm softwood plain edge flooring and bearers in making good where partitions removed or openings formed (boards running at right angles to partitions removed)

150 mm wide	m	8·00
225 mm „	„	10·00
300 mm „	„	13·00
450 mm „	„	19·50
600 mm „	„	26·00

Fit existing softwood skirting to new frame or architrave

75 mm high	No.	0·45
150 mm „	„	0·60
225 mm „	„	0·75

Short length of salvaged softwood skirting to new reveal, one end mitred to existing other end fitted to new frame, including grounds plugged to brickwork and bringing skirting forward for redecoration

75 mm high	„	1·65
150 mm „	„	2·35
225 mm „	„	3·00

25 mm softwood skirtings moulded to match existing including grounds plugged to brickwork

75 mm high	m	3·45
150 mm „	„	5·95
225 mm „	„	8·15

Weatherboard; screwed to bottom rail of existing door

762 mm long	No.	3·10
838 mm „	„	3·20

REPAIRS AND RENEWALS

Cut out infected or decayed structural timbers, shore up adjacent work and replace with new 'treated' timbers

Floor or roof joists

50 × 125 mm	m	3·15
50 × 150 mm	„	3·60
50 × 175 mm	„	4·05
50 × 200 mm	„	4·50
50 × 225 mm	„	4·95

Pitched roof members

38 × 100 mm	„	2·70
50 × 100 mm	„	3·15
50 × 125 mm	„	3·60
50 × 150 mm	„	4·05

Kerbs, bearers, plates and the like

50 × 75 mm	„	2·05
50 × 100 mm	„	2·45
75 × 100 mm	„	3·10

WOODWORK

REPAIRS AND RENEWALS – *continued*

	Unit	£
Scarfed joints of new to existing timber, sectional area over 450 mm²	No.	5·00
Scarfed and bolted joint of new to existing timber including letting in head and nut flush; sectional area over 450 mm²	„	8·70
Remove dust, cobwebs and roof insulation and treat roof timbers with proprietary insecticide by spray application and afterwards replace insulation – per m³ of roof volume	m³	3·30
Lift necessary floorboards and treat floors with proprietary insecticide by spray application and afterwards replace boards	m²	3·10
Treat individual timbers with proprietary insecticide and fungicide by spray application		
Boarding	„	1·40
Structural timbers	„	1·80
Treat individual timbers with proprietary insecticide and fungicide by brush application		
Boarding	„	1·95
Structural timbers	„	2·65
Skirtings	m	0·90
Treat surfaces of concrete or brickwork adjoining infected areas with two coats of fungicide by brush application	m²	1·45
Remove or punch in projecting nails and refix loose boards in softwood or hardwood floors	„	0·75
Take up broken flooring and replace with new 25 mm plain edge softwood boards		
Individual boards	m	3·30
Small areas	m²	16·35
6 mm Plywood laid on joisted floors to level surface	„	4·85
Piece in new softwood skirting to match existing where old removed	m	5·00
„ „ „ „ „ „ „ „ „ switch or socket outlet removed	No.	3·00
Re-surface softwood or hardwood floors, prepare body in with shellac and wax polish	m²	2·45
Cut away defective foot of 75 × 100 mm external door frame provide and scarf in new piece 300 mm long and bring forward for redecorations including bedding in cement mortar and pointing one side	No.	8·60
Cut out part of defective window sill, provide and scarf in new piece 83 × 65 mm weathered on top and throated on underside and bring forward for redecorations	·,	12·85
Take off damaged door stop and provide and screw new 25 × 38 mm softwood stop	m	1·80
Ease and adjust hanging of softwood door, oil furniture and bring forward affected parts for redecoration	No.	3·10
Take out mortice lock, piece out softwood door and bring forward affected parts for redecoration	„	5·80
Take down softwood door, plane 12·5 mm off bottom edge and re-hang		7·85
Take down softwood door and rehang on opposite style, piece out and rebate frames, ease, adjust and bring forward affected parts for redecoration.	„	18·35
Fix only salvaged door	„	6·25
Take down softwood door, remove deadlock and prepare door for fire precaution upgrading, replace existing beads with new 25 × 38 mm hardwood beads screwed on, repair frame where damaged and later rehang upgraded door on wider butts, adjust all ironmongery, bring forward affected parts for redecoration and seal all round frame in cement mortar	„	37·65
Face up one side and upgrade flush door 830 × 2000 mm with 9 mm Superlux screwed on	„	17·10
Face up one side and upgrade panelled door 830 × 2000 mm with 9 mm Superlux screwed on and plasterboard infilling in recesses	„	26·35

WOODWORK

REPAIRS AND RENEWALS – *continued*

	Unit	£
Ease and adjust softwood casement sash and bring forward affected parts for redecoration	No.	3·90
Take off softwood casement sash, repair corner with angle repair plate let in flush, rehang and bring forward affected parts for redecoration . .	,,	13·50
Overhaul double hung sash window, ease adjust and oil pulley wheels, replace parting beads, rehang sashes on new best-quality hemp sash lines, re-assemble and bring forward affected parts for redecoration . . .	,,	20·90
Strip off damaged formica from worktop and replace with new . .	m²	14·40
Strip off damaged formica edging 25 mm wide and replace with new .	m	1·55
Cut out and remove rusted waterbar and make good concrete . . .	,,	1·40

PROTECTION AND TEMPORARY SCREENS

	Unit	£
Protect floors with softwood sawdust	m²	0·25
Protect handrails with hair felt wrapped and tied on	m	1·80
Protect steps of timber staircase with temporary softwood treads . .	No.	0·90
Temporary screens to openings formed with 50 × 50 mm sawn softwood framing covered on one side with 13 mm insulating board and on other with single layer of polythene sheet	m²	12·60

STEELWORK AND METALWORK

	Unit	£
Rolled steel joist beams and stanchions painted one coat primer and hoisting and fixing at first floor level within existing building	t	0·90
Take down tubular handrailing and brackets and make good . .	m	1·20
Remove metal balustrades	,,	2·15
Take down metal corrugated partition	m²	1·00
Take down lightweight steel mesh security screen	,,	3·00
Carefully dismantle metal demountable solid partitioning . . .	,,	3·25
,, ,, ,, ,, ,, ,, and glazed partitioning .	,,	3·85
Take off metal bars from windows	No.	1·70
Remove plates and bolts from concrete floor and make good . .	,,	4·00
Take down and set aside steel cat ladder 4 m long	,,	12·00
Take down metal shutter door and track		
6·20 × 4·60 m (track 12·6 m long)	,,	60·00
12·40 × 4·60 m (,, 16·4 m ,,)	,,	75·00
Overhaul and repair metal casement window, ease, adjust and oil ironmongery and bring forward affected parts for redecoration	,,	6·00

PLUMBING AND ENGINEERING INSTALLATIONS

REMOVAL INCLUDING REMOVING DEBRIS FROM SITE

	Unit	£
Take down 100 mm eaves gutter and supports		
PVC or asbestos	m	0·60
Cast iron	,,	0·90
Take down rainwater head and supports		
PVC or asbestos	No.	1·40
Cast iron	,,	1·70
Take down 100 mm rainwater stacks and supports		
PVC or asbestos	m	1·10
Cast iron	,,	1·45
Take down 100 mm cast iron jointed soil stacks and supports	,,	2·00
Remove 100 mm cast iron rainwater shoe	No.	0·30
Take out sanitary fittings complete with brackets and cap off wastes		
Sink or lavatory basin	,,	8·75
Bath	,,	12·50
W.C. Suite	,,	15·00
Take out sanitary fittings complete with brackets, overflow, services, waste and soil-pipe connections, remove from site and make good holes, facings and plaster and bring forward affected work for redecoration as applicable		
Sink or lavatory basin	,,	26·85
Range of lavatory basins	,,	48·50
Bath	,,	33·45
W.C. Suite	,,	42·00
2 stall urinal	,,	116·80
3 ,, ,,	,,	152·50
4 ,, ,,	,,	188·20
Take down toilet roll holder	,,	1·35
Take down splashback, mirror, towel or soap dispenser and make good	,,	2·85
Remove cold water tank and disconnect services		
900 × 500 × 900 mm	,,	11·25
1540 × 900 × 900 mm	,,	16·15
Remove cold water tank and housing on roof and strip out all associated pipework and make good		
1540 × 900 × 900 mm	,,	62·50
Take out redundant water or gas pipework and make good where disturbed	m	0·65
Take out 'Ascot' water heater and disconnect and cap off services	No.	30·75
Take out gas fire and disconnect and cap off services	,,	10·50
Take out redundant cast iron radiator and make good	,,	25·00
Remove extract fan from window	,,	3·40
Take off light fitting or switch and make good plaster	,,	2·75
Take off fluorescent batten ceiling fitting	,,	3·15
Take off switch socket outlet and make good skirting	,,	3·40
Remove fuse box or control panel and make good plaster	,,	5·60
Remove redundant surface wiring and make good plaster	m	0·23

INCIDENTAL WORKS

	Unit	£
Cut into existing cast iron soil stack for and provide and insert single branch		
75 mm diameter	No.	53·00
75 mm ,, with bolted access door	,,	60·10
90 mm ,,	,,	70·40
90 mm ,, with bolted access door	,,	78·35
100 mm ,,	,,	73·28
100 mm ,, with bolted access door	,,	81·25

PLUMBING AND ENGINEERING INSTALLATIONS

INCIDENTAL WORKS – *continued*

	Unit	£
Cut hole for new or cut out small pipe or conduit from wall and make good		
Half brick wall	No.	1·70
One brick wall	,,	3·00
One and a half brick wall	,,	4·30
Add for making good facings per side	,,	0·60
Add for making good plaster per side	,,	1·00
Cut hole for new or cut out large pipe from wall and make good		
Half brick wall	,,	2·85
One brick wall	,,	5·00
One and a half brick wall	,,	7·15
Add for making good facings per side	,,	0·80
Add for making good plaster per side	,,	1·15
Cut horizontal or vertical chase in plastered brick wall for, and make good plaster		
Small pipe, cable or conduit	m	6·20
Large pipe	,,	8·35
Take up softwood single floor board, notch or bore softwood joists for small pipe, cable or conduit and refix board with screws	,,	2·10

REPAIRS AND RENEWALS

Clean out existing rainwater gutters	,,	0·25
Remove old balloon grating and provide and fix new galvanized wire grating in gutter outlet		
75 mm	No.	2·80
100 mm	,,	3·25
Overhaul 100 mm cast iron gutter, cut out existing joints, adjust brackets to correct falls and remake joints including new gutter bolts	m	4·20
Clean out and repair crack in existing lead gutter with solder	m	5·80
Overhaul leaking joint in 100 mm cast iron gutter and remake including new gutter bolt	No.	3·85
Cut off lower portion of 100 mm cast iron rainwater pipe, disconnect shoe and rejoint shoe to discharge 500 mm higher	,,	8·55

FLOOR, WALL AND CEILING FINISHINGS

REMOVAL INCLUDING REMOVING DEBRIS FROM SITE

	Unit	£
Take up floor finishings		
Carpet and underfelt	m²	0·45
Linoleum flooring	”	0·35
Take up floor finishings and prepare screed to receive new		
Carpet and underfelt	”	2·40
Vinyl or thermoplastic tiles	”	3·15
Hack up flooring		
Floor screed	”	1·85
Granolithic flooring and screed	”	2·30
Terrazzo or ceramic floor tiles and screed	”	4·80
Hack in situ or tile skirting from wall	m	1·25
Take up woodblock flooring	m²	2·65
Hack from walls		
Plaster	”	1·25
Cement rendering or pebbledash	”	1·75
Wall tiling and screed	”	2·25
Remove plasterboard plastered lining from wall	”	1·45
Take down ceilings		
Suspended ceilings	”	3·10
Plasterboard and skim from joists, and withdraw nails.	”	1·85
Wood lath and plaster ,, ,, ,, ,, ,,	”	2·15
Plaster moulded cornice per 25 mm girth	m	0·70
Remove part of plasterboard ceiling for new steel beam	”	4·40
Take out fireplace surround and interior		
Tiled	No.	8·00
Cast iron, and set aside	”	11·50
Stone, and set aside	”	30·00

INCIDENTAL WORKS

Protect tile flooring or carpeting	m²	0·50
Clean and protect vinyl flooring	”	1·30
Clean and protect ceramic tile flooring	”	3·10
Prepare, sand down and apply two coats of sealer to hardwood block flooring	”	2·80
Cut back finishings to allow new partition to be built and afterwards make good to same		
Plaster on walls	m	6·20
Plasterboard and skim to ceilings	”	9·50

REPAIRS AND RENEWALS

Refix individual loose floor tiles or blocks	No.	1·85
Level and repair floor screed with 'Ardit K10'	m²	6·10
Render and set on brick or block walls		
Lightweight plaster	”	4·40
Limelight plaster	”	4·80
Sirapite plaster	”	5·50
Dubbing out in cement and sand (1:3) average 13 mm thick	”	4·00
Brush down and apply two coats of bonding agent on existing walls	”	1·90

FLOOR, WALL AND CEILING FINISHINGS

REPAIRS AND RENEWALS – *continued*

	Unit	£
Joint new to existing plaster	m	0·70
Trim back existing plaster and render and set in gypsum plaster in making good where cross walls removed including jointing new to existing		
150 mm wide		
225 mm „	„	2·75
300 mm „	„	3·45
	„	4·15
Trim back existing lath and plaster and render float and set in gypsum plaster in making good lath and plaster ceilings where partitions removed including jointing new plaster to existing		
150 mm wide		
225 mm „	„	5·45
300 mm „	„	6·85
	„	8·20
Render and set in gypsum plaster to brick reveals or returns and joint new plaster to existing		
75 mm wide including narrow return and arris	„	6·85
150 mm „ „ „ „	„	7·55
225 mm „ „ „ „	„	8·30
300 mm „ „ „ „	„	9·05
Render and set in gypsum plaster to faces and soffits of concrete lintels including trimming and jointing new plaster to existing		
75 mm wide	„	3·05
150 mm „	„	3·70
225 mm „	„	4·45
300 mm „	„	5·20
Labour arris	„	1·15
Hack off area of defective or damaged wall plaster and prepare and finish in two coats of Carlite plaster in making good to walls including dubbing out		
Small areas	m²	10·70
Small areas not exceeding 0·50 m²	No.	7·30
„ „ „ 1·00 m²	„	13·40
Hack off area of damp wall plaster, investigate and treat wall, prepare and finish in two coats of Limelight plaster including dubbing out		
Small areas	m²	14·05
Cut out and repair cracks in existing wall plaster	m	1·85
Newtonlite lath fixed to walls with nails		
over 300 mm wide	m²	5·60
Hack off area of defective or damaged ceiling plaster and prepare and fix new plasterboard and skim in making good		
Small areas	„	12·30
Cut out and repair cracks in existing ceiling plaster	m	2·30
Make good plaster where items removed		
Small pipe or conduit	No.	0·65
Large pipe	„	0·75
Extra large pipe	„	0·85
Light switch	„	0·75
Make good plasterboard and skim where items removed		
Small pipe or conduit	„	2·10
Large pipe	„	2·45
Extra large pipe	„	2·85
Light switch	„	2·75
Make good plaster to small holes in ceilings	„	2·75
Refix loose tiles in suspended ceilings	„	2·65

GLAZING

Hack out, prepare rebates and reglaze to wood with linseed oil putty in
squares up to 0·40 m²

The following prices are for totally reglazing windows. Prices for renewing   Unit   £
single squares in repairs will be much higher, dependent on the quantity
involved.

	Unit	£
3 mm clear sheet ordinary quality	m²	17·40
4 „ float	„	18·35
3 „ patterned	„	18·25
5 „ rough cast	„	21·35
6 „ Georgian wired cast	„	23·40
6 „ „ „ polished	„	41·80
6 „ float in squares up to 4 m²	„	26·80

PREPARATION FOR REDECORATION

NOTE. The prices for painting and decorating in 'Prices for Measured Work' will apply for redecoration of existing surfaces, and the prices given hereunder are for the preparation only of existing decorated surfaces in fair condition.

	Unit	£
Prepare decorated plastered walls, including washing surfaces, cutting out cracks and making good with Keene's cement		
Emulsion painted surfaces and bring forward bare patches . . .	m²	0·45
Strip wallpaper	,,	0·80
Prepare plastered ceilings, including washing surfaces, cutting out cracks and making good with Keene's cement		
Wash off distemper	,,	0·50
Emulsion painted surfaces and bring forward bare patches . . .	,,	0·60
Strip distempered lining paper	,,	1·00
Prepare plaster moulded cornices, including washing surfaces, cutting out cracks and making good with Keene's cement		
Wash off distemper	,,	0·75
Emulsion painted surfaces and bring forward bare patches . . .	,,	0·90
Prepare painted metalwork, including washing and rubbing down, priming bare patches and bringing forward general surfaces, pipes, etc., internally		
Over 300 mm girth	,,	0·80
Not exceeding 150 mm girth	m	0·20
Exceeding 150 mm not exceeding 300 mm girth	,,	0·30
Metal windows in average medium squares	m²	1·50
Prepare painted metalwork, including washing and rubbing down, priming bare patches and bringing forward general surfaces, pipes, etc., externally		
Over 300 mm girth	,,	1·00
Not exceeding 150 mm girth	m	0·25
Exceeding 150 mm not exceeding 300 mm girth	,,	0·40
Metal windows in average medium squares	m²	1·85
Prepare painted woodwork, including washing and rubbing down, priming bare patches and bringing forward general surfaces, frames, skirtings, margins of treads and risers, etc., internally		
Over 300 mm girth	,,	1·10
Not exceeding 150 mm girth	m	0·30
Exceeding 150 mm, not exceeding 300 mm girth	,,	0·45
Casements in average medium squares	m²	2·10
Prepare painted woodwork, including washing and rubbing down, priming bare patches and bringing forward general surfaces, frames, etc., externally		
Over 300 mm girth	,,	1·40
Not exceeding 150 mm girth	m	0·40
Exceeding 150 mm, not exceeding 300 mm girth	,,	0·60
Casements in average medium squares	m²	2·60
Prepare painted woodwork, including burning off, rubbing down, knotting and priming general surfaces, frames, etc., externally		
Over 300 mm girth	,,	5·30
Not exceeding 150 mm girth	m	1·40
Exceeding 150 mm, not exceeding 300 mm girth	,,	2·10
Casements in average medium squares	m²	9·80

SPOT ITEMS

Few exactly similar composite items of alteration work are encountered on different schemes, and for this reason it is considered more accurate for the reader to build up the value of such items from individual prices in the preceeding sections. However, for estimating purposes, the following 'spot' items have been prepared.
Prices do not include for shoring, scaffolding or redecoration.

DEMOLITION

	Unit	£
Demolish single storey brick out-buildings		
50 m³	m³	3·85
200 m³	,,	2·75
500 m³	,,	1·55
Demolish two storey brick out-buildings with timber first floor		
200 m³	,,	2·00
Demolish three storey bay window on front of remaining building	No.	1375·00

Pull down brick chimney stacks to 300 mm below roof level, seal flues with two courses of slates in cement mortar, piece in 'treated' sawn softwood rafters and make good roof coverings to match existing over

680 × 680 × 900 mm high above roof	,,	92·00
Add for each additional 300 mm of height	,,	13·80
680 × 1030 × 900 mm high above roof	,,	136·00
Add for each additional 300 mm of height	,,	20·40
1030 × 1030 × 900 mm high above roof	,,	202·00
Add for each additional 300 mm of height	,,	30·30

Carefully take off existing chimney pots and set aside, pull down defective chimney stack to roof level and rebuild using 25 % new facing bricks to match existing, parge and core flues and reset chimney pots including flaunching in cement mortar (scaffolding costs excluded)

680 × 680 × 900 mm high above roof	,,	180·80
Add for each additional 300 mm of height	,,	36·15
680 × 1030 × 900 mm high above roof	,,	267·60
Add for each additional 300 mm of height	,,	53·50
1030 × 1030 × 900 mm high above roof	,,	397·80
Add for each additional 300 mm of height	,,	79·60

FORMING OPENINGS

The requirement for shoring and strutting for the formation of large openings are dependent on a number of factors, for example, the weight of the super-imposed structure to be supported, number, if any, of windows and, number of floors and the roof to be strutted, whether raking shores are required, and the depth to a load-bearing surface.

Prices would best be built up by assessing the use and waste of materials and the labour involved, including getting timber from and returning to yard, cutting away and making good, overheads and profit. This method is considered the more practical way of pricing than endeavouring to price the work on a metre cube basis of timber used, and has been adopted in preparing the prices for the examples which follow.

The shoring and strutting for the formation of an opening 6100 mm wide, 2440 mm high through a one-brick external wall at ground floor of a normal two-storey structure with solid ground-floor construction, timber joisted suspended floor and timber pitched roof would cost about as follows:

	Unit	£
Strutting to window openings over proposed new opening	No.	3·20
Plates, struts, braces and hardwood wedges in supports to floors and roof	Metre of opening	8·45
Dead shore and needle using die square timber with sole plates, braces, hardwood wedges and steel dogs	No.	101·00
Cut holes through one brick wall for die square needle and make good, including facings externally and plaster internally	,,	31·00
Set of two raking shores using die square timber with 50 mm thick wall piece, hardwood wedges and steel dogs, including forming holes for needles and making good	Set	132·00

SPOT ITEMS

FORMING OPENINGS – *continued*

	Unit	£
Form opening through 100 mm softwood stud partition for single door and frame including studwork framing around, making good boarding and any plaster either side and extending floor finish through opening (new door and frame measured elsewhere)	No.	92·50

Form opening through internal plastered wall for single door and frame including cutting structure, quoining or making good jambs, cutting and pinning in suitable precast concrete plate lintel/s, making good plasterwork up to new frame both sides and extending floor finish through opening (new door and frame measured elsewhere)

150 mm reinforced concrete wall	,,	176·00
225 mm ,, ,, ,, ...	,,	218·00
Half brick	,,	133·00
One brick	,,	176·00
One and a half brick	,,	215·00
Two brick	,,	260·00
100 mm block	,,	145·00
215 mm ,, ...	,,	195·00

Form opening through internal plastered wall for pair of doors and frame including cutting structure, quoining or making good jambs, cutting and pinning in suitable precast concrete plate lintel/s, making good plasterwork up to new frame both sides and extending floor finish through opening (new door and frame measured elsewhere)

150 mm reinforced concrete wall	,,	235·00
250 mm ,, ,, ,, ...	,,	300·00
Half brick	,,	193·00
One brick	,,	226·00
One and a half brick	,,	270·00
Two brick	,,	325·00
100 mm block	,,	186·00
215 mm block	,,	255·00

Form opening through faced wall for new window including cutting structure, quoining up jambs, cutting and pinning in suitable precast concrete boot lintel with angle bolted on to support outer brick soldier course in facing bricks to match existing (new window and frame measured elsewhere)

One brick wall or two half brick skins	m²	80·00
One and a half brick wall	,,	105·00
Two brick wall	,,	126·00
Cut and form 700 × 1100 mm opening in existing slated, boarded and timbered roof for new rooflight including trimming timbers in rafters and making good roof coverings up to new kerb and units (measured separately)	No.	132·50

FILLING OPENINGS

Take out door and frame, make good plaster and skirtings across reveal and leave as blank opening

Single door	,,	38·60
Pair of doors	,,	46·40
Take out single door and frame in 100 mm softwood stud partition, block up opening with timber covered on both sides with boarding to match existing and extend skirting both sides	,,	68·00

Take out single door and frame in internal wall, brick or block up opening, plaster walls and extend skirting both sides

Half brick	,,	84·00
One brick	,,	120·00
One and a half brick	,,	153·00
Two brick	,,	187·00
100 mm block	,,	81·00
215 mm ,, ...	,,	103·00

SPOT ITEMS

FILLING OPENINGS – *continued*   Unit   £

Take out pair of doors and frame in internal wall, brick or block up opening,
plaster walls and extend skirting both sides

Half brick	No.	137·00
One brick	,,	192·00
One and a half brick	,,	236·00
Two brick	,,	290·00
100 mm block	,,	129·00
215 mm ,,	,,	161·00

Take out sash window and frame in external faced wall, brick up opening
with facing bricks on outside to match existing and common bricks on
inside plastered internally

One brick or two half brick skins	m²	76·00
One and a half brick	,,	100·00
Two brick	,,	121·00

Take out curved headed sash window in external stuccoed wall, brick up
opening with common bricks, stucco on outside and plaster on inside to
match existing

One brick	m²	73·00
One and a half brick	,,	97·00
Two brick	,,	118·00

Take out curved headed sash window in external masonry faced brick wall,
brick up opening with masonry on outside and common bricks on inside
plastered internally

350 mm wall	,,	210·00
500 mm ,,	,,	235·00
600 mm ,,	,,	255·00

Take out fireplace surround and interior, break up kerb and hearth, block
up fireplace opening with half brick wall including building in air-brick,
plaster wall to finish flush with existing, fix fibrous plaster ventilator and
extend skirting to match existing across front of fireplace

Tiled	No.	75·00
Cast iron, and set aside	,,	67·00
Stone and set aside	,,	122·00

PART III

Approximate Estimating

This part of the book contains the following sections:

Building Prices Per Square Metre

Prices given under this heading are average prices, on a 'fluctuating basis', for typical buildings at November 1980 (forecast tender price level index = 480) and, unless otherwise stated, they do not allow for external works, other than those adjacent to the building, furniture, loose or special equipment and are, of course, exclusive of fees for professional services.

Prices are based upon the total floor area of all storeys, measured between external walls and without deduction for internal walls.

As in previous editions it is emphasised that the prices must be treated with reserve in that they represent the average of prices from our records and cannot provide more than a rough guide to the probable cost of a building.

In many instances normal commercial pressures together with a limited range of available specifications ensure that a single rate is sufficient to indicate the prevailing average price. However, where such restriction do not apply a range has been given; this is not to suggest that figures outside this range will not be encountered, but simply that the calibre of such a type of building can itself vary significantly.

For assistance with the compilation of a closer estimate, or of a 'Cost Plan' the reader is directed to the 'Approximate Estimates' and 'Elemental Cost Plan' sections.

Transport and Utility Buildings (Cl/SfB 1)	Square metre £
Multi-storey car parks	125
Underground car parks:	
partially underground under buildings	160
completely underground under buildings	210
completely underground with landscaped roof	240
Bus and Coach stations	320
Bus garages	330
Garages	250
Airport Passenger Terminal Buildings (excluding aprons):	
national standard	600 to 700
international standard	800 to 1000
Airport facility buildings:	
large hangars	700
workshops and small hangars	360
Mortuaries	700

Industrial Buildings (Cl/SfB 2)	
Farm sheds	50 to 100
Factories:	
for letting	170
workshops	220
maintenance/motor transport workshops	400 to 450
owner occupation – for light industrial use	220 to 280
owner occupation – for heavy industrial use	400 to 450

	Square metre £
Printing works	600
Pumping station excluding specialist equipment	850
Distilleries	480
Laboratory workshops and offices	400 to 500
High technology laboratory workshop centres, air conditioned	1200 to 1500
Warehouses:	
low bay for letting	120
low bay for owner occupation	180
high bay for owner occupation	240

Administrative, Public, Commercial and Office Buildings; General (Cl/SfB 3)

Embassies	600 to 950
County Courts	650 to 750
Magistrates Courts	500 to 600
Civic Offices:	
non air conditioned	490 to 550
fully air conditioned	600 to 650
Administrative buildings:	
low rise	340
medium rise	450
medium rise, air conditioned	600
Probation/Registrar Offices	360
Offices for letting:	
non air conditioned	390
air conditioned	500 to 610
Offices for owner occupation:	
low rise	420
medium rise	520
medium rise air conditioned	800
high rise air conditioned	1100
Two storey ancillary office accommodation to warehouses/factories	330
Fitting out offices	250 to 350
High quality office renovation	480
Banks:	
local	600
city centre	1100
Branch Post Offices	500
Postal delivery offices	450
Telephone exchanges	400
Shop shells:	
small	230
large including department stores and supermarkets	190
Fitting out shell for small shop (including shop fittings)	250
Fitting out shell for department store or supermarket (including shop fittings)	400 to 500

Hypermarkets, single store (including fitting out)	330
Central area redevelopments (excluding fitting out):	
open malls	200
enclosed malls – medium quality	350 to 450
*Ambulance stations	400
Fire Stations	380
Police Stations	360

Health and Welfare Facilities (Cl/SfB 4)

*District General Hospitals	550
*Health Centres	330 to 400
*Day Centres	330
*Day Nurseries	380
Group Practice Surgeries	420
*Homes for the physically handicapped:	
houses	250
residential centres	400
*Homes for the mentally handicapped	350
Geriatric Day Hospital	480
Convalescent Homes:	
dormitory and welfare blocks	320
nursing homes	460
*Children's Homes	300
*Homes for the aged	300
*Observation and Assessment units	340
*Sheltered housing block with wardens' accommodation . .	330

Refreshment, Entertainment, Recreation Buildings (Cl/SfB 5)

Public Houses	450
Dining blocks and canteens in shop and factory	350
Restaurants	450 to 500
Community Centres	340
Tenants' Clubrooms	450
Youth Clubs	350
Theatres, including seating and stage equipment:	
large – over 500 seats	650 to 850
small – under 500 seats	500 to 650
Conference Centres	900
Exhibition Centres	500
Swimming pools:	
international standard	650
local authority standard	550
school standard	400

* *Refer also to Cost Limits and Allowances in next Section*

	Square metre £
Leisure centre:	
dry	425 to 525
wet	550 to 650
Sports halls	350
School gymnasiums	300
Squash courts	470
Sports pavilions:	
social and changing	280
changing only	360
changing and public toilets	460
Grandstands	400
Clubhouses	525

Religious Buildings (Cl/5fB 6)

Churches	380 to 460
Crematoria	750

Educational, Scientific and Information Buildings (Cl/SfB 7)

*Primary/junior school	300
*Secondary/middle schools	340
*Extensions to schools:	
classrooms	360
residential	400
laboratories	500
*Special schools	350
*Polytechnic:	
Students Union Buildings	350
Arts building	320
Scientific Laboratories	400
*Training Colleges	330
Management Training Centres	525
*Universities:	
Arts buildings	320
Science Buildings – Chemistry	420
– Biology	400
– Physics	350
Computer buildings	1000 to 1250
Museums:	
local	550
national	1200
Libraries:	
city centre	550
branch	450

* *Refer also to Cost Limits and Allowances in next section*

Square metre
£

Residential Buildings (Cl/SfB 8)

*Local authority and housing association housing schemes:

4 person low rise flats excluding lifts	300
4 person medium rise flats including lifts	350
4 person two storey houses	250
Private houses, detached including central heating	450 to 650
Houses in private developments	250 to 300
Flats in private developments	250 to 350
Terraced blocks of garages	180
Hotels:	
city centre	850 to 950
commercial	500 to 600
motels	350
*Students' residences	300 to 400

* *Refer also to Cost Limits and Allowances in next section*

Cost Limits and Allowances

Information given under this heading is based upon the cost targets currently in force for buildings financed out of public funds, i.e. Hospitals, Schools, Universities and Public Authority Housing. The information enables the cost limit for a scheme to be calculated and is not intended to be a substitute for estimates prepared from drawings and specifications.

The cost limits are generally set as target costs based upon the user accommodation, i.e. in the case of schools they are given per place. However ad hoc additions can be agreed with the relevant Authority in exceptional circumstances.

The documents setting out cost targets are almost invariably complex and cover a range of differing circumstances. They should be studied carefully before being applied to any scheme; this study should preferably be undertaken in consultation with a Chartered Quantity Surveyor.

The cost limits for Public Authority Housing are contained in the Department of the Environment's Circular, generally known as the Housing Cost Yardstick. This covers a wide range of possibilities and consequently they are not included in detail; however, the principles embodied in the Housing Cost Yardstick have been summarized and a method of budgeting to these limits is given.

HOSPITAL BUILDINGS

The information and tables which follow are contained in Section A and Appendices 7 and 9 of Hospital Building Procedure Note No. 6 – Cost Control, published by H.M. Stationery Office and included here by kind permission of the Controller. The tables, which should be read in conjunction with the Building Notes concerned, provide the preliminary cost limit for a project but are not a substitute for estimates prepared from drawings and specifications.

The cost allowance figures given in the following tables are those set out in Appendix 7 to HBPN 6 including revisions up to January 1980 and should be used when submitting new schemes.

C.I.S. note 1039 dated March 1980 states that from 1 April 1980 these cost allowances will be increased by approximately 11·5%

The reader should check that the cost allowance figures and any increases that relate to them are valid at the time they are being used.

DEPARTMENTAL COSTS

The Departmental Cost section provides tables from which the total departmental cost of a scheme or project can be estimated from the functional content.

Cost Allowance figures (col. 6) include an allowance over and above the cost of the basic accommodation appropriate to the individual Building Notes (where applicable) to cover the cost of both the additional accommodation and additional engineering for medical and other user requirements.

In interpolating for functional units which fall between two of the figures given in the tables the cost should be rounded to the nearest £1000.

HOSPITAL BUILDINGS

DEPARTMENTAL AREAS

These are areas of basic accommodation (col. 4a). The areas of basic accommodation include internal circulation space. Their status for planning purposes also depends upon the current applicability of guidance material, some of which was issued some while ago. To arrive at the gross floor area of a project it will be necessary to make an addition for communication space. The amount of this addition can vary considerably, but at preliminary stages a token figure of 17% should be made.

ON-COSTS

These are costs additional to the Departmental Costs for the particular scheme arising from the site and its use. At stage 1 50% should be added to the cost of accommodation to cover on-costs unless a broad assessment of proposed sites or phasing costs already indicates a significantly different figure. At stage 2 these items will be measured and costed. Items included under this heading are:

Communications
Space which provides access to, or between, the various departments of the hospital. Plant rooms and major structural ducts and shafts for engineering services between departments. Lifts and other forms of vertical moving access.

External works
Site clearance and levelling, roads, paths and paved areas, vehicle parking, drainage, gates and fences, landscaping and builders' work in connection with engineering services. External engineering services to buildings and internal distribution between departments within buildings, street lighting, water storage, stand-by generators, etc.

Height factor
Additional cost factors due to the need to build in blocks over four storeys high.

Auxiliary buildings
Separate minor buildings.

Abnormals
Exceptional factors of cost such as piling due to adverse soil conditions and air-conditioning due to density of building on a restricted site [and which normally would not be required for clinical purposes].

PARTIAL DEVELOPMENT OR EXTENSION OF AN EXISTING HOSPITAL

Where it is intended to add or alter departmental or service provision of an existing hospital, the stage 1 cost should be taken as the accommodation cost of the functional units to be newly provided (derived from the Costing Section of the table) plus 50% to cover demolition costs, costs of temporary accommodation and on-costs as defined above.

HOSPITAL BUILDINGS

UPGRADING OF EXISTING ACCOMMODATION

The cost limit for upgrading work should be fixed in each case in relation to the improvement in standards it is desired to achieve; but in no case should it exceed 60% of the cost of the relevant functional units if these were provided through new building. The calculation for this purpose should be made on the cost of the functional units alone, without any addition for 'on-costs'. At least 20 years service should be expected from any upgrading scheme costed at more than £75,000.

HOSPITAL BUILDINGS

DEPARTMENTAL COST AND AREA GUIDE

NOTES. (1) *The areas of basic accommodation in Column 4a are appropriate to the individual Building Notes or Design Guides (where applicable) and include circulation space. Their status for planning purposes depends on the current applicability of guidance material, some of which was issued some time ago.*

SERVICE Department (1) and (2)	Functional units (3)	Areas of basic accommodation appropriate to B.N. (where applicable, to nearest 5 m^2) (m^2) (4a)
Administration Services		
General administration	9 points/300 beds	640
	12 „ /450 „	795
	16 „ /600 „	955
	20 „ /800 „	1,110
Non-Resident Staff changing (excluding medical staff)	220 places	345
	250 „	375
	300 „	420
	400 „	515
	500 „	605
	600 „	700
Main entrance accommodation	9 points/300 beds	135
	12 „ /450 „	145
	16 „ /600 „	160
	20 „ /800 „	170
Medical records	9 points/300 beds	265
	12 „ /450 „	330
	16 „ /600 „	390
	20 „ /800 „	455
Group accommodation	19 points	435
	30 „	605
	40 „	765
In-patients' Services		
General acute wards	30 beds	710
	60 „	1,410
	120 „	2,815
	180 „	4,225
	240 „	5,630
	300 „	7,040
	360 „	8,445
	420 „	9,855
	480 „	11,260
	540 „	12,670
	600 „	14,075
	660 „	15,485
	720 „	16,890

HOSPITAL BUILDINGS

(2) *The cost allowance figures shown in Column 6 include an allowance over and above the cost of the basic accommodation appropriate to the individual Building Notes or Design Guides (where applicable) to cover the cost of both the additional accommodation and additional engineering for medical and other user requirements.*

Hospital B.N. number (where applicable) (5)	*Cost allowance (£)* (6)	*Remarks* (7)
	188,000	
	221,000	
	262,000	
	303,000	
	96,500	**Non-resident staff changing**
	103,000	At Stage A1 it may not be possible to accurately
	115,000	assess proportion of N–R Staff, and adjustment
	138,000	at Stage 2 may be necessary
	161,000	
	184,000	
18	39,000	For Central Telephone Installations (i.e.
	42,500	accommodation and equipment) see
	43,500	'Service Facilities.'
	48,000	
	89,500	
	110,000	
	135,000	
	159,000	
	101,000	
	138,000	
	172,000	
	238,000	Piped medical gases and plated tray service
	455,000	included
	911,000	
	1,366,000	
	1,823,000	
	2,278,000	
4	2,733,000	
	3,189,000	
	3,644,000	
	4,099,000	
	4,555,000	
	5,010,000	
	5,467,000	

HOSPITAL BUILDINGS

DEPARTMENTAL COST AND AREA GUIDE

NOTES. (1) *The areas of basic accommodation in Column* 4a *are appropriate to the individual Building Notes or Design Guides (where applicable) and include circulation space. Their status for planning purposes depends on the current applicability of guidance material, some of which was issued some time ago.*

SERVICE Department (1) *and* (2)	Functional units (3)	Areas of basic accommodation appropriate to B.N. (*where applicable, to nearest 5 m^2*) (m^2) (4a)
In-patients' Services – *continued*		
Children's wards	20 beds	790
	40 ,,	1,575
	60 ,,	2,365
	80 ,,	3,155
Geriatric wards (assessment / acute)	28 ,,	765
	56 ,,	1,425
	112 ,,	2,845
Geriatric wards (long stay)	28 ,,	765
	56 ,,	1,425
	112 ,,	2,845
	168 ,,	4,270
	224 ,,	5,690
	280 ,,	7,115
	336 ,,	8,535
Intensive therapy	8 ,,	460
Anaesthetic services	DGH	185
Maternity department clinic	10 Doctor sessions	590
Maternity department (Reception, administration, education and central delivery suite)	Serving a population of:	
	150,000	1,685
	200,000	2,005
	250,000	2,315
Maternity department (Ward accommodation)	75 beds	2,160
	100 beds	2,860
	125 ,,	3,555
Maternity department (Special care baby nursery)	20 cots	420
	30 ,,	530
Maternity department (Milk kitchen/store; sub-stores and flying squad store)	Serving a population of:	
	150,000	105
	200,000	115
	250,000	125

HOSPITAL BUILDINGS

(2) *The cost allowance figures shown in Column 6 include an allowance over and above the cost of the basic accommodation appropriate to the individual Building Notes or Design Guides (where applicable) to cover the cost of both the additional accommodation and additional engineering for medical and other user requirements.*

Hospital B.N. number (where applicable) (5)	Cost allowance (£) (6)	Remarks (7)
23	201,000 397,000 592,000 787,000	Piped medical gases and plated tray service included
N/A	241,000 430,000 855,000	Ditto
	235,000 419,000 838,000 1,258,000 1,676,000 2,095,000 2,514,000	Ditto
27	149,000	Piped medical gases included
N/A	46,500	See Design Guide, October 1971
	176,000	
21 and Interim guidance	522,000 617,000 710,000	Piped medical gases included
	656,000 884,000 1,115,000	Ditto
	166,000 222,000	Ditto
	33,000 35,500 37,500	

HOSPITAL BUILDINGS

DEPARTMENTAL COST AND AREA GUIDE

NOTES. (1) *The areas of basic accommodation in Column* 4a *are appropriate to the individual Building Notes or Design Guides (where applicable) and include circulation space. Their status for planning purposes depends on the current applicability of guidance material, some of which was issued some time ago.*

SERVICE Department (1) *and* (2)	Functional units (3)	Areas of basic accommodation appropriate to B.N. (where applicable, to nearest 5 m^2) (m^2) (4a)
Main operating facilities		
Operating suites and related rooms	1 theatre	⎫
	2 theatres	⎬ N/A
	3 ,,	⎭
	4 ,,	1,045
	5 ,,	1,250
	6 ,,	1,475
	7 ,,	1,685
	8 ,,	1,890
T.S.S.U. (for departments with no nearby C.S.S.D.)	Serving 4 theatres	150
	,, 5 ,,	165
	,, 6 ,,	185
	,, 7 ,,	205
	,, 8 ,,	220
Sterile store (where no T.S.S.U. provided)	Serving 4 theatres	45
	,, 5 ,,	55
	,, 6 ,,	60
	,, 7 ,,	70
	,, 8 ,,	80
Diagnostic and treatment facilities		
X-ray department	1 R/D room	⎫
	2 ,, rooms	⎬ N/A
	3 ,, ,,	⎭
	4 ,, ,,	615
	5 ,, ,,	800
	6 ,, ,,	960
	7 ,, ,,	1,100
	8 ,, ,,	1,175
Radioisotopes		270
Pathology department	Area Laboratory (39 L.S.U.s)	1,895
	Public Health Service Laboratory (Supplementary) (6.5 L.S.U.s)	320
	Reference Laboratories serving 500,000 population (Supplementary)	
	Cytology (1 L.S.U.)	45
	Trace Elements ($\frac{1}{2}$ L.S.U.)	25
	Toxicology ($\frac{1}{2}$ L.S.U.)	25

HOSPITAL BUILDINGS

(2) *The cost allowance figures shown in Column 6 include an allowance over and above the cost of the basic accommodation appropriate to the individual Building Notes or Design Guides (where applicable) to cover the cost of both the additional accommodation and additional engineering for medical and other user requirements.*

Hospital B.N. number (where applicable) (5)	*Cost allowance* (£) (6)	*Remarks* (7)
N/A	268,000	Plant rooms excluded, cooling, piped medical gases and improved ventilation included
	393,000	
	514,000	
	638,000	
	764,000	
	896,000	
	1,024,000	
	1,155,000	
26	99,500	Cooling and emergency sterilizer included
	106,000	
	113,000	
	120,000	
	128,000	
	20,000	
	23,000	
	26,000	
	28,500	
	31,500	
N/A	121,000	Plantrooms excluded. Ultrasonics included in departments with 5–8 R/D rooms. Piped medical gases included
	142,000	
	164,000	
	195,000	
6	247,000	
	295,000	
	337,000	
	359,000	
N/A	88,500	See Design Guide, October 1973
15	909,000	
	115,000	
	13,500	
	12,000	
	12,000	

HOSPITAL BUILDINGS

DEPARTMENTAL COST AND AREA GUIDE

NOTES. (1) *The areas of basic accommodation in Column 4a are appropriate to the individual Building Notes or Design Guides (where applicable) and include circulation space. Their status for planning purposes depends on the current applicability of guidance material, some of which was issued some time ago.*

SERVICE		
Department (1) *and* (2)	*Functional units* (3)	*Areas of basic accommodation appropriate to B.N.* (*where applicable, to nearest 5 m²*) (*m²*) (4a)
Diagnostic and treatment facilities – *continued*		
Pathology – *continued*	Immunology (1½ L.S.U.s)	70
	Automation (2 L.S.U.s)	95
	Neuropathology (2 L.S.U.s)	95
	Reference Laboratories serving 1,000,000 population (Supplementary)	
	Cytology (2 L.S.U.s)	95
	Trace Elements (1 L.S.U.)	45
	Toxicology (1 L.S.U.)	45
	Immunology (3 L.S.U.s)	140
	Chromosomes (1 L.S.U.)	45
	Automation (4 L.S.U.s)	185
	Neuropathology (3 L.S.U.s)	140
Pharmaceutical department	300–500 beds	535
	501–1,000 beds	735
	1,001–1,200 ,,	865
Mortuary and post-mortem room	300 beds	215
	500 ,,	230
	800 ,,	290
Rehabilitation	Rehabilitation Services serving population of:	
	150,000	1,585
	200,000	1,845
	250,000	2,105
	Hydrotherapy section	185
Medical photography	Department	130

HOSPITAL BUILDINGS

(2) *The cost allowance figures shown in Column 6 include an allowance over and above the cost of the basic accommodation appropriate to the individual Building Notes or Design Guides (where applicable) to cover the cost of both the additional accommodation and additional engineering for medical and other user requirements.*

Hospital B.N. number (where applicable) (5)	Cost allowance (£) (6)	Remarks (7)
	23,500	
	42,500	
	34,500	
	27,000	
15	24,000	
	24,000	
	46,500	
	17,500	
	85,000	
	52,000	
	272,000	Plant and Plant Room for medical gases under
29	338,000	communications
	383,000	
	81,000	
20	95,000	
	115,000	
	434,000	See Design Guide, August 1974
N/A	483,000	
	533,000	
	74,000	Deduct if pool not required
19	33,000	

HOSPITAL BUILDINGS

DEPARTMENTAL COST AND AREA GUIDE

NOTES. (1) *The areas of basic accommodation in Column 4a are appropriate to the individual Building Notes or Design Guides (where applicable) and include circulation space. Their status for planning purposes depends on the current applicability of guidance material, some of which was issued some time ago.*

| SERVICE | | |
| | | Areas of basic accommodation appropriate to B.N. (where applicable, to nearest 5 m^2) |
Department (1) *and* (2)	Functional units (3)	(m^2) (4a)
Out-patients' services		
O.P.D. (Consulting suite)	36 Doctor sessions	595
	72 ,, ,,	1,020
	106 ,, ,,	1,450
	144 ,, ,,	1,880
O.P.D. (Operating theatre suite)	1 theatre	285
	2 theatres	455
O.P.D. (Day ward)	8 beds	240
	12 ,,	305
	16 ,,	370
	20 ,,	435
O.P.D. (Dental Department)	1 Surgery (up to 9 sessions)	80
	2 Surgeries (10–18 sessions)	100
	3 Surgeries (19–27 sessions)	260
	4 Surgeries (28–36 sessions)	295
	5 Surgeries (37–45 sessions)	325
O.P.D. (Hospital Dermatology Services Supplementary)	Serving population of 250,000–300,000	55
Orthodontic Supplementary	1 Surgery (up to 9 sessions)	115
	2 Surgeries (10–18 sessions)	145
	3 Surgeries (19–27 sessions)	180
Ear, Nose and Throat Services	Serving population of:	
	150,000	540
	200,000	590
	250,000	640
Special Treatment Clinic	Serving population of 300,000	435

HOSPITAL BUILDINGS

(2) *The cost allowance figures shown in Column 6 include an allowance over and above the cost of the basic accommodation appropriate to the individual Building Notes or Design Guides (where applicable) to cover the cost of both the additional accommodation and additional engineering for medical and other user requirements.*

Hospital B.N. number (where applicable) (5)	*Cost allowance (£)* (6)	*Remarks* (7)
	187,000	Piped medical gases included
	310,000	
	434,000	
	557,000	
12	117,000	Plant rooms excluded. Cooling and piped medical gases included
	187,000	
	68,000	Piped medical gases included
	87,000	
	106,000	
	126,000	
	35,000	
	48,500	
28	109,000	
	125,000	
	140,000	
N/A	19,500	See Design Guide, October 1976
	43,000	
28	55,000	
	70,500	
N/A	160,000	See Design Guide, February 1974
	171,000	
	182,000	
N/A	157,000	

HOSPITAL BUILDINGS

NOTES. (1) *The areas of basic accommodation in Column* 4a *are appropriate to the individual Building Notes or Design Guides (where applicable) and include circulation space. Their status for planning purposes depends on the current applicability of guidance material, some of which was issued some time ago.*

SERVICE Department (1) *and* (2)	Functional units (3)	Areas of basic accommodation appropriate to B.N. (where applicable to nearest 5 m²) (m²) (4a)
Out-patients' services – *continued*		
Accident and emergency (Departmental accommodation)	Patients in a 3-hour peak period	
	60 patients	595
	110 ,,	680
	160 ,,	765
Accident and emergency department (Minor operating theatre suites)	Patients in a 3-hour peak period:	
	60 patients	315
	110 ,,	325
	160 ,,	330
Accident and emergency department (Recovery and short stay unit)	8 beds	245
	10 ,,	275
	12 ,,	305
Accident and emergency department (Orthopaedic and fracture clinic)	18 Doctor sessions	240
	27 ,, ,,	325
	36 ,, ,,	410
	45 ,, ,,	490
Service facilities		
C.S.S.D.	Department	330
Central Telephone Installation	180 Extensions	60
	(300 beds)	
	200 ,, (300 ,,)	60
	300 ,, (450 ,,)	70
	400 ,, (600 ,,)	80
	500 ,, (800 ,,)	95
	600 ,, (1,000 ,,)	105
Central kitchens	300 meals	400
	400 ,,	455
	500 ,,	520
	750 ,,	665
	1,000 ,,	830
	1,250 ,,	1,000
	1,500 ,,	1,100

HOSPITAL BUILDINGS

(2) *The cost allowance figures shown in Column 6 include an allowance over and above the cost of the basic accommodation appropriate to the individual Building Notes or Design Guides (where applicable) to cover the cost of both the additional accommodation and additional engineering for medical and other user requirements.*

Hospital B.N. number (where applicable) (5)	*Cost allowance* (£) (6)	*Remarks* (7)
	154,000	
	174,000	
	195,000	
	178,000	Plant rooms, excluded. Cooling and piped medical gases included
	180,000	
22	186,000	
	67,000	Piped medical gases included
	76,000	
	84,000	
	73,000	
	97,000	Ditto
	121,000	
	144,000	
13	157,000	
	69,500	
	78,500	
N/A	114,000	
	138,000	
	171,000	
	205,000	

Reduction to be made where Area or Bulk Food Store exists.

		sq. metres	£
	236,000	30	9,500
	266,000	35	12,500
	310,000	45	15,500
10	381,000	70	23,500
	472,000	95	31,500
	566,000	115	39,000
	618,000	140	47,000

Centralized tray service, central wash-up, hospital bulk food store, milk store, dairy and butcher's shop included.

HOSPITAL BUILDINGS

DEPARTMENTAL COST AND AREA GUIDE

NOTES. (1) *The areas of basic accommodation in Column* 4a *are appropriate to the individual Building Notes or Design Guides* (*where applicable*) *and include circulation space. Their status for planning purposes depends on the current applicability of guidance material, some of which was issued some time ago.*

SERVICE Department (1) *and* (2)	Functional units (3)	Areas of basic accommodation appropriate to B.N. (where applicable to nearest 5 m^2) (m^2) (4a)
Service facilities – *continued*		
Laundries	Articles per week	
Basic accommodation	55,000	1,130
	120,000	2,140
	180,000	3,075
	250,000	4,030
Supplementary accommodation	Articles per week	
Infants' napkins section	over 1,000	60
	Kilogrammes per week	
Dry cleaning section	600	55
	950	70
	1,450	100
	2,400	150
	Articles per week	
Special personal clothing section	3,000/3,500	65
	4,500/5,000	85
	6,000/7,000	110
Storage at sending hospital	300 beds	75
	400 ,,	110
	600 ,,	155
	800 ,,	175
	1,000 ,,	230
	1,200 ,,	265
Teaching facilities		
Nurses' training school	100 students	460
	200 ,,	630
	300 ,,	820
	400 ,,	1,030
	20 pupils	165
	40 ,,	225
	60 ,,	280
	80 .,	300
Post graduate medical centre	Centre	555
Education	800 bed D.G.H.	150

HOSPITAL BUILDINGS

(2) *The cost allowance figures shown in Column 6 include an allowance over and above the cost of the basic accommodation appropriate to the individual Building Notes or Design Guides (where applicable) to cover the cost of both the additional accommodation and additional engineering for medical and other user requirements.*

Hospital B.N. number (where applicable) (5)	*Cost allowance* (£) (6)	*Remarks* (7)
	728,000	Engineering Services account for up to 67% of costs
	1,374,000	of additional accommodation
	1,951,000	
	2,545,000	
	32,000	
	40,500	
	50,500	
	69,000	
25	99,000	
	42,000	
	56,000	
	77,500	
	15,000	
	21,000	
	28,500	
	32,500	
	42,000	
	47,500	
	99,000	
	136,000	
	173,000	
	211,000	
14	42,500	
	51,000	
	59,500	
	68,000	
N/A	144,000	See Design Guide, October 1968
N/A	43,000	Facilities for staff other than doctors or nurses

HOSPITAL BUILDINGS

DEPARTMENTAL COST AND AREA GUIDE

NOTES. (1) *The areas of basic accommodation in Column 4a are appropriate to the individual Building Notes or Design Guides (where applicable) and include circulation space. Their status for planning purposes depends on the current applicability of guidance material, some of which was issued some time ago.*

SERVICE Department (1) *and* (2)	Functional units (3)	Areas of basic accommodation appropriate to B.N. (where applicable, to nearest 5 m²) (m²) (4a)
Staff facilities		
Dining rooms	125 meals	125
	250 ,,	215
	500 ,,	410
	750 ,,	605
Residential accommodation for staff	100 points	
	150 ,,	
	200 ,,	
	250 ,,	
	500 ,,	
	For each additional point after the first 500 allow £500	
Occupational health centre for staff	1,000 beds	165
Hospital engineering and works services		
Boiler houses and fuel storage	Solid fuel or oil fuel	
	1,760 KW	190
	4,400 ,,	320
	6,150 ,,	365
	8,800 ,,	460
	11,700 ,,	610
	20,500 ,,	975
Works department	Group with:	
	300– 600 beds	460
	601– 800 ,,	605
	801–1,200 ,,	755
	1,201–2,000 ,,	890
	2,001–3,000 ,,	1,035
	3,001–4,000 ,,	1,165
	Lock-up store/workshop Group with:	
	300– 600 beds	20
	601– 800 ,,	30
	801–1,200 ,,	40
	1,201–2,000 ,,	45
	2,001–3,000 ,,	50
	3,001–4,000 ,,	55

HOSPITAL BUILDINGS

(2) *The cost allowance figures shown in Column 6 include an allowance over and above the cost of the basic accommodation appropriate to the individual Building Notes or Design Guides (where applicable) to cover the cost of both the additional accommodation and additional engineering for medical and other user requirements.*

Hospital B.N. number (where applicable) (5)	Cost allowance (£) (6)	Remarks (7)
11	40,500	Wash up not included (see Central Kitchen)
	69,000	
	131,000	
	194,000	
24	69,500	
	97,500	
	125,000	
	151,000	
	290,000	
N/A	44,000	
16	136,000	Area in column 4a is averaged for the two types of installations
	208,000	Variations are in the order of:
	255,000	Solid fuel: 6–17% above mean
	327,000	Oil fuel: 6–17% below mean
	410,000	
	588,000	
34	101,000	Compound, garages and separate lock-up store are included
	132,000	
	163,000	
	196,000	
	228,000	
	254,000	
34	3,000	To serve hospital/Group remote from Works Department
	5,000	
	6,500	
	7,500	
	8,500	
	9,000	

HOSPITAL BUILDINGS

DEPARTMENTAL COST AND AREA GUIDE

NOTES. (1) *The areas of basic accommodation in Column 4a are appropriate to the individual Building Notes or Design Guides (where applicable) and include circulation space. Their status for planning purposes depends on the current applicability of guidance material, some of which was issued some time ago.*

SERVICE Department (1) *and* (2)	Functional units (3)	Areas of basic accommodation appropriate to B.N. (*where applicable to nearest 5 m²*) (m²) (4a)
Psychiatric patients' services		
Psychiatry (mental illness) Ward unit	60 beds	1,555
	90 „	2,330
	120 „	3,110
Psychiatry (mental illness) Annexe	10 beds	385
Psychiatry (mental illness) Day hospital	80 places	1,570
	120 „	1,900
	160 „	2,285
Small mental handicap hospital units	Children's residential unit: 8 places	
	16 „	
	24 „	
	Adult residential unit: 12 places	
	24 „	
	Adult day care unit: 60 places	
	115 „	

HOSPITAL BUILDINGS

(2) *The cost allowance figures shown in Column 6 include an allowance over and above the cost of the basic accommodation appropriate to the individual Building Notes or Design Guides (where applicable) to cover the cost of both the additional accommodation and additional engineering for medical and other user requirements.*

Hospital B.N. number (where applicable) (5)	*Cost allowance* (£) (6)	*Remarks* (7)
	464,000	
	690,000	
	921,000	
35	126,000	
	416,000	
	506,000	
	590,000	
	79,000	
	134,000	
	186,000	
	96,500	
	164,000	
	205,000	
	335,000	

HOSPITAL BUILDINGS

DEPARTMENTAL COST AND AREA GUIDE

NOTES. (1) *The areas of basic accommodation in Column* 4a *are appropriate to the individual Building Notes or Design Guides* (*where applicable*) *and include circulation space. Their status for planning purposes depends on the current applicability of guidance material, some of which was issued some time ago.*

SERVICE Department (1) *and* (2)	Functional units (3)	*Areas of basic accommodation appropriate to B.N.* (*where applicable to nearest 5 m²*) (*m²*) (4a)
Community Health Service Health Centre	Primary care services General Medical Practitioner Services 3 General Medical Practitioners 6 „ „ „ 9 „ „ „ 12 „ „ „ Community Health Services, Health Visiting, District Nursing and Midwifery; and Welfare Foods Expected to serve a population of: 7,500 15,000 22,500 30,000 Other services Chiropody 1 Chiropodist 2 „ Speech Therapy and/or Child Health Assessment 1 Therapist 2 „ Social Services 1 Social worker 2 „ „ 3 „ „	

HOSPITAL BUILDINGS

(2) *The cost allowance figures shown in Column 6 include an allowance over and above the cost of the basic accommodation appropriate to the individual Building Notes or Design Guides (where applicable) to cover the cost of both the additional accommodation and additional engineering for medical and other user requirements.*

Hospital B.N. number (where applicable) (5)	Cost allowance (£) (6)	Remarks (7)
	38,500	
	56,500	
	88,000	
	110,000	
36	41,500	
	57,000	
	73,500	
	91,500	
	5,500	
	11,500	
	9,000	
	13,500	
	3,500	
	3,500	
	7,500	

Cost Limits and Allowances

HOSPITAL BUILDINGS

DEPARTMENTAL COST AND AREA GUIDE

NOTES. (1) *The areas of basic accommodation in Column* 4a *are appropriate to the individual Building Notes or Design Guides* (*where applicable*) *and include circulation space. Their status for planning purposes depends on the current applicability of guidance material, some of which was issued some time ago.*

SERVICE Department (1) *and* (2)	Functional units (3)	Areas of basic accommodation appropriate to B.N. (*where applicable to nearest 5 m²*) (m²) (4a)
Community Health Service (*contd*) Health Centre (*contd*)	Shared facilities Related to General Medical Practitioners 3 General Medical Practitioners 6　　" 　　" 　　" 9　　" 　　" 　　" 12　　" 　　" 　　" Related to Community Health Services Expected to serve a population of: 7,500 15,000 22,500 30,000 Dental Services School and Priority Dental Services 1 Dentist 2　　" 3　　" 4　　" General Dental Services 1 Dentist 2　　" 3　　" 4　　"	

HOSPITAL BUILDINGS

(2) *The cost allowance figures shown in Column 6 include an allowance over and above the cost for the basic accommodation appropriate to the individual Building Notes or Design Guides (where applicable) to cover the cost of both the additional accommodation and additional engineering for medical and other user requirements.*

Hospital B.N. number (where applicable) (5)	*Cost allowance (£)* (6)	*Remarks* (7)
	46,500	
	68,000	
	90,500	
	112,000	
	56,000	
	72,000	
36	85,000	
	103,000	
	31,000	
	40,000	
	50,500	
	59,500	
	20,500	
	29,500	
	47,500	
	57,000	

HOSPITAL BUILDINGS

DEPARTMENTAL COST AND AREA GUIDE

NOTES. (1) *The areas of basic accommodation in Column* 4a *are appropriate to the individual Building Notes or Design Guides (where applicable) and include circulation space. Their status for planning purposes depends on the current applicability of guidance material, some of which was issued some time ago.*

SERVICE			Areas of basic accommodation appropriate to B.N. (where applicable to nearest 5 m²)
	Department (1) *and* (2)	*Functional units* (3)	*(m²)* (4a)
Community Health Service (*contd*)			
Health Centre (*contd*)		Hospital Consultant Services	
		9 Doctor sessions per week	180
		18 ,, ,, ,, ,,	190
		Pharmaceutical Services	
		Expected to serve a population of:	
		Less than 15,000	200
		More than 15,000	210
Health Authority Clinic		Clinic to serve population of:	
		7,000–10,000	220
		10,000–20,000	230
		20,000–30,000	240
Ambulance Stations			

HOSPITAL BUILDINGS

(2) *The cost allowance figures shown in Column 6 include an allowance over and above the cost of the basic accommodation appropriate to the individual Building Notes or Design Guides (where applicable) to cover the cost of both the additional accommodation and additional engineering for medical and other user requirements.*

Hospital B.N. number (where applicable) (5)	Cost allowance (£) (6)	Remarks (7)
36	6,000 12,000	
LABN3	19,500 26,000	
LABN7	79,500 89,000 101,000	Cost information available from Department on request

LOCAL AUTHORITY BUILDINGS

Following the introduction of block allocations the use of Departmental cost allowance will no longer be mandatory except for children's projects containing secure accommodation. However in the interests of economy and good cost planning Departmental cost guidance will still be made available by providing updated functional unit costs under the arrangements set out in the Department of Health and Social Security Local Authority Social Services Letter LASSL (78) 6: Revised Building Procedural Guidance and Cost Allowances dated 12th May 1978. The tables which follow are based upon extracts from LASSL (78) 6 and C.I.S. Note 508 dated 1 January 1980. The tables should be read in conjunction with the appropriate design guidance published by the DHSS.

Basic Accommodation

Brief area guidance is contained within Circular LASSL (78) 6.

Functional Unit Costs

These include for engineering costs.

On Costs

These include external works and abnormal requirements. Exceptionally higher costs may be considered by the Department.

The reader should check that the information and tables are valid at the time they are being used.

Type of accommodation	Functional Unit	Cost allowance	On Costs	Engineering Services Costs
Adult residential accommodation				
Residential accommodation for elderly people	30 residents for each additional resident	£261,600 £6,100	22½%	Engineering content should normally be contained within 30% to 33% of the functional unit cost.
Residential accommodation for mentally handicapped adults. Residential accommodation for mentally ill	16 residents for each additional resident	£146,800 £6,100	22½%	Ditto
Residential accommodation for physically handicapped people	20 residents for each additional resident	£254,300 £7,800	22½%	Ditto
Community homes				
Community homes without specialist facilities	8 residents 9–20 residents 21 residents 22–30 residents 31 residents 32–40 residents 41 residents 42–50 residents	£87,200 £6,650/resident £179,500 £5,650/resident £241,200 £5,100/resident £302,250 £4,800/resident	22½%	Ditto

LOCAL AUTHORITY BUILDINGS

Type of accommodation	Functional Unit	Cost allowance	On Costs	Engineering Services Costs
Community homes with observation and assessment facilities	20–30 residents	£40,000		
	31–40 residents	£42,900		
	41–50 residents	£49,400		
	Education provision	£1,950/ place		
	Day attenders	£1,900/ day attender		
	Small secure unit for			
	2 places	£46,200	22½%	Engineering content should normally be contained within 20% to 22% of the functional unit cost.
	3 places	£53,800		
	4 places	£63,500		
	5 places	£72,100		
	6 places	£84,000		
	7 places	£90,400		
	8 places	£95,500		
	Special bedsitting room with soft finishes	£1,200		
Adult training centres	50 trainees	£188,900	33⅓%	Engineering content should normally be contained within 25% to 27% of the functional unit cost.
	for each additional trainee	£2,180		
Day centres				
Multi-purpose day centres equipped largely for occupational activities	40 day attenders	£188,900	33⅓%	Engineering content should normally be contained within 30% to 33⅓% of the functional unit cost.
	for each additional day attended up to a maximum of 150	£2,180		
Day centres providing cultural, educational and recreational activities	40 day attenders	£171,450		
	for each additional day attender up to a maximum of 150	£1,745		
Day nurseries	50 place day nursery	£130,800	33⅓%	Ditto
Residential staff accommodation Houses	4 persons	£14,500		
	5 persons	£16,400		
	6 persons	£18,500		
Flats (including storage)	1 person	£6,200		
	2 persons	£9,000		
	3 persons	£11,400		
	4 persons	£14,000		
	5 persons	£15,700		
	6 persons	£17,100		
Bedsitting accommodation (including storage)	1 person	£4,800		
	2 persons	£7,600		

UNIVERSITY BUILDINGS

A revised procedure for the assessment of grants for university building projects has been introduced by the University Grants Committee and is now operative. Details appear in Notes on Procedure 1977 – Capital Grants. This procedure involves a university submitting a Building Project Form which in essence asks for permission to design a project that will have the capacity to accommodate certain numbers of units. Examples of units are – undergraduate, postgraduate course or research students studying specific subjects; accommodation for administrative staff, maintenance services, health services or sports facilities on a per student basis.

When the Committee have agreed the capacity of a project with a university, the officers of the Committee provide the university with a provisional expenditure limit based on current costs together with notional and minimum usable areas. Development of the sketch design follows and when this is approved the provisional expenditure limit is adjusted to allow for current costs and fully substantiated abnormal costs, and restated as a final expenditure limit.

The university is free to design the project entirely as it wishes subject only to the constraints that the agreed unit capacity is provided, at least the minimum usable area is provided and costs do not exceed the expenditure limit.

In accordance with current Government policy no standard rates per square metre or unit costs are now published but Notes on Procedure 1977 – Capital Grants does contain the following list of notional usable areas which are for use by universities in design work and by the Committee's officers when calculating expenditure limits for individual projects.

UNIVERSITY BUILDINGS

Notional Unit Areas

ACADEMIC

Subject	Usable area all students (m^2)

LECTURE THEATRES

All subjects except mathematics	0·50
Mathematics	0·80

ARTS

Basic arts and social studies	2·64
Social psychology	3·44
Languages	3·44
Geography (traditional)	5·34
Archaeology	5·34
Education	5·34
Geography (scientific)	8·14
Experimental Psychology	9·54
Music	
Standard base unit (Constant)	300·00
plus for every student up to 50 FTE	4·14
plus for every student over 50 FTE	5·64

	Usable area	
	UG and PGC (m^2)	PGR (m^2)

SCIENCES

Mathematics	3·80	3·15
Physics	10·05	18·60
Chemistry	10·15	19·05
Biology	10·15	20·00
Architecture	9·35	8·60
Electronics and Engineering Science		
Equipment dominated space	ad hoc	—
plus for every FTE student	10·05	18·60
All other engineering including Heavy, Electrical, Mechanical, Civil Aeronautical, Chemical and Nuclear		
Equipment dominated space	ad hoc	—
plus for every FTE student	10·85	18·80
Medical and Dental-Pre-clinical	13·98	20·25
Medical-clinical	7·50	23·15
Dental-clinical	11·40	17·05

UNIVERSITY BUILDINGS

Notional Unit Areas

NON-ACADEMIC

	unit per	*usable area* (m^2)
LIBRARY		
Basic provision (based on planned FTE numbers at the estimated date of completion of project)	each	1·25
Addition where necessary for each law student . . .	each	0·80
Expansion provision (based on planning horizon) . . .	each	0·20
Special collections	ad hoc	—
RESERVE BOOK STORE		
Standard base unit (constant)	each	50·00
plus for every 1000 volumes	1000	3·50
ADMINISTRATION		
up to 3000 students	each	0·55
over 3000 students	each	0·35
MAINTENANCE SERVICES		
up to 3000 students	each	0·25
over 3000 students	each	0·15
HEALTH SERVICES		
(Consultancy only)		
up to 3000 students	each	0·03
over 3000 students	each	0·015
SPORTS FACILITIES		
(1) Indoor		
up to 3000 students	each	0·47
3000 to 6000 students	each	0·13
over 6000 students	each	0·25
(2) Outdoor		
up to 3000 students	each	0·18
over 3000 students	each	0·10

EDUCATIONAL BUILDINGS

On 31st December 1974 the Department of Education and Science issued Circular 13/74 of which the following is an extract:

'CONTROL AND APPROVAL PROCEDURES – GENERAL'

3. The annual resources available for educational building outside the university sector will be determined as national cash allocations for starts in four main sectors:

(*a*) primary and secondary schools;

(*b*) nursery education;

(*c*) special schools, including child guidance clinics and hospital schools;

(*d*) institutions for further education and teacher training.

4. These sectors and their scope may subsequently be varied in accordance with Government policies. The procedures for each sector differ in detail as set out below, notably in that allocations for primary and secondary schools, and for nursery education, will be made as lump sum authorisations within which all work in each sector must be contained; in other sectors individual projects will be authorised.

5. Main features common to all sectors are:

(*a*) The restoration of a three stage rolling programme to promote the systematic processing of work. The aim will be to give local education authorities progressively firmer allocations (either as lump sums or for specific projects) at regular intervals before the beginning of the starts year, viz. $2\frac{1}{2}$ years (Provisional), $1\frac{1}{4}$ years (Planning) and 6 months (Final) beforehand. (New terms have been introduced to avoid confusion with the previous system.) At each stage allocations may be adjusted in the light of available resources. Final allocations will take account of prospective cost movements and will not be subject to re-adjustment on this count after they have been announced.

(*b*) There will be no formal cost limits, instead, the building work for each sector will be contained within nationally determined cash allocations for the total value of starts. Thus the individual allocations made subsequently as lump sums or on a project basis will be totals within which the cost of building work (other than abnormal costs in primary and secondary schools of paragraph 6) must be contained.

EDUCATIONAL BUILDINGS

(c) All major projects (that is, schools projects costing £120,000 or more, and further education and teacher training projects costing £120,000 or more) will need to be submitted to Department of Education and Science at tender stage. No other formal approval will be required though it is envisaged that, as in the past, informal exchanges will take place between departments and authorities throughout the inception and planning of the project. It will continue to be necessary to meet minimum statutory and other requirements. The Department of Education and Science and Welsh Education Office retain the right to disallow an individual project at tender stage on cost grounds; but their aim will be to secure, through close and continuing contact with authorities and other providing bodies in all the preceding stages, that potential difficulties are identified and eliminated so that the formal approval process can be carried out smoothly and quickly.

The Department of Education and Science issue annual guidelines on costs to local authorities.

PUBLIC AUTHORITY HOUSING

INTRODUCTION

The Housing Cost Yardstick for local authority housing will be discontinued in April 1981, however a decision as to the method of establishing cost limits for Housing Associations and the New Towns has still to be made.

In March 1978 the Department of the Environment issued Circular 24/78 – The Housing Cost Yardstick which set out the new cost limits for public authority housing and consolidated a number of previous circulars. The standards and other data upon which these cost limits are based are generally recorded in other circulars which are referred to in Circular 24/78.

The cost limits in the circular have been reviewed quarterly in October, January, April and July each year, to take account of inflation. Readers are reminded of the need to ensure that they are in possession of the latest reviews of the yardstick. In order to provide a common base this section of the price book uses the limit set in Circular 24/78.

The Yardstick contains sufficient data for calculating the basic cost limit for any proposed housing scheme by means of a series of tables setting out the cost limits for a wide range of situations and these are not repeated here. The following explanatory notes have been prepared by the editors as a guide to the structure of the Yardstick. A method is suggested for planning the cost of a scheme in line with the cost limits.

GENERAL APPLICATION

For the purposes of the Circular the items which make up the cost of a housing scheme will generally fall into one of three categories, details of which are given in the 'Manual on Local Authority Housing Subsidies and Accounting' published by H.M.S.O. (referred to hereafter as 'The Manual'). The three categories are broadly as follows:

(1) Items which relate to the dwellings themselves and which qualify for a Treasury subsidy; generally these comprise the substructures, superstructures and external works forming part of the dwellings including approved car accommodation not exceeding one per dwelling. The cost of these items must not exceed the limits set by the cost tables in the Yardstick or any *ad hoc* addition thereto agreed with the Department. Where these limits are exceeded the excess cost must be considered as part of (2) below.

(2) Items which relate to the dwellings themselves but which do not qualify for a Treasury subsidy; generally these comprise higher standards than laid down in the Parker Morris report and such items as cookers, refrigerators, furniture, radio and T.V. aerials, contingency sums and any excess costs referred to under category 1 and category 3. The total cost of all these items should not exceed 10% of the agreed yardstick, or loan sanctions will be refused for the whole scheme.

(3) Major site development work such as roads and sewers; transformer buildings and electricity substations, work to existing buildings and items not directly associated with the housing, such as shops, restaurants, factories, community halls and the like. The 10% limit mentioned above does not apply to these items nor, except in the case of roads and sewers, do they qualify for a Treasury subsidy. Where the cost limits for car accommodation are exceeded the excess must be considered under (2) above; the cost of other items must be shown to represent value for money.

The basic cost limit is expressed in £ per person; an indicative sub-division into substructure, superstructure and external works is given in an Appendix to the Yardstick.

There are three major cost variables recognized by the Yardstick; these are the average number of persons per dwelling, the density of the development and regional variations in building prices.

PUBLIC AUTHORITY HOUSING

Average number of persons per dwelling

The size of a dwelling is expressed as the number of persons (bed-spaces) per dwelling; the Yardstick is based upon the average for the whole of the development, i.e. the total number of bed-spaces divided by the total number of dwellings.

The Yardstick tables are expressed in costs per person, and these costs reduce as the average number of bed-spaces per dwelling increases; this is because the area of a six-person dwelling is not twice that of a three-person dwelling as much of the accommodation, e.g. bathroom and kitchen, is common to both.

Density

The density of a scheme is the number of bed-spaces required by the brief per hectare of site. The definition of the site area is given on page 62 of 'The Manual' and generally covers all the land which is available for building, including half the width of any roads bordering the site (except major highways from which vehicular access to the site is prohibited). As the density of a scheme rises so the use of higher buildings may become necessary if the accommodation is to be fitted on to the site and still maintain light angles, adequate space around the buildings, etc. The higher the building, then generally the more expensive it becomes. Thus the total cost limits given in the Yardstick increase with the density, although the costs of substructure and external works may reduce.

Regional Variations

The Yardstick contains a list of percentages to be added to the basic cost limit dependent upon the geographical location of the site. These percentages range from nil to 35% for Variation of Price Contracts (see Table 6) but are subject to adjustment for inflation.

SUPPLEMENTARY ITEMS

There are a number of items supplementary to the Yardstick:

Accommodation designed for one, two or three persons – an allowance of £300 per dwelling is added to the yardstick for these dwellings.

Accommodation designed for old people – see Circulars 82/69, 1/80 and 8/80.

Accommodation designed for disabled people – appropriate standards are set out in Circular 92/75.

Hostels – subtractions from the total basic yardstick are set out according to storey height and standard of accommodation designed – see Circular 170/74.

Children's Play-space – an allowance of £60 per child bed-space is allowed. The number of child bed-spaces is in effect defined as the number in excess of two bed-spaces per dwelling – see Circular 79/72.

Accommodation for single working people – appropriate standards are set out in Circular 12/76.

Small schemes – a range of allowances is given for a variety of schemes up to 20 dwellings per scheme. Schemes of 25 dwellings or less are generally exempt from full scrutiny – see Circular 3/79.

Accommodating the motor car – the limit for the cost of constructing a hardstanding is given at £160 in the yardstick and a series of cost limits for other forms of car accommodation (if approved) will be made available by the Department's regional offices.

The Yardstick also recognizes the fact that in some instances the car accommodation

PUBLIC AUTHORITY HOUSING

will occupy so much of the site that the housing will have to be constructed in higher (and therefore more expensive) buildings than might otherwise be the case. This 'extra housing cost' is given in a supplementary table to the Yardstick. This gives an equivalent higher density – see Table 1 – to be used for calculating the housing yardstick where more than 75 car spaces per hectare are required.

TABLE 1

Average number of car spaces/hectare	% increase on actual density to give equivalent higher density
75	Nil
100	6
125	13
150	21
175 and over	31

All these supplementary allowances are subject to the regional variations mentioned earlier.

BUDGETING FOR THE COST YARDSTICK

The Cost Yardstick for a project can be calculated as soon as the client has established the area of site available, the number of dwellings and the number of bed-spaces. The Yardstick aims to provide reasonable cost limits by recognizing the cost effects of different dwelling sizes and of density. As the density rises above a certain base point the cost limit also rises to finance the more expensive building forms that become necessary. The actual solution is left to the Architect, but the Yardstick remains the same whatever solution is chosen.

Initial Strategy

The introductory notes to Circular No. 36/67 – the original mandatory cost yardstick – stated that medium rise flats and high rise flats were 20% and 50% more expensive respectively than low rise dwellings which were generally the cheapest form of housing.

Until the June 1975 revision of the yardstick the cost limits were based upon a theoretical mix of low, medium and high rise buildings. However, the last two revisions to the yardstick take a more general and generous view of the cost implications of increasing the density which should enable designers to be less restricted particularly in the middle range of densities than hitherto.

Therefore the table of mixes that was the basis of building up the old yardstick (which was also used to check the cost feasibility of an embryo proposal) technically no longer applies. However, in practical terms, some initial guidance as to the likely viable sub-division between low, medium and high rise buildings is invaluable and the table, which appeared in earlier editions of the price book, has been revised and is reproduced at Table 2; for the purpose of this table low rise can be considered to be one and two storey buildings, medium rise to be three and four storeys and high rise five storeys and above.

In applying this test it is important to base calculations on the proportion of floor area, not the number of bed-spaces or dwellings.

The density used in the table is the equivalent higher density after allowing for the effects of car accommodation referred to earlier.

PUBLIC AUTHORITY HOUSING

TABLE 2. – % DIVISION OF FLOOR AREAS

Density in bedspaces per hectare	2 persons			3 persons			4 persons			5 persons		
	Low rise	Medium rise	High rise	Low rise	Medium rise	High rise	Low rise	Medium rise	High rise	Low rise	Medium rise	High rise
100	100			100			100			100		
125	92	8		100			100			100		
150	57	43		87	13		100			100		
175	27	73		57	43		74	26		86	14	
200	2	98		31	69		48	52		59	41	
225		84	16	9	91		26	74		36	64	
250		69	31		92	8	6	94		16	84	
275		55	45		78	22		90	10		99	1
300		43	57		65	35		78	22		86	14
325		33	67		54	46		65	35		74	26
350		23	77		43	57		55	45		64	36
375		15	85		34	66		46	54		54	46
400		8	92		26	74		37	63		45	55
425		2	98		18	82		29	71		37	63
450			100		12	88		22	78		29	71
475			100		6	94		16	84		23	77
500			100		1	99		10	90		17	83
525			100			100		5	95		11	89
550			100			100		1	99		6	94
575			100			100			100		2	98
600			100			100			100			100

The division of floor area given in Table 2 is a target only; it is not mandatory. However, if the proportion of the floor area shown in this table as being accommodated in the lower of the two building types is not achieved, then it may be difficult if not impossible to meet the basic cost limits in the Yardstick. Where, on the other hand, this proportion is exceeded, then it may often be possible to achieve the cost limits with a margin to spare for increasing the amenities or standards of the scheme or for financing the more expensive building forms that may become necessary.

Whilst the table only admits two forms at any one density, in practice it may be possible to introduce a proportion of all three forms, particularly at the middle range of densities. In deviating from this two form mix it must be recognized that high rise is 20–25% more expensive than medium rise but covers much less ground area.

Checking the feasibility of a proposal

Once the initial strategy has been determined and outline proposals prepared it will be found that the mix tables above will be too crude a guide to judge the likely financial picture with precision. Moreover, whilst the mix tables might provide useful initial guidance they are no longer the basis of the Yardstick.

The new Yardstick commences for any dwelling size at a minimum level that will finance the cheapest form, e.g. all low rise building. As the density increases it contains a growing margin to finance higher (and more expensive) building forms. Thus the simplest way to check the feasibility of a proposal is to calculate the margin above the low rise figure that is allowed in the Yardstick for the scheme and compare this with the extra cost that is likely to be generated by the proposed dwelling forms above an all low rise solution; i.e. compare the budget index with the estimated cost index of a proposal.

PUBLIC AUTHORITY HOUSING

The following paragraphs elaborate upon this approach; an example is then given of how the data can be applied to a specific scheme. In both cases costs are related to the basic yardstick limits in Circular 24/78 and are therefore subject to adjustment for regional variations and inflation.

Calculating the budget index

The budget index for any scheme can be found by comparing the actual amount given in Yardstick tables with the minimum Yardstick figure for the same dwelling size. For example, at a density of 260 persons per hectare and an average dwelling size of 3·6 persons per dwelling the yardstick is £2,632 compared with £2,172 – the minimum figure quoted at that dwelling size; the index is therefore £2,632 ÷ £2,172 × 100, i.e. 121·18 when an all two-storey house scheme is 100·00. (This budget index can be deduced merely from the density and dwelling size – if, for example, Yardstick tables are not available – by using Tables 3 and 4 as described later, but this is slightly more laborious than the method described above.)

Where the budget index is 121·18 a solution that comprises dwellings having an average cost index of up to 121·18 should be within the Yardstick.

Comparative costs of dwelling forms

In general the cheapest form of housing per unit is two-storey. The margin above this level for alternative forms will vary considerably but Table 3 indicates the likely variants.

It will be appreciated that single storey dwellings (e.g. bungalows) absorb a greater site area than two or three storey dwellings and their use could create the need for higher density on other parts of the site.

TABLE 3. COMPARATIVE COSTS OF DIFFERING DWELLING FORMS

Dwelling Form	Cost Index
(a) Terraced bungalows	98
(b) Two storey houses	100
(c) Two storey flats	111
(d) Three storey houses	112
(e) Three storey flats	118
(f) Three storey maisonettes/flats	120
(g) Four storey maisonettes	129
(h) Four storey flats	133
(i) Five storey flats	140
(j) Other higher mix forms	144+

The above indices are largely indicative and are for the purpose of demonstrating this particular cost checking technique and have changed marginally over the years on better data becoming available. In practice they can vary considerably depending upon such factors as the plan shape, the number of units per terrace, whether staircase or balcony access is provided to the flats, the configuration of the dwellings; the quantity surveyor should have more precise data available for use on any particular scheme.

It is possible to deduce from this table the cost index for any set of proposals and this can then be compared with the budget index referred to above.

For example a three storey block of maisonettes and flats at 120 would be within the limit index of 121·18 used above. The following complete example illustrates how this process might work in practice.

PUBLIC AUTHORITY HOUSING

Example:

Basic data

Site area 1·50 hectares

Number of dwellings	40 No. with 2 persons each	=	80 persons
	60 No. „ 3 „ „	=	180 „
	50 No. „ 5 „ „	=	250 „
Total	150 No. dwellings	=	510 persons

Car accommodation   1 space per dwelling

Average dwelling size $= \dfrac{510 \text{ persons}}{150 \text{ dwellings}}$ $= 3\cdot4$ persons per dwelling

Housing density $= \dfrac{510 \text{ persons}}{1\cdot5 \text{ hectares}}$ $= 340$ persons per hectare

Car spaces per hectare $= \dfrac{150}{1\cdot50}$ $= 100$

Percentage increase to
give Equivalent Higher
Density $= 6\%$ (See Table 1)

Therefore the 'Equivalent Higher Density' $= 340 + 6\% = 360$ persons per hectare.

Initial strategy

By interpolating the data in Table 2 it will be clear that a solution comprising 44% medium rise and 56% high rise (by floor area) would be a reasonable solution; to increase the proportion of high rise would probably mean exceeding the cost limit, reducing might give a contingency within which to work but the extent of such a movement is limited by the physical problem of fitting the housing on the site.

Checking the feasibility of a proposed solution

The Yardstick tables for this brief of 360 persons per hectare and 3·4 persons per dwelling is £2997 per person (before regional allowances) and the minimum figure at 3·4 persons per dwelling is £2247; the budget index therefore is $\dfrac{2997}{2247} \times 100 = 133\cdot38$, i.e. dwellings which are on average 33·38% more expensive than all low rise can be afforded with the budget.

The Architects' proposals, reflecting as far as possible the initial strategy, might be to accommodate the majority of units in four storey blocks with the 5 person units in ground level maisonettes with 2 levels of 3 person and 2 person flats over; the residue of small flats being provided in a five storey block, i.e.

45% of floor area × 129 (Table 3, item g) =	58·05
33% of floor area × 133 (Table 3, item h) =	43·89
22% of floor area × 140 (Table 3, item i) =	30·80
	132·74

PUBLIC AUTHORITY HOUSING

As this is to be compared with a budget index of 133·38 this proposal should be within the Yardstick.

The above paragraphs set down the guide lines for the initial development of a scheme and indicate how a budget can be established so that the Yardstick total is not exceeded. As the scheme is developed cost checks, preferably in elements, should be undertaken to ensure that the shape of the dwelling, the intricacy of the design and the standard of specification are such that these targets can be achieved.

It is possible to calculate the budget index without the yardstick tables. For any average dwelling size there is a density above which it is assumed dwelling forms higher than two storey will be necessary; this appears as the starting density on the yardstick tables and for present purposes we have called this the 'threshold density'. The threshold densities are recorded in Table 4.

For a given percent increase in density beyond this threshold the increase in the yardstick allowance will be the same; for example at double the threshold density the budget index will always be 126·3 whatever the average dwelling size. This common pattern of increase in the budget index is given in Table 5.

For any particular brief therefore one can work out the actual density, compare it with the threshold density and then read off the budget increase. Thus if the brief gives a density of 330 persons per hectare for a scheme with an average dwelling size of 2·5 persons the threshold density will be seen from Table 4 to be 132. However, the actual density is 211 which is an increase of 60% on the threshold of 132. From Table 5 it will be seen that the budget index for this brief is 118·2 (100 + 18·2); i.e. dwelling forms which are on average up to 18·2% more expensive than two storey forms should fall within the yardstick limit.

PUBLIC AUTHORITY HOUSING

TABLE 4

Average persons per dwelling	Threshold density* in yardstick tables persons/hectares	Average persons per dwelling	Threshold density* in yardstick tables persons/hectares	Average persons per dwelling	Threshold density* in yardstick tables persons/hectares
1·0	84	3·0	141	5·0	163
1·1	89	3·1	143	5·1	164
1·2	94	3·2	144	5·2	165
1·3	98	3·3	146	5·3	165
1·4	102	3·4	147	5·4	166
1·5	106	3·5	148	5·5	167
1·6	109	3·6	150	5·6	167
1·7	112	3·7	151	5·7	168
1·8	113	3·8	152	5·8	168
1·9	118	3·9	153	5·9	169
2·0	121	4·0	154	6·0	170
2·1	123	4·1	155		
2·2	126	4·2	156		
2·3	128	4·3	157		
2·4	130	4·4	158		
2·5	132	4·5	159		
2·6	134	4·6	160		
2·7	136	4·7	161		
2·8	138	4·8	162		
2·9	140	4·9	162		

* Density above which the yardstick tables increase progressively.

TABLE 5. – PATTERN OF INCREASES IN YARDSTICK COST TABLES

% Increase in density above threshold	% Margin in cost limit in excess of minimum figure	Budget Index	% increase in density above threshold	% Margin in cost limit in excess of minimum figure	Budget Index
10	3·9	103·9	140	32·6	132·6
20	7·3	107·3	150	34·0	134·0
30	10·4	110·4	160	35·2	135·2
40	13·2	113·2	170	36·4	136·4
50	15·8	115·8	180	37·5	137·5
60	18·2	118·2	190	38·6	138·6
70	20·4	120·4	200	39·6	139·6
80	22·5	122·5	210	40·5	140·5
90	24·6	124·6	220	41·3	141·3
100	26·3	126·3	230	42·2	142·2
110	28·0	128·0	240	42·9	142·9
120	29·6	129·6	250	43·6	143·6
130	31·1	131·1	260 and over	44·1	144·1

PUBLIC AUTHORITY HOUSING

TABLE 6. – REGIONAL VARIATIONS OF YARDSTICK COSTS

Economic Planning Region	Counties and Metropolitan Counties	Yardstick plus Variation of price contracts	Firm price contracts
Eastern	Bedfordshire, Buckinghamshire, Cambridgeshire, Essex, Hertfordshire, Norfolk and Suffolk, excluding the Districts and any Housing Sites set out in Schedules A and B.	2½%	8½%
North West	Cheshire, Greater Manchester, Lancashire and Merseyside.		
East Midlands	Derbyshire, Leicestershire, Lincolnshire, Northamptonshire and Nottinghamshire.	NIL	6%
South West	Avon, Cornwall, Devon, Dorset, Gloucestershire, Somerset and Wiltshire.		
Yorkshire and Humberside	Humberside, North Yorkshire, South Yorkshire, and West Yorkshire.		
Northern	Cleveland, Cumbria, Durham, Northumberland and Tyne and Wear.	7½%	14%
Wales			
South East	Berkshire, East Sussex, Hampshire, Isle of Wight, Kent, Oxfordshire, Surrey and West Sussex, excluding the Districts and any Housing Sites set out in Schedules A and B.	5%	11½%
West Midlands	Hereford and Worcester, Salop, Staffordshire, Warwickshire and West Midlands.		
Eastern and South East – *continued*	District Authorities wholly within 35 miles of Charing Cross as Schedule A. Any housing site wholly or partly within 35 miles of Charing Cross in the District Authorities as Schedule B.	12½%	19½%
London	London Boroughs as Schedule C.	20%	27%
	City of London and London Boroughs as Schedule D.	35%	43%

SCHEDULE A
Variation of price contracts – Yardstick plus 12½%
Firm price contracts   – Yardstick plus 19½%
District Authorities wholly within 35 miles (56·327 kilometres) of Charing Cross:

Basildon
Beaconsfield
Bracknell
Brentwood
Broxbourne
Castle Point
Chiltern
Crawley
Dartford
East Hertfordshire
Elmbridge
Epping Forest
Epsom and Ewell

Gillingham
Gravesham
Guildford
Harlow
Hertsmere
Luton
Mole Valley
Reigate and Banstead
Runnymede
Rushmoor
St Albans
Sevenoaks
Slough

Spelthorne
Stevenage
Surrey Heath
Tandridge
Three Rivers
Thurrock
Tonbridge and Malling
Watford
Welwyn Hatfield
Windsor and Maidenhead
Woking

SCHEDULE B
Variation of price contracts – Yardstick plus 12½%
Firm price contracts   – Yardstick plus 19½%
Any housing sites wholly or partly within 35 miles (56·327 kilometres) of Charing Cross in the following District Authorities:

Aylesbury Vale
Braintree
Chelmsford

Chichester
Dacorum
Hart

Horsham
Maidstone
Maldon

PUBLIC AUTHORITY HOUSING

TABLE 6. – *Continued.*

SCHEDULE B – *continued*

Medway	South Bedfordshire	Uttlesford
Mid Bedfordshire	Southend-on-Sea	Waverley
Mid Sussex	South Oxfordshire	Wealden
North Hertfordshire	Swale	Wokingham
Rochford	Tunbridge Wells	Wycombe

SCHEDULE C

Variation of price contracts – Yardstick plus 20%
Firm price contracts   – Yardstick plus 27%
The London Boroughs of:

Barking	Enfield	Merton
Barnet	Haringey	Newham
Bexley	Harrow	Redbridge
Brent	Havering	Richmond-upon-Thames
Bromley	Hillingdon	Sutton
Croydon	Hounslow	Waltham Forest
Ealing	Kingston-upon-Thames	

SCHEDULE D

Variation of price contracts – Yardstick plus 35%
Firm price contracts   – Yardstick plus 43%
The City of London and the London Boroughs of:

Camden	Islington	Southwark
Greenwich	Kensington and Chelsea	Tower Hamlets
Hackney	Lambeth	Wandsworth
Hammersmith	Lewisham	City of Westminster

Approximate Estimates

Estimating by means of priced approximate quantities is always more accurate than by using prices per square metre. The prices in this section, which is arranged in elements, are based upon the 'Prices for Measured Work' contained in Part II but allow for incidentals which normally would be measured separately in a full Bill of Quantities. They do not include for preliminaries, details of which are given in Part II and which may amount to approximately 15% of the value of the measured work. Further alternatives to the items given here may be interpolated, if required, from 'Comparative Prices'.

Item No.		Unit	Excavation Hand £	Mechanical £
1·0	**SUBSTRUCTURE**			
	Strip foundations			
	Excavating trenches 1 m deep in heavy soil; levelling; compacting; planking and strutting; back filling; disposal of surplus material from site; concrete 11·50 N/mm² foundations 300 mm thick; common bricks *p.c.* £45·00 1000 in cement mortar (1:3) walls to 150 mm above ground level; bitumen hessian-based horizontal damp-proof course; facing bricks *p.c.* £122·00 1000 300 mm high one side; weather pointing as work proceeds one side			
1·0·1	for one brick walls	m	44·00	36·00
1·0·2	Extra; each additional 300 mm in depth . .	,,	11·00	8·70
1·0·3	for hollow walls; two half-brick skins; 50 mm cavity with ties; filling cavity with concrete 21·00 N/mm² . .	,,	50·00	40·35
1·0·4	Extra; each additional 300 mm in depth . . .	,,	13·00	10·10
1·0·5	for one and a half brick walls . . .	,,	59·60	49·20
1·0·6	Extra; each additional 300 mm in depth . . .	,,	15·35	12·00
	Ground beams			
	Excavating trenches 500 mm deep in heavy soil; levelling; compacting; planking and strutting; backfilling; disposal of surplus material from site; hand packing hardcore to battering face; concrete 25·50 N/mm²; reinforcement four 25 mm high tensile steel bars; sawn formwork to vertical sides of foundation			
1·0·7	construction; ground beam over 0·05 m² not exceeding 0·10 m²	,,	16·45	
	Column foundations			
	Excavating pit 1·50 m deep in heavy soil; levelling; compacting; planking and strutting; backfilling; disposal of surplus material from site; concrete (1:12) foundation to stanchion 850 mm thick; concrete 25·50 N/mm² thick; reinforcement 215 Kg of 25 mm high tensile steel bars; sawn formwork to vertical sides of foundation			
1·0·8	construction; foundation to stanchion 3500 × 3500 × 450 mm	No.	1004·00	

ω̲e̲l̲ ̲x̲2̲

Item No.		Unit	Excavation Hand £	Mechanical £
	SUBSTRUCTURE – *continued*			
	Solid ground-floor construction			
	Excavating to reduce levels in heavy soil; disposal of surplus material from site; hardcore filling 150 mm thick; concrete 11·50 N/mm^2 bed 150 mm thick; mesh reinforcement 2·22 kg/m^2; two coats Synthaprufe			
1·0·9	construction; excavating to reduce levels 225 mm deep	m^2	14·40	11·85
1·0·10	Extra; excavating to reduce levels each additional 150 mm deep	,,	2·65	0·84
1·0·11	Thickening 150 mm concrete bed to 250 mm; 450 mm wide; bitumen hessian-based horizontal damp-proof course 113 mm wide	m	1·95	1·45
	Excavating to reduce levels in heavy soil; levelling; compacting; disposal of surplus material from site; hardcore filling 150 mm thick; levelling blinding; concrete (1:12) bed 50 mm thick; concrete 25·50 N/mm^2 bed 200 mm thick; mesh reinforcement 3·02 Kg/m^2; 0·13 mm polythene sheet			
1·0·12	construction; excavating to reduce levels 300 mm deep	m^2	17·00	
	Hollow ground-floor construction			
	Excavating to reduce levels in heavy soil; disposal of surplus material from site; hardcore filling 100 mm thick; concrete 11·50 N/mm^2 bed 150 mm thick; common bricks *p.c.* £45·00 1000 half brick walls 300 mm in cement mortar (1:3), 300 mm high, honeycomb bond at 2 m centres; two courses slates horizontal damp-proof course; 100 × 50 mm softwood plates; bedding in cement mortar (1:3); 100 × 50 softwood joists at 400 mm centres			
1·0·13	construction; excavating to reduce levels 375 mm deep	m^2	20·50	16·25
1·0·14	Extra; excavating to reduce levels and half-brick walls each additional 150 mm deep	,,	3·30	1·53

Item No.		Unit	£
2·0	**SUPERSTRUCTURE**		
2·1	**FRAME**		
	Reinforced-concrete construction		
	In situ concrete mix 25·50 N/mm^2; reinforcement high tensile steel bars; sawn formwork		
2·1·1	construction; beam over 0·05 m^2 not exceeding 0·10 m^2; reinforcement 50 Kg/m	m	40·75
2·1·2	construction· column over 0·05 m^2 not exceeding 0·10 m^2; reinforcement 40 Kg/m	,,	42·00
2·1·3	Extra; wrought formwork	m^2	1·90

Item No.		Unit	£
2·2	**UPPER FLOORS**		
	Reinforced-concrete construction		
	In situ concrete mix 21 N/mm²; reinforcement high tensile steel bars 20 Kg/m²; formwork to soffit and exposed edges; suitable for span of 4 m		
2·2·1	construction 125 mm thick	m²	26·45
	In situ concrete mix 25·50 N/mm²; reinforcement high tensile steel bars 36Kg/m²; formwork to soffit and exposed edges; suitable for span of 4·80 m		
2·2·2	construction 225 mm thick	,,	38·15
	Hollow clay blocks 150 mm thick; clay filler tiles; concrete mix 25·50 N/mm² ribs and topping 50 mm thick; reinforcement high tensile steel bars 8·15 Kg/m² and mesh 2·22 Kg/m²; formwork to soffit and exposed edges; suitable for span of 4·80 m		
2·2·3	construction 200 mm thick	,,	26·00
	Precast, prestressed concrete construction		
	Precast prestressed units; fixed direct from lorry to hard level mountings; one hour fire grading; total loading 5·00 kN/m² in addition to self weight of units;		
2.2.4	construction 150 mm thick for spans up to 3 m . . .	,,	14·05
2.2.5	,, ,, ,, ,, ,, 3 to 6 m . . .	,,	15·65
2.2.6	,, 200 ,, ,, ,, ,, 6 to 7·5 m . . .	,,	17·45
	total loading 8·50 kN/m² in addition to self weight of units;		
2.2.7	construction 150 mm thick, for spans up to 3 m . . .	,,	14·25
2.2.8	,, 200 ,, ,, ,, ,, 3 to 6 m . . .	,,	18·20
2.2.9	,, 250 ,, ,, ,, ,, 6 to 7·5 m . . .	,,	20·55
	total loading 12·50 kN/m² in addition to self weight of units;		
2.2.10	construction 150 mm thick, for spans up to 3m . . .	,,	15·55
2.2.11	,, 250 ,, ,, ,, ,, 3 to 6 m .	,,	20·65
	Softwood construction		
	Joists at 400 mm centres; ends built in; 50 × 25 mm herringbone strutting; 50 × 25 mm fillets to perimeters; trimming to openings		
2·2·12	construction with 175 × 50 mm joists	,,	9·00
2·2·13	,, ,, 200 × 50 ,, ...	,,	10·00
2·2·14	,, ,, 225 × 50 ,, ...	,,	11·00
2·3	**ROOF**		
	Flat roof		
	Reinforced-concrete construction; coverings		
	In situ concrete mix 21 N/mm²; mesh reinforcement; formwork to soffit and exposed edges; 25 mm (average) cement and sand (1 : 3) screed to falls; 19 mm two-coat asphalt B.S. 988 coverings; suitable for span of 4 m and superimposed load of 145 kg/m²		
2·3·1	construction 125 mm thick; screed; coverings . . .	,,	29·50

Item No.	ROOF – *continued*	Unit	£

Flat roof – *continued*

Softwood construction; coverings

Joists at 400 mm centres; firrings; 100 × 50 mm plates; bedding; 50 × 25 mm herringbone strutting; 25 mm roof boarding; butt joints; to falls; three-layer one-ply bitumen-felt coverings

2·3·2	construction with 175 × 50 mm joists; coverings . . .	m²	22·75
2·3·3	,, ,, 200 × 50 ,, ,, . . .	,,	23·55
2·3·4	,, ,, 225 × 50 ,, ,, . . .	,,	24·40

Joists at 400 mm centres; firrings; 100 × 50 mm plates; bedding; 50 × 25 mm herringbone strutting; 50 mm thick wood-wool slab to falls; three-layer one-ply bitumen-felt coverings

2·3·5	construction with 175 × 50 mm joists; coverings . . .	,,	22·75
2·3·6	,, ,, 200 × 50 ,, ,, . . .	,,	23·60
2·3·7	,, ,, 225 × 50 ,, ,, . . .	,,	24·45

Joists at 400 mm centres; firrings; 100 × 50 mm plates; bedding; 50 × 25 mm herringbone strutting; 25 mm roof boarding; butt joints; to falls; 19 mm two-coat asphalt B.S. 988 coverings

2·3·8	construction with 175 × 50 mm joists; coverings . . .	,,	23·55
2·3·9	,, ,, 200 × 50 ,, ,, . . .	,,	24·35
2·3·10	,, ,, 225 × 50 ,, ,, . . .	,,	25·20

Joists at 400 mm centres; firrings; 100 × 50 mm plates; bedding; 50 × 25 mm herringbone strutting; 50 mm thick wood-wool slab to falls; 19 mm two-coat asphalt B.S. 988 coverings

2·3·11	construction with 175 × 50 mm joists; coverings . . .	,,	23·55
2·3·12	,, ,, 200 × 50 ,, ,, . . .	,,	24·40
2·3·13	,, ,, 225 × 50 ,, ,, . . .	,,	25·25

Steel construction; coverings

Plain girders in roof beams; all fabrication; placing in position; fixing; one coat primer

2·3·14	construction rolled steel joist roof beams . . .	kg	0·60

Reinforced wood-wool to falls; mild steel clips at 600 mm centres; 19 mm two-coat asphalt B.S. 1162 coverings; felt underlay

2·3·15	construction; 50 mm wood-wool; coverings . . .	m²	20·00

Mansard roof

Softwood construction; coverings

2·3·16	Treated sawn softwood; plates; joists; rafters . . .	m³	243·00

Coverings; mansard; 70° pitch

2·3·17	Asbestos cement slates; blue; 508 × 254 mm; fixing with two copper nails and one copper rivet per slate; to 75 mm lap; 19 × 50 mm battens; 50 mm glass fibre quilt . . .	m²	19·75

Coverings; vertical

2·3·18	Asbestos cement slates; blue; 508 × 254 mm; fixing with two copper nails and one copper rivet per slate to 75 mm lap; 19 mm blockboard	,,	30·65

Eaves

Softwood fascia; to backgrounds requiring plugging; 75 × 50 mm softwood splayed fillet; three-layer one-ply bitumen felt coverings; aluminium edge trim; one coat primer, two undercoats, one coat full gloss finish on exposed woodwork surfaces

2·3·19	eaves; 280 × 25 mm fascia	m	9·75

Softwood fascia; to backgrounds requiring plugging; cast-iron gutter and brackets; No. 4 lead drip dressed between layers of roofing and into gutter; one coat primer, two undercoats, one coat full-gloss finish on exposed woodwork surfaces; touch up primer, two undercoats, one full-gloss finish on gutters and brackets

2·3·20	eaves; 230 × 25 mm fascia; 100 mm half-round gutter .	,,	18·90
2·3·21	,, 230 × 25 ,, 150 ,, ,, ,, .	,,	23·85

Item No.		Unit	£

ROOF – *continued*

Pitched roofs

NOTE. *Prices are for roofs measured on the flat plan area over all external walls and not the area on slope*

Softwood construction; gable ends 35° pitch; clear span not exceeding 8 m; coverings

75 × 40 mm plates; T.R.A.D.A. type trussed rafters at 2 m centres; 100 × 32 mm rafters and ceiling joists at 500 mm centres; 150 × 50 mm purlins; 125 × 50 mm binders; 150 × 25 mm ridge; 40 × 19 mm battens; sarking felt

2·3·22	construction; natural slates; B.S. 680; Welsh blue; uniform size; 508 × 254 mm; centre fixing with two copper nails per slate .	m²	39·80
2·3·23	construction; concrete interlocking tiles B.S. 550 type A; 413 × 330 mm; fixing every second course with two aluminium nails per tile	”	19·20

NOTE. *Brickwork to gables not included.*

Softwood construction; hipped ends; 40° pitch; clear span not exceeding 8 m; coverings

75 × 40 mm plates; T.R.A.D.A. type trussed rafters at 2 m centres; 100 × 50 mm rafters and 100 × 40 mm ceiling joists at 500 mm centres; 150 × 50 mm purlins; 125 × 50 mm binders; 150 × 25 mm ridge; 175 × 32 mm hips; 40 × 19 mm battens; sarking felt

2·3·24	construction; clayware Broseley machine-made plain tiles; B.S. 402; 267 × 165 mm; hanging by nibs; fixing every fourth course with two aluminium nails per tile; to 64 mm lap .	m²	42·00
2·3·25	construction; clayware best hand-made sand-faced plain tiles; B.S. 402; 267 × 165 mm; hanging by nibs; fixing every fourth course with two aluminium nails per tile; to 64 mm lap .	”	55·20

Eaves

Double course of slates or tiles; softwood fascia; cast-iron gutters and brackets; one coat primer, two undercoats, one full-gloss finish on exposed woodwork surfaces; touch up primer, two undercoats, one coat full-gloss finish on gutters and brackets

2·3·26	eaves; 200 × 25 mm fascia; 6 mm flat asbestos-cement sheet soffit 225 mm wide; 100 mm half-round gutter . . .	m	15·15
2·3·27	eaves; 200 × 25 mm fascia; 6 mm flat asbestos-cement sheet soffit 225 mm wide; 150 mm half-round gutter . . .	”	19·85
2·3·28	eaves; 200 × 25 mm fascia; 225 × 19 mm softwood soffit; 100 mm half-round gutter	”	17·75
2·3·29	eaves; 200 × 25 mm fascia; 225 × 19 mm softwood soffit; 150 mm half-round gutter	”	22·45

Verges

Undercloak course; bedding and pointing in cement mortar (1 : 4); softwood bargeboards; one coat primer, two undercoats, one full-gloss finish on exposed woodwork surfaces

2·3·30	verges; 250 × 25 mm bargeboard; 6 mm flat asbestos-cement sheet soffit 150 mm wide	”	6·15
2·3·31	verges; 250 × 25 mm bargeboard; 150 × 19 mm softwood soffit	”	7·80

		Unit	£
Item No.	**ROOF** – *continued*		
	Hips		
	Cappings to hips		
2·3·32	machine-made clayware half-round hip tiles; bedding and jointing in cement mortar (1 : 4); ends; mitred intersections; cutting roof tiles prior to covering	m	12·65
2·3·33	hand-made clayware bonnet hip tiles; supplementary fixing with copper nails; coursed and bonded with general tiling; bedding and jointing in cement mortar (1 : 4); ends; mitred intersections	„	14·40
	Ridges		
	Cappings to ridges		
2·3·34	concrete half-round ridge tiles; butt jointed; bedding and jointing in cement mortar (1 : 4); ends; cutting on roof tiles prior to covering	„	5·85
2·3·35	hand-made clayware half-round ridge tiles; butt jointed; bedding and jointing in cement mortar (1 : 4); ends; cutting on roof tiles prior to covering	„	7·00
	Rainwater pipes		
	6 m length of cast-iron rainwater pipe; one offset; one shoe; red lead joints; fixing with galvanized nails and galvanized pipe distance pieces to backgrounds requiring plugging; touch up primer, two undercoats, one coat full-gloss finish		
2·3·36	75 mm diameter	No.	78·35
2·3·37	Add or deduct for each 300 mm variation in length . .	„	3·10
2·3·38	100 mm diameter	„	99·00
2·3·39	Add or deduct for each 300 mm variation in length . .	„	3·85
	12 m length of pvc rainwater pipe; one roof outlet; one connection to vitrified clay pipe; fixing with pvc brackets with screws to backgrounds requiring plugging		
2·3·40	100 mm diameter	„	108·00
2·3·41	Add or deduct for each 300 mm variation in length . .	„	2·05

2·4   STAIRS

In situ reinforced-concrete construction

900 mm wide; treads and risers; half space landing; granolithic one-coat work to treads, risers and landing; steel trowelled; carborundum grains trowelled in to treads and landings; fair concrete soffits; mild steel plain balustrade one side; one coat primer, two undercoats, one coat full-gloss finish on fair concrete soffits and metal balustrades

2·4·1	one storey high; rising 2600 mm	„	734·50
2·4·2	Add or deduct for each 300 mm variation in storey height .	„	73·30

900 mm wide; treads and risers; half space landing; terrazzo finish to treads, risers and landing; plastered soffits; mild-steel ornamental balustrade one side; one coat primer, two undercoats, one coat full-gloss finish on plaster soffits and metal balustrades

2·4·3	one storey high; rising 2600 mm	„	1703·00
2·4·4	Add or deduct for each 300 mm variation in storey height .	„	159·00

Softwood construction B.S. 585

900 mm wide; treads and risers; winders; balustrade one side; plasterboard B.S. 1230 type b, fixing with nails and one coat gypsum plaster to soffit; one coat primer, two undercoats, one coat full-gloss finish on plaster and woodwork surfaces

2·4·5	one storey high; rising 2600 mm	„	250·00

Item No.		Unit	£

STAIRS – *continued*

Oak construction

900 mm wide; 32 mm treads and winders; 32 mm quarter space landing; 25 mm risers; 40 mm strings; 100 × 100 mm turned newels; 32 × 32 mm turned balusters; moulded handrails; 75 × 100 mm softwood carriages and bearers; plasterboard B.S. 1230 type b, fixing with nails and one coat gypsum plaster to soffits; one coat primer, two undercoats, one coat full-gloss finish to plaster surfaces; staining and wax polishing hardwood surfaces

Item No.		Unit	£
2·4·6	one storey high; rising 2600 mm	No.	2070·00
2·4·7	Add or deduct for each 300 mm variation in storey height	,,	222·00

Metal construction

Straight flight; one storey high; cast-iron perforated treads and landing plates (no risers); mild-steel strings, plain balusters and handrails; one coat primer, two undercoats, one coat full-gloss finish

2·4·8	760 mm wide; rising 2600 mm	,,	915·00
2·4·9	Add or deduct for each 300 mm variation in storey height	,,	94·15
2·4·10	900 mm wide; rising 2600 mm	,,	942·00
2·4·11	Add or deduct for each 300 mm variation in storey height	,,	100·50
2·4·12	1070 mm wide; rising 2600 mm	,,	914·00
2·4·13	Add or deduct for each 300 mm variation in storey height	,,	113·35

Spiral staircase; one storey high; cast-iron perforated treads and landing plates (no risers); mild-steel strings, plain balusters and handrails both sides; one coat primer, two undercoats, one coat full-gloss finish

2·4·14	1070 mm diameter; rising 2600 mm	,,	772·00
2·4·15	Add or deduct for each 300 mm variation in storey height	,,	83·60
2·4·16	1370 mm diameter; rising 2600 mm	,,	825·00
2·4·17	Add or deduct for each 300 mm variation in storey height	,,	93·70
2·4·18	1830 mm diameter; rising 2600 mm	,,	1034·00
2·4·19	Add or deduct for each 300 mm variation in storey height	,,	104·50

2·5	**EXTERNAL WALLS**		

In situ concrete mix 25·50 N/mm² wall; reinforcement 13 kg/m²; sawn formwork

2·5·1	construction 150 mm thick	m²	36·00
2·5·2	,, 240 ,,	,,	40·35

Common bricks *p.c.* £45·00 1000 in cement–lime mortar (1 : 2 : 9)

2·5·3	one brick walls	,,	21·25
2·5·4	Add for fair face one side; flush pointing as work proceeds	,,	1·48
2·5·5	hollow walls; two half-brick skins; 50 mm cavity with ties	,,	22·70
2·5·6	Add for fair face one side; flush pointing as work proceeds	,,	1·48
2·5·7	one and a half brick walls	,,	30·40
2·5·8	Add for fair face one side; flush pointing as work proceeds	,,	1·48

Common bricks *p.c.* £45·00 1000 in cement–lime mortar (1 : 2 : 9); facing bricks *p c.* £122·00 1000

2·5·9	one brick walls; facings and flush pointing one side as work proceeds; Flemish bond	,,	32·85
2·5·10	Add or deduct for each variation of £1·00 1000 in *p.c.* for facing bricks	,,	0·09
2·5·11	Add for weather pointing as a separate operation	,,	2·95
2·5·12	hollow walls; one skin commons, one skin facings and flush pointing one side as work proceeds; 50 mm cavity with ties; stretcher bond	,,	31·00
2·5·13	Add or deduct for each variation of £1·00 1000 in p.c. for facing bricks	,,	0·07
2·5·14	Add for weather pointing in cement mortar as a separate operation	,,	2·54

Item No.		Unit	£
	EXTERNAL WALLS – *continued*		
2·5·15	one and a half brick walls; facings and flush pointing one side as work proceeds; Flemish bond	m²	42·30
2·5·16	Add or deduct for each variation of £1·00 1000 in *p.c.* for facing bricks	„	0·09
2·5·17	Add for weather pointing in cement mortar as a separate operation	„	2·95
	Lightweight concrete blocks; facing bricks *p.c.* £122·00 1000		
2·5·18	hollow walls; one skin 100 mm lightweight concrete blocks; one skin facings and flush pointing one side as the work proceeds; 50 mm cavity with ties, stretcher bond	„	29·25
	Facing bricks *p.c.* £122·00 1000		
2·5·19	half brick wall	„	20·00
2·5·20	„ „ against concrete	„	20·85

2·6 WINDOWS AND EXTERNAL DOORS

Windows

Prices are for normal sizes; suitable ironmongery; glazing with 3 mm clear sheet glass

2·6·1	Module '100' metal windows; opening lights; bedding frame in cement mortar; one coat primer, two undercoats, one coat full-gloss finish, both sides	„	50·15
2·6·2	Stock pattern or standard unit softwood casement windows, B.S. 644 Part 1; casements without glazing bars; cramps to jambs; bedding frame in cement mortar; pointing one side; one coat primer, two undercoats, one coat full-gloss finish, both sides .	„	44·20
2·6·3	Purpose-made West African mahogany; selected; casement windows; cramps to jambs; bedding frame in cement mortar; pointing one side; staining and wax polishing, both sides . . .	„	97·00
	Window openings		
	Precast reinforced-concrete mix 21 N/mm² boot shaped lintel, fair finish on exposed surfaces; artificial stone sill; softwood window board, one coat primer, two undercoats, one coat full-gloss finish		
2·6·4	in one brick walls	m	28·60
2·6·5	in hollow walls; two half brick skins; 50 mm cavity; bitumen hessian based cavity gutter	„	38·35
2·6·6	in one and a half brick walls	„	33·80
	Facings to reveals		
2·6·7	of one brick walls	„	0·37
2·6·8	of hollow walls; two half brick skins; 50 mm cavity; closing cavities at jambs; bitumen hessian-based vertical damp-proof courses	„	3·27
2·6·9	of one and a half brick walls	„	0·37

External doors

Softwood

44 mm (f) two panelled doors B.S. 459 Part 1 type 2XG; bottom edge rebated; weather fillet; frame and sill; cramps to jambs; bedding frame and sill in cement mortar; pointing one side; ironmongery *p.c.* £20·00; one coat primer, two undercoats, one coat full-gloss finish both sides

2·6·10	838 × 1981 mm door	No.	122·00
	44 mm (f) flush doors with opening for glass, B.S. 459 Part 2 figure 1; plywood facing both sides; hardwood lipping both long edges; bottom edge rebated; weather fillet; 100 × 50 mm once rebated frame; 150 × 75 mm oak sunk weathered threshold; cramps to jambs; bedding frame and threshold in cement mortar; pointing one side; ironmongery *p.c.* £20·00; 4 mm clear sheet glass; one coat primer; two undercoats, one coat full-gloss finish both sides		
2·6·11	826 × 2040 mm door	„	136·00

Item No.		Unit	£

WINDOWS AND EXTERNAL DOORS – *continued*

External doors – *continued*

West African mahogany; selected

44 mm (f) panelled doors; two open panels for glass; 150 mm wide stiles and top rail; 225 mm wide middle and bottom rails in one width; 125 × 75 mm once rebated frame; 150 × 75 mm sunk weathered threshold; cramps to jambs; bedding frame and threshold in cement mortar; pointing one side; ironmongery *p.c.* £35·00; 4 mm clear sheet glass; staining and wax polishing both sides

2·6·12	826 × 2040 mm door	No.	274·00

Door openings

Precast reinforced concrete mix 21 N/mm² boot shaped lintels; fair finish on exposed surfaces; oak threshold; staining and wax polishing

2·6·13	in one brick walls	m	20·75
2·6·14	in hollow walls; two half brick skins; 50 mm cavity; bitumen hessian based cavity gutter	,,	37·75
2·6·15	in one and a half brick walls	,,	25·00

Facings to reveals

2·6·16	of one brick walls	,,	0·37
2·6·17	of hollow walls; two half-brick skins; 50 mm cavity; closing cavities at jambs; bitumen hessian-based vertical damp-proof courses	,,	3·27
2·6·18	of one and a half brick walls	,,	0·37

2·7 INTERNAL WALLS AND PARTITIONS

Common bricks *p.c.* £45·00 1000 in cement–lime mortar (1 : 2 : 9)

2·7·1	half brick walls	m²	11·10
2·7·2	one brick walls	,,	21·20

Lightweight concrete blocks; keyed both sides in cement–lime mortar (1 : 2 : 9)

2·7·3	75 mm thick walls or partitions	,,	7·50
2·7·4	100 mm ,, ,, ,, ...	,,	9·10

2·8 INTERNAL DOORS

Softwood

40 mm (f) flush doors B.S. 459, Part 2; hardboard facing both sides; softwood lippings two long edges; 100 × 25 mm linings; 40 × 13 mm stops; 45 × 14 mm (f) architraves B.S. 584 type TF2/2 both sides; ironmongery *p.c.* £7·00 one coat primer, two undercoats, one coat full-gloss finish both sides

2·8·1	726 × 2040 mm door and frame	No.	88·25

44 mm (f) panelled doors; six 13 mm cross-tongued panels; mouldings worked on solid both sides; 150 × 40 mm once rebated linings 75 × 32 mm moulded architraves both sides; ironmongery *p.c.* £13·00; one coat primer, two undercoats, one coat full-gloss finish both sides

2·8·2	838 × 1981 mm door and frame	,,	240·00

Item No.		Unit	£

INTERNAL DOORS – *continued*

West African mahogany; selected

50 mm (f) panelled doors; one 25 mm cross-tongued panel; mouldings worked on solid both sides; once rebated; mouldings worked on solid one side; 150 × 50 mm once rebated, once grooved frame; 75 × 32 mm moulded architraves both sides; ironmongery *p.c.* £40·00; 4 mm clear sheet glass; staining and wax polishing both sides

Item No.		Unit	£
2·8·3	726 × 2040 mm door and frame	No.	287·00
	Door openings		
	Precast reinforced concrete mix 21 N/mm² lintels		
2·8·4	in half brick wall	,,	3·83
2·8·5	in one brick wall	,,	5·53
2·8·6	75 mm partition	,,	3·00
2·8·7	100 mm ,, ...	,,	4·20

IRONMONGERY

Included with elements for windows and doors, etc.

3·1   WALL FINISHES

NOTE. *Allowance has been made in prices hereafter for reveals, but quantities should be taken on the face areas of walls and excluding the areas of reveals*

3·1·1	Plaster; first and finishing coat of light-weight gypsum plaster; steel trowelled; to brickwork or blockwork base; two coats emulsion paint	m²	4·30
3·1·2	Extra; one coat primer, two undercoats, one coat full-gloss finish in lieu of two coats emulsion paint	,,	2·10
3·1·3	108 × 108 × 4 mm white glazed ceramic tiles, B.S. 1281; fixing with approved adhesive; grouted with white cement; butt joints straight both ways; plaster base	,,	13·60

3·2   FLOOR FINISHES

3·2·1	25 mm thick softwood board flooring; 150 mm widths; tongued-and-grooved joints; surface prepared to receive finishings after laying	,,	7·60
3·2·2	25 mm thick Iroko strip flooring 75 mm widths; tongued-and-grooved joints; tongued-and-grooved heading joints; secret nailed; surface sanded after laying; one coat sealer, one coat wax polish	,,	19·75
3·2·3	25 mm one coat granolithic paving; steel trowelled; 25 mm cement and sand (1 : 3) bed; screeded	,,	6·10
3·2·4	16 mm thick in situ terrazzo; cement and white Sicilian marble aggregate; polished; 34 mm cement and sand (1 : 3) bed; screeded	,,	15·20
3·2·5	3 mm thermoplastic tiles; butt joints straight both ways; fixing with adhesive; 47 mm cement and sand (1 : 3) bed, trowelled with a plain surface	,,	8·40
3·2·6	3·2 mm linoleum tiles, B.S. 810, marbled patterns; butt joints straight both ways; fixing with adhesive; 47 mm cement and sand (1 : 3) bed, trowelled with a plain surface	,,	10·95

Item No.		Unit	£
	FLOOR FINISHES – *continued*		
3·2·7	200 × 200 × 19 mm brown quarry tiles B.S. 1286, joints straight both ways; bedding in cement mortar (1:3); 30 mm cement and sand (1:3) bed	m²	13·85
3·2·8	25 mm thick Iroko blocks; to regular pattern; fixing with adhesive; surfaces sanded after laying; one coat sealer, one coat wax polish; 25 mm cement and sand (1:3) bed; trowelled with a plain surface	,,	20·35
3·2·9	Add to or deduct from above items for each 18 mm variation in thickness of cement and sand beds	,,	0·80
3·2·10	75 × 25 mm softwood skirting; once splayed; softwood grounds; plugging; one coat primer, two undercoats, one coat full-gloss finish	m	4·00
3·2·11	100 × 25 mm West African mahogany skirting; once splayed; softwood grounds; plugging; staining, wax polishing . .	,,	5·00
3·2·12	13 × 75 mm granolithic; rounded edge; coved junction with paving; to brickwork or blockwork base	,,	3·30
3·2·13	10 × 75 mm terrazzo; rounded edge; coved junction with paving; 13 mm cement and sand (1:3) backing; to brickwork or blockwork base	,,	9·55
3·2·14	150 × 150 × 12·5 mm brown quarry tile; rounded; coved junction with paving; 13 mm cement and sand (1:3) backing; to brickwork or blockwork base	,,	1·87
3·3	**CEILING FINISHES**		
3·3·1	Plaster; first coat of gypsum bonding plaster; finishing coat of gypsum plaster; steel trowelled; to concrete base; two coats emulsion paint	m²	3·82
3·3·2	Extra; one coat primer, two undercoats, one coat full-gloss finish in lieu of two coats emulsion paint . . .	,,	2·14
3·3·3	Plasterboard, B.S. 1230 type b; fixing with nails; joints filled with plaster and scrimmed; 5 mm one coat gypsum plaster; steel trowelled; two coats emulsion paint	,,	5·10
3·3·4	Extra; one coat primer, two undercoats, one coat full-gloss finish in lieu of two coats emulsion paint . . .	,,	2·14
	Suspended from structure; steel tees 600 mm centres; aluminium angle trim		
3·3·5	15 mm mineral fibre tiles flame-proofed class I . . .	,,	11·90
	DECORATIONS		
	Included with elements for windows, doors, wall and ceiling finishes, etc.		
4·0	**FITTINGS AND FURNISHINGS**		
	Softwood		
4·0·1	25 mm thick shelves; cross-tongued joints; 50 × 25 mm bearers; plugging	,,	21·75
4·0·2	25 mm thick shelves; slatted; 50 mm wide longitudinal slats at 75 mm centres; 50 × 25 mm bearers; plugging . . .	,,	19·50
4·0·3	25 mm thick; 225 mm wide in one width; black japanned steel brackets at 450 mm centres; screwed	m	5·05

Item No.		Unit	£
	FITTINGS AND FURNISHINGS – *continued* **Softwood**		
	Stock pattern or standard cupboard units; floor units with plinths; framed construction, plywood facings; doors 16 mm thick plywood or framed construction with plywood facing both sides; hung on butt hinges and fitted with catches and pulls; drawers fitted with pulls; fixing to background requiring plugging; plastic laminate finish (B.S. references to pattern only)		
4·0·4	1200 × 900 × 600 mm sink unit; without sink top	No.	70·00
4·0·5	500 × 900 × 600 mm floor unit; reference 5B6; blockboard work-top; laminated plastic facing one side	,,	50·00
4·0·6	600 × 600 × 300 mm wall unit; reference 6W	,,	32·00
4·0·7	500 × 600 × 300 mm wall unit; reference 5W	,,	29·00
4·0·8	1950 × 600 × 500 mm store cupboard; reference 6TS	,,	81·00
5·1	**SANITARY APPLIANCES**		
5·1·1	Sinks; white glazed fireclay, waste fitting, cantilever brackets and pair of taps (*p.c.* £31·00 complete); trap; copper waste pipe	,,	121·50
5·1·2	Sinks; stainless steel, waste fitting and pair of taps (*p.c.* 73·15 complete); trap; copper waste pipe	,,	161·00
5·1·3	Lavatory basins; white glazed vitreous china, waste fitting, cantilever brackets and pair of taps (*p.c.* £29·00 complete); trap; copper waste pipe	,,	89·00
5·1·4	Baths; white Acrylic, waste and overflow fittings, pair of taps, side and end panels (*p.c.* £127·00 complete); trap; copper waste pipe	,,	216·00
5·1·5	W.C. High level suites (on ground floor); cistern, flush pipe and seat (*p.c.* £60·00 complete); connection to drain	,,	100·00
5·1·6	W.C. high level suites (on upper floor); cistern, flush pipe and seat (*p.c.* £60·00 complete); 6·10 m cast iron soil pipe, connection to drain	,,	285·00
5·1·7	W.C. low level suites (on upper floor); cistern, flush pipe and seat (*p.c.* £46·00 complete); 1 m cast iron soil pipe	,,	165·00
5·1·8	Soil and ventilating pipe to upper floors; single branches with access; connection to drain; 3·15 m storey height; cast iron soil pipe balloon grating	,,	240·00
5·1·9	Add to above for additional floors	,,	84·00
5·4	**WATER INSTALLATION** Cold water Services for sink on ground floor; bath, w.c. suite and lavatory basin on first floor; *p.c.* £100 for connection to water main, 10 m of water main externally; rising main galvanized storage cistern: down services; lagging pipes and cistern		
5·4·1	services in copper pipes	,,	455·00
5·4·2	services in galvanized steel pipes	,,	475·00
	Add to above for services to additional w.c. suite or lavatory basin on ground floor		
5·4·3	services in copper pipes	,,	28·00
5·4·4	services in galvanized steel pipes	,,	29·00
	Hot water		
5·4·5	Copper pipe primary and secondary services for sink on ground floor and bath and lavatory basin on first floor; indirect cylinder; lagging pipes	,,	420·00
5·4·6	*Add* to above for services to additional lavatory basin on ground floor	,,	28·00

Item No.		Unit	£
5·5	**HEAT SOURCE**		
5·5·1	Gas fired boiler 13 Kw rating, lined flue; expansion tank; immersion heater (electric point not included)	No.	460·00
5·6	**SPACE HEATING**		
5·6·1	Tiled hearth and surround, firebrick back and slow-burning fire *p.c.* £120·00 complete; chimney breast projection and stack; fuel storage (heating to first floor not included)	„	570·00
5·8	**ELECTRICAL INSTALLATIONS**.		
	Wiring complete; PVC insulated and sheathed cable installed in floor and roof voids; protected by light-gauge conduits where buried (excluding cost of lamps and light fittings)		
5·8·1	consumer control unit (100 amp.)	„	100·00
5·8·2	lighting point	„	21·50
5·8·3	single 13-amp. switched socket outlet wired in a ring-main circuit	„	25·50
5·8·4	double 13-amp. switched socket outlet wired in a ring-main circuit	„	29·50
5·8·5	cooker point; control unit	„	50·00
5·8·6	immersion heater point (excluding heater)	„	31·50
5·9	**GAS INSTALLATIONS**		
5·9·1	Carcassing meter to capped-off points; builder's work in connection (based on a Gas Board's contributory charge of £45·00 for the installation to houses)	„	50·00
5·10	**LIFT INSTALLATIONS**		
	Electric passenger lifts to carry ten persons at approximately 1·50 m/sec		
5·10·1	6 stops	„	34,000·00
5·10·2	8 „	„	38,000·00
5·10·3	9 „	„	39,000·00
5·10·4	10 „	„	40,000·00
5·10·5	13 „	„	46,000·00
	Electric passenger lifts to carry twenty persons at approximately 2·50 m/sec		
5·10·6	18 stops	„	62,000·00

Item No.		Unit	£
6·1	**SITE WORKS**		
	Preserving vegetable soil; excavating average 225 mm deep; depositing in temporary spoil heaps; spreading; on site		
6·1·1	by hand	m²	2·10
6·1·2	by machine	„	0·48
	Excavating to reduce levels; not exceeding 300 mm deep; depositing in temporary spoil heaps; filling in making up levels; depositing and compacting in layers		
6·1·3	by hand	„	2·98
6·1·4	by machine	„	0·74
	Add for each 300 mm additional depth of excavation and disposal		
6·1·5	by hand	„	2·98
6·1·6	by machine	„	0·74
	Pavings to pavements; to falls, crossfalls or slopes not exceeding 15° from horizontal; disposing of excavated material by filling in making up levels; depositing and compacting in layers		
6·1·7	50 mm gravel pavings; hardcore filling in making up levels average 100 mm thick; excavating to reduce levels average 150 mm deep	„	3·90
6·1·8	50 mm tar pavings; hardcore filling in making up levels average 100 mm thick; excavating to reduce levels average 150 mm deep .	„	7·75
6·1·9	750 × 600 × 50 mm concrete flags B.S. 368 natural finish; bedding in lime mortar; jointing and pointing in cement mortar; 6 mm joints straight both ways; ash filling in making up levels average 100 mm thick; excavating to reduce levels average 150 mm deep	„	8·95
6·1·10	Pavings to surface car parks, to falls, crossfalls or slopes not exceeding 15° from horizontal; 75 mm tarmacadam B.S. 802; hardcore filling 225 mm thick (excluding excavation and disposal)	„	7·70
	Pavings to roads, to falls, crossfalls or slopes not exceeding 15° from horizontal; 75 mm tarmacadam B.S. 802; hardcore filling 225 mm thick; precast concrete B.S. 340, 225 × 125 mm kerbs and channels; both sides; in situ concrete mix 11·50 N/mm² foundations (excluding excavation and disposal)		
6·1·11	4·90 m wide	m	56·60
6·1·12	6·10 „	„	66·00
6·1·13	7·32 „	„	75·00
	In situ reinforced concrete; mix 21 N/mm² roads; 150 mm thick; laid in bays; mesh reinforcement; 15 mm thick impregnated fibreboard expansion joints; polysulphide based sealant; hardcore filling 150 mm thick; blinding; precast concrete B.S. 340, 225 × 125 mm kerbs and channels; both sides; in situ concrete mix 11·50 N/mm² foundations (excluding excavation and disposal)		
6·1·14	4·90 m wide	„	70·00
6·1·15	6·10 „	„	84·50
6·1·16	7·32 „	„	100·00
	Fencing; cleft chestnut pale; pales 50 mm apart; two lines galvanized wire; galvanized tying wire; 65 mm diameter posts at 3 m centres; 75 × 50 mm struts; driving posts and struts 610 mm into ground		
6·1·17	1220 mm high	„	4·60
	Fencing; chain link, 10½ swg (3 mm) 2 in. (51 mm) mesh; galvanized mild-steel mesh, line wires and tying wires; galvanized steel components; posts at 3050 mm centres; line wires threaded through posts; setting posts and struts 610 mm below ground		
6·1·18	1220 mm high; mild-steel angle posts and struts . . .	„	8·50
6·1·19	1220 „ concrete posts and struts	„	9·10

Item No.		Unit	£
6·2	**DRAINAGE**		
	Drain pipes; excavating trenches average 1 m deep in heavy soil; grading bottoms planking and strutting; filling with excavated material and compacting; disposal of surplus spoil by spreading on site average 50 m from excavation; normal concrete mix 11·50 N/mm² beds 150 mm thick and benchings in similar concrete; vitrified clay pipes and fittings; extra strength quality; flexible joints		
6·2·1	100 mm internal diameter	m	13·00
6·2·2	150 mm ,, ,, ...	,,	15·65
	Cast iron pipes; B.S. 1211 Class B; caulked lead joints		
6·2·3	100 mm internal diameter	,,	20·50
6·2·4	150 mm ,, ,, ...	,,	26·20
	Add for each 250 mm additional trench depth		
6·2·5	not exceeding 1·50 m deep	,,	1·47
6·2·6	over 1·50 m not exceeding 3 m deep	,,	3·80
	Manholes; excavation in heavy soil; backfillings; disposal of surplus material; levelling and compacting bottoms; planking and strutting; normal concrete mix 11·50 N/mm² bed 150 mm thick; normal concrete mix 21 N/mm² benchings; one brick wall; commons *p.c.* £45·00 1000 in cement mortar (1 : 3); 13 mm cement and sand (1 : 3) rendering; building in ends of pipes; 610 × 457 mm access cover and frame; bedding frame in cement mortar; cover in grease and sand; 100 mm half-section vitrified clay channel		
6·2·9	686 × 457 × 600 mm deep internally; suitable for not more than two branches per side	No.	136·00
6·2·10	Add for each 300 mm additional depth up to 1·50 m deep internally	,,	43·20
6·2·11	980 × 686 × 600 mm deep internally; normal concrete mix 21 N/mm² cover slab 100 mm thick, reinforced with mesh; aperture for cover and frame; suitable for not more than three branches per side	,,	205·00
6·2·12	Add for each 300 mm additional depth up to 1·50 m deep internally	,,	57·00
6·2·13	1370 × 800 × 1500 mm deep internally; normal concrete mix 21 N/mm² cover slab 100 mm thick, reinforced with mesh; aperture for cover and frame; suitable for not more than four branches per side	,,	455·00
6·2·14	Add for 685 × 685 mm shaft for each additional 300 mm depth up to 3 m deep internally	,,	57·00
6·2·15	Extra; 100 mm three-quarter section branch bends . .	,,	3·30
6·2·16	Extra; 150 mm ,, ,, ,, . .	,,	5·20

Elemental Cost Plan

The Elemental Cost Plans which follow are examples of how priced approximate quantities may be used to compile detailed estimates for a building using a known or assumed specification. They are based on items contained in the Approximate Estimates Section and referenced accordingly.

Detached House

This example is for a two-storey detached dwelling having a floor area of approximately 90 m^2 (measured within the external walls and over all internal walls and partitions) and storey heights of 2·44 m in the clear. The building is of traditional construction, having solid concrete ground-floor slab and strip foundations, load-bearing brick cavity walls externally and brick and block internal walls and partitions, with timber joisted upper floor and flat roof, felt roof coverings and a reasonable standard of doors, windows, finishings and sanitary fittings.

Item No.	Element	Total cost of element £	£	Cost of element per m^2 of floor area £
1·0	**Substructure**			
1·0·3	Strip foundation, 27 m @ £50·00	1,350·00		
1·0·9	Ground floor slab, 44 m^2 @ £14·40	633·60		
1·0·11	Thickening for partitions, 13 m @ £1·95	25·35	2,008·95	22·32
2·2	**Upper floors**			
2·2·12	175 mm deep joists, 41 m^2 @ £9·00	369·00	369·00	4·10
2·3	**Roof**			
2·3·5	175 mm flat roof joists with wood-wool and felt, 51 m^2 @ £22·75	1,160·25		
2·3·19	Flush eaves, 22 m @ £9·75	214·50		
2·3·20	Ditto with 100 mm gutter, 7 m @ £18·90	132·30		
2·3·36	75 mm rainwater pipe, 2 No. @ £78·35	156·70	1,663·75	18·49
2·4	**Stairs**			
2·4·5	Staircase to B.S. 585, 1 No. @ £250·00	250·00	250·00	2·78
2·5	**External walls**			
2·5·12	Hollow wall of two half brick skins with outer skin in facings, 113 m^2 @ £31·00	3,503·00	3,503·00	38·92
2·6	**Window and external doors**			
2·6·1	Standard metal windows in standard sub-frames, 33 m^2 @ £50·15	1,654·95		
2·6·5	Lintels, etc., 18 m @ £38·35	690·30		
2·6·8	Reveals, etc., 53 m @ £3·27	173·31		
2·6·11	Flush doors 826 × 2040 mm, complete, 3 No. @ £136·00	408·00		
2·6·14	Lintels, etc., 3 m @ £37·75	113·25		
2·6·17	Reveals, etc., 12 m @ £3·27	39·24	3,079·05	34·21
	Continued		10,873·75	120·82

Item No.	Element		Total cost of element £	Cost of element per m² of floor area £
		Continued	10,873·75	120·82
2·7	**Internal walls and partitions**			
2·7·1	Half brick wall, 27 m² @ £11·10		299·70	
2·7·2	One brick wall, 12 m² @ £21·20		254·40	
2·7·3	75 mm partition, 21 m² @ £7·50 . . .		157·50	
2·7·4	100 mm partition, 7 m² @ £9·10 . . .		63·70	775·30	8·61
2·8	**Internal doors**			
2·8·1	Flush doors 726 × 2040 mm complete, 9 No. @ £88·25		794·25	
2·8·4	Lintels, 5 m @ £3·83		19·15	
2·8·5	Lintels, 3 m @ £5·53		16·59	
2·8·6	Lintels, 4 m @ £3·00		12·00	841·99	9·36
	Ironmongery			
	Included with windows and doors, etc. . . .			
3·1	**Wall finishes**			
3·1·1	Two coat plaster and emulsion paint, 181 m² @ £4·30		778·30	
3·1·3	Wall tiling, 6 m² @ £13·60		81·60	859·90	9·55
3·2	**Floor finishes**			
3·2·1	25 mm softwood flooring 42 m² @ £7·60 . .		319·20	
3·2·5	3 mm thermoplastic tile paving 43 m² @ £8·40 .		361·20	
3·2·10	75 × 25 mm softwood skirtings, 92 m @ £4·00 .		368·00	1,048·40	11·65
3·3	**Ceiling finishes**			
3·3·3	Plasterboard and setting coat and emulsion paint, 84 m² @ £5·10		428·40	428·40	4·76
	Decorations			
	Included with windows, doors and wall and ceiling finishes, etc.			
4·0	**Fittings**			
4·0·1	25 mm softwood shelving, 1 m² @ £21·75 . . .		21·75	
4·0·2	Slat shelving, 2 m² @ £19·50		39·00	
4·0·3	225 × 25 mm softwood shelving, 4 m @ £5·05 . .		20·20	
4·0·4	Sink unit 1 No. @ £70·00 . . .		70·00	
4·0·5	Floor cupboard No. B51, 1 No. @ £50·00 . .		50·00	
4·0·6	Wall cupboard No. W66, 1 No. @ £32·00 . .		32·00	
4·0·7	Ditto No. W56, 1 No. @ £29·00 . . .		29·00	
4·0·8	Broom cupboard No. T60/67, 1 No. @ £81·00 .		81·00	342·95	3·81
5·1	**Sanitary appliances**			
5·1·2	Sink top complete, 1 No. @ £161·00		161·00	
5·1·3	Basin complete, 1 No. @ £89·00 . . .		89·00	
5·1·4	Bath complete, 1 No. @ £216·00 . . .		216·00	
5·1·6	W.C. complete, 1 No. @ £285·00		285·00	751·00	8·35
		Continued	15,921·69	176·91

Item No.	Element	Total cost of element £	Cost of element per m² of floor area £
	Continued	15,921·69	176·91
5·4	**Water installation**		
5·4·1	Cold services in copper to bath, W.C., sink and basin .	455·00	
5·4·5	Hot services in copper, to bath, sink and basin . .	420·00	875·00	9·72
5·5	**Heat source**		
5·5·1	Boiler, tank, immersion heater	460·00	460·00	5·11
5·6	**Space heating**		
5·6·1	Hearth, surround, chimney breast and stack and fuel storage	570·00	570·00	6·33
5·8	**Electrical installations**		
5·8·1	Consumer control unit, 1 No @ £100·00	100·00	
5·8·2	Lighting point, 11 No. @ £21·50	236·50	
5·8·3	Single socket outlet, 6 No. @ £25·50 . . .	153·00	
5·8·4	Double ditto, 2 No. @ £29·50	59·00	
5·8·5	Cooker point, 1 No. @ £50·00	50·00	
5·8·6	Immersion heater point, 1 No. @ £31·50 . . .	31·50	630·00	7·00
5·9	**Gas installations**		
5·9·1	Carcassing meter to points	50·00	50·00	0·56
6·1	**Site works**		
	Precast concrete paving serving external doors, 1220 mm chain link fencing to one side and rear of plot and dwarf brick wall to front and comprising 24 m² of paving, 43 m of fencing and 8 m of wall . . .	1,057·00	1,057·00	11·74
6·2	**Drainage**		
	Drainage from soil and waste pipe including 3 No. manholes and 14 m of (100 mm) stoneware drain and provision for sewer connection (*p.c.* £60) . . .	725·00	725·00	8·06
	Preliminaries		
	15% of the value of work, say	3·041·31	3,041·31	33·79
	Contingencies		
	Allowance, say £100		100·00	1·11
	TOTAL COST OF DWELLING	£23,430·00	
	TOTAL COST PER m² OF FLOOR AREA		£260·33

Office Block

This example is for a four storey office block having a floor area of 2,400 m² (measured within the external walls and over all internal walls and partitions) and a car park of 500 m² forming part of the ground floor. The building is of reinforced concrete frame construction having a storey height of 3·15 m, solid concrete ground floor slab with ground beams and column bases, precast concrete cladding panels externally with brick inner skins, brick and block internal walls and partitions, in situ reinforced concrete and hollow block with structural topping upper floors, mansard roof with asphalt roof coverings on reinforced wood-wool supported on rolled steel joists and asbestos cement slate coverings on timber roof joists, anodized aluminium windows, hardwood doors and a reasonable standard of finishings and sanitary fittings.

The sums included for Mechanical and Electrical Services are based upon rates for this type of work included in the companion volume of 'Spon's Mechanical and Electrical Services Price Book' and a detailed analysis of these sums is included in that publication.

Item No.	Element	Total cost of element £	£	Cost of element per m² of floor area £
1·0	**Substructure**			
1·0·7	Ground beam, 142 m @ £16·45	2,335·90		
1·0·8	Column base, 21 No. @ £1004·00	21,084·00		
1·0·12	Ground floor slab, 265 m² @ £17·00	4,505·00	27,924·90	11·64
2·0	**Superstructure**			
2·1	**Frame**			
2·1·1	Beams, 1307 m @ £40·75	53,260·25		
2·1·2	Columns, 391 m @ £42·00	16,422·00		
2·1·3	Extra for wrought formwork, 454 m² @ £1·90	862·60	70,544·85	29·39
2·2	**Upper floors**			
2·2·2	225 mm reinforced concrete, 830 m² @ £38·15	31,664·50		
2·2·3	200 mm hollow block, 1617 m² @ £26·00	42,042,00	73,706·50	30·71
2.3	**Roof**			
2·3·14	Rolled steel joist roof beams, 20,200 kg @ £0·60	12,120·00		
2·3·15	50 mm reinforced wood-wool and asphalt, 675 m² @ £20·00	13,500·00		
2·3·16	Treated sawn softwood roof members, 25 m³ @ £243·00	6,075·00		
2·3·17	Asbestos cement slates and 50 mm quilt, 460 m² @ £19·75	9,085·00		
2·3·18	Asbestos cement slates on 20 mm blockboard, 100 m² @ £30·65	3,065·00		
2·3·40 } 2·3·41 }	100 mm rainwater pipe with roof outlet, 12·60 m long 7 No. @ £110·05	770·35	44,615·35	18·59
2·4	**Stairs**			
2·4·1 } 2·4·2 }	Reinforced concrete stairs with granolithic finish rising 3·15 m, 3 No. @ £869·25	2,607·75		
2·4·3 } 2·4·4 }	Reinforced concrete stairs with terrazzo finish rising 3·15 m, 3 No. @ £1,994·50	5,983·50	8,591·25	3·58
	Continued		225,382·85	93·91

Item No.	Element	Total cost of element £	Cost of element per m² of floor area £
	Continued	225,382·85	93·91
2·5	**External Walls**		
2·5·1	150 mm reinforced concrete, 85 m² @ £36·00 . .	3,060·00	
2·5·2	240 mm reinforced concrete, 25 m² @ £40·35 . .	1,008·75	
2·5·9	One brick wall faced one side, 4 m² @ £32·85 . .	131·40	
2·5·12	Hollow wall of two half brick skins with outer skin in facings, 8 m² @ £31·00	248·00	
2·5·18	Hollow wall of 100 mm blockwork inner skin with outer skin in facings, 300 m² @ £29·25 . . .	8,775·00	
2·5·19	Half brick wall in facings, 24 m² @ £20·00 . .	480·00	
2·5·20	Half brick wall in facings against concrete. 140 m² @ £20·85	2,919·00	
2·7·1	Half brick wall in commons, 2 m² @ £11·10 . .	22·20	
2·7·2	One brick in commons, 100 m² @ £21·20 . . .	2,120·00	
2·7·4	100 mm block partition, 32 m² @ £9·10 . .	291·20	
	Budget estimate for precast concrete cladding panels, 450 m² @ £51·25	23,062·50	42,118·05 17·55
2·6	**Windows and external doors**		
	Budget estimate for aluminium windows supplied fixed and glazed, 615 m² @ £90·70	55,780·50	
2.6.8	Reveals, 79 m @ £3·27	258·33	
2·6·12	Two panel hardwood doors 826 × 2040 mm, 16 No. @ £274·00	4,384·00	
2·6·14	Lintels, 16 m @ £37·75	604·00	
2·6·17	Reveals, 65 m @ £3·27	212·55	61,239·38 25·52
2·7	**Internal walls and partitions**		
2·5·1	150 mm reinforced concrete, 70 m² @ £36·00 . .	2,520·00	
2·7·1	Half brick wall in commons, 155 m² @ £11·10 . .	1,720·50	
2·7·2	One brick wall in commons, 1 m² @ £21·20 . .	21·20	
2·7·3	75 mm block partition, 13 m² @ £7·50 . .	97·50	
2·7·4	100 mm block partition, 62 m² @ £9·10 . .	564·20	
	Budget estimate for prefabricated W.C. partitions, 100 m² @ £20·50	2,050·00	6,973·40 2·91
2·8	**Internal doors**		
2·8·1	Flush doors 726 × 2040 mm, 28 No. @ £88·25 . .	2,471·00	
2·8·4	Lintels, 44 m @ £3·83	168·52	2,639·52 1·10
	Ironmongery		
	Included with windows and doors, etc. . . .		
3·1	**Wall finishes**		
3·1·1	Two coat plaster and emulsion paint, 1135 m² @ £4·30	4,880·50	
3·1·3	Wall tiling, 5 m² @ £13·60	68·00	
3·3·3	Plasterboard and setting coat and emulsion paint, 475 m² @ £5·10	2,422·50	7,371·00 3·07
3·2	**Floor finishes**		
3·2·3	25 mm granolithic paving, 75 m² @ £6·10 . .	457·50	
3·2·5	3 mm thermoplastic tiles, 2300 m² @ £8·40 . .	19,320·00	
3·2·7	19 mm quarry tiles, 75 m² @ £13·85 . .	1,038·75	20,816·25 8·67
	Continued	366,540·45	152·73

Item No.	Element	Total cost of element £	£	Cost of element per m² of floor area £
	Continued		366,540·45	152·73
3·3	**Ceiling finishes**			
3·3·1	Two coat plaster and emulsion paint, 15 m² @ £3·82	57·30		
3·3·3	Plasterboard and setting coat and emulsion paint, 85 m² @ £5·10	433·50		
3·3·5	19 mm insulating board tile suspended ceiling, 2350 m² @ £11·90	27,965·00	28,455·80	11·86
	Decorations			
	Included with windows, doors, and wall and ceiling finishes, etc.			
4·0	**Fittings**			
4·0·1	25 mm softwood shelving, 3 m² @ £21·75	65·25	65·25	0·03
5·1	**Sanitary appliances**			
5·1·1	Sink complete, 4 No. @ £121·50	486·00		
5·1·3	Basin complete, 24 No. @ £89·00	2,136·00		
5·1·7	W.C. complete, 24 No. @ £240·00	5,760·00		
5·1·8} 5·1·9}	Disposal installation, soil and ventilating pipes to four storey building, 2 No. @ £408·00	816·00	9,198·00	3·83
5·4	**Mechanical services**			
5·5 5·6	Budget estimate for heating, hot and cold water, gas and fire fighting installations	139,000·00		
	Allow for builders work in connection with mechanical services, say 15% on £139,000·00	20,850·00		
	Allow for general contractor's profit and attendance, say 5% on £139,000·00	6,950·00	166,800·00	69·50
5·8	**Electrical Services**			
	Budget estimate for sub-mains, lighting and power, supply to lift, mechanical services, telephone conduit and fire alarm installation	75,700·00		
	Allow for builders' work in connection with electrical services, say 12½% on £75,700·00	9,462·50		
	Allow for general contractor's profit and attendance say 5% on £75,700·00	3,785·00	88,947·50	37·06
5·10	**Lift installation**			
5·10·1	Budget estimate for electric passenger lift to carry ten persons, 4 stops	34,600·00		
	Allow for builder's work in connection with lift installation, say 10% on £34,600·00	3,460·00		
	Allow for general contractor's profit and attendance, say 5% on £34,600·00	1,730·00	39,790·00	16·58
6·1	**Site works**			
6·1·10	75 mm tarmacadam paving, 880 m² @ £7·70	6,776·00	6,776·00	2·82
	Continued		706,573·00	294·41

Item No.	Element		Total cost of element £	Cost of element per m² of floor area £
		Continued	706,573·00	294·41
6·2	**Drainage** Budget estimate for drainage from soil and waste pipes, including 9 No. manholes, 120 m of 100 mm and 45 m of 150 mm stoneware drain and provision for sewer connection (*p.c.* £100)		4,200·00 4,200·00	1·75
	Preliminaries 15% of the value of work, say		106,630·84 106,630·84	44·43
	Contingencies and design risk Allowance, 4% on £817,403·84, say		32,696·16 32,696·16	13·62
	TOTAL COST OF OFFICE BLOCK .		£850,100·00	
	TOTAL COST PER m² OF FLOOR AREA			£354·21

Comparative Prices

Under this heading is given a priced list of some of the more commonly used forms of construction and types and qualities of finishings. The list is intended as a guide to Architects and others when considering alternative materials and constructions. The prices are based upon the 'Prices for Measured Work' contained in Part II, but some items (e.g. block partitions) allow for incidental labours to make comparisons more valid. Preliminaries are not included.

In some cases it will be necessary to combine two or more of the items to give a truer comparison; for example, if the choice is between a self-finished partition and one requiring plastering and painting (on both sides). Again, the extra cost of additional screeding must be added when comparing solid floor finishings of different thicknesses.

It should be stressed that although comparisons of this kind may be a useful indication of cost differences, they can never be exact, and they can never take all the relevant factors into account. Differences in thermal conductivity, for example, may affect the heating system, and differences in weight may affect the design of the foundations or even of the structure. Speed of erection, and its effect upon other trades, is another factor which it is difficult to take into account in a direct analysis of comparative costs.

	Cost per m^2 £
1·0 SUBSTRUCTURE	
Solid ground floor with 225 mm excavation, 150 mm bed of hardcore, 150 mm concrete surface bed, fabric reinforcement and damp-proof membrane	14·40
Hollow ground floor with 375 mm excavation, 100 mm bed of hardcore, 150 mm concrete surface bed, half brick sleeper walls 300 mm high, slate damp-proof course, 100 × 50 mm softwood plates and joists	20·50

	Cost per m^2 £
2·2 UPPER FLOORS	
Timber, 150 × 50 mm joists with herringbone strutting	8·00
„ 175 × 50 „ „ „	9·00
„ 200 × 50 „ „ „	10·00
„ 225 × 50 „ „ „	11·00
„ 250 × 50 „ „ „	12·00
„ 275 × 40 „ „ „	13·00
Hollow tile, 140 mm total thickness	25·15
Reinforced concrete 125 mm thick, 3·65 m span, 300 kg/m² loading	26·15
„ „ 125 „ „ 2·75 „ 800 „ „	26·15
Hollow tile 190 mm total thickness	26·65
„ „ 240 „ „ „	28·35
Reinforced concrete 150 mm thick, 3·35 m span, 800 kg/m² loading	31·90
„ „ 150 „ „ 4·25 „ 300 „ „	31·90
„ „ 185 „ „ 4·25 „ 800 „ „	33·45

Cost per m²
£

2·3 ROOF

Flat roofs

Construction

	Cost per m² £
Timber, 150 × 50 mm joists on 100 × 50 mm plates with herringbone strutting	8·00
„ 175 × 50 „ 100 × 50 „ „ „	9·00
„ 200 × 50 „ 100 × 50 „ „ „	10·00
„ 225 × 50 „ 100 × 50 „ „ „	11·00
„ 250 × 50 „ 100 × 50 „ „ „	12·00
„ 275 × 50 „ 100 × 50 „ „ „	13·05
Hollow tile, 140 mm total thickness	25·15
Reinforced concrete, 125 mm thick, 3·65 m span, 300 kg/m² loading	26·15
„ „ 125 „ 2·75 „ 800 „	26·15
Prestressed concrete beams 175 mm thick at 460 mm centres, 6 m span with lightweight infilling blocks	26·45
Hollow tile, 190 mm total thickness	26·65
„ „ 240 „ „ „	28·35
Prestressed concrete beams, 230 mm thick at 510 mm centres, 6 m span with lightweight infilling blocks	30·85
Reinforced concrete, 150 mm thick, 4·25 m span, 300 kg/m² loading	31·90
Reinforced concrete, 150 mm thick, 3·35 m span, 800 kg/m² loading	31·90
Reinforced concrete, 185 mm thick, 4·25 m span, 800 kg/m² loading	33·45

Screed, boarding and insulation

	Cost per m² £
Layer of building paper laid on boarding (not included)	0·75
Layer of double-sided reflective insulation paper draped over joists (not included)	1·30
38 mm expanded polystyrene fixed with adhesive	2·00
80 mm glass fibre quilt over joists (not included)	2·30
32 mm cement and sand screed to receive felt roofing	2·60
50 mm cement and sand screed to receive felt roofing	3·40
25 mm cork slabs laid on boarding (not included)	3·55
25 mm softwood roof boarding nailed to joists	6·15
75 mm lightweight bituminous screed laid on concrete on and including felt vapour barrier to receive roof coverings	8·30
50 mm wood wool slabs laid on joists (not included) including mild-steel ties at 600 mm centres	8·85
100 mm lightweight bituminous screed laid on concrete on and including felt vapour barrier to receive roof coverings	10·50
125 mm lightweight bituminous screed laid on concrete on and including felt vapour barrier to receive roof coverings	12·70

Finishes

	Cost per m² £
19 mm mastic asphalt to B.S. 988 on and including felt underlay	6·00
Three layer glass fibre based bitumen felt roofing with granite chipping finish	6·70
Three layer asbestos based bitumen felt roofing with granite chipping finish	7·00
19 mm mastic asphalt to B.S. 1162 on and including felt underlay	7·95
0·91 mm commercial quality aluminium	19·25
0·81 mm zinc	19·45
0·56 mm copper	27·70
0·61 mm copper	29·80
No. 4 sheet lead	48·30
No. 6 sheet lead	66·60

Cost per m²
£

2·3 ROOFS – *continued*

Flat roofs – *continued*

Composite roofs

	Cost per m² £
Roof decking complete with 13 mm fibre board and three-layer felt roofing, 0·71 mm galvanized steel for spans up to 2 m	19·00
Ditto, ,, ,, ,, 3 m	20·00
Ditto, ,, ,, ,, 3·50 m	20·70
Ditto, ,, ,, ,, 4·25 m	22·00
Ditto, 0·91 mm aluminium for spans up to 1·50 m	23·00
Ditto, ,, ,, ,, 2·50 m	25·00
Ditto, ,, ,, ,, 3 m	26·00
Ditto, ,, ,, ,, 3·50 m	28·00

Pitched roofs

NOTE. *Cost per m² of plan area measured to outside of external walls and based on a span of 7·60 m. The extra cost of hipped or gable ends is not included in the rates.*

Pitch

	20° £	30° £	35° £	40° £
Construction				
Timber, trussed rafters at 600 mm centres, 75 × 50 mm rafters, 75 × 50 mm ceiling joists, 75 × 25 mm struts, 75 × 38 mm plate and 75 × 50 mm ridge	5·45	—	—	—
Timber, trussed rafters at 1800 mm centres similar to T.D.A. pattern, 100 × 38 mm common rafters, 100 × 38 mm ceiling joists, 150 × 50 mm purlins, 75 × 38 mm hangers and ties, 75 × 38 mm plate and 150 × 25 mm ridge	—	8·65	8·90	9·85
Timber, 100 × 50 mm rafters, 100 × 50 mm ceiling joists at 375 mm centres, 150 × 50 mm purlins, 50 × 32 mm hangers, 100 × 75 mm plate and 150 × 25 mm ridge	—	10·30	10·55	10·85
Boarding and insulation (fixed to rafters)				
Underlining felt	2·05	2·50	2·80	2·90
19 mm softwood boarding	6·10	7·50	8·35	8·60
50 mm unreinforced wood wool slabs	6·45	7·90	8·80	9·05
25 mm softwood boarding	5·40	6·65	7·40	7·65
Finishes (including battening, eaves courses and ridges)				
Two-layer self finished bituminous felt	—	—	7·15	7·38
Three-layer ,, ,, ,,	7·30	8·85	—	—
Concrete interlocking tiles, 413 × 330 mm laid to 76 mm lap	—	7·00	7·70	7·95
Red plain pantiles, 337 × 241 mm, laid to 76 mm head and 38 mm side laps	—	15·10	16·60	17·10
Concrete plain tiles, 267 × 165 mm, laid to 64 mm lap	—	19·70	21·65	22·30
Machine-made plain tiles 267 × 165 mm	—	20·70	22·80	23·50
Welsh slates, 405 × 255 mm, laid to 76 mm lap	—	23·55	26·00	26·80
0·91 mm commercial quality aluminium fixed to boarding (not included)	19·10	23·40	26·00	26·85
Welsh slates, 510 × 255 mm laid to 76 mm lap	—	27·10	29·85	30·75
Hand-made sandfaced plain tiles, 267 × 165 mm, laid to 64 mm lap	—	31·00	34·30	35·50
0·56 mm copper fixed to boarding (not included)	28·40	34·80	38·65	39·90
0.61 mm copper fixed to boarding (not included)	30·70	37·65	41·80	43·20
Westmorland Green slates, random sizes, laid to 76 mm lap	—	46·40	51·15	52·70

2·5 EXTERNAL WALLS

Walls

Construction

Timber framing of 125 × 50 mm studs at 600 mm centres with 125 × 50 mm head and plate	4·60
Ditto with layer of double-sided building paper	5·70
Solid wall in 140 mm Thermalite blocks	11·15
Cavity wall with inner skin of 75 mm hollow clay blocks and outer skin of half brick fletton brickwork	19·30
Cavity wall with inner skin of 75 mm Thermalite blocks and outer skin of half brick fletton brickwork	19·30
Solid wall in one brick fletton brickwork	22·60
Cavity wall with inner and outer skins of half brick fletton brickwork . . .	24·15
Solid wall of one and a half brick fletton brickwork	32·40
Solid wall of 150 mm reinforced concrete	36·15
Solid wall of 225 mm reinforced concrete	39·80

Finishes

Two coats cement paint on brick walls	1·05
Building fletton brickwork with fair face and flush pointing	1·50
Extra; over concrete wall for wrought formwork one side including rubbing down concrete	1·90
Extra; over concrete wall for shot blasting to expose aggregate, price depending on quality of concrete and depth of penetration . . . . . . to	2·75 to 4·10
Cement and sand plain face rendering	4·50
Extra; over concrete wall for bush hammering to expose aggregate . . .	4·95
Grey corrugated asbestos, 'Big Six' or similar, fixed to timber framing (not included) .	7·50
Three-coat Tyrolean rendering on and including backing	8·95
Extra; over cavity wall in flettons for outer skin in facing bricks, *p.c.* £96·00 per 1000 and pointing	7·80
25 mm tongued and grooved softwood boarding fixed vertically on and including battens at 600 mm centres	10·80
Machine-made tiles hung vertically on and including battens	16·20
25 mm tongued and grooved Western Red cedar boarding fixed vertically on and including battens at 600 mm centres	19·05
Best hand made sand-faced tiles hung vertically on and including battens . .	23·40
20 × 20 mm mosaic glass or ceramic in common colours and fixing on prepared surface	30·00
75 mm Portland stone facing slabs and fixing, including cramps	112·00
75 mm Ancaster stone facing slabs ditto	117·50

Curtain Walls

Construction

Galvanized steel standard grid curtain walling containing a proportion of opening lights but excluding glazing and infill panels	78·40
Anodized aluminium ditto	88·50

Finishes

5 mm rough cast glass glazed with metal beads (not included)	14·75
6 mm float ditto	18·30
Double glazing unit of two skins of 6 mm float, in copper channel and fixing with metal beads (not included)	47·65
Double glazing unit of inner skin of 6 mm float and outer skin of 6 mm Antisun ditto	62·40

Cost per m²
£

2·7 INTERNAL WALLS AND PARTITIONS

Load-bearing Walls

75 mm hollow clay block	7·90
75 mm Thermalite block	7·90
100 mm Thermalite block	9·46
Half brick thick fletton brickwork	8·50
140 mm Thermalite block	12·25
One brick thick fletton brickwork	21·50

Partitions

Stud partition of 100 × 38 mm studs at 400 mm centres, 100 × 38 mm head and plate and 50 × 25 mm wedges.	3·80
57 mm Paramount dry partition	7·10
50 mm laminated plaster board partition	7·30
64 mm Paramount dry partition	8·00
65 mm laminated plaster board partition	9·50
50 mm demountable steel partition, self finish	26·40
50 mm ditto fully glazed	28·50
75 mm ditto firecheck	42·90

2·8 INTERNAL DOORS

	Size of door	
	726 × 2040 mm No. £	826 × 2040 mm No. £

Doors

40 mm finished flush door, cellular core, hardboard faced, lippings to edges, 102 mm steel butts, sealed and painted three coats . . .	25·35	27·00
40 mm finished flush door, cellular core, plywood faced, lippings to edges, 102 mm steel butts, primed and painted three coats	30·90	32·50
40 mm finished flush door, solid core, hardboard faced, lippings to edges, 102 mm steel butts, sealed and painted three coats	37·60	39·20
40 mm finished flush door, cellular core, sapele faced hardboard faced, lippings to edges, 102 mm steel washered brass butts and wax polished . . .	41·60	43·30
40 mm finished flush door, solid core, plywood faced, lippings to edges, 102 mm steel butts, primed and painted three coats	41·90	43·60
44 mm finished flush door, half hour fire check, sapele faced hardboard faced, lippings to edges, 102 mm steel washered brass butts and wax polished .	52·75	54·50
40 mm finished flush door, solid core, sapele faced hardboard faced, lippings to edges, 102 mm steel washered brass butts and wax polished . .	54·60	56·40
44 mm finished softwood purpose made four panel door, 102 mm steel butts, knotted, primed, stopped and painted three coats	60·20	65·75
54 mm finished flush door, one hour fire check, sapele veneered plywood faced, lippings to edges, 102 mm steel washered brass butts and wax polished . .	83·70	85·50
50 mm finished mahogany purpose made four panel door, 102 mm steel washered brass butts and wax polished	95·45	104·00

Frames and Linings

100 × 25 mm softwood lining on and including softwood grounds, 50 × 19 mm softwood stop, knotted, primed, stopped and painted three coats (for 75 mm partition plastered both sides)	29·45	30·00
150 × 25 mm softwood lining on and including softwood grounds, 50 × 19 mm softwood stop, knotted, primed, stopped and painted three coats (for half brick wall plastered both sides)	34·90	35·60
100 × 25 mm mahogany lining on and including softwood grounds, 50 × 19 mm mahogany stop, wax polished (for 75 mm partition plastered both sides).	38·50	39·40
250 × 25 mm softwood lining on and including softwood grounds, 50 × 19 mm softwood stop, knotted, primed, stopped and painted three coats (for one brick wall plastered both sides)	41·75	42·50
100 × 75 mm softwood rebated and rounded frame, knotted, primed, stopped and painted three coats (including cramps and dowels)	41·10	42·85
150 × 25 mm mahogany lining on and including softwood grounds, 50 × 19 mm mahogany stop, wax polished (for half brick wall plastered both sides).	47·50	48·60
100 × 75 mm mahogany rebated and rounded frame, wax polished (including cramps and dowels)	54·10	55·00
250 × 25 mm mahogany lining on and including softwood grounds, 50 × 19 mm mahogany stop, wax polished (for one brick wall plastered both sides).	56·00	57·30
375 × 25 mm softwood lining on and including softwood grounds, 50 × 19 mm softwood stop, knotted, primed, stopped and painted three coats (for one and a half brick wall plastered both sides)	70·80	72·15
375 × 25 mm mahogany lining on and including softwood grounds, 50 × 19 mm mahogany stop, wax polished (for one and a half brick wall plastered both sides).	75·70	77·40

	Size of door	
	726 × 2040 *mm* No. £	826 × 2040 *mm* No. £

2·8 INTERNAL DOORS – *continued*

Architraves

50 × 25 mm softwood splayed and rounded architrave on and including softwood grounds to both sides of opening, knotted, primed, stopped and painted three coats	27·05	27·45
50 × 25 mm mahogany splayed and rounded architrave on and including softwood grounds to both sides of opening and wax polished	38·70	39·25

Cost per m²
£

3·1 WALL FINISHES

Building fletton brickwork with fair face and flush pointing	1·50
Render and set in lightweight gypsum on brick walls	2·90
Render in cement-sand (1:3) and set in gypsum on brick walls	4·05
Skim coat of gypsum on and including 9·5 mm gypsum lath fixed to softwood studs (not included)	4·20
12·5 mm insulation board on and including battens plugged to wall . .	4·30
9·5 mm gypsum wallboard on and including battens plugged to wall . .	4·35
6·4 mm medium-quality hardboard on and including battens plugged to wall . .	5·45
6 mm birch-faced plywood ditto	7·30
6 mm W.A.M.-faced plywood ditto	9·45
4 mm white glazed wall tiles on and including 13 mm cement-sand (1:3) backing . .	11·75
4 mm coloured wall tiles ditto	11·90 to 15·75
6 mm terrazzo wall lining on and including backing	17·10
12 mm softwood wall lining on and including battens plugged to wall . . .	18·00
25 mm softwood wall lining on and including battens plugged to wall . . .	21·80

Cost per m²
£

3·4 FLOOR FINISHES
Solid floors
Beds

	£
3 mm one-coat latex cement	2·40
32 mm cement-sand (1:3) screeded bed	2·60
32 ,, ,, (1:3) floated bed	2·85
32 ,, ,, (1:3) trowelled bed	2·85
50 mm cement-sand (1:3) screeded bed	3·40
50 ,, ,, (1:3) floated bed	3·65
50 ,, ,, (1:3) trowelled bed	3·65
25 mm Synthanite on and including building paper	3·90
5 mm two-coat latex cement	4·20
75 mm cement-sand (1:3) screeded bed	5·00
75 ,, ,, (1:3) floated bed	5·15
75 ,, ,, (1:3) trowelled bed	5·25
50 mm Synthanite on and including building paper	6·10
75 mm Synthanite on and including building paper	8·30

Finishes

	£
32 mm cement and sand paving, surface hardened.	3·15
25 mm granolithic paving	3·50
2 mm vinyl asbestos tiles	4·65
3 mm thermoplastic tiles. 'B' range	4·70
2 mm vinyl tiles	4·80
6 mm jointless composition flooring	4·90
2·50 mm vinyl tiles	5·00
15 mm red pitchmastic	5·15
2·50 vinyl asbestos tiles	5·15
20 mm red pitchmastic	5·90
25 mm red pitchmastic	6·60
3·20 mm coloured linoleum tiles	7·25
5 mm cork tile flooring and polishing	8·60
6 mm cork tile flooring and polishing	9·70
25 mm softwood tongued-and-grooved flooring including fillets	9·85
8 mm cork tile flooring and polishing	11·20
19 mm brown floor tile paving	11·25
3·75 mm rubber tile flooring	13·00
25 mm Gurjun block flooring and polishing	14·50
5 mm rubber tile flooring	15·50
25 mm Maple strip flooring including fillets and polishing	16·55
25 mm Iroko block flooring and polishing	17·50
25 mm Redwood block flooring and polishing	17·50
25 mm Merbau block flooring and polishing	17·50
16 mm terrazzo paving divided into squares with ebonite strip	17·55
28 mm terrazzo tiles	17·90
25 mm Oak block flooring and polishing	18·70
25 mm Oak strip flooring including fillets and polishing	19·25
25 mm Iroko strip flooring including fillets and polishing	22·00
25 mm Sapele strip flooring including fillets and polishing	25·75

3·2 FLOOR FINISHES – *continued*

Joisted floors

Boarding

25 mm straight edge softwood flooring	6·75
25 mm tongued-and-grooved ditto	7·55

Finishes

3·20 mm coloured sheet linoleum on and including underlay	6·65
25 mm straight-edge softwood flooring, cleaned off and polishing	7·20
25 mm tongued-and-grooved softwood flooring, cleaned off and polishing . . .	8·00
5 mm cork tile flooring and polishing	8·60
6 mm cork tile flooring and polishing	9·70
8 mm cork tile flooring and polishing	11·20
25 mm Maple strip flooring nailed to joists and polishing	14·25
25 mm Oak strip flooring nailed to joists and polishing	16·95
Pile carpet (*p.c.* £10·00 m²) jointed on and including underlay	17·25
25 mm Iroko strip flooring nailed to joists and polishing	19·65
25 mm Sapele strip flooring nailed to joists and polishing	23·45

3·3 CEILING FINISHES

Render and set in lightweight gypsum on concrete soffit	2·75
Skim coat of gypsum on and including 9·7 mm gypsum lath fixed to soffit of joists .	4·20
Skim coat of gypsum on and including 9·7 mm insulating gypsum lath fixed to soffit of joists	4·35
50 mm wood-wool slabs laid on formwork as permanent lining	4·50
16 mm tongued-and-grooved softwood boarding fixed to soffit of joists . . .	7·30
Suspended ceiling of 15 mm mineral fibre tiles flame proofed to Class 1 B.S. 476 fixed in steel tees suspended from structure, tiles chamfered and grooved, aluminium angle trim at perimeter	11·50
Render in gypsum-sand (1:1½) and set in gypsum including gypsum sand (1:2) pricking up coat on and including expanded metal lathing fixed to and including counter battens	11·80

DECORATIONS

Prepare and two coats emulsion paint on plastered walls or ceilings . . .	0·97
Prepare walls and hang wallpaper *p.c.* £4·00 per piece	1·82
Prepare, size, stain and two coats varnish on woodwork	2·12
Prepare, prime and three coats oil paint on plastered walls or ceilings . . .	3·11
Prepare, prime and three coats oil paint on metalwork	3·12
Knot, prime, stop and three coats oil paint on woodwork	3·27
Seal, body-in and wax polish on hardwood	3·57
Stain, body-in and French polish on hardwood	10·70

Cost per
point
£

5·8 ELECTRICAL INSTALLATIONS

Prices for wiring complete, including mains cost on consumer's side of supply point including builder's work but excluding cost of lamps and light fittings

Lighting points

P.V.C. insulated and sheathed cable in houses and flats, installed in floor cavities and roof voids, protected by light gauge conduits where buried	25·80
P.V.C. insulated cables contained in screwed welded conduit in commercial property	30·25
M.I.C.C. cables in commercial property	44·50
M.I.C.C. cables in industrial property	56·80
P.V.C. insulated cables contained in screwed welded conduit in industrial property	57·25

Switched Socket Outlets

13-amp. point with P.V.C. insulated and sheathed cables, wired as a ring-main circuit, installed in the floor cavity and ceiling void protected by light-gauge conduit where buried in houses and flats	29·65
13-amp. point with P.V.C. insulated cable contained in screwed welded conduit on a ring-main circuit in commercial property	34·35
13-amp. point with M.I.C.C. cable on a ring-main circuit in houses and flats .	44·20
13-amp. point with M.I.C.C. cable on a ring-main circuit in commercial property .	50·00
13-amp. point with P.V.C. insulated cable contained in screwed welded conduit on a ring main circuit in industrial property	54·50
13-amp. point with M.I.C.C. cable on a ring-main circuit in industrial property .	63·00

Cost per m²
£

6·1 SITE WORKS

Preparatory work

75 mm bed of hardcore spread, levelled and well rammed	0·75
50 mm bed of sand, spread and levelled	0·95
100 mm bed of hardcore, spread, levelled and well rammed	1·00
75 mm bed of sand, spread and levelled	1·05
75 mm bed of clinker, ditto	1·30
Excavate over site to remove top soil and vegetable matter, average 150 mm deep . .	1·60
50 mm concrete (1:12) 40 mm aggregate as blinding layer	1·75
75 mm concrete (1:3:6) 40 mm aggregate spread and levelled	2·95

Finishes

50 mm (after rolling) gravel paving rolled to cambers and falls	0·86
63 mm ditto	1·13
Cold bitumen emulsion paving, total thickness 25 mm spread and rolled on prepared bed (not included) in three layers	2·00
50 mm tarmacadam paving in two layers, finished with limestone or granite chippings .	4·75
*50 mm precast concrete paving slabs to B.S. 368 bedded in lime mortar . . .	5·35
*42 mm Noelite paving slabs, ditto	8·35
*50 mm Yorkstone paving slabs, ditto	8·60
*38 mm Colourstone paving slabs, ditto	9·50
229 × 114 × 38 mm paving bricks laid flat, bedded and pointed in cement mortar .	10·15
229 × 114 × 38 mm paving bricks laid to herringbone pattern bedded and pointed in cement mortar	11·50
75 mm cobble paving	15·00

NOTE *Prices are dependent on colour and sizes.*

6·2 DRAINAGE

100 mm dia. drain pipes laid and jointed in trench including excavation average 1 m deep in normal soils

Pitch fibre pipe including bed of sand 50 mm thick	9·50
Surface water quality vitrified stoneware pipe including 150 mm concrete bed and benching	12·50
'B.S.' tested quality vitrified stoneware pipe ditto	13·20
Spun-iron pipe to B.S. 1211 ditto	21·30

PART IV

European Section

This part of the book contains a summary of tendering procedures and costs of labour and materials for ten European countries.

TENDERING PROCEDURES AND CONSTRUCTION COSTS
IN EUROPE

This ninth issue of the European Section of Spon's contains up-dated summaries of tendering procedures and the costs of labour and materials for Belgium, Denmark, France, West Germany, Italy, The Netherlands, Norway, Sweden, the Republic of Ireland and Luxembourg, Spon's now includes all members within the enlarged Economic Community. In this issue for seven of the countries (Belgium, Denmark, France, West Germany, Republic of Ireland, Italy and The Netherlands) is included a detailed approximate estimate for a four storey office block. This has been based on a typical scheme built in the United Kingdom and priced in the countries concerned at prices ruling in May 1980. The specification, being based on English design and practice, is not common in all cases to other countries and in such cases the item has been priced for the nearest comparable specification.

At the end of the section the results of these estimates have been summarized and compared with the current cost of similar work in this country. Schedules of rates have again been included for the remaining countries together with approximate building prices for most of the countries.

The principal difference in tendering procedures between the United Kingdom and the other nine European countries (excluding the Republic of Ireland) is that only in West Germany and Luxembourg and less frequently in Denmark, France and Sweden is any form of bill of quantities supplied to the tenderers. In most other countries each tendering contractor usually takes off his own quantities from the drawings and specification and thus adds to the risk and cost of tendering. Another major difference is that, especially in France and Luxembourg and often in West Germany, the employer enters into separate contracts with specialist and trades contractors; co-ordination is the responsibility of the architect or of one of the contractors or a contractor employed for the purpose. Belgium, The Netherlands and Norway use general contractors, as in Britain, and this system is also gaining ground in Sweden, West Germany and to some extent in Italy.

No European country has such a sophisticated system of cost targets, cost planning and cost control procedures as in Britain. Perhaps because there is no independent profession (except in the Republic of Ireland) equivalent to quantity surveying and little use of bills of quantities. Information about the cost of building is less organized and more difficult to ascertain than in Britain. In particular there is very little published information on the cost of specific types of building. Approximate prices of building types for some countries have been included in this year's edition, but it is unwise to make comparisons between countries because of differences in the economy of each country, in productivity and in building standards and regulations. Anyone who becomes involved with a building project is advised that much detailed research will be necessary to establish applicable cost data. Reference to E.E.C. Directives 64/427, 64/429, 71/304, 71/305, 72/277 (and any subsequent) is also recommended.

The summaries of tendering and contract procedures in each country are an outline of the current practice. To date there has been little standardization of procedures between the countries within the E.E.C., although there is a directive that invitations to tender for public works contracts of more than one million accounting units (currently construed in the United Kingdom as £416,000) should be notified in all the community languages in the official journal of the European Communities (J.O.C.E.). A standard form of contract for use in E.E.C. countries is also under investigation; this would be a considerable achievement as at present few European countries have a form which is accepted generally even within their own territory.

The wage rates given apply to the districts stated. Regional variations are sometimes quite wide, especially in the larger countries, and there may also be regional differences in the charges on labour. The prices for materials are also subject to regional variations, largely as the result of haulage costs. In addition, at the time of preparing this section many countries were anticipating substantial increases in material costs due to sharply rising transport charges; intending users of this section are therefore advised to check prices wherever possible at the time of use. The schedules of rates and approximate building prices are all average figures and should only be used as a guide to establish an approximate 'order of cost'. It should be noted that the 'sterling equivalent' given against each rate is a direct conversion of the foreign price into British currency at 26th June 1980. Because of the changing economic situation in Europe the exchange rates given should be compared with current rates.

The Editors gratefully acknowledge the help they have received in compiling this section from many organizations and individuals including the British Embassies in the countries concerned, the Building Research Station and offices of the Department of the Environment in England and West Germany. Considerable assistance in the compilation of the prices sections has been received from: C.B.C. Byggeadministration A.-S. (Denmark and Norway), Enterprise Francis Bouygues (France), Boyd and Creed (Ireland), Signor Fabozzi (Italy), Berenschot Osborne B.V. (The Netherlands), D. McColl A.I.Q.S. (Norway and Sweden), Herr. W. Weiss (West Germany) and E. R. Skoyles Esq., A.R.I.C.S., A.I.Q.S.

REFERENCES

The following publications may be useful for more detailed information on procedures in the countries concerned.

Barclay, M., Happold, E., Martin, J. & Watt, B.
Working in France.
The Structural Engineer 52 (1) January 1974, pp. 3–16.

Bindslev, Bjorn.
Contract procedures and techniques in Europe and Scandinavia.
Chartered Surveyor 104 (3) 1971, pp. 122–127.

Building Centre.
Construction into Europe.
Building Centre Forum, 1973.

Cibula, Evelyn.
Building control in West Germany.
(Building Research Station Current Paper 10/70).
Watford: BRS, 1970.

Coates, M.
Construction Contracts. France.
Chartered Surveyor 105 (7) January 1973, pp. 317–319.

Cox, L. V.
The construction industry in the Federal Republic of West Germany.
Building 218 (6618) 20 March 1970, p. 121.

Eade, J. M.
The construction industry in the Federal Republic of Germany Parts 1 & 2.
The Quantity Surveyor 30 (4) January/February 1974, pp. 90–94.
                30 (5) March/April   1974, pp. 113–117.

Feigl, B. C.
The quantity surveyor in West Germany.
Chartered Surveyor 107 (10) May 1975, pp. 273–274.

Forder, J. N.
Quantity surveying in The Netherlands.
Parts I–III.
Quantity Surveyor – 26 (5) March/April 1970, pp. 105–108.
                27 (1) July/August 1970, pp. 9–16.
                27 (3) November/December 1970, pp. 56–62.

Glendinning, N. J.
European practice in controlling construction costs –
Federal Republic of Germany.
Chartered Surveyor. BQS Quarterly 1 (3) March 1974, pp. 54–58.

Grafton P. W. & Branson, A. J. W.
The German construction industry and the quantity surveyor.
Chartered Surveyor. BQS Quarterly 1 (2) December 1973, pp. 25–28.

Tendering procedures arising from entry into the European Common Market,
Greater London Council Bulletin 65 May 1973. Item 2.

Groves, P.
   Architects in the Common Market. Building Exhibition Conference.
   Architects Journal 158 (46) 14 November 1973, pp. 1187–1192.

House Information Services Ltd.
   Construction Industry Europe.
   London: 1974.

Houtsma, G. O.
   Dutch post contract administration 1.
   Building 225 (6800) 28 September 1973.

Keyte, M.
   Working in France. Building Exhibition Conference.
   Architects Journal 158 (46) 14 November 1973, pp. 1193–1196.

Madden, L. W.
   Building in Europe.
   Building 224 (6762) 5 January 1973, pp. 51–66.

Ministry of Public Building & Works,
   The French construction industry.
   (MPBW Research and Development Paper).
   London: MPBW, 1967.

National Building Agency.
   Housing and the European Community.
   European Studies Part 2 Volume 1.
   The Netherlands, Belgium, Luxembourg.
   London: October 1972.

National Building Agency.
   Construction in the six – how it works.
   Department of the Environment, January 1973.

Osborne, J. G.
   European practice in controlling construction costs: Netherlands.
   Chartered Surveyor, BQS Quarterly 106 (2) 1973, pp. 3–7.

Royal Institute of British Architects.
   Architectural Practice in Europe.
      1. France.
      2. Federal Republic of Germany.
      3. Italy.
      4. Benelux.
   London: 1975.

Skoyles, E. R.
   Norway. A short conspectus of its building practice and industry.
   Quantity Surveyor 32 (8) March 1976, pp. 141–143.

Steele, Gordon.
   Building in Sweden.
   Building 222 (6718) 25 February 1972, pp. 79–82.

Sturtewagen, D.
   European practice in controlling construction costs: Belgium.
   Chartered Surveyor, BQS Quarterly 1 (2) December 1973, pp. 23–24.

Taylor, G. D.
   Dutch post-contract administration 2.
   Building 225 (6801) 5 October 1973, pp. 123–124.

The quantity surveyor in The Netherlands.
   Quantity Surveyor 32 (2) September 1975, p. 34.

Trentham, J. M. E.
   Chartered quantity surveyor practising in Belgium.
   Chartered Surveyor 106 (13) August 1974, pp. 17–19.

The Royal Institution of Chartered Surveyors.
   The French Building Industry.
   London: 1975.

Westminster Chamber of Commerce Building in the E.E.C.
   (Report of the Building Group Mission to Europe, April 1972).
   London: 1972.

BELGIUM

TENDERING PROCEDURES

There are three basic types of contract in use in Belgium, which can be defined according to the method of establishing the cost of the works. These are, fixed lump sum, lump sum with provision for variations and a remeasurement contract.

The fixed lump sum contract (*Marché à Forfait Absolu*) is a rigorous contract with no provision for variations. The contractor is legally bound to construct the building for the agreed sum irrespective of any unforeseen difficulties which may occur; however a variation may be effected by the mutual agreement of both parties. Tendering is on the basis of drawings and specification only and the contractor must carry out all necessary work whether specified or not.

The other lump sum contract (more widely used) is the *Marché à Forfait Relatif*. This is similar to the *Marché à Forfait Absolu* except that it includes variation and fluctuation clauses. A schedule of quantities and rates must also be provided by the contractor in support of his tender and this forms the basis of valuation for variations.

The remeasurement contract (*Marché à Borderaux de Prix*) is based upon a schedule of rates; the rates are fixed at the time of tender but the extent of the work is undefined. The work is measured upon completion and the agreed rates are used for valuing the measured items.

The controlled (*en régie*) or American contract is a cost plus contract rarely used nowadays.

Contractors are normally sent the following tender documents:

1 Drawings (*Série de plans*)

2 Contract Form (*cahier général des charges première partie*)

3 General specification of workmanship and materials (*cahier général des charges deuxième partie*)

4 Particular specification (*cahier spécial des charges*) which takes precedence over the form of contract

These documents together with a copy of the tender and the letter of instruction will become the contract documents and signed copies of each are held by both parties. In addition, if required, a schedule of rates supplied by the contractor may be included.

There is no generally accepted standard form of contract in use in Belgium as in England, although a standard form has been issued recently by the *Fédération Royale des Sociétés d'Architectes* and agreed by the *Confédération Nationale de la Construction*, but it is not yet in general use. There is also a Government standard form of contract for public works and a standard form for dwellings built with government subsidies.

Selective tendering is still the usual method for private work though the amount of negotiated tendering is on the increase. Another form of tendering used in the private sector is where contractors submit unit prices for labour, material, plant, etc. with a percentage addition for on-costs, risk and profit. During the contract valuations are based on assessment of total consumption made by the successful contractor.

Public tendering is used for Government work and in this case, the Government is bound to pay damages if the lowest tenderer's offer is not accepted. The *appel d'offres* method is encouraged (see section on France), and the conditions under which negotiated contracts may be used have been extended.

BELGIUM

As in The Netherlands, contractors cover the risks involved in open tendering by a well established system of price adjustment before tenders are submitted; tenders are adjusted to the arithmetical average and allowance is also made for the cost of tendering. There is also once again often a considerable amount of post tender negotiation.

Interim payments are made in various ways. For public works the work executed is measured at agreed periods of time. For private development the same method may be used or more often the payments are made at various stages, e.g. foundations, at the completion of each floor, roof, etc. or alternatively in stages of say 10%. Retention of 10% is normally held on interim payments, of which 5% is paid on practical completion of the contract and the remaining 5% is paid on final completion six months after practical completion, provided the building is satisfactory.

It is normal to employ a general contractor in Belgium who may sublet work to specialist trades; however, this does not relieve him of any of his obligations to the building owner and he is directly responsible for their work. Contracts are not infrequently let trade by trade in which case it is the architect's duty to act as co-ordinator.

Contracts which include a fluctuations clause are normally adjusted by a pre-agreed formula based upon monthly price indices published by the Statistical Office.

The final account is prepared by the contractor and agreed by the architect. Both the contractor and the architect are jointly responsible for the condition of the building, excluding fair wear and tear, for a ten year period following its completion.

BELGIUM

PRICE DATA

The prices for materials are ex factory, quarry, etc. and do not include haulage to site. Unless otherwise stated all prices and rates exclude Value Added Tax; this is currently 18% on supply only items and 16% on supply and fix items. These percentages are added as a separate item to the total of the building tender.

For the purposes of comparing Belgian prices with those in the United Kingdom it has been assumed that 1 Franc = 1·5p approximately.

BASIC WAGE RATES (May 1980)

	Francs per hour	Sterling equivalent £
Labourers	192·50	2·89
Semi-skilled Workers	209·90	3·15
Skilled Workers		
First Grade	227·20	3·41
Second Grade	244·60	3·67

ADDITIONS TO BASIC WAGE RATES

Allowance is usually made to the basic rates for the following:

Unemployment Insurance.
Health Service Contributions.
Health Indemnities.
Insurance against illness caused through nature of work.
Family allowances.
Pensions.
Accidents at work.
Holidays with pay.
Bank Holidays with pay.
Paid leave for family reasons (births, deaths, marriages, etc.).
Guaranteed week.
Contribution to fares.
Civil Liabilities.
Inclement Weather.
O.N.S.S. Subscriptions.
Life Insurance.
Working Clothes.
Bonus payments.

The effect of the above is to add approximately 112% to the basic rates. In addition, a contractor tendering adds his own on-costs and profit.

BELGIUM

MATERIALS PRICES (Second Quarter 1979)

	Unit	Price *Francs*	Sterling equivalent £
Ready mix concrete, quality 31 N/mm²	m³	1,660·00	24·90
Cement Type P300	tonne	1,650·00	24·75
Sand for mortar and plastering	,,	60·00	0·90
Fine aggregate for concrete	,,	60·00	0·90
Belgian river gravel grade 5/30	,,	93·00	1·40
Hardcore (broken stone/brick) filling	m³	870·00	13·05
Mild steel bar reinforcement	tonne	1,020·00	15·30
Mild steel fabric reinforcement (weighing 3 kg/m²) . . .	,,	14,900·00	223·50
25 mm softwood boarding for formwork	m²	157·00	2·36
100 mm Exposed aggregate precast concrete wall panels . .	,,	1,840·00	27·60
Asphalt	tonne	8,300·00	124·50
Low quality bricks size 190 × 90 × 65 mm	1000	2,200·00	33·00
Good quality facing bricks	,,	4,010·00	60·15
100 mm Lightweight large building blocks	m²	160·00	2·40
4 mm Blue/black asbestos slates size 500 × 250 mm . .	1000	14,360·00	215·40
Northern softwood for framing type IV 64 × 178 mm . .	m	87·00	1·31
50 × 50 mm Softwood batten	,,	24·00	0·36
100 × 35 mm Softwood door frame	,,	166·00	2·49
20 mm Blockboard	m²	400·00	6·00
Hardwood (Afrormosia)	m³	24,000·00	360·00
76 × 114 mm Afrormosia door frame	m	605·00	9·08
40 mm Afrormosia faced solid core flush door size 762 × 1981 mm	No.	2,980·00	44·70
50 mm wood wool slabs	m²	205·00	3·08
Structural steelwork	tonne	13,000·00	195·00
Aluminium	kg	70·00	1·05
Black anodized aluminium windows	m²	3,130·00	46·95
100 mm Diameter PVC pipe	m	140·00	2·10
15 mm Diameter copper pipe	,,	60·00	0·90
Low level coloured WC suite	No.	1,780·00	26·70
Gypsum plaster	tonne	2,740·00	41·10
12 mm Plasterboard	m²	82·00	1·23
3 mm Vinyl floor tiles	,,	173·00	2·60
150 × 150 × 6 mm Quarry (clay) floor tiles . . .	,,	200·00	3·00
50 mm Precast concrete paving slabs	,,	157·00	2·36
4 mm Clear sheet glass	,,	420·00	6·30
Emulsion paint	litre	68·00	1·02
Oil paint	,,	136·00	2·04
Polyurethane lacquer	,,	140·00	2·10
100 mm Diameter clay drain pipes	m	55·00	0·83
150 mm Diameter cast iron drain pipes (3 m lengths) . .	,,	530·00	7·95

BELGIUM

ESTIMATE OF COST FOR OFFICE TYPE BUILDING

The following estimate relates to a reinforced concrete framed four storey office block with a total floor area of 2400 m² and a car park of 500 m² forming part of the ground floor.

It has been assumed that the work would take about 15 months to complete and that access to the site is good.

	Quantity	Unit	Rates Francs	Total Francs	Sterling equivalent £
SUBSTRUCTURE					
Excavate					
to reduce levels	585	m³	185·00	108,225·00	1,623·38
to form trenches	413	,,	340·00	140,420·00	2,106·30
Extra for breaking out brickwork or concrete	90	,,	1,830·00	164,700·00	2,470·50
Remove surplus material from site .	846	,,	130·00	109,980·00	1,649·70
Hardcore (stone/brick)					
infilling	84	,,	1,200·00	100,800·00	1,512·00
in bed 150 mm thick . . .	220	m²	245·00	53,900·00	808·50
Plain concrete 8·00 N/mm²					
in foundations . . .	220	m³	2,725·00	599,500·00	8,992·50
in bed 50 mm thick . . .	603	m²	153·00	92,259·00	1,383·89
Reinforced concrete 26·00 N/mm²					
in foundations . . .	190	m³	2,920·00	554,800·00	8,322·00
in bed 200 mm thick . . .	265	m²	610·00	161,650·00	2,424·75
Mild steel bar reinforcement					
in foundations . . .	16,730	kg	36·00	602,280·00	9,034·20
Mild steel fabric reinforcement					
3·00 kg/m² in bed . . .	258	m²	132·00	34,056·00	510·84
Formwork					
to sides of foundations . . .	270	,,	850·00	229,500·00	3,442·50
SUPERSTRUCTURE					
Reinforced concrete 31 N/mm²					
in beams	196	m³	3,160·00	619,360·00	9,290·40
in columns	50	,,	3,160·00	158,000·00	2,370·00
Mild steel bar reinforcement					
in beams and columns . .	81,850	kg	36·00	2,946,600·00	44,199·00
Formwork					
in beams	900	m²	1,014·00	912,600·00	13,689·00
in columns	550	,,	1,014·00	557,700·00	8,365·50
Wrot formwork					
in beams	400	,,	1,100·00	440,000·00	6,600·00
in columns	54	,,	1,120·00	60,480·00	907·20
Structural steelwork in roof beams etc. including all fabrication, placing in position, fixing and protective coat of paint .	20,200	kg	45·00	909,000·00	13,635·00
			C/F	9,555,810·00	143,337·16

BELGIUM

	Quantity	Unit	Rates Francs	Total Francs	Sterling equivalent £
		B/F	9,555,810·00		143,337·16

225 mm thick concrete 31 N/mm² trough suspended floor consisting of 165 mm (av.) thick × 175 mm deep ribs at 600 mm centres one way, between 600 mm wide × 175 mm deep special trough forms supported as necessary, with a 50 mm concrete topping, the whole reinforced with 8·15 kg of steel bar reinforcement per m² and 2·20 kg of fabric reinforcement per m², including all necessary formwork to soffit
OR:
Suitable equivalent concrete floor spanning 4·80 m with a dead load of 4·57 N/mm² and a superimposed load of 4·60 N/mm² including all necessary reinforcement and formwork

	745	m²	2,200·00	1,639,000·00	24,585·00

200 mm thick concrete 31 N/mm² hollow pot suspended floor consisting of 295 × 295 × 150 mm deep clay blocks at 400 mm centres with clay filler tiles between and a 50 mm concrete topping
OR:
Suitable equivalent concrete floor spanning 4·80 m with a dead load of 4·57 N/mm² and a superimposed load of 4·60 N/mm²

	1,617	,,	700·00	1,131,900·00	16,978·50
Reinforced concrete 31 N/mm² in 200 mm suspended floor slab	85	,,	700·00	59,500·00	892·50
Mild steel bar reinforcement in suspended floors	13,900	kg	37·00	514,300·00	7,714·50
Mild steel fabric reinforcement 2·2 kg/m² in suspended slab	1,702	m²	88·00	149,776·00	2,246·64
Formwork to soffit of slab	1,450	,,	920·00	1,334,000·00	20,010·00
External rendering to underside of exposed slab and beams	1,060	,,	360·00	381,600·00	5,724·00
20 mm mastic asphalt roofing on felt on 50 mm thick lightweight channel reinforced composite decking spanning 4·80 m	675	,,	1,015·00	685,125·00	10,276·88
20 mm asphalt skirtings average 175 mm girth	215	m	180·00	38,700·00	580·50
4 mm blue black asbestos cement slates size 500 × 250 mm laid to slope on and including battens and 25 mm insulation quilt	460	m²	740·00	340,400·00	5,106·00
Ditto but fixed vertically on and including 20 mm blockboard	100	,,	894·00	89,400·00	1,341·00
15 mm blue black asbestos cement sheet, 300 mm wide fixed to softwood soffit	130	m	415·00	53,950·00	809·25
Aluminium flashings	1,350	,,	570·00	769,500·00	11,542·50
Treated sawn softwood in roof members, plates etc.	25	m³	18,700·00	467,500·00	7,012·50
		C/F		17,210,461·00	258,156·93

BELGIUM

	Quantity	Unit	Rates Francs B/F	Total Francs 17,210,461·00	Sterling equivalent £ 258,156·93
100 mm diameter PVC flat roof outlet with PVC pipe 12 m long including all fixings	7	No.	8,900·00	62,300·00	934·50
Dog leg stairs and half landing rising three levels (concrete 6 m³, reinforcement 440 kg, formwork 50 m²)	2	,,	100,000·00	200,000·00	3,000·00
Mild steel balustrade with plastic handrail to six flights of stairs (handrail 17 m, standards 36 No.)	1	,,	42,000·00	42,000·00	630·00
Mild steel balustrading with hardwood handrail to six flights of stairs, string to profile of treads and risers and wall string (handrail 19 m, standards 26 No.)	1	,,	108,000·00	108,000·00	1,620·00
Granolithic finish 25 mm thick to treads and risers	30	m²	800·00	24,000·00	360·00
Vinyl tiles 3 mm thick ditto	22	,,	600·00	13,200·00	198·00
Exposed aggregate pre-cast concrete panels 100 mm thick as wall cladding including all fixings and joint sealing	345	,,	3,670·00	1,266,150·00	18,992·25
Ditto but coping panels 520 mm overall width	97	,,	4,950·00	480,150·00	7,202·25
Ditto but parapet stones 620 mm wide × 470 mm deep including damp proof course	15	m	3,130·00	46,950·00	704·25
75 × 500 mm pre-cast concrete (26·00 N/mm²) cill or head finished fair on all exposed faces	52	,,	1,100·00	57,200·00	858·00
Reinforced concrete 26·00 N/mm² in walls 150 mm thick (155 m²) and 240 mm thick (25 m²)	30	m³	3,270·00	98,100·00	1,471·50
Mild steel bar reinforcement in walls	2,000	kg	35·50	71,000·00	1,065·00
Formwork to walls	230	m²	960·00	220,800·00	3,312·00
Wrot formwork to walls	110	,,	1,100·00	121,000·00	1,815·00
Brickwork, low quality bricks 115 mm wall	175	,,	960·00	168,000·00	2,520·00
240 mm wall	115	,,	1,650·00	189,750·00	2,846·25
Brickwork, good quality facings 115 mm wall	560	,,	1,700·00	952,000·00	14,280·00
Lightweight building blocks 100 mm wall	650	,,	840·00	546,000·00	8,190·00
Hardwood two panel door, 762 × 1981 × 50 mm, with glass in top panel and louvres in panel with and including hardwood frame 76 × 114 mm, including all fixing and ironmongery	5	No.	14,000·00	70,000·00	1,050·00
Hardwood window board 400 mm wide fixed to concrete cill	26	m	1,420·00	36,920·00	553·80
Plastic faced blockboard window lining 25 × 150 mm fixed to concrete	500	,,	440·00	220,000·00	3,300·00
			C/F	22,203,981·00	333,059·73

	Quantity	Unit	Rates Francs	Total Francs	Sterling equivalent £
			B/F	22,203,981·00	333,059·73
Black anodized aluminium windows, 50% horizontal sliding, 1700 × 1600 mm including all fixing glass and ironmongery	440	m²	5,600·00	2,464,000·00	36,960·00
Ditto but 12,250 × 675 mm . .	70	,,	6,200·00	434,000·00	6,510·00
Black anodized aluminium doors and screens including all fixing, 6 mm obscured glass and ironmongery . .	50	,,	8,100·00	405,000·00	6,075·00
Range of 6 No. prefabricated WC compartments 4·80 m long × 1·96 m high × 1·5 m deep consisting of 5 No. partitions and 6 No. doors . . .	4	No.	38,750·00	155,000·00	2,325·00
100 × 150 mm pre-cast concrete (26·00 N/mm²) lintel reinforced with one 12 mm mild steel rod	44	m	250·00	11,000·00	165·00
Hardwood faced solid core flush door, 762 × 1981 × 40 mm, with and including softwood frame, 100 × 35 mm, including all fixing and ironmongery . .	28	No.	11,500·00	322,000·00	4,830·00
Pair of hardwood two panel doors, 512 × 1981 mm, one hour fire resisting, panels glazed fire resisting glass, with and including hardwood frame, 135 × 90 mm, including all fixing and ironmongery	11	,,	32,400·00	356,400·00	5,346·00
Treated wrot softwood fillet fixed to concrete	2,300	m	86·00	197,800·00	2,967·00
Two-coat plaster, 12 mm thick to walls, columns and soffits . .	1,150	m²	465·00	534,750·00	8,021·25
Foil backed plaster board fixed to softwood, with and including one-coat plaster	560	,,	545·00	305,200·00	4,578·00
Two coats emulsion paint on walls and soffits	1,500	,,	174·00	261,000·00	3,915·00
Two coats oil paint ditto . .	400	,,	209·00	83,600·00	1,254·00
Three coats oil paint on wood or metal .	70	,,	251·00	17,570·00	263·55
Two coats polyurethane lacquer on aluminium or hardwood . .	270	,,	197·00	53,190·00	797·85
Insulation quilt 25 mm thick covered with cement and sand screed 48 mm thick reinforced with chicken wire . .	750	,,	436·00	327,000·00	4,905·00
Cement and sand screed 48 mm thick .	1,700	,,	283·00	481,100·00	7,216·50
Granolithic finish 50 mm thick to floors	70	,,	436·00	30,520·00	457·80
Vinyl tiles 3 mm thick to floors . .	2,300	,,	390·00	897,000·00	13,455·00
Brown quarry tiles 150 × 150 × 16 mm to floors	75	,,	610·00	45,750·00	686·25
OR: Terrazzo tiles 300 × 300 × 25 mm .	75	,,	1,360·00	—	—
Hardwood skirting 100 × 25 mm on and including softwood grounds fixed to wall	820	m	390·00	319,800·00	4,797·00
Mineral board ceiling tiles 9·5 mm thick laid in and including aluminium tee section grid supported by metal hangers 400 mm long fixed to soffit, including all decorations	2,350	m²	925·00	2,173,750·00	32,606·25
			C/F	32,079,411·00	481,191·18

BELGIUM

	Quantity	Unit	Rates Francs	Total Francs	Sterling equivalent £
			B/F	32,079,411·00	481,191·18
Lavatory basin, coloured, 560 × 400 mm, including fittings but excluding associated pipework	24	No.	3,600·00	86,400·00	1,296·00
Sink, white, 455 × 380 × 205 mm ditto	4	„	2,800·00	11,200·00	168·00
WC suite, low level, coloured, excluding associated pipework	24	„	4,630·00	111,120·00	1,666·80
Towel dispensers	8	„			
Soap dispensers	24	„		70,000·00	1,050·00
Toilet roll holders	24	„			
Mirrors 1750 × 450 × 6 mm . .	8	„			
Disposal installation to ground level for sanitary fittings listed above including all builders work in connection . .				445,000·00	6,675·00
Hot and cold water installation for sanitary fittings listed above including all builders work in connection . .				1,050,000·00	15,750·00
Hot water and wall radiator heating system for offices (2400 m² of floor area) including all builders work in connection OR:				4,175,000·00	62,625·00
Low cost full air conditioning (6000 m³)				(8,700,000·00)	(130,500·00)
Electrical power and lighting installation including mains, transformers and all builders work in connection to the following: 90 lighting points without fittings 10 „ „ with fittings 50 power points 5 control points				5,400,000·00	81,000·00
Passenger lift, electric, eight persons, four stops, travelling through 9·50 m. Including all builders work in connection but excluding cost of shaft . . .				1,770,000·00	26,550·00
Allow for 7 No. 30 m hose reels and all builders work in connection . .				155,000·00	2,325·00
Car park slab consisting of 175 mm bed of hardcore and 75 mm tarmacadam paving in two coats	880	m²	950·00	836,000·00	12,540·00
Total building cost				46,189,131·00	692,836·98
Add for site establishment, supervision and head office overheads where not included in above rates .			10%	4,618,913·10	69,283·70
				50,808,044·10	762,120·68
Add for any other estimated expenditure				—	—
				50,808,044·10	762,120·68
Add for V.A.T. or equivalent tax . .			16%	8,129,287·06	121,939·30
TOTAL ESTIMATED TENDER FIGURE				58,937,331·16	884,059·98

BELGIUM

AVERAGE BUILDING PRICES PER SQUARE METRE

(Second Quarter 1980)

	Price Francs	Approximate Sterling equivalent £
Offices for letting, including lifts and central heating	26,000·00	390·00
Offices for owner occupation, including lifts and air-conditioning . .	29,500·00	442·50
Shops (without finishings)	26,000·00	390·00
Flats, four storey	27,000·00	405·00
Houses, two storey	28,000·00	420·00
Detached houses	29,500·00	442·50
Hotels	28,350·00	425·25

N.B. The above prices include Value Added Tax and an engineering fee.

BELGIUM

BUILDING COST INDEX

				Annual Average 1965 = 100
Year				
1963	.	.	.	.	85
1964	.	.	.	.	98
1965	.	.	.	.	100
1966	.	.	.	.	106
1967	.	.	.	.	113
1968	.	.	.	.	116
1969	.	.	.	.	122
1970	.	.	.	.	129
1971	.	.	.	.	138
1972	.	.	.	.	146
1973	.	.	.	.	187
1974	.	.	.	.	200
1975	.	.	.	.	210
1976	.	.	.	.	232
1977	.	.	.	.	252
1978	.	.	.	.	270
1979	.	.	.	.	284
1980	.	.	.	.	310 (First quarter)

N.B.: The above indices exclude Value Added Tax.

DENMARK

TENDERING PROCEDURES

The principal tender documents comprise a comprehensive specification, drawings, form of agreement, form of tender and general conditions of contract. The building industry's standard form of building contract, *Almindelige Betingelser for Arbejder og Leverancer* (1972) appears to be as widely used as our own J.C.T. standard form of building contract.

Although cost reimbursement and negotiated contracts are known, the majority of contracts are let on a lump-sum basis. Most Government contracts are placed as a result of public advertisement, but occasionally a single or limited number of contractors are approached. Until recently, it was standard practice for tenderers to submit details of their proposed tenders to their trade organizations for checking and approval: the trade organization could prevent the submission or acceptance of a tender they considered to be too low and the only recourse of the building owner was through a special form of arbitration. However, Parliament has now prohibited tendering contractors from 'adjusting' their tenders. Trade organizations are still permitted to hold meetings of tenderers to discuss the work but the building owner must be invited and it is prohibited for prices to be discussed at such meetings.

Trade Associations are very prominent in Denmark and many of them have prepared standard schedules of prices for their members. General contractors are not very common and usually separate trade contracts are entered into. These trade contractors prepare their own unit quantities and often base their prices on standard schedules prepared by their particular association. Several Danish Building Institutions have advocated the wider use of bills of quantities but to date they are used only occasionally, mainly in association with advanced techniques of computer analysis and planning. Co-ordination of the various trades on site is usually the responsibility of the supervising architect but the complexity of planning required on some large schemes has led to the introduction of project supervisors and other special overall planning consultants.

Industrialized building systems are becoming increasingly popular in Denmark, as well as in other Scandinavian countries; initially these were mainly prefabricated timber systems but systems based on other materials, especially pre-cast concrete, are now more common. This has resulted in many projects with independent architects and consulting engineers employed by contractors to design their 'package' offers to the client. Housing is one such field with two-thirds stemming from the private sector and from non-profit making housing associations; it is claimed by some (and denied by others) that industrialized house building systems are 10% to 15% cheaper than traditional building methods.

All Government financed and state subsidized building contracts are let on a firm price–fixed time basis. Exceptions to the firm price rule are for expenses incurred by the contractor due to *force majeure* or circumstances occasioned by the authority.

Many private contracts are let under the same conditions and even though the term 'a firm price contract' implies that there should be no change in price from the date of acceptance of the contract to the date of final completion, modifications to the agreed work can be made and payment is made for these.

Statements for interim payments submitted by the contractor are usually subject to a 10% retention. The final account is prepared by the contractor.

DENMARK

PRICE DATA

Unless otherwise stated all prices and rates exclude Value Added Tax, which is currently levied as a 22% addition to tenders. In certain circumstances, e.g. housing, the eventual property owner may be entitled to claim a partial rebate of the Tax.

The prices for both labour and materials are approximate and exclude delivery charges.

For the purposes of comparing Danish prices with those in the United Kingdom it has been assumed that 1 Danish Kroner = 7·8p approximately.

AVERAGE EARNINGS

As in other Scandinavian countries a large percentage of earnings are calculated on a piece work system. Average earnings for the first quarter 1980, excluding overtime, shift pay and holiday supplements, were as follows:

	Copenhagen Kroner per hour	District Sterling equivalent £
Skilled Workers:		
Bricklayers	73·00	5·69
Carpenters	68·00	5·30
Joiners	66·00	5·15
Plumbers	70·00	5·46
Unskilled Workers (varies)	61·50	4·80

ADDITIONS TO AVERAGE EARNINGS

Overtime, shift pay and holiday supplements tend to add about Kr. 4·50 (£0·35) to the above figures. Allowance is usually made to the above rates for the following:

Social Security.
Public Health insurance.
Supplementary pension A.T.P.
Statutory accident/disability insurance.
Annual Holidays.
Public Holidays falling on weekdays. } (Most people receive 15 to 20 working days annual holiday plus a further 12 days for public holidays.)

The effect of the above is to add approximately 16% to the average earnings. In addition a contractor tendering adds his own on-costs and profit.

DENMARK

MATERIALS PRICES (First Quarter 1980)

	Unit	*Price Kroner*	*Sterling equivalent* £
Hardcore filling (broken brick)	m³	60·00	4·68
Portland cement	tonne	460·00	35·88
Aggregates			
Fine gravel 0–8 mm ⎱ including delivery	m³	71·00	5·54
Coarse gravel 16–32 mm ⎰ within 5 km	,,	70·00	5·46
Ready mixed concrete including delivery within 15 km	,,	430·00	33·54
Reinforcement in lengths exceeding			
10 m and not exceeding 12 m delivered in loads in excess of 200 kg			
Mild steel rods – 10 mm diameter	100 kg	370·00	28·86
High tensile steel rods – ,,	,,	440·00	34·32
Mild steel fabric reinforcement (weighing 3 kg/m²)	m²	18·00	1·40
Clay bricks 230 × 110 × 55 mm	1000	850·00	66·30
Machine made facings	,,	1,900·00	148·20
Hand made facings	,,	570·00	44·46
Plain backing bricks for internal use	,,	600·00	46·80
Porous insulation bricks ,, ,,	,,	850·00	66·30
Hard burnt clinker bricks			
Cellular concrete blocks			
59 × 19 × 10 cm	No.	5·50	0·43
59 × 19 × 30 cm	,,	16·00	1·25
Mortar materials			
Slaked lime	H1	81·00	6·32
Beach sand	m³	57·50	4·49
Ready mixed lime mortar	,,	200·00	15·60
Building timber – Danish Spruce in lengths 4·5 m to 6·0 m			
38 × 75 mm	m	5·00	0·39
50 × 100 mm	,,	9·00	0·70
75 × 150 mm	,,	19·50	1·52
75 × 75 mm	,,	10·50	0·82
100 × 100 mm	,,	18·00	1·40
150 × 150 mm	,,	41·00	3·20
Imported hardwoods			
Afrormosia	cu. ft	140·00	10·92
African Mahogany	,,	100·00	7·80
Teak	,,	190·00	14·82
		(average)	
20 mm blockboard	m²	115·00	8·97
50 mm wood wool slabs	,,	25·50	1·99
4 mm blue/black asbestos slates	1000	3,465·00	270·27
Steel Sections			
IPE profiles			
IPE 100–100 × 55 mm	100 kg	447·00	34·87
IPE 200–200 × 100 mm	,,	420·00	32·76
IPE 400–400 × 180 mm	,,	503·00	39·23
HEB profiles			
HE 100B–100 × 100 mm	,,	420·00	32·76
HE 200B–200 × 200 mm	,,	400·00	31·20
HE 400B–400 × 300 mm	,,	510·00	39·78
White glazed wall tiles 150 × 150 mm	No.	2·00	0·16
Clay floor tiles 150 × 150 × 16 mm	m²	230·00	17·94
WC suite complete (average)	No.	1,090·00	85·02
Lavatory basin complete (average)	,,	1,000·00	78·00
4 mm clear sheet glass	m²	121·00	9·44
Cast iron drain/waste pipe in 3 m lengths			
100 mm diameter	m	127·00	9·91
150 mm diameter	,,	90·00	7·02
100 mm pvc pipe	,,	24·00	1·87

DENMARK

ESTIMATE OF COST FOR OFFICE TYPE BUILDING

The following estimate relates to a reinforced concrete framed four storey office block with a total floor area of 2400 m² and a car park of 500 m² forming part of the ground floor.

It has been assumed that the work would take about 15 months to complete and that access to the site in the Copenhagen area is good.

Substructure	Quantity	Unit	Rate Kroner	Total Kroner	Sterling equivalent £
Excavate					
to reduce levels	585	m³	30·00	17,550·00	1,368·90
to form trenches	413	„	43·00	17,759·00	1,385·20
Extra for breaking out:					
brickwork or concrete	90	„	29·50	2,655·00	207·09
Remove surplus material from site . .	846	„	24·00	20,304·00	1,583·71
Hardcore (stone/brick)					
in filling	84	„	83·00	6,972·00	543·82
in bed 150 mm thick . . .	220	m²	30·00	6,600·00	514·80
Plain concrete 8·00 N/mm²					
in foundations	220	m³	440·00	96,800·00	7,550·40
in bed 50 mm thick	603	m²	28·00	16,884·00	1,316·95
Reinforced concrete 26·00 N/mm²					
in foundations	190	m³	585·00	111,150·00	8,669·70
in bed 200 mm thick . . .	265	m²	110·00	29,150·00	2,273·70
Mild steel bar reinforcement					
in foundations	16,730	kg	7·20	120,456·00	9,395·57
Mild steel fabric reinforcement					
3·00 kg/m² in bed	258	m²	24·00	6,192·00	482·98
Formwork					
to sides of foundations . . .	270	„	125·00	33,750·00	2,632·50
Superstructure					
Reinforced concrete 31 N/mm²					
in beams	196	m³	600·00	117,600·00	9,172·80
in columns	50	„	600·00	30,000·00	2,340·00
Mild steel bar reinforcement					
in beams and columns . . .	81,850	kg	7·50	613,875·00	47,882·25
Formwork					
in beams	900	m²	190·00	171,000·00	13,338·00
in columns	550	„	190·00	104,500·00	8,151·00
Wrot formwork					
in beams	400	„	260·00	104,000·00	8,112·00
in columns	54	„	260·00	14,040·00	1,095·12
Structural steelwork in roof beams etc. including all fabrication, placing in position, fixing and protective coat of paint . .	20,200	kg	12·00	242,400·00	18,907·20
			C/F	1,883,637·00	146,923·69

	Quantity	Unit	Rate Kroner	Total Kroner	Sterling equivalent £
			B/F	1,883,637·00	146,923·69
225 mm thick concrete 31 N/mm² trough suspended floor consisting of 165 mm (av.) thick × 175 mm deep ribs at 600 mm centres one way, between 600 mm wide × 175 mm deep special trough forms supported as necessary, with a 50 mm concrete topping, the whole reinforced with 8·15 kg of steel bar reinforcement per m² and 2·20 kg of fabric reinforcement per m², including all necessary formwork to soffit. OR: Suitable equivalent concrete floor spanning 4·80 m with a dead load of 4·57 N/mm² and a superimposed load of 4·60 N/mm² including all necessary reinforcement and formwork	745	m²	510·00	379,950·00	29,636·10
200 mm thick concrete 31 N/mm² hollow pot suspended floor consisting of 295 × 295 × 150 mm deep clay blocks at 400 mm centres with clay filler tiles between and a 50 mm concrete topping OR: Suitable equivalent concrete floor spanning 4·80 m with a dead load of 4·57 N/mm² and a superimposed load of 4·60 N/mm²	1,617	,,	117·50	189,997·50	14,819·81
Reinforced concrete 31 N/mm² in 200 mm suspended floor slab	85	,,	117·50	9,987·50	779·03
Mild steel bar reinforcement in suspended floors	13,900	kg	7·50	104,250·00	8,131·50
Mild steel fabric reinforcement 2·2 kg/m² in suspended slab	1,702	m²	19·00	32,338·00	2,522·36
Formwork to soffit of slab	1,450	,,	142·00	205,900·00	16,060·20
External rendering to underside of exposed slab and beams	1,060	,,	35·00	37,100·00	2,893·80
20 mm mastic asphalt roofing on felt on 50 mm thick lightweight channel reinforced composite decking spanning 4·80 m	675	,,	370·00	249,750·00	19,480·50
20 mm asphalt skirtings average 175 mm girth	215	m	70·00	15,050·00	1,173·90
4 mm blue black asbestos cement slates size 500 × 250 mm laid to slope on and including battens and 25 mm insulation quilt	460	m²	160·00	73,600·00	5,740·80
Ditto but fixed vertically on and including 20 mm blockboard	100	,,	380·00	38,000·00	2,964·00
15 mm blue black asbestos cement sheet, 300 mm wide fixed to softwood soffit	130	m	88·00	11,440·00	892·32
Aluminium flashings, 300 mm girth	1,350	,,	96·00	129,600·00	10,108·80
Treated sawn softwood in roof members, plates, etc.	25	m³	5,950·00	148,750·00	11,602·50
100 mm diameter PVC flat roof outlet with PVC pipe 12 m long including all fixings	7	No.	1,035·00	7,245·00	565·11
			C/F	3,516,595·00	274,294·42

DENMARK

	Quantity	Unit	Rate Kroner	Total Kroner	Sterling equivalent £
			B/F	3,516,595·00	274,294·42
Dog leg stairs and half landings rising three levels (concrete 6 m³, reinforcement 440 kg, formwork 50 m²)	2	No.	15,300·00	30,600·00	2,386·80
Mild steel balustrade with plastic handrail to six flights of stairs (handrail 17 m, standards 36 No.)	1	„	7,750·00	7,750·00	604·50
Mild steel balustrading with hardwood handrail to six flights of stairs, string to profile of treads and risers and wall string (handrail 19 m, standards 26 No.)	1	„	13,800·00	13,800·00	1,076·40
Granolithic finish 25 mm thick to treads and risers	30	m²	1,280·00	38,400·00	2,995·20
Vinyl tiles 3 mm thick ditto	22	„	345·00	7,590·00	592·02
Exposed aggregate pre-cast concrete panels 100 mm thick as wall cladding including all fixings and joint sealing	345	„	530·00	182,850·00	14,262·30
Ditto but coping panels 520 mm overall width	97	„	530·00	51,410·00	4,009·98
Ditto but parapet stones 620 mm wide × 470 mm deep including damp proof course	15	m	400·00	6,000·00	468·00
75 × 500 mm pre-cast concrete (26·00 N/mm²) cill or head finished fair on all exposed faces	52	„	240·00	12,480·00	973·44
Reinforced concrete 26·00 N/mm² in walls 150 mm thick (155 m²) and 240 mm thick (25 m²)	30	m³	560·00	16,800·00	1,310·40
Mild steel bar reinforcement in walls	2,000	kg	7·75	15,500·00	1,209·00
Formwork to walls	230	m²	142·00	32,660·00	2,547·48
Wrot formwork to walls	110	„	200·00	22,000·00	1,716·00
Brickwork, low quality bricks					
115 mm wall	175	„	205·00	35,875·00	2,798·25
240 mm wall	115	„	360·00	41,400·00	3,229·20
Brickwork, good quality facings 115 mm wall	560	„	280·00	156,800·00	12,230·40
Lightweight building blocks 100 mm wall	650	„	160·00	104,000·00	8,112·00
Hardwood two panel door, 762 × 1981 × 50 mm, with glass in top panel and louvres in bottom panel with and including hardwood frame 76 × 114 mm, including all fixing and ironmongery	5	No.	2,945·00	14,725·00	1,148·55
Hardwood window board 400 mm wide fixed to concrete cill	26	m	485·00	12,610·00	983·58
Plastic faced blockboard window lining 25 × 150 mm fixed to concrete	500	„	160·00	80,000·00	6,240·00
Black anodized aluminium windows, 50% horizontal sliding, 1700 × 1600 mm including all fixing glass and ironmongery	440	m²	1,775·00	781,000·00	60,918·00
Ditto but 12,250 × 675 mm	70	„	1,100·00	77,000·00	6,006·00
Black anodized aluminium doors and screens including all fixing, 6 mm obscured glass and ironmongery	50	„	2,160·00	108,000·00	8,424·00
Range of 6 No. prefabricated WC compartments 4·80 m long × 1·96 m high × 1·5 m deep consisting of 5 No. partitions and 6 No doors	4	No.	13,490·00	53,960·00	4,208·88
			C/F	5,419,805·00	422,744·80

DENMARK

	Quantity	Unit	Rate Kroner	Total Kroner	Sterling equivalent £
			B/F	5,419,805·00	422,744·80
100 × 150 mm pre-cast concrete (26·00 N/mm²) lintel reinforced with one 12 mm mild steel rod	44	m	56·25	2,475·00	193·05
Hardwood faced solid core flush door, 762 × 1981 × 40 mm, with and including softwood frame 100 × 35 mm including all fixing and ironmongery	28	No.	1,370·00	38,360·00	2,992·08
Pair of hardwood two panel doors, 512 × 1981 mm, one hour fire resisting, panels glazed fire resisting glass, with and including hardwood frame, 135 × 90 mm, including all fixing and ironmongery	11	,,	2,740·00	30,140·00	2,350·92
Treated wrot softwood fillet fixed to concrete	2,300	m	12·00	27,600·00	2,152·80
Two coat plaster, 12 mm thick to walls, columns and soffits	1,150	m²	66·00	75,900·00	5,920·20
Foil backed plasterboard fixed to softwood, with and including one coat plaster	560	,,	62·00	34,720·00	2,708·16
Two coats emulsion paint on walls and soffits	1,500	,,	35·00	52,500·00	4,095·00
Two coats oil paint ditto . . .	400	,,	60·00	24,000·00	1,872·00
Three coats oil paint on wood or metal . .	70	,,	74·00	5,180·00	404·04
Two coats polyurethane lacquer on aluminium or hardwood	270	,,	48·00	12,960·00	1,010·88
Insulation quilt 25 mm thick covered with cement and sand screed 48 mm thick reinforced with chicken wire . . .	750	,,	120·00	90,000·00	7,020·00
Cement and sand screed 48 mm thick . .	1,700	,,	79·00	134,300·00	10,475·40
Granolithic finish 50 mm thick to floors .	70	,,	265·00	18,550·00	1,446·90
Vinyl tiles 3 mm thick to floors . . .	2,300	,,	167·00	384,100·00	29,959·80
Brown quarry tiles 150 × 150 × 16 mm to floors	75	,,	420·00	31,500·00	2,457·00
OR: Terazzo tiles 300 × 300 × 25 mm . .	75	,,	—	—	—
Hardwood skirting 100 × 25 mm on and including softwood grounds fixed to wall .	820	m	52·00	42,640·00	3,325·92
Mineral board ceiling tiles 9·5 mm thick laid in and including aluminium tee section grid supported by metal hangers 400 mm long fixed to soffit, including all decoration . .	2,350	m²	204·00	479,400·00	37,393·20
Lavatory basin, coloured, 560 × 400 mm, including fittings but excluding associated pipework	24	No.	1,170·00	28,080·00	2,190·24
			C/F	6,932,210·00	540,712·39

DENMARK

	Quantity	Unit	Rate Kroner	Total Kroner	Sterling equivalent £
			B/F	6,932,210·00	540,712·39
Sink, white, 455 × 380 × 205 mm ditto	4	No.	725·00	2,900·00	226·20
WC suite, low level, coloured, excluding associated pipework	24	„	2,785·00	66,840·00	5,213·52
Towel dispensers	8	„			
Soap dispensers	24	„			
Toilet roll holders	24	„	}	17,500·00	1,365·00
Mirrors 1750 × 450 × 6 mm	8	„			
Disposal installation to ground level for sanitary fittings listed above including all builders work in connection				292,000·00	22,776·00
Hot and cold water installations for sanitary fittings listed above including all builders work in connection				254,000·00	19,812·00
Hot water and wall radiator heating system for offices (2400 m² of floor area) including all builders work in connection				1,556,000·00	121,368·00
OR:					
Low cost full air conditioning (6000 m³)				—	
Electric power and lighting installation including mains, transformers and all builders work in connection to the following:					
90 lighting points without fittings					
10 „ „ with fittings					
10 power points					
5 control points				125,000·00	9,750·00
Passenger lift, electric, eight persons, four stops, travelling through 9·50 m. Including all builders work in connection but excluding cost of shaft				194,000·00	15,132·00
Allow for 7 No. 30 m hosereels and all builders work in connection				14,000·00	1,092·00
Car park slab consisting of 175 mm bed of hardcore and 75 mm tarmacadam paving in two coats	880	m²	80·00	70,400·00	5,481·20
Total building cost				9,524,850·00	742,938·31
Add for site establishment, supervision and head office overheads where not included in above rates			7%	666,739·50	52,005·68
				10,191,589·50	794,943·99
Add for any other estimated expenditure				—	
				10,191,589·50	794,943·99
Add for V.A.T. or equivalent tax			22%	2,242,149·69	174,887·68
TOTAL ESTIMATED TENDER FIGURE				12,433,739·19	969,831·67

DENMARK

BUILDING COST INDEX
1955 = 100

Year	Annual average
1964	. . .	139
1965	. . .	154
1966	. . .	164
1967	. . .	175
1968	. . .	190
1969	. . .	205
1970	. . .	229
1971	. . .	250
1972	. . .	267
1973	. . .	300
1974	. . .	333
1975	. . .	358
1976	. . .	447
1977	. . .	485
1978	. . .	538
1979	. . .	618
1980	. . .	723

(First quarter)

N.B. The above index includes for the effect of Value Added Tax, introduced in 1967.

FRANCE

TENDERING PROCEDURES

The main systems of tendering in use are negotiated contracts (*de gré à gré*) a restricted form of competitive tendering for private development known as the *appel d'offres* and competitive tendering for government and other public work which can be either restricted (*adjudication restreinte*) or unrestricted (*adjudication*). Negotiated contracts are far more popular than they are in England, and even the *appel d'offres* contains an element of negotiation. In this latter system the architect frequently bargains with the tenderers after the receipt of tenders, setting one contractor off against another until a reasonably low price and a reasonably high standard of specification are obtained.

Negotiated contracts are also fairly common in the public sector: when competitive tenders are invited it is the rule for the lowest tender to be accepted; the tenderer can withdraw only within twenty-four hours of the tender being opened providing he pays the difference between his tender and the next lowest.

General contractors are becoming more common though they often undertake only the main structural work (*gros oeuvre*) and sub-let the remaining work to specialist sub-contractors. They also manage the project, co-ordinate the work of sub-contractors, order materials and supply all plant services etc., recovering the cost of these preliminary items from sub-contractors in proportion to their share of the total contract value. In those cases where no general contractor is employed the client enters into direct contracts with several specialist contractors, overall co-ordination becoming the responsibility of the Architect, Engineer or *Bureau d'étude* concerned.

Except in the Lille and Lyon regions, bills of quantities are not used as tender documents, tendering normally being based upon drawings and specification. With the *adjudication* system these are fully detailed documents but with the *appel d'offres* the drawings and specification only indicate the architectural solution and the quality required in general terms. Under the *appel d'offres* system the selected contractor often produces a large proportion of the working drawings himself to the approval of the architect; frequently concrete designs and drawings, and joinery details are prepared by contractors.

The standard form of contract, most commonly used, for private work is the *norme PO3·001* published by *l'Association Française de Normalisation* (similiar to the British Standards Institute). Unfortunately this document is not as widely used as the Standard Form in the United Kingdom and it is not uncommon for architects to draw up their own clauses. For public work, the major building ministries have their own forms.

Contract documents for each trade contractor usually comprise the following:

1. Drawings (*série de plans*).
2. Specification
    (*a*) For workmanship (*le dévis descriptif*).
    (*b*) For materials (*le cahier des prescriptions techniques particulières*).
    (*c*) National Code of Practice (*prescriptions techniques générales*).
3. Contract conditions, embodying
    (*a*) Standard Form of Contract (*cahiers des conditions et charges générales*).
    (*b*) Special conditions of contract (*cahier des charges particulières*).
4. Contract agreement (*la lettre de commande*).
5. The agreed progress chart (*le planning*).
6. A detailed break-down of the tender price to provide rates for the purpose of pricing variations. Occasionally this might take the form of a bill of quantities in respect of rates only.
7. Form of tender (*soumission*).

FRANCE

Standard published schedules of rates known as the *séries des prix* are often incorporated in the contract for the purpose of measuring variations. There are a number of priced schedules published for various purposes. The most widely used, including by Government Departments, is that known as *La Série Centrale* prepared by the Society of Architects in conjunction with building employers. A monthly supplement is issued with coefficients to bring the prices for each trade up to date, for all parts of France.

Although lump sum contracts (*marché à forfait*) are the most popular form of contract in France, re-measurement contracts are also used. In negotiated tenders the rates are often agreed between the parties, but in competitive tendering the tenderers quote percentages on or off (usually off) the *série des prix*.

All contracts contain a fluctuations clause and an agreed formula based upon price indices published by the Ministry of Construction is used to arrive at the additional payments due. The price indices are issued monthly and cover individual trades, thus enabling the formula to be applied to interim payments. Interim payments are normally monthly and subject to a 10% retention except in the case of very large contracts where a maximum retention sum may be fixed. Half of the retention fund is released on practical completion of the contract, the remainder being paid to the builder at the end of the six months maintenance period. However, both the contractor and the architect are jointly responsible for the condition of the building, excluding fair wear and tear, for a ten year period following its completion.

Variations can be authorized to the work described in the contract documents and their value is agreed if possible on contract rates on a lump sum basis before the work is put in hand, omissions being valued at 85% of the contract rates. The final account is prepared by the contractor and checked by the architect.

FRANCE

PRICE DATA

Wage rates and certain materials prices in France vary according to the region. The wage rates given are those applicable in the Paris region.

The prices for materials are ex factory, quarry, etc. and exclude haulage to site.

Unless otherwise stated all prices and rates exclude Value Added Tax. The Standard rate of Value Added Tax is 17·60% but certain goods can qualify for a maximum tax of 33⅓% whilst others may be subject to a reduced tax at 7%. In addition to Value Added Tax timber carries a special forestry tax amounting to 5·90% – similarly concrete goods are subject to an additional professional tax of 0·30%.

For the purposes of comparing French prices with those in the United Kingdom it has been assumed that 1 Franc = 10·4p approximately.

BASIC WAGE RATES (June 1980)

	Francs per hour	Sterling equivalent £
Labourer	11·82	1·23
Specialized Workers:		
First Grade	12·80	1·33
Second Grade	13·80	1·44
Third Grade	14·80	1·54
Qualified Workers:		
First Grade	15·76	1·64
Second Grade	16·75	1·74
Third Grade	18·00	1·87
Highly Qualified Workers	19·70	2·05
Foreman Grade 1	20·70	2·15
Foreman Grade 2	22·20	2·31

ADDITIONS TO BASIC WAGE RATES

Allowance is usually made to the basic rates for the following:

Sickness Insurance.
Old Age Insurance.
Accident Insurance.
Family Allowances.
Holidays with pay and holiday bonus.
Public holidays with pay and authorized absences.
Inclement weather insurance.
OPPBTP Contributions.
Trade Contributions.
Decennial Liability Insurance.
Apprenticeship Scheme Contribution.
May Day holiday.
Apprenticeship Tax.
Supplementary Benefits.
First Aid.
Professional Contribution (payable when number of employees exceeds 9).
Unemployment Insurance.

FRANCE

Workman's Supplementary Retirement Pension.
Provident Society Contributions.
Unemployment pay for less than 90 days of work resulting from sickness or accident.
Contribution to Fares (Applicable to large towns only).
Dwelling Allowance.
Redundancy Indemnity.
National fund of guaranteed wages.

The effect of the above is to add approximately 77% to the basic rate. In addition, a contractor tendering adds his own on-costs and profit.

MATERIALS PRICES (Second Quarter 1980)	Unit	Price Francs	Sterling equivalent £
Hardcore filling (broken stone/brick)	m³	42·00	4·37
Ready mixed concrete quality 31 N/mm²	,,	180·00	18·72
Portland Cement (type CPAL 325) in 10 tonne loads . .	tonne	173·00 (average)	17·99
Hydraulic Lime (types XHA and XHN 100) in 10 tonne loads	,,	140·00 (average)	14·56
Coarse Sand grade 0/25	m³	60·00	6·24
Fine Sand grade 0/6	,,	45·00	4·68
Gravel grade 5/25	,,	53·00	5·51
,, ,, 5/15	,,	54·00	5·62
,, ,, 15/25	,,	55·00	5·72
Stone grade 20/63	,,	44·00	4·58
,, ,, 20/40	,,	44·00	4·58
,, ,, 40/70	,,	44·00	4·58
Steel reinforcing rods 22 to 28 mm diameter . . .	tonne	2,400·00	249·60
Mild steel fabric reinforcement (weighing 3 kg/m²) . .	,,	2,800·00	291·20
Bricks and Blocks (clay)			
Partition blocks size:			
4 × 25 × 48 cm	1000	1,160·00	120·64
5 × 25 × 40 ,, ...	,,	1,040·00	108·16
5 × 25 × 48 ,, ...	,,	1,265·00	131·56
8 × 20 × 40 ,, ...	,,	1,300·00	135·20
10 × 20 × 40 ,, ...	,,	1,474·00	153·30
11 × 20 × 40 ,, ...	,,	1,575·00	163·80
Ordinary hollow blocks size:			
15 × 20 × 40 cm	,,	1,960·00	203·84
15 × 20 × 48 ,, ...	,,	2,375·00	247·00
20 × 20 × 40 ,, ...	,,	2,700·00	280·80
Ordinary solid bricks size 6 × 10·5 × 22 cm . . .	,,	464·00	48·26
Hollow concrete blocks size:			
5 × 20 × 50 cm	,,	1,576·00	163·90
10 × 20 × 50 ,, ...	,,	1,730·00	179·92
15 × 20 × 50 ,, ...	,,	2,685·00	279·24
Solid concrete blocks size:			
5 × 20 × 50 cm	,,	1,455·00	151·32
10 × 20 × 50 ,, ...	,,	2,330·00	242·32
15 × 20 × 50 ,, ...	,,	3,675·00	382·20
Clay Roof Tiles			
Large flat red tiles	,,	1,240·00	128·96
Roman tiles	,,	1,247·00	129·69
Flat antique style size 17 × 24 cm	,,	780·00	81·12

FRANCE

MATERIALS PRICES (Second Quarter 1980) – *continued*

	Unit	Price Francs	Sterling equivalent £
Softwood			
For carpentry	m³	625·00 to 670·00	65·00 to 69·68
Tongued and grooved boarding 24/27 mm thick first quality .	m²	40·00 to 42·00	4·16 to 4·37
Tongued and grooved boarding 15/16 mm thick first quality .	„	35·00 to 38·00	3·64 to 3·95
Oak for joinery	m³	2,900·00 to 4,100·00	301·60 to 426·40
20 mm blockboard	m²	65·00	6·76
50 mm wood wool slabs	„	35·00	3·64
Structural steel	kg	3·50	0·36
Plaster in bags	tonne	150·00	15·60
Sheet lead	100 kg	750·00	78·00
Vitreous enamelled steel bath 1·60 m long (excluding fittings) .	No.	1,050·00 (average)	109·20
WC Suite (excluding fitting)	„	235·00 (average)	24·44
Polished plate glass			
6 mm thick	m²	175·00	18·20
8 mm „	„	265·00	27·56
10 mm „	„	315·00	32·76
Paint			
Anti rust paint for metalwork	kg	12·00	1·25
Paint for building	„	12·00	1·25
Rigid PVC drain pipes 160 mm diameter	m	25·00	2·60

FRANCE

ESTIMATE OF COST FOR OFFICE TYPE BUILDING

The following estimate relates to a reinforced concrete framed four storey office block with a total floor area of 2400 m² and a car park of 500 m² forming part of the ground floor.

It has been assumed that the work would take about 16 months to complete and that access to the site in the Paris area is good.

	Quantity	Unit	Rates Francs	Total Francs	Sterling equivalent £
SUBSTRUCTURE					
Excavate					
to reduce levels	585	m³	9·50	5,557·50	577·98
to form trenches	413	,,	37·60	15,528·80	1,615·00
Extra for breaking out:					
brickwork	45	,,	60·00	2,700·00	280·80
concrete	45	,,	250·00	11,250·00	1,170·00
Remove surplus material from site	846	,,	35·00	29,610·00	3,079·44
Hardcore (stone/brick)					
infilling	84	,,	55·00	4,620·00	480·48
in bed 150 mm thick	220	m²	13·50	2,970·00	308·88
Plain concrete 8·00 N/mm²					
in foundations	220	m³	390·00	85,800·00	8,923·20
in bed 50 mm thick	603	m²	46·00	27,738·00	2,884·75
Reinforced concrete 26·00 N/mm²					
in foundations	190	m³	525·00	99,750·00	10,374·00
in bed 200 mm thick	265	m²	95·00	25,175·00	2,618·20
Mild steel bar reinforcement					
in foundations	16,730	kg	8·50	142,205·00	14,789·32
Mild steel fabric reinforcement					
3·00 kg/m² in bed	258	m²	22·00	5,676·00	590·30
Formwork					
to sides of foundations	270	,,	105·00	28,350·00	2,948·40
SUPERSTRUCTURE					
Reinforced concrete 31 N/mm²					
in beams	196	m³	440·00	86,240·00	8,968·96
in columns	50	,,	460·00	23,000·00	2,392·00
Mild steel bar reinforcement					
in beams and columns	81,850	kg	9·00	736,650·00	76,611·60
Formwork					
in beams	900	m²	185·00	166,500·00	17,316·00
in columns	550	,,	107·00	58,850·00	6,120·40
Wrot formwork					
in beams	400	,,	200·00	80,000·00	8,320·00
in columns	54	,,	117·00	6,318·00	657·07
Structural steelwork in roof beams etc. including all fabrication, placing in position, fixing and protective coat of paint	20,200	kg	6·00	121,200·00	12,604·80
			C/F	1,765,688·30	183,631·58

FRANCE

	Quantity	Unit	Rates Francs	Total Francs	Sterling equivalent £
			B/F	1,765,688·30	183,631·58
225 mm thick concrete 31 N/mm² trough suspended floor consisting of 165 mm (av.) thick × 175 mm deep ribs at 600 mm centres one way, between 600 mm wide × 175 mm deep special trough forms supported as necessary, with a 50 mm concrete topping, the whole reinforced with 8·15 kg of steel bar reinforcement per m² and 2·20 kg of fabric reinforcement per m², including all necessary formwork to soffit OR: Suitable equivalent concrete floor spanning 4·80 m with a dead load of 4·57 N/mm² and a superimposed load of 4·60 N/mm² including all necessary reinforcement and formwork	745	m²	270·00	201,150·00	20,919·60
200 mm thick concrete 31 N/mm² hollow pot suspended floor consisting of 295 × 295 × 150 mm deep clay blocks at 400 mm centres with clay filler tiles between and a 50 mm concrete topping OR: Suitable equivalent concrete floor spanning 4·80 m with a dead load of 4·57 N/mm² and a superimposed load of 4·60 N/mm²	1,617	,,	85·00	137,445·00	14,294·28
Reinforced concrete 31 N/mm² in 200 mm suspended floor slab	85	,,	94·00	7,990·00	830·96
Mild steel bar reinforcement in suspended floors	13,900	kg	10·00	139,000·00	14,456·00
Mild steel fabric reinforcement 2·2 kg/m² in suspended slab	1,702	m²	14·00	23,828·00	2,478·11
Formwork to soffit of slab	1,450	,,	110·00	159,500·00	16,588·00
External rendering to underside of exposed slab and beams	1,060	,,	85·00	90,100·00	9,370·40
20 mm mastic asphalt roofing on felt on 50 mm thick lightweight channel reinforced composite decking spanning 4·80 m	675	,,	230·00	155,250·00	16,146·00
20 mm asphalt skirtings average 175 mm girth	215	m	23·00	4,945·00	514·28
4 mm blue black asbestos cement slates size 500 × 250 mm laid to slope on and including battens and 25 mm insulation quilt	460	m²	145·00	66,700·00	6,936·80
Ditto but fixed vertically on and including 20 mm blockboard	100	,,	210·00	21,000·00	2,184·00
15 mm blue black asbestos cement sheet, 300 mm wide fixed to softwood soffit	130	m	70·00	9,100·00	946·40
Aluminium flashings	1,350	,,	170·00	229,500·00	23,868·00
Treated sawn softwood in roof members, plates etc.	25	m³	3,500·00	87,500·00	9,100·00
100 mm diameter PVC flat roof outlet with PVC pipe 12 m long including all fixings	7	No.	720·00	5,040·00	524·16
			C/F	3,103,736·30	322,788·57

FRANCE

	Quantity	Unit	Rates Francs	Total Francs	Sterling equivalent £
			B/F 3,103,736·30		322,788·57
Dog leg stairs and half landing rising three levels (concrete 6 m³, reinforcement 440 kg, formwork 50 m²)	2	No.	19,800·00	39,600·00	4,118·40
Mild steel balustrade with plastic handrail to six flights of stairs (handrail 17 m, standards 36 No.)	1	,,	5,290·00	5,290·00	550·16
Mild steel balustrading with hardwood handrail to six flights of stairs, string to profile of treads and risers and wall string (handrail 19 m, standards 26 No.) . . .	1	,,	10,400·00	10,400·00	1,081·60
Granolithic finish 25 mm thick to treads and risers	30	m²	325·00	9,750·00	1,014·00
Vinyl tiles 3 mm thick ditto . . .	22	,,	165·00	3,630·00	377·52
Exposed aggregate pre-cast concrete panels 100 mm thick as wall cladding including all fixings and joint sealing	345	,,	875·00	301,875·00	31,395·00
Ditto but coping panels 520 mm overall width	97	,,	990·00	96,030·00	9,987·12
Ditto but parapet stones 620 mm wide × 470 mm deep including damp proof course . .	15	m	320·00	4,800·00	499·20
75 × 500 mm pre-cast concrete (26·0 N/mm²) cill or head finished fair on all exposed faces .	52	,,	240·00	12,480·00	1,297·92
Reinforced concrete 26·00 N/mm² in walls 150 mm thick (155 m²) and 240 mm thick (25 m²)	30	m³	445·00	13,350·00	1,388·40
Mild steel bar reinforcement in walls . .	2,000	kg	10·00	20,000·00	2,080·00
Formwork to walls	230	m²	73·00	16,790·00	1,746·16
Wrot formwork to walls	110	,,	73·00	8,030·00	835·12
Brickwork, low quality bricks 115 mm wall	175	,,	150·00	26,250·00	2,730·00
240 mm wall	115	,,	256·00	29,440·00	3,061·76
Brickwork, good quality facings 115 mm wall	560	,,	217·00	121,520·00	12,638·08
Lightweight building blocks 100 mm wall	650	,,	72·00	46,800·00	4,867·20
Hardwood two panel door, 762 × 1981 × 50 mm, with glass in top panel and louvres in panel with and including hardwood frame 76 × 114 mm, including all fixing and ironmongery	5	No.	955·00	4,775·00	496·60
Hardwood window board 400 mm wide fixed to concrete cill	26	m	105·00	2,730·00	283·92
Plastic faced blockboard window lining 25 × 150 mm fixed to concrete . . .	500	,,	74·00	37,000·00	3,848·00
Black anodized aluminium windows, 50% horizontal sliding, 1700 × 1600 mm including all fixing glass and ironmongery . . .	440	m²	980·00	431,200·00	44,844·80
Ditto but 12,250 × 675 mm . . .	70	,,	600·00	42,000·00	4,368·00
Black anodized aluminium doors and screens including all fixing, 6 mm obscured glass and ironmongery	50	,,	1,575·00	78,750·00	8,190·00
			C/F	4,466,226·30	464,487·53

FRANCE

	Quantity	Unit	Rates Francs	Total Francs	Sterling equivalent £
			B/F	4,466,226·30	464,487·53
Range of 6 No. prefabricated WC compartments 4·80 m long × 1·96 m high × 1·5 m deep consisting of 5 No. partitions and 6 No. doors	4	No.	27,700·00	110,800·00	11,523·20
100 × 150 mm pre-cast concrete (26·00 N/mm²) lintel reinforced with one 12 mm mild steel rod	44	m	63·00	2,772·00	288·29
Hardwood faced solid core flush door, 762 × 1981 × 40 mm, with and including softwood frame 100 × 35 mm including all fixing and ironmongery	28	No.	870·00	24,360·00	2,533·44
Pair of hardwood two panel doors, 512 × 1981 mm, one hour fire resisting, panels glazed fire resisting glass, with and including hardwood frame, 135 × 90 mm, including all fixing and ironmongery	11	,,	1,575·00	17,325·00	1,801·80
Treated wrot softwood fillet fixed to concrete	2,300	m	24·50	56,350·00	5,860·40
Two coat plaster, 12 mm thick to walls, columns and soffits	1,150	m²	33·00	37,950·00	3,946·80
Foil backed plasterboard fixed to softwood, with and including one coat plaster	560	,,	71·00	39,760·00	4,135·04
Two coats emulsion paint on walls and soffits	1,500	,,	18·00	27,000·00	2,808·00
Two coats oil paint ditto	400	,,	26·50	10,600·00	1,102·40
Three coats oil paint on wood or metal	70	,,	30·50	2,135·00	222·04
Two coats polyurethane lacquer on aluminium or hardwood	270	,,	24·50	6,615·00	687·96
Insulation quilt 25 mm thick covered with cement and sand screed 48 mm thick reinforced with chicken wire	750	,,	74·00	55,500·00	5,772·00
Cement and sand screed 48 mm thick	1,700	,,	37·00	62,900·00	6,541·60
Granolithic finish 50 mm thick to floors	70	,,	165·00	11,550·00	1,201·20
Vinyl tiles 3 mm thick to floors	2,300	,,	45·00	103,500·00	10,764·00
Brown quarry tiles 150 × 150 × 16 mm to floors	75	,,	235·00	17,265·00	1,833·00
OR:					
Terrazzo tiles 300 × 300 × 25 mm	75	,,	145·00	—	—
Hardwood skirting 100 × 25 mm on and including softwood grounds fixed to wall	820	m	27·00	22,140·00	2,302·56
Mineral board ceiling tiles 9·5 mm thick laid in and including aluminium tee section grid supported by metal hangers 400 mm long fixed to soffit, including all decorations	2,350	m²	95·00	223,250·00	23,218·00
Lavatory basin, coloured, 560 × 400 mm, including fittings but excluding associated pipework	24	No.	560·00	13,440·00	1,397·76
Sink, white, 455 × 380 × 205 mm ditto	4	,,	370·00	1,480·00	153·92
WC suite, low level, coloured, excluding associated pipework	24	,,	570·00	13,680·00	1,422·72
Towel dispensers	8	,,	}		
Soap dispensers	24	,,	}	7,900·00	821·60
Toilet roll holders	24	,,	}		
Mirrors 1750 × 450 × 6 mm	8	,,	}		
			C/F	5,334,858·30	554,825·26

FRANCE

	Quantity	Unit	Rates Francs	Total Francs	Sterling equivalent £
			B/F	5,334,858·30	554,825·26
Disposal installations to ground level for sanitary fittings listed above including all builders work in connection				82,000·00	8,528·00
Hot and cold water installation for sanitary fittings listed above including all builders work in connection					
Hot water and wall radiator heating system for offices (2400 m² of floor area) including all builders work in connection . . .				350,000·00	36,400·00
OR:					
Low cost full air conditioning (6000 m³) .				(1,424,000·00)	(148,096·00)
Electric power and lighting installation including mains, transformers and all builders work in connection to the following:					
90 lighting points without fittings					
10 ,, ,, with fittings					
50 power points					
5 control points				442,000·00	45,968·00
Passenger lift, electric, eight persons, four stops, travelling through 9·50 m. Including all builders work in connection but excluding cost of shaft				111,000·00	11,544·00
Allow for 7 No. 30 m hose reels and all builders work in connection . . .				21,900·00	2,277·60
Car park slab consisting of 175 mm bed of hardcore and 75 mm tarmacadam paving in two coats.	880	m²	82·00	72,160·00	7,504·64
Total building cost				6,413,918·30	667,047·50
Add for site establishment, supervision and head office overheads where not included in above rates . .				—	—
				6,413,918·30	667,047·50
Add for any other estimated expenditure .				—	—
				6,413,918·30	667,047·50
Add for V.A.T. or equivalent tax . .			17·60%	1,128,849·62	117,400·36
TOTAL ESTIMATED TENDER FIGURE				7,542,767·92	784,447·86

FRANCE

BUILDING COSTS INDEX

1965 = 100

Year						Annual average
1965	.	.	.	.	.	100
1966	.	.	.	.	.	104
1967	.	.	.	.	.	105
1968	.	.	.	.	.	110
1969	.	.	.	.	.	117
1970	.	.	.	.	.	120
1971	.	.	.	.	.	130
1972	.	.	.	.	.	138
1973	.	.	.	.	.	160
1974	.	.	.	.	.	193
1975	.	.	.	.	.	217
1976	.	.	.	.	.	246
1977	.	.	.	.	.	272
1978	.	.	.	.	.	293
1979	.	.	.	.	.	325

N.B. The above index includes for the effect of Value Added Tax, introduced in 1968.

WEST GERMANY

TENDERING PROCEDURES

The Federal Republic has published rules covering contract procedure for all building work, the '*Verdingungsordnung für Bauleistungen*' '*V.O.B.*' (Regulations for Building Contracts), and this document is the main compendium of regulations and a good practice guide for the industry, the basis of many public work contracts, and though considerably amended, often used for private work as well.

The '*V.O.B.*' also contains standard specification clauses and recommends that a bill (schedule) of quantities (*Leistungsverzeichnis*) is prepared. This bill or schedule is usually based on approximate quantities and is prepared either by an employee of the architect or by the engineer. Part C of the '*V.O.B.*' gives broad guidelines for the measurement of quantities and ensures some uniformity of tender documents. Items of work are often loosely collected together for lump sum pricing or may be individually described, giving the location and requiring the contractor to give details of wage costs, material costs, plant and overheads, etc. for each item. This latter format encourages tendering contractors to balance their costs within each item and so arrive at lower overheads.

The specification together with a bill of approximate quantities (if prepared) all necessary drawings, the '*V.O.B.*' and any other special conditions are collected together and form the tender documents; these, supplemented by an exchange of letters placing and accepting the contract, also become the contract documents. The conditions of contract may be as '*V.O.B.*' Part B, but these are frequently supplemented with amended or additional clauses to suit the particular circumstances of each scheme.

For Government schemes, the majority of contracts are placed as a result of public advertisement, but for private schemes it is more usual to invite tenders from a limited number of contractors, followed up by further negotiation with the lowest tenderer.

The type of contract most commonly entered into is a lump sum contract based on a specification and bills of approximate 'unit' quantities (subject to remeasurement on completion); though in recent years, the growing advent of general contracting and the subsequent introduction of 'package deals' is gaining popularity.

However, main contractors remain in the minority and the majority of work is still let directly to individual trade contractors. In these cases, site co-ordination remains the responsibility of the Architect, though the introduction of small independent project management firms is becoming more frequent.

For large contracts, it is quite common for several contractors to form a joint company to tender for and construct individual projects. About one-third of all construction work in West Germany is undertaken as a 'joint-venture'. Many clients favour the system as it helps to share the financial responsibility; if one member of the venture goes into liquidation the other members will usually support the job through to completion.

Most contracts are 'firm' price though alterations to rates are usually permitted when there are quantity adjustments in excess of ten per cent. Interim payments are usually paid to a contractor upon receipt of a detailed claim from him subject to a minimum value payment laid down by the client.

Authorized variations are permitted providing they do not change the scope of the work; the contractor prepares details of the cost of the alteration for agreement with the architect based on rates in the contract documents, as far as possible but new rates are agreed when necessary. It is normal for the cost of variations to be agreed before the work is undertaken on site.

Most contracts make provision for retention, usually 10% of the contract sum, which is released upon agreement of the final account. Each contractor, however, is liable for the

WEST GERMANY

maintenance of his part of the work for anything between two and five years after its completion.

It is the duty of the contractor to prepare the final account which in the case of Government work is subject to audit by the Federal Auditor which has to be carried out within five years and the account may be re-opened within this time.

WEST GERMANY

PRICE DATA

Wage rates and material prices vary in the Federal Republic not only from one federal *Land* to another, but also from town to town. The following information is based on prices, etc. current at the time stated in North Rhine-Westphalia and should only be used as a guide to those applicable in the other states.

Unless otherwise stated all prices and rates exclude Value Added Tax, for which an allowance of 13% is currently added as a separate item to tenders.

For the purposes of comparing German prices with those in the United Kingdom it has been assumed that 1 Deutsch Mark = 24·4p approximately.

BASIC WAGE RATES (First Quarter 1980)

	Deutsch Marks per hour	*Sterling equivalent* £
Bricklayer foreman	14·70	3·59
Bricklayer	13·70	3·34
Reinforcement fixer	13·08	3·19
Bricklayers' labourer	11·28	2·75
General labourer	11·20	2·73

ADDITIONS TO BASIC WAGE RATES

Overtime and piece work can add between 30% and 70% to basic wage rates. Allowance to the basic rates must also be made for the following:

Unemployment Assistance and Insurance.
Health Insurance.
Old Age Pensions Insurance.
Industrial Injuries Insurance.
Seriously Injured Persons Insurance.
Family Compensation.
Sick Pay Scheme.
Inclement Weather Insurance.
Building Compensation on Tariff Wages.
Annual Holidays (normally three weeks).
Public Holidays (eleven days).

The effect of the above is to add approximately 80% to the basic rate. In addition, a contractor tendering would add his own on-costs, e.g. subsistence, plus rates, profit, etc.

WEST GERMANY

MATERIALS PRICES (First Quarter 1980)

	Unit	Price Deutsch Marks	Sterling equivalent £
Hardcore (broken stone/brick)	m³	18·00	4·39
Cement	50 kg bag	8·75	2·14
Coarse Aggregate ⎤ for concrete	m³	32·00	7·81
Fine Aggregate ⎦	,,	32·00	7·81
Sand for plastering	,,	44·00	10·74
Steel reinforcing rods (over 10 mm)	tonne	1,215·00	296·46
Mild steel fabric reinforcement weighing 3 kg/m²	,,	1,265·00	308·66
Common bricks (KSL silicate)	1000	530·00	129·32
Facing bricks	m²	50·00	12·20
100 mm lightweight building blocks 49 × 24 cm	1000	1910·00	466·04
Felt damp proof courses	m²	2·75	0·67
Clay roofing tiles	1000	970·00	236·68
Interlocking clay roofing tiles	,,	1,130·00	275·72
Softwood	m³	615·00	150·06
Hardwood – oak	,,	2,460·00	600·24
– beech	,,	980·00	239·12
Structural steel (standard sections)	100 kg	370·00	90·28
Gypsum plaster	50 kg bag	10·50	2·56
Copper pipe (15 mm diameter)	m	2·80	0·68
Galvanized piping (15 mm diameter)	,,	4·00	0·98
Clear sheet glass (varies according to thickness and size of pane)	m²	18·50 (average)	4·51
Emulsion paint	kg	6·00	1·46
Oil Paint (varies)	,,	10·75	2·62
Stoneware drain pipes (100 mm diameter)	m	10·75	2·62
Cast Iron drain pipes (100 mm diameter)	,,	18·50	4·51

WEST GERMANY

ESTIMATE OF COST FOR OFFICE TYPE BUILDING

The following estimate relates to a reinforced concrete framed four storey office block with a total floor area of 2400 m² and a car park of 500 m² forming part of the ground floor.

It has been assumed that the work would take about 15 months to complete and that access to the site in the Dusseldorf area is good.

	Quantity	Unit	Rates Deutsch Marks	Total Deutsch Marks	Sterling equivalent £
SUBSTRUCTURE					
Excavate					
to reduce levels	585	m³	5·00	2,925·00	713·70
to form trenches	413	,,	45·00	18,585·00	4,534·74
Extra for breaking out:					
brickwork or concrete	90	,,	124·00	11,160·00	2,723·04
Remove surplus material from site (ne 5 kms)	846	,,	14·50	12,267·00	2,993·15
Hardcore (stone/brick)					
infilling	84	,,	22·00	1,848·00	450·91
in bed 150 mm thick	220	m²	7·25	1,595·00	389·18
Plain concrete 8·00 N/mm²					
in foundations	220	m³	157·00	34,540·00	8,427·76
in bed 50 mm thick	603	m²	9·10	5,487·30	1,338·90
Reinforced concrete 26·00 N/mm²					
in foundations	190	m³	170·00	32,300·00	7,881·20
in bed 200 mm thick	265	m²	28·00	7,420·00	1,810·48
Mild steel bar reinforcement					
in foundations	16,730	kg	2·45	40,988·50	10,001·19
Mild steel fabric reinforcement					
3·00 kg/m² in bed	258	m²	7·30	1,883·40	459·55
Formwork					
to sides of foundations	270	,,	38·00	10,260·00	2,503·44
SUPERSTRUCTURE					
Reinforced concrete 31 N/mm²					
in beams	196	m³	182·00	35,672·00	8,703·97
in columns	50	,,	193·50	9,675·00	2,360·70
Mild steel bar reinforcement					
in beams and columns	81,850	kg	2·50	204,625·00	49,928·50
Formwork					
in beams	900	m²	61·00	54,900·00	13,395·60
in columns	550	,,	73·00	40,150·00	9,796·60
Wrot formwork					
in beams	400	,,	73·00	29,200·00	7,124·80
in columns	54	,,	80·00	4,320·00	1,054·08
Structural steelwork in roof beams etc. including all fabrication, placing in position, fixing and protective coat of paint	20,200	kg	5·75	116,150·00	28,340·60
			C/F	675,951·20	164,932·09

WEST GERMANY

	Quantity	Unit	Rates Deutsch Marks	Total Deutsch Marks	Sterling equivalent £
			B/F	675,951·20	164,932·09
225 mm thick concrete 31 N/mm² trough suspended floor consisting of 165 mm (av.) thick × 175 mm deep ribs at 600 mm centres one way, between 600 mm wide × 175 mm deep special trough forms supported as necessary, with a 50 mm concrete topping, the whole reinforced with 8·15 kg of steel bar reinforcement per m² and 2·20 kg of fabric reinforcement per m², including all necessary formwork to soffit OR:					
Suitable equivalent concrete floor spanning 4·80 m with a dead load of 4·57 N/mm² and a superimposed load of 4·60 N/mm² including all necessary reinforcement and formwork	745	m²	146·00	108,770·00	26,539·88
200 mm thick concrete 31 N/mm² hollow pot suspended floor consisting of 295 × 295 × 150 mm deep clay blocks at 400 mm centres with clay filler tiles between and a 50 mm concrete topping OR:					
Suitable equivalent concrete floor spanning 4·80 m with a dead load of 4·57 N/mm² and a superimposed load of 4·60 N/mm²	1,617	,,	43·00	69,531·00	16,965·56
Reinforced concrete 31 N/mm² in 200 mm suspended floor slab	85	,,	42·00	3,570·00	871·08
Mild steel bar reinforcement in suspended floors	13,900	kg	2·50	34,750·00	8,479·00
Mild steel fabric reinforcement 2·2 kg/m² in suspended slab	1,702	m²	5·35	9,105·70	2,221·79
Formwork to soffit of slab	1,450	,,	46·00	66,700·00	16,274·80
External rendering to underside of exposed slab and beams	1,060	,,	51·00	54,060·00	13,190·64
20 mm mastic asphalt roofing on felt on 50 mm thick lightweight channel reinforced composite decking spanning 4·80 m	675	,,	177·00	119,475·00	29,151·90
20 mm asphalt skirtings average 175 mm girth	215	m	48·00	10,320·00	2,518·08
4 mm blue/black asbestos cement slates size 500 × 250 mm laid to slope on and including battens and 25 mm insulation quilt	460	m²	157·00	72,220·00	17,621·68
Ditto but fixed vertically on and including 20 mm blockboard	100	,,	140·00	14,000·00	3,416·00
15 mm blue/black asbestos cement sheet, 300 mm wide fixed to softwood soffit	130	m	133·00	17,290·00	4,218·76
Aluminium flashings	1,350	,,	67·50	91,125·00	22,234·50
Treated sawn softwood in roof members, plates etc.	25	m³	2,610·00	65,250·00	15,921·00
100 mm diameter PVC flat roof outlet with PVC pipe 12 m long including all fixings	7	No.	360·00	2,520·00	614·88
Dog leg stairs and half landing rising three levels (concrete 6 m³, reinforcement 440 kg, formwork 50 m²)	2	,,	14,340·00	28,680·00	6,997·92
Mild steel balustrade with plastic handrail to six flights of stairs (handrail 17 m, standards 36 No.)	1	,,	4,250·00	4,250·00	1,037·00
			C/F	1,447,567·90	353,206·56

WEST GERMANY

	Quantity	Unit	Rates Deutsch Marks	Total Deutsch Marks	Sterling equivalent £
		B/F		1,447,567·90	353,206·56
Mild steel balustrading with hardwood handrail to six flights of stairs, string to profiles of treads and risers and wall string (handrail 19 m, standards 26 No.)	1	No.	6,440·00	6,440·00	1,571·36
Granolithic finish 25 mm thick to treads and risers	30	m²	236·00	7,080·00	1,727·52
Vinyl tiles 3 mm thick ditto . . .	22	,,	50·00	1,100·00	268·40
Exposed aggregate pre-cast concrete panels 100 mm thick on wall cladding including all fixings and joint sealing	345	,,	160·00	55,200·00	13,468·80
Ditto but coping panels 520 mm overall width	97	,,	190·00	18,430·00	4,496·92
Ditto but parapet stones 620 mm wide × 470 mm deep including damp proof course .	15	m	152·00	2,280·00	556·32
75 × 500 mm pre-cast concrete (26·00 N/mm²) cill or head finished fair on all exposed faces	52	,,	105·00	5,460·00	1,332·24
Reinforced concrete 26·00 N/mm² in walls 150 mm thick (155 m²) and 240 mm thick (25 m²)	30	m³	170·00	5,100·00	1,244·40
Mild steel bar reinforcement in walls	2,000	kg	2·50	5,000·00	1,220·00
Formwork to walls	230	m²	41·60	9,568·00	2,334·59
Wrot formwork to walls	110	,,	55·00	6,050·00	1,476·20
Brickwork, low quality bricks 115 mm wall	175	,,	62·00	10,850·00	2,647·40
240 mm wall	115	,,	105·00	12,075·00	2·946·30
Brickwork, good quality facings 115 mm wall	560	,,	135·00	75,600·00	18,446·40
Lightweight building blocks 100 mm wall	650	,,	53·00	34,450·00	8,405·80
Hardwood two panel door, 762 × 1981 × 50 mm, with glass in top panel and louvres in bottom panel with and including hardwood frame 76 × 114 mm, including all fixings and ironmongery	5	No.	675·00	3,375·00	823·50
Hardwood window board 400 mm wide fixed to concrete cill	26	m	60·00	1,560·00	380·64
Plastic faced blockboard window lining 25 × 150 mm fixed to concrete . . .	500	,,	35·00	17,500·00	4,270·00
Black anodized aluminium windows, 50% horizontal sliding, 1700 × 1600 mm, including all fixing glass and ironmongery . .	440	m²	934·00	410,960·00	100,274·24
Ditto but 12,250 × 675 mm . . .	70	,,	975·00	68,250·00	16,653·00
Black anodized aluminium doors and screens including all fixing, 6 mm obscured glass and ironmongery	50	,,	945·00	47,250·00	11,529·00
Range of 6 No. prefabricated WC compartments 4·80 m long × 1·96 m high × 1·5 m deep consisting of 5 No. partitions and 6 No. doors	4	No.	4,570·00	18,280·00	4,460·32
		C/F		2,269,425·90	553,739·91

WEST GERMANY

	Quantity	Unit	Rates Deutsch Marks	Total Deutsch Marks	Sterling equivalent £
			B/F	2,269,425·90	553,739·91
100×150 mm pre-cast concrete ($26·00$ N/mm²) lintel reinforced with one 12 mm mild steel rod	44	m	43·15	1,898·60	463·26
Hardwood faced solid core flush door, $762 \times 1981 \times 40$ mm, with and including softwood frame 100×35 mm including all fixing and ironmongery	28	No.	580·00	16,240·00	3,962·56
Pair of hardwood two panel doors, 512×1981 mm, one hour fire resisting panels glazed fire resisting glass, with and including hardwood frame, 135×90 mm, including all fixing and ironmongery	11	,,	970·00	10,670·00	2,603·48
Treated wrot softwood fillet fixed to concrete	2,300	m	30·00	69,000·00	16,836·00
Two coat plaster, 12 mm thick to walls, columns and soffits	1,150	m²	23·00	26,450·00	6,453·80
Foil backed plasterboard fixed to softwood, with and including one coat plaster .	560	,,	70·00	39,200·00	9,564·80
Two coats emulsion paint on walls and soffits	1,500	,,	8·50	12,750·00	3,111·00
Two coats oil paint ditto	400	,,	13·40	5,360·00	1,307·84
Three coats oil paint on wood or metal .	70	,,	26·00	1,820·00	444·08
Two coats polyurethane lacquer on aluminium or hardwood	270	,,	16·50	4,455·00	1,087·02
Insulation quilt 25 mm thick covered with cement and sand screed 48 mm thick reinforced with chicken wire	750	,,	25·00	18,750·00	4,575·00
Cement and sand screed 48 mm thick . .	1,700	,,	14·75	25,075·00	6,118·30
Granolithic finish 50 mm thick to floors .	70	,,	91·00	6,370·00	1,554·28
Vinyl tiles 3 mm thick to floors . . .	2,300	,,	31·50	72,450·00	17,677·80
Brown quarry tiles $150 \times 150 \times 16$ mm to floors . . . . OR:	75	,,	87·50	6,562·50	1,601·25
Terrazzo tiles $300 \times 300 \times 25$ mm . .	75	,,	84·00	—	—
Hardwood skirting 100×25 mm on and including softwood grounds fixed to wall	820	m	9·00	7,380·00	1,800·72
Mineral board ceiling tiles 9·5 mm thick laid in and including aluminium tee section grid supported by metal hangers 400 mm long fixed to soffit, including all decoration . .	2,350	m²	57·00	133,950·00	32,683·80
Lavatory basin, coloured, 560×400 mm, including fittings but excluding associated pipework	24	No.	610·00	14,640·00	3,572·16
Sink, white, $455 \times 380 \times 205$ mm ditto .	4	,,	365·00	1,460·00	356·24
WC suite, low level, coloured, excluding associated pipework	24	,,	650·00	15,600·00	3,806·40
Towel dispensers	8	,,			
Soap dispensers	24	,,		5,850·00	1,427·40
Toilet roll holders	24	,,			
Mirrors $1750 \times 450 \times 6$ mm . . .	8	,,			
Disposal installations to ground level for sanitary fittings listed above including all builders work in connection				21,500·00	5,246·00
			C/F	2,786,857·00	679,993·10

WEST GERMANY

	Quantity	Unit	Rates Deutsch Marks	Total Deutsch Marks	Sterling equivalent £
			B/F	2,786,857·00	679,993·10
Hot and cold water installation for sanitary fittings listed above including all builders work in connection				27,900·00	6,807,60
Hot water and wall radiator heating system for offices (2400 m² of floor area) including all builders work in connection . . .				102,000·00	24,888·00
OR:					
Low cost full air conditioning (6000 m³) .				(365,000·00)	(89,060·00)
Electric power and lighting installation including mains, transformers and all builders work in connection to the following: 90 lighting points without fittings 10 ,, ,, with fittings 50 power points 5 control points				48,000·00	11,712·00
Passenger lift, electric, eight persons, four stops, travelling through 9·50 m. Including all builders work in connection but excluding cost of shaft				67,000·00	16,348·00
Allow for 7 No. 30 m hose reels and all builders work in connection				14,000·00	3,416·00
Car park slab consisting of 175 mm bed of hardcore and 75 mm tarmacadam paving in two coats	880	m²	45·00	39,600·00	9,662·40
Total building cost				3,085,357·00	752,827·10
Add for site establishment, supervision and head office overheads where not included in above rates			10%	308,535·70	75,282·71
				3,393,892·70	828,109·81
Add for any other estimated expenditure .				—	—
				3,393,892·70	828,109·81
Add for V.A.T. or equivalent tax . .			13%	441,206·05	107,654·28
TOTAL ESTIMATED TENDER FIGURE				3,835,098·75	935,764·09

WEST GERMANY

AVERAGE BUILDING PRICES PER SQUARE METRE

(First Quarter 1980)

	Price Deutsch Marks	Approximate Sterling equivalent £
Offices for letting including lifts and central heating	1,560·00	380·64
Offices for owner occupation, including lifts and air-conditioning . .	1,950·00	475·80
Departmental stores	1,220·00	297·68
Flats, four storeys		
State assisted schemes	1,220·00	297·68
Private	1,220·00	297·68
Private houses	1,285·00	313·54
Hotels	1,770·00	431·88

N.B The above prices include Value Added Tax.

BUILDING COST INDEX

Second Quarter 1962 = 100

Year	First quarter	Second quarter	Third quarter	Fourth quarter	Annual average
1962 . . .	—	100	101	102	101 (for three quarters only)
1963 . . .	103	105	106	106	105
1964 . . .	107	110	111	112	110
1965 . . .	112	115	116	116	115
1966 . . .	117	119	119	119	119
1967 . . .	117	116	116	115	116
1968 . . .	120	120	122	122	121
1969 . . .	123	126	128	133	128
1970 . . .	141	147	150	152	148
1971 . . .	156	163	165	165	162
1972 . . .	169	174	175	177	174
1973 . . .	180	188	190	190	187
1974 . . .	194	202	204	203	201
1975 . . .	204	207	207	207	206
1976 . . .	209	214	216	217	214
1977 . . .	219	225	227	227	225
1978 . . .	234	243	245	245	242
1979 . . .	255	264	270	279	267

N.B. The above index reflects the effect of Value Added Tax, introduced in 1968. The figures are derived from statistics compiled by the Statistisches Bundesamt, Wiesbaden.

REPUBLIC OF IRELAND

TENDERING PROCEDURES

Tendering and contract procedures in the Republic of Ireland are very similar to those in the United Kingdom. The Royal Institute of the Architects of Ireland publish Forms of Contract both with and without Quantities which are generally used throughout the country; in addition to these, Government Departments have their own Forms of Contract and occasionally the Standard Form of Contract published in the United Kingdom is used.

PRICE DATA

Wage rates and certain materials prices in the Republic of Ireland vary slightly according to region. The wage rates given are those applicable in the Dublin area. The prices for materials are approximate and include delivery to site in the Dublin area in full loads.

Unless otherwise stated all prices and rates exclude Value Added Tax. In relation to building contracts, Value Added Tax is charged at 10% on 30% of the value of the contract. In practice, an effective rate of 3% is applied to the value of the contract. Construction, repair maintenance and improvement of roads, harbours and sewage works by the State, Local Authorities or Harbour Authorities is zero rated.

For the purpose of comparing Irish prices with those in the United Kingdom it has been assumed that 1 Irish Punt = 91·0p approximately.

BASIC WAGE RATES (June 1980)

	Punt per hour	£ per hour
Tradesmen	2·09	1·90
Labourers	1·90	1·73

ADDITIONS TO BASIC WAGE RATES

Allowance is usually made to the basic rates for the following:

Holidays.
Social Services (Including Occupational Injuries Act).
'Wet-Time'.
Employer's Liability Insurance.
Public Liability Insurance.
Cash in Transit.
Clerical Work.
Pension Scheme.
Guaranteed Week.
Redundancy Scheme.
Sick Pay Scheme.
ANCO Levy/Grant.
Pay Related Benefit Scheme.

The effect of the above is to add approximately 41% to the basic rate. In addition a Contractor tendering adds his own on-costs and profit.

REPUBLIC OF IRELAND

MATERIALS PRICES (Second Quarter 1980)	Unit	Price Punt	Sterling Equivalent £
All prices delivered Dublin area in full loads			
Hardcore (broken stone/brick)	m³	5·60	5·10
Ready mixed concrete quality 31 N/mm²	,,	30·00	27·30
Portland cement in 50 kg bags, under 10 bags	Bag	2·75	2·50
Ditto over 10 bags	,,	2·35	2·14
Portland cement in bulk	tonne	40·00	36·40
Aggregates for concrete	,,	5·60	5·10
Mild steel reinforcing rods (10 mm)	,,	295·00	268·45
High tensile steel ditto	,,	300·00	273·00
Clay bricks size 225 × 112 × 75 mm			
Machine made facings	1000	135·00	122·85
Concrete blocks size 450 × 225 × 112 mm			
Hollow	,,	274·00	249·34
Solid	,,	310·00	282·10
Felt damp proof courses	m²	1·10	1·00
Concrete interlocking roof tiles size 418 × 330 mm	1000	260·00	236·60
Imported hardwoods			
Afrormosia (25 mm to 100 mm)	m³	740·00	673·40
African mahogany (ditto)	,,	480·00	436·80
Teak (ditto)	,,	1,400·00	1,274·00
Iroko	,,	380·00	345·80
Imported softwoods			
White deal 115 × 35 mm	,,	235·00	213·85
175 × 35 ,,	,,	236·00	214·76
Red deal 115 × 35 mm	,,	270·00	245·70
175 × 35 ,,	,,	275·00	250·25
20 mm blockboard	m²	5·75	5·23
12 mm plasterboard	,,	1·05	0·96
Gypsum perlited plaster, undercoat (50 kg bags)	tonne	65·00	59·15
Ditto finishing coat (ditto)	,,	50·00	45·50
R.S. Joists, basis price, ordinary sections	,,	350·00	318·50
Copper pipes (15 mm) in 5 m lengths	m	1·65	1·50
Clear sheet glass 4 mm thick	m²	8·00	7·28
Paint			
Primer for metalwork	5 litre	11·00	10·01
Emulsion paint	,,	10·00	9·10
Oil paint	,,	11·00	10·01
Stoneware drain pipes			
100 mm diameter	m	1·10	1·00
150 ,, ,,	,,	2·20	2·00
PVC drain pipes			
100 mm diameter	,,	1·50	1·37
150 ,, ,,	,,	2·95	2·68
Cast iron drain pipes 150 mm diameter	,,	13·00	11·83
White glazed wall tiles 150 × 150 × 6 mm	m²	6·70	6·10
Quarry floor tiles 100 × 100 × 18 mm	,,	8·45	7·69
WC Suite complete (average)	No.	80·00	72·80
Lavatory basin complete (ditto)	,,	37·00	33·67

REPUBLIC OF IRELAND

ESTIMATE OF COST FOR OFFICE TYPE BUILDING

The following estimate relates to a reinforced concrete framed four storey office block with a total floor area of 2400 m² and a car park of 500 m² forming part of the ground floor.

It has been assumed that the work would take about 15 months to complete and that access to the site in the Dublin area is good.

	Quantity	Unit	Rate Punt	Total Punt	Sterling Equivalent £
SUBSTRUCTURE					
Excavate					
to reduce levels	585	m³	1·50	877·50	798·53
to form trenches	413	,,	2·90	1,197·70	1,089·91
Extra for breaking out:					
brickwork or concrete	90	,,	19·50	1,755·00	1,597·05
Remove surplus material from site	846	,,	2·18	1,844·28	1,678·29
Hardcore (stone/brick)					
infilling	84	,,	8·60	722·40	657·38
in bed 150 mm thick	220	m²	1·70	374·00	340·34
Plain concrete 8·00 N/mm²					
in foundations	220	m³	38·00	8,360·00	7,607·60
in bed 50 mm thick	603	m²	2·18	1,314·54	1,196·23
Reinforced concrete 26·00 N/mm²					
in foundations	190	m³	40·00	7,600·00	6,916·00
in bed 200 mm thick	265	m²	8·15	2,159·75	1,965·37
Mild steel bar reinforcement					
in foundations	16,730	kg	0·45	7,528·50	6,850·94
Mild steel fabric reinforcement					
3·00 kg/m² in bed	258	m²	2·25	580·50	528·26
Formwork					
to sides of foundations	270	,,	8·00	2,160·00	1,965·60
SUPERSTRUCTURE					
Reinforced concrete 31 N/mm²					
in beams	196	m³	46·70	9,153·20	8,329·41
in columns	50	,,	51·00	2,550·00	2,320·50
Mild steel bar reinforcement					
in beams and columns	81,850	kg	0·45	36,832·50	33,517·58
Formwork					
in beams	900	m²	13·05	11,745·00	10,687·95
in columns	550	,,	13·05	7,177·50	6,531·53
Wrot formwork					
in beams	400	,,	15·80	6,320·00	5,751·20
in columns	54	,,	15·80	853·20	776·41
Structural steelwork in roof beams etc. including all fabrication, placing in position, fixing and protective coat of paint	20,200	kg	0·85	17,170·00	15,624·70
			C/F	128,275·57	116,730·78

REPUBLIC OF IRELAND

	Quantity	Unit	Rate Punt	Total Punt	Sterling Equivalent £
			B/F	128,275·57	116,730·78
225 mm thick concrete 31 N/mm² trough suspended floor consisting of 165 mm (av.) thick × 175 mm deep ribs at 600 mm centres one way, between 600 mm wide × 175 mm deep special trough forms supported as necessary, with a 50 mm concrete topping, the whole reinforced with 8·15 kg of steel bar reinforcement per m² and 2·20 kg of fabric reinforcement per m², including all necessary formwork to soffit OR: Suitable equivalent concrete floor spanning 4·80 m with a dead load of 4·57 N/mm² and a superimposed load of 4·60 N/mm² including all necessary reinforcement and formwork . . .	745	m²	26·00	19,370·00	17,626·70
200 mm thick concrete 31 N/mm² hollow pot suspended floor consisting of 295 × 295 × 150 mm deep clay blocks at 400 mm centres with clay filler tiles between and a 50 mm concrete topping OR: Suitable equivalent concrete floor spanning 4·80 m with a dead load of 4·57 N/mm² and a superimposed load of 4·60 N/mm² . . .	1,617	,,	8·65	13,987·05	12,728·22
Reinforced concrete 31 N/mm² in 200 mm suspended floor slab	85	,,	10·00	850·00	773·50
Mild steel bar reinforcement in suspended floors	13,900	kg	0·45	6,255·00	5,692·05
Mild steel fabric reinforcement 2·2 kg/m² in suspended slab	1,702	m²	2·05	3,489·10	3,175·08
Formwork to soffit of slab . . .	1,450	,,	10·10	14,645·00	13,326·95
External rendering to underside of exposed slab and beams	1,060	,,	5·75	6,095·00	5,546·45
20 mm mastic asphalt roofing on felt on 50 mm thick lightweight channel reinforced composite decking spanning 4·80 m	675	,,	19·50	13,162·50	11,977·88
20 mm asphalt skirting average 175 mm girth	215	m	4·50	967·50	880·43
4 mm blue/black asbestos cement slates size 500 × 250 mm laid to slope on and including battens and 25 mm insulation quilt . . .	460	m²	14·85	6,831·00	6,216·21
Ditto but fixed vertically on and including 20 mm blockboard	100	,,	36·00	3,600·00	3,276·00
15 mm blue/black asbestos cement sheet. 300 mm wide fixed to softwood soffit . .	130	m	2·60	338·00	307·58
Aluminium flashings	1,350	,,	5·00	6,750·00	6,142·50
Treated sawn softwood in roof members, plates etc.	25	m³	400·00	10,000·00	9,100·00
100 mm diameter PVC flat roof outlet with PVC pipe 12 m long including all fixings . .	7	No.	130·00	910·00	828·10
Dog leg stairs and half landing rising three levels (concrete 6 m³, reinforcement 440 kg, formwork 50 m²)	2	,,	1,300·00	2,600·00	2,366·00
			C/F	238,125·72	216,694·43

REPUBLIC OF IRELAND

	Quantity	Unit	Rate Punt	Total Punt	Sterling Equivalent £
B/F				238,125·72	216,694·43
Mild steel balustrade with plastic handrail to six flights of stairs (handrail 17 m, standards 36 No.)	1	No.	1,250·00	1,250·00	1,137·50
Mild steel balustrading with hardwood handrail to six flights of stairs, string to profiles of treads and risers and wall string (handrail 19 m, standards 26 No.)	1	,,	1,985·00	1,985·00	1,806·35
Granolithic finish 25 mm thick to treads and risers	30	m²	8·65	259·50	236·15
Vinyl tiles 3 mm thick ditto . . .	22	,,	7·25	159·50	145·15
Exposed aggregate pre-cast concrete panel 100 mm thick on wall cladding including all fixings and joint sealing	345	m²	43·25	14,921·25	13,578·34
Ditto but coping panels 520 mm overall width	97	,,	56·00	5,432·00	4,943·12
Ditto but parapet stones 620 mm wide × 470 mm deep including damp proof course .	15	m	43·00	645·00	586·95
75 × 500 mm pre-cast concrete (26·00 N/mm²) cill or head finished fair on all exposed faces .	52	,,	19·00	988·00	899·08
Reinforced concrete 26·00 N/mm² in walls 150 mm thick (155 m²) and 240 mm thick (25 m²) .	30	m³	48·00	1,440·00	1,310·40
Mild steel bar reinforcement in walls . .	2,000	kg	0·50	1,000·00	910·00
Formwork to walls	230	m²	13·00	2,990·00	2,720·90
Wrot formwork to walls	110	,,	16·00	1,760·00	1,601·60
Brickwork, low quality bricks 115 mm wall	175	,,	16·00	2,800·00	2,548·00
240 mm wall	115	,,	35·00	4,025·00	3,662·75
Brickwork, good quality facings 115 mm wall	560	,,	20·00	11,200·00	10,192·00
Lightweight building blocks 100 mm wall	650	,,	8·00	5,200·00	4,932·00
Hardwood two panel door, 762 × 1981 × 50 mm, with glass in top panel and louvres in bottom panel with and including hardwood frame 76 × 114 mm, including all fixing and ironmongery	5	No.	240·00	1,200·00	1,092·00
Hardwood window board 400 mm wide fixed to concrete cill	26	m	23·00	598·00	544·18
Plastic faced blockboard window lining 25 × 150 mm fixed to concrete	500	,,	6·00	3,000·00	2,730·00
Black anodized aluminium windows, 50% horizontal sliding, 1700 × 1600 mm including all fixing, glass and ironmongery . . .	440	m²	115·00	50,600·00	46,046·00
Ditto but 12,250 × 675 mm . . .	70	,,	120·00	8,400·00	7,644·00
Black anodized aluminium doors and screens including all fixing, 6 mm obscured glass and ironmongery	50	,,	145·00	7,250·00	6,597·50
Range of 6 No. prefabricated WC compartments 4·80 m long × 1·96 m high × 1·5 m deep consisting of 5 No. partitions and 6 No. doors.	4	No.	1,305·00	5,220·00	4,750·20
100 × 150 mm pre-cast concrete (26·00 N/mm²) lintel reinforced with one 12 mm mild steel rod	44	m	6·00	264·00	240·24
			C/F	370,712·97	337,348·84

REPUBLIC OF IRELAND

	Quantity	Unit	Rate Punt	Total Punt	Sterling Equivalent £
			B/F	370,712·97	337,348·84
Hardwood faced solid core flush door, 762 × 1981 × 40 mm, with and including softwood frame 100 × 35 mm including all fixing and ironmongery	28	No.	190·00	5,320·00	4,841·20
Pair of hardwood two panel doors, 512 × 1981 mm, one hour fire resisting, panels glazed fire resisting glass, with and including hardwood frame, 135 × 90 mm, including all fixing and ironmongery	11	,,	375·00	4,125·00	3,753·75
Treated wrot softwood fillet fixed to concrete .	2,300	m	2·25	5,175·00	4,709·25
Two coat plaster, 12 mm thick to walls, columns and soffits	1,150	m²	5·00	5,750·00	5,232·50
Foil backed plaster board fixed to softwood, with and including one coat plaster	560	,,	6·00	3,360·00	3,057·60
Two coats emulsion paint on walls and soffits .	1,500	m²	2·00	3,000·00	2,730·00
Two coats oil paint ditto . . .	400	,,	2·20	880·00	800·80
Three coats oil paint on wood or metal . .	70	,,	3·50	245·00	222·95
Two coats polyurethane lacquer on aluminium or hardwood	270	,,	2·20	594·00	540·54
Insulation quilt 25 mm thick covered with cement and sand screed 48 mm thick reinforced with chicken wire	750	,,	6·50	4,875·00	4,436·25
Cement and sand screed 48 mm thick . .	1,700	,,	4·35	7,395·00	6,729·45
Granolithic finish 50 mm thick to floors .	70	,,	7·35	514·50	468·20
Vinyl tiles 3 mm thick to floors . .	2,300	,,	6·40	14,720·00	13,395·20
Brown quarry tiles 150 × 150 × 16 mm to floors	75	,,	20·00	1,500·00	1,365·00
OR:					
Terrazzo tiles 300 × 300 × 25 mm . .	75	,,	27·50	—	—
Hardwood skirting 100 × 25 mm on and including softwood grounds fixed to wall . .	820	m	4·50	3,690·00	3,357·90
Mineral board ceiling tiles 9·5 mm thick laid in and including aluminium tee section grid supported by metal hangers 400 mm long fixed to soffit, including all decoration . . .	2,350	m²	8·65	20,327·50	18,498·03
Lavatory basin, coloured, 560 × 400 mm, including fittings but excluding associated pipework	24	No.	102·00	2,448·00	2,227·68
Sink, white, 455 × 380 × 205 mm ditto . .	4	,,	95·00	380·00	345·80
WC suite, low level, coloured, excluding associated pipework	24	,,	135·00	3,240·00	2,948·40
Towel dispensers	8	,,			
Soap dispensers	24	,,			
Toilet roll holders	24	,,	}	1,650·00	1,501·50
Mirrors 1750 × 450 × 6 mm . . .	8	,,			
Disposal installation to ground level for sanitary fittings listed above including all builders work in connection				3,600·00	3,276·00
Hot and cold water installation for sanitary fittings listed above including all builders work in connection				10,100·00	9,191·00
			C/F	473,601·97	430,977·84

REPUBLIC OF IRELAND

	Quantity	Unit	Rate Punt B/F	Total Punt	Sterling Equivalent £
				473,601·97	430,977·84
Hot water and wall radiator heating system for offices (2400 m² of floor area) including all builders work in connection				65,000·00	59,150·00
OR					
Low cost full air conditioning (6000 m³) . .				(175,000·00)	(159,250·00)
Electric power and lighting installation including mains, transformers and all builders work in connection to the following: 90 lighting points without fittings 10 ,, ,, with fittings 50 power points 5 control points				7,250·00	6,597·50
Passenger lift, electric, eight persons, four stops, travelling through 9·50 m. Including all builders work in connection but excluding cost of shaft .				24,500·00	22,295·00
Allow for 7 No. 30 m hose reels and all builders work in connection				1,500·00	1,365·00
Car park slab consisting of 175 mm bed of hardcore and 75 mm tarmacadam paving in two coats	880	m²	8·00	7,040·00	6,406·40
Total building cost				578,891·97	526,791·74
Add for site establishment, supervision and head office overheads where not included in above rates			15%	86,833·80	79,018·76
Add for any other estimated expenditure (contingencies)				28,000·00	25,480·00
				693,725·77	631,290·50
Add for V.A.T. or equivalent tax . . .			3%	20,811·77	18,938·72
TOTAL ESTIMATED TENDER FIGURE				714,537·54	650,229·22

REPUBLIC OF IRELAND

AVERAGE BUILDING PRICES PER SQUARE METRE
(Second Quarter 1980)

	Price Punt	Sterling Equivalent £
Offices for letting, including lifts and central heating	340·00	309·40
Offices for owner occupation, including lifts and air-conditioning . .	500·00	455·00
Shops (excluding shop fronts)	225·00	204·75
Flats, four storey, private	320·00	291·20
Hotels	450·00	409·50
Banks	600·00	546·00

N.B. The above prices include Value Added Tax.

BUILDING COST INDEX

January 1968 = 100

	January
1969	104
1970	120
1971	134
1972	142
1973	156
1974	205
1975	246
1976	291
1977	345
1978	407
1979	460
1980	588

N.B. The above index includes Value Added Tax.

ITALY

TENDERING PROCEDURES

There are three main methods of obtaining tenders in Italy – by public advertisement, by selective tendering and by negotiation. The first two methods are those normally adopted by Government Departments and Local Authorities whilst negotiation is usually confined to private work.

In Italy contractors are registered with the National Register of Building Contractors which lists the type and financial size of work each firm is capable of carrying out. When tendering, contractors must demonstrate, usually by reference to the Register, that the contract under consideration lies within their scope.

Prior to submitting tenders the contractors often have to pay a deposit to the employer which is refunded to unsuccessful tenderers less a nominal tax. With public authority work, a contractor may compile his own tender figure but it is not uncommon for contractors to tender by offering a percentage adjustment to a sum set by the client.

There is no widely accepted standard form of building contract as we know it in the United Kingdom and each client tends to draw up his own clauses based upon 'regional' standard forms published by engineering organizations though many of their contract clauses are nationally recognized. The contract documents normally comprise the following:

1. Drawings.
2. General Contract and Specification Clauses.
3. Special specification clauses, applicable to the particular contract.
4. Form of Tender.

Three copies of the contract documents are made; one set is held by the employer and contractor respectively and the third is deposited with the Public Registrar. Contracts may be let to a general contractor who will be responsible to the client for the whole project whether or not he employs sub-contractors. On the other hand, contracts are sometimes let on a trade by trade system with the client's consultants being responsible for co-ordination.

Most contracts are of a firm price type. In some instances fluctuations are permitted, with a formula for valuing any adjustment set out in the Special Specification Clauses, usually related to a published National Index.

Variations are normally permitted and providing the cost of these falls within certain limits they are valued at rates pro rata to the contract. The final account is prepared jointly by the contractor and the employer. Interim payments are made to the contractor and are subject to a retention of 10%, this being held until the satisfactory completion of the contract. Any disputes arising out of the contract are referred to a panel of three arbitrators; one being appointed by the client, one by the builder and one by the Courts or a similar body as provided for in the contract documents. The arbitrators make a majority decision which is binding on both parties.

ITALY

PRICE DATA

Unless otherwise stated, all prices and rates exclude Value Added Tax. Materials are generally subject to a standard rate of 14% apart from timber and quarry products subject only to a reduced rate of 6%, 14% is added to the final tender for industrial, commercial and luxury residential buildings but only 3% is added for other residential development.

The prices for materials are approximate and include delivery to site in reasonable loads. The following information is based on prices and rates current at the time stated in and around Milan, in northern Italy, and should be only used as a guide to those applicable in other areas of the country.

For the purposes of comparing Italian prices with those in the United Kingdom it has been assumed that 1,000 lire = 51p approximately.

BASIC WAGE RATES (January 1980)

	Lire per hour	*Sterling equivalent*
Foreman	5,160·00	2·63
Specialized worker . . .	4,700·00	2·40
Qualified worker . . .	4,420·00	2·25
Specialized labourer . .	4,100·00	2·09

ADDITIONS TO BASIC WAGE RATES

Addition is usually made to the basic rates for the following:

Social charges.
Industrial injury insurance.
Sickness insurance.
Accident insurances.
Family allowance.
etc.

The effect of this is to add approximately 65% to the basic rate. In addition, a contractor, tendering, adds his own on-costs and profit.

ITALY

MATERIALS PRICES (January 1980)

	Unit	Price Lire	Sterling equivalent £
Hardcore (broken stone/brick)	m³	6,600·00	3·37
Ready mixed concrete quality 31 N/mm²	,,	25,100·00	12·80
Cement	100 kg	4,925·00	2·51
Lime	,,	5,820·00	2·97
Coarse aggregate for concrete	m³	5,500·00	2·81
Fine aggregate for concrete	,,	5,750·00	2·93
Sand for mortar and plastering	,,	7,400·00	3·77
Steel reinforcing rods	kg	360·00	0·18
Common bricks (230 × 110 × 60 mm)	1000	57,000·00	29·07
Facing bricks	,,	106,000·00	54·06
Clay cellular blocks			
500 × 250 × 80 mm thick	m²	5,300·00	2·70
500 × 250 × 120 ,, ,,	,,	6,600·00	3·37
500 × 250 × 200 ,, ,,	,,	8,600·00	4·39
Concrete blocks (500 × 250 × 80 mm)	,,	4,600·00	2·35
7 mm blue/black asbestos slates	,,	5,200·00	2·65
Carcassing timber	m³	400,000·00	204·00
Afrormosia hardwood	,,	570,000·00	290·70
Softwood internal door 43 mm thick, including frame and iron-mongery	No.	80,500·00	41·06
Structural steel (standard sections)	kg	59,500·00	30·35
Plaster	,,	6,000·00	3·06
13 mm plasterboard	m²	2,650·00	1·35
2·5 mm vinyl floor tiles	,,	6,600·00	3·37
11 mm clay floor tiles	,,	8,000·00	4·08
4 mm clear sheet glass	,,	10,000·00	5·10
Low level WC suite	No.	66,000·00	33·66
16 mm diameter copper pipe	m	1,650·00	0·84
100 mm diameter PVC pipe	,,	2,000·00	1·02
Emulsion paint	kg	2,000·00	1·02
Oil paint	,,	2,650·00	1·35
100 mm diameter clay drain pipes	m	4,600·00	2·35
150 mm diameter cast iron drain pipes	kg	900·00	0·46

ITALY

ESTIMATE OF COST FOR OFFICE TYPE BUILDING

The following estimate relates to a reinforced concrete framed four storey office block with a total floor area of 2400 m² and a car park of 500 m² forming part of the ground floor.

It has been assumed that the work would take about 14 months to complete and that access to the site in the Milan area is good.

	Quantity	Unit	Rate Lire	Total Lire	Sterling equivalent £
SUBSTRUCTURE					
Excavate					
to reduce levels . . .	585	m³	2,000·00	1,170,000·00	596·70
to form trenches . . .	413	,,	3,400·00	1,404,200·00	716·14
Extra for breaking out:					
brickwork or concrete . .	90	,,	17,000·00	1,530,000·00	780·30
Remove surplus material from site .	846	,,	1,700·00	1,438,200·00	733·48
Hardcore (stone/brick)					
infilling	84	,,	6,600·00	554,400·00	282·74
in bed 150 mm thick . . .	220	m²	1,300·00	286,000·00	145·86
Plain concrete 8·00 N/mm²					
in foundations . . .	220	m³	45,000·00	9,900,000·00	5,049·00
in bed 50 mm thick . . .	603	m²	2,200·00	1,326,600·00	676·57
Reinforced concrete 26·00 N/mm²					
in foundations . . .	190	m³	44,000·00	8,360,000·00	4,263·60
in bed 200 mm thick . . .	265	m²	8,000·00	2,120,000·00	1,081·20
Mild steel bar reinforcement					
in foundations	16,730	kg	475·00	7,946,750·00	4,052·84
Mild steel fabric reinforcement					
3·00 kg/m² in bed . . .	258	m²	2,000·00	516,000·00	263·16
Formwork					
to sides of foundations . .	270	,,	9,200·00	2,484,000·00	1,266·84
SUPERSTRUCTURE					
Reinforced concrete 31 N/mm²					
in beams	196	m³	53,000·00	10,388,000·00	5,297·88
in columns	50	,,	55,000·00	2,750,000·00	1,402·50
Mild steel bar reinforcement					
in beams and columns . .	81,850	kg	400·00	32,740,000·00	16,697·40
Formwork					
in beams	900	m²	12,000·00	10,800,000·00	5,508·00
in columns	550	,,	10,000·00	5,500,000·00	2,805·00
Wrot formwork					
in beams	400	,,	13,000·00	5,200,000·00	2,652·00
in columns	54	,,	13,000·00	702,000·00	358·02
Structural steel work in roof beams etc. including all fabrication, placing in position, fixing and protective coat of paint	20,200	kg	1,000·00	20,200,000·00	10,302·00
			C/F	127,316,150·00	64,931·23

	Quantity	Unit	Rate Lire	Total Lire	Sterling equivalent £
			B/F	127,316,150·00	64,931·23
225 mm thick concrete 31 N/mm² trough suspended floor consisting of 165 mm (av.) thick × 175 mm deep ribs at 600 mm centres one way, between 600 mm wide × 175 mm deep special trough forms supported as necessary, with a 50 mm concrete topping, the whole reinforced with 8·15 kg of steel bar reinforcement per m² and 2·20 kg of fabric reinforcement per m². including all necessary formwork to soffit OR: Suitable equivalent concrete floor spanning 4·80 m with a dead load of 4·57 N/mm² and a superimposed load of 4·60 N/mm² including all necessary reinforcement and formwork	745	m²	15,600·00	11,622,000·00	5,927·22
200 mm thick concrete 31 N/mm² hollow pot suspended floor consisting of 295 × 295 × 150 mm deep clay blocks at 400 mm centres with clay filler tiles between and a 50 mm concrete topping OR: Suitable equivalent concrete floor spanning 4·80 m with a dead load of 4·57 N/mm² and a superimposed load of 4·60 N/mm² . . .	— 1,617	— ,,	— 12,000·00	— 19,404,000·00	— 9,896·04
Reinforced concrete 31 N/mm² in 200 mm suspended floor slab .	85	,,	13,500·00	1,147,500·00	585·23
Mild steel bar reinforcement in suspended floors	13,900	kg	400·00	5,560,000·00	2,835·60
Mild steel fabric reinforcement 2·2 kg/m² in suspended slab . .	1,702	m²	2,100·00	3,574,200·00	1,822·84
Formwork to soffit of slab . .	1,450	,,	9,000·00	13,050,000·00	6,655·50
External rendering of underside of exposed slab and beams . .	1,060	,,	4,200·00	4,452,000·00	2,270·52
20 mm mastic asphalt roofing on felt on 50 mm thick lightweight channel reinforced composite decking spanning 4·80 m	675	,,	3,000·00	2,025,000·00	1,032·75
20 mm asphalt skirtings average 175 mm girth	215	m	3,300·00	709,500·00	361·85
			C/F	188,860,350·00	96,318·78

ITALY

	Quantity	Unit	Rate Lire	Total Lire	Sterling equivalent £
			B/F	188,860,350·00	96,318·78
4 mm blue/black asbestos cement slates size 500 × 250 mm laid to slope on and including battens and 25 mm insulation quilt . . .	460	m²	29,000·00	13,340,000·00	6,803·40
Ditto but fixed vertically on and including 20 mm blockboard . .	100	,,	29,000·00	2,900,000·00	1,479·00
15 mm blue/black asbestos cement sheet, 300 mm wide fixed to softwood soffit	130	m	6,200·00	806,000·00	411·06
Aluminium flashings . . .	1,350	,,	6,200·00	8,370,000·00	4,268·70
Treated sawn softwood in roof members, plates etc. .	25	m²	316,000·00	7,900,000·00	4,029·00
100 mm diameter PVC flat roof outlet with PVC pipes 12 m long including all fixings . . .	7	No.	31,500·00	220,500·00	112·46
Dog leg stairs and half landing rising three levels (concrete 6 m³, reinforcement 440 kg, formwork 50 m²) .	2	,,	165,000·00	330,000·00	168·30
Mild steel balustrade with plastic handrail to six flights of stairs (handrail 17 m, standards 36 No.) .	1	,,	169,000·00	169,000·00	86·19
Mild steel balustrading with hardwood handrail to six flights of stairs, string to profile of treads and risers and wall string (handrail 19 m, standards 26 No.)	1	,,	173,000·00	173,000·00	88·23
Granolithic finish 25 mm thick to treads and risers	30	m²	18,000·00	540,000·00	275·40
Vinyl tiles 3 mm thick ditto . .	22	,,	13,000·00	286,000·00	145·86
Exposed aggregate pre-cast concrete panels 100 mm thick on wall cladding including all fixings and joint sealing	345	,,	24,000·00	8,280,000·00	4,222·80
Ditto but coping panels 520 mm overall width	97	,,	26,450·00	2,565,650·00	1,308·48
Ditto but parapet stones 620 mm wide × 470 mm deep including damp proof course . . .	15	m	60,000·00	900,000·00	459·00
75 × 500 mm pre-cast concrete (26·00 N/mm²) cill or head finished fair on all exposed faces . .	52	,,	9,200·00	478,400·00	243·98
Reinforced concrete 26·00 N/mm² in walls 150 mm thick (155 m²) and 240 mm thick (25 m²) . . .	30	m³	75,000·00	2,250,000·00	1,147·50
Mild steel bar reinforcement in walls	2,000	kg	450·00	900,000·00	459·00
Formwork to walls	230	m²	10,000·00	2,300,000·00	1,173·00
Wrot formwork to walls	110	,,	14,000·00	1,540,000·00	785·40
Brickwork, low quality bricks 115 mm wall	175	,,	10,000·00	1,750,000·00	892·50
240 mm wall	115	,,	16,000·00	1,840,000·00	938·40
			C/F	246,698,900·00	125,816·44

ITALY

	Quantity	Unit	Rate Lire	Total Lire	Sterling equivalent £
			B/F	246,698,900·00	125,816·44
Brickwork, good quality facings 115 mm wall	560	m²	10,750·00	6,020,000·00	3,070·20
Lightweight building blocks 100 mm wall	650	„	9,500·00	6,175,000·00	3,149·25
Hardwood two panel door, 762 × 1981 × 50 mm, with glass in top panel and louvres in bottom panel with and including hardwood frame 76 × 114 mm, including all fixing and ironmongery . . .	5	No.	244,000·00	1,220,000·00	622·20
Hardwood window board 400 mm wide fixed to concrete cill . .	26	m	28,750·00	747,500·00	381·23
Plastic faced blockboard window lining 25 × 150 mm fixed to concrete	500	„	9,550·00	4,775,000·00	2,435·25
Black anodized aluminium windows, 50% horizontal sliding, 1700 × 1600 mm including all fixing glass and ironmongery	440	m²	135,000·00	59,400,000·00	30,294·00
Ditto but 12,250 × 675 mm . .	70	„	138,000·00	9,660,000·00	4,926·60
Black anodized aluminium doors and screens including all fixing, 6 mm obscured glass and ironmongery .	50	„	165,000·00	8,250,000·00	4,207·50
Range of 6 No. pre-fabricated WC compartments 4·80 m long × 1·96 m high × 1·5 m deep consisting of 5 No. partitions and 6 No. doors .	4	No.	3,300,000·00	13,200,000·00	6,732·00
100 × 150 mm pre-cast concrete (26·00 N/mm²) lintel reinforced with one 12 mm mild steel rod . .	44	m	9,500·00	418,000·00	213·18
Hardwood faced solid core flush door, 762 × 1981 × 40 mm, with and including softwood frame 100 × 35 mm including all fixing and iron-mongery	28	No.	2,513,000·00	70,364,000·00	35,885·64
Pair of hardwood two panel doors, 512 × 1981 mm, one hour fire resisting, panels glazed fire resisting glass, with and including hardwood frame, 135 × 90 mm, including all fixing and ironmongery	11	„	185,000·00	2,035,000·00	1,037·85
Treated wrot softwood fillet fixed to concrete	2,300	m	5,200·00	11,960,000·00	6,099·60
Two coat plaster, 12 mm thick to walls, columns and soffits . .	1,150	m²	6,600·00	7,590,000·00	3,870·90
Foil backed plaster board fixed to softwood, with and including one coat plaster	560	„	10,000·00	5,600,000·00	2,856·00
Two coats emulsion paint on walls and soffits	1,500	„	1,100·00	1,650,000·00	841·50
Two coats oil paint ditto . .	400	„	4,500·00	1,800,000·00	918·00
Three coats oil paint on wood or metal	70	„	4,750·00	332,500·00	169·58
			C/F	457,895,900·00	233,526·92

ITALY

	Quantity	Unit	Rate Lire	Total Lire	Sterling equivalent £
			B/F	457,895,900·00	233,526·92
Two coats polyurethane lacquer on aluminium or hardwood . .	270	m²	3,500·00	945,000·00	481·95
Insulation quilt 25 mm thick covered with cement and sand screed 48 mm thick reinforced with chicken wire .	750	,,	11,000·00	8,250,000·00	4,207·50
Cement and sand screed 48 mm thick	1,700	,,	4,500·00	7,650,000·00	3,901·50
Granolithic finish 50 mm thick to floors	70	,,	8,000·00	560,000·00	285·60
Vinyl tiles 3 mm thick to floors .	2,300	,,	14,000·00	32,200,000·00	16,422·00
Brown quarry tiles 150 × 150 × 16 mm to floors	75	,,	15,000·00	1,125,000·00	573·75
OR:					
Terrazzo tiles 300 × 300 × 25 mm.	75	,,	16,000·00	—	—
Hardwood skirting 100 × 25 mm on and including softwood grounds fixed to wall	820	m	7,500·00	6,150,000·00	3,136·50
Mineral board ceiling tiles 9·5 mm thick laid in and including aluminium tee section grid supported by metal hangers 400 mm long fixed to soffit, including all decoration . .	2,350	m	12,650·00	29,727,500·00	15,161·03
Lavatory basin, coloured, 560 × 400 mm, including fittings but excluding associated pipework . .	24	No.	210,000·00	5,040,000·00	2,570·40
Sink, white, 455 × 380 × 205 mm ditto	4	,,	140,000·00	560,000·00	285·60
WC suite, low level, coloured, excluding associated pipework . .	24	,,	142,000·00	3,408,000·00	1,738·08
Towel dispensers	8	,,	13,000·00	104,000·00	53·04
Soap dispensers	24	,,	16,000·00	384,000·00	195·84
Toilet roll holders . . .	24	,,	20,000·00	480,000·00	244·80
Mirrors 1750 × 450 × 6 mm .	8	,,	26,000·00	208,000·00	106·08
Disposal installation to ground level for sanitary fittings listed above including all builders work in connection				360,000·00	183·60
Hot and cold water installation for sanitary fittings listed above including all builders work in connection.				22,000,000·00	11,220·00
Hot water and wall radiator heating system for offices (2400 m² of floor area) including all builders work in connection				20,000,000·00	10,200·00
OR:					
Low cost full air conditioning (6000 m³)				(28,000,000·00)	(14,280·00)
			C/F	597,047,400·00	304,494·19

ITALY

	Quantity	Unit	Rate Lire	Total Lire	Sterling equivalent £
			B/F	597,047,400·00	304,494·19
Electric power and lighting installation including mains, transformers and all builders work in connection to the following: 90 lighting points without fittings 10 ,, ,, with fittings 50 power points 5 control points . . .				14,500,000·00	7,395·00
Passenger lift, electric, eight persons, four stops, travelling through 9·50 m. Including all builders work in connection but excluding cost of shaft .				18,000,000·00	9,180·00
Allow for 7 No. 30 m hose reels and all builders work in connection .				2,500,000·00	1,275·00
Car park slab consisting of 175 mm bed of hardcore and 75 mm tarmacadam paving in two coats . .	880	m²	16,000·00	14,080,000·00	7,180·80
Total building costs . . .				646,127,400·00	329,524·99
Add for site establishment, supervision and head office overheads where not included in above rates .			23%	148,609,302·00	75,790·75
				794,736,702·00	405,315·74
Add for any other estimated expenditure			—	—	—
				794,736,702·00	405,315·74
Add for V.A.T. or equivalent tax .			14%	111,263,138·28	56,744·20
TOTAL ESTIMATED TENDER FIGURE				905,999,840·28	462,059·94

ITALY

AVERAGE BUILDING PRICES PER SQUARE METRE

(First Quarter 1980)	Price Lire	Approximate sterling equivalent £
Offices for letting, including lifts and central heating with shops at ground level	352,000·00	179·52
Offices for owner occupation including lifts and air-conditioning	450,000·00	229·50
Flats, four storey:		
State assisted schemes	250,000·00	127·50
Private houses	420,000·00	214·20
Hotels	560,000·00	285·60

N.B. The above prices include Value Added Tax.

BUILDING COST INDEX

1965 = 100

Year	Annual average
1965	100
1966	101
1967	104
1968	108
1969	118
1970	136
1971	143
1972	151
1973	188
1974	218
1975	257
1976	270
1977	293
1978	340
1979	391
1980	450

(first quarter)

N.B. The index reflects the effect of Value Added Tax introduced in 1973.

LUXEMBOURG

TENDERING PROCEDURES

There are two main types of contract in use in Luxembourg:

Marché à Forfait Relatif – A lump sum contract which always includes fluctuation clauses for labour and sometimes for materials. The Contractor allows a provisional sum in his tender for price changes, which are later reimbursed according to an official government price index.

Marché à Bordereau de Prix – A remeasurement contract based upon a schedule of rates; the rates are fixed at the time of tender but the extent of the work is undefined, and only measured upon completion of the job.

Other types of contract less commonly used are a firm price contract (*Marché à Forfait Absolu*) with no provisions for variations and a cost-plus contract (*Contract en Régie*) for works where it is difficult to estimate likely costs. With the latter contract an agreed percentage is added to Contractors basic costs to cover general expenses and profit.

Negotiated contracts are only occasionally used in the public sector, most contracts being placed by way of public advertisement. However, they are increasingly gaining popularity in the private sector as an alternative to selective tendering. General Contractors are uncommon and usually individual trade contractors are employed by the Architect/Engineer, who pays them direct, co-ordinates them on site and generally manages the project. Preliminary items are normally the responsibility of the Contractor for earthworks, foundation works and the main structure, and he recovers any costs due from other Contractors.

Contractors usually receive the following tender documents:

1. Drawings.
2. General conditions of contract, descriptions and specification.
3. Schedule of quantities.
4. Tender form.

These documents, together with the letter of instruction from the Architect/Engineer eventually become the contract documents and signed copies are held by all parties concerned.

In Luxembourg the only standard form of building contract is a government one for public works, contract forms therefore are usually prepared by the Architect according to the individual requirements of each project. The law (civil code) establishes and safeguards the legal relationship between all parties entering into a contract.

When selective and open tendering methods are used it is not always the lowest tenderer who is awarded the contract, a considerable amount of emphasis being placed upon a contracting firm's reputation and financial standing.

Interim payments are either paid at agreed periods of time (usually monthly) or as stage payments and retention amounting to 10% is held from sums certified. If there are any supposed defects that the Contractor refuses to remedy then the Architect/Engineer has to refer his claim either through the civil courts (i.e. the local magistrate) or through the Trades Council (an arbitration body) before he can exercise his proper rights against the Contractor.

LUXEMBOURG

Retention monies are usually held for about one year, but the final account is normally expected from the Contractor within four months of contract completion. The civil code places responsibility for the safety of the project on the designer and contractors for a period of ten years.

LUXEMBOURG

PRICE DATA

The prices for materials are ex factory, quarry, etc. and do not include haulage to site. Unless otherwise stated all prices and rates exclude Value Added Tax which is currently 10% but this is not levied on all materials and labour costs and in fact only adds between 2·6% and 4% to total building costs.

For the purposes of comparing Luxembourg prices with those in the United Kingdom it has been assumed that 1 Franc = 1·5p approximately.

BASIC WAGE RATES (First Quarter 1980)

	Francs per hour	*Sterling equivalent* £
Qualified Workers		
First Grade	176·75	2·65
Second Grade.	146·00	2·19
Third Grade	128·50	1·93
Semi-Qualified Workers . . .	116·00	1·74
Labourers	109·00	1·64

ADDITIONS TO BASIC WAGE

Health insurance.
Insurance against illness through nature of work.
Accident insurance.
Family allowances.
Pensions.
Holidays with pay.
Bank holidays with pay.
Inclement weather.

The effect of the above is to add approximately 45% to the basic rates. In addition, a contractor tendering adds his own on-costs and profit.

LUXEMBOURG

MATERIALS PRICES (First Quarter 1980)

	Unit	Price Francs	Sterling equivalent £
Cement			
Up to 70 bags	50 kg	105·00	1·58
Beyond 70 bags	,,	95·00	1·43
Aggregates	m³	430·00	6·45
Sand			
Ordinary quality	,,	395·00	5·93
Washed Moselle	,,	670·00	10·05
Ready mixed concrete (depending on quantity and specification)	,,	170·00	2·55
Steel reinforcing bars			
14 mm diameter	100 kg	1,100·00	16·50
21 ,, ,,	,,	1,060·00	15·90
8 ,, ,,	,,	1,350·00	20·25
Bricks			
Common	1000	3,825·00	57·38
Red facing	,,	6,800·00	102·00
Handmade	,,	8,500·00	127·50
Concrete blocks	each	42·00	0·63
Clay roofing tiles	,,	30·00	0·45
Timber			
Carcassing quality	m³	8,000·00	120·00
Wrot softwood battens	,,	10,000·00	150·00
Structural steelwork (basis price only)	100 kg	1,470·00	22·05
Plaster	40 kg bag	117·00	1·76
Sand for plastering	m³	735·00	11·03
Hydrated Lime	40 kg bag	105·00	1·58
Plastic paint			
Medium quality	kg	82·00	1·23
High quality	,,	105·00	1·58
Oil paint	litre	140·00	2·10
Varnish	,,	140·00	2·10
Drainage pipes			
100 mm diameter clay	m	75·00	1·13
120 ,, ,, ,,	,,	108·00	1·62
150 ,, ,, ,,	,,	130·00	1·95
150 mm diameter cast iron	,,	530·00	7·95

BUILDING COST INDEX
1968 = 100

Year	Annual average
1968	100
1969	106
1970	125
1971	141
1972	150
1973	164
1974	176
1975	194
1976	211
1977	230
1978	245
1979	272

N.B. The above index reflects the effect of Value Added Tax, introduced in 1970.

THE NETHERLANDS

TENDERING PROCEDURES

Control of new building projects in the Netherlands is exercised through the Physical Planning Act. The National Planning Agency in the Hague is responsible for approving all Government schemes and also for approving any large private projects which are likely to have a direct influence on the economy. At a lower level, local authorities have a strong controlling effect on private building developments and developers find that it is a long and tedious process to obtain planning permission.

The Government finances about half of the annual amount spent in the building and construction industry and employs its own architects and engineers who form the backbone of building teams within the various local government departments. Initial estimates are prepared but are often little more than approximate rates per m³ of building volume.

In the private sector, professional architects are employed on a fee basis and with the approval of the client they often use the services of other specialist engineers if required.

Tender documents usually consist of a specification and drawings. From these the Contractor is expected to prepare his tender and to do this he often has to measure quantities. As specifications sometimes relate to performance requirements rather than a precise specification he may have to prepare 'working' details to arrive at his final estimate. Once a contract is underway either the Architect or the Contractor prepares detailed working drawings.

A standard form of contract, the U.A.V 68 (*Uniforme Administratieve, Voorwaarden voor de Uitvoering van Werken*) also exists and this is widely used in the Netherlands.

There are three main methods of obtaining tenders:

by public advertisement
by selective tendering (after public announcement of intention to let a contract, or by invitation or by direct selection)
by negotiation with a single contractor

Public advertisement is still largely used for most Government and local authority contracts but selective tendering (even with the extra cost of tendering due to the practice of reimbursing tendering costs) is steadily gaining favour. Normally, selective and negotiated tendering methods are used in private development. In both open and selective tendering the contract is awarded after consideration of the tender sum as well as achievements of the firm on previous contracts and its financial standing at the time.

With negotiated tenders the contractor is brought into the design team during the planning process, at the end of which he is asked to state his price for the work. If this cannot be negotiated successfully the contractor is paid for the preliminary work which he has done and further tenders are sought.

When compiling the tender, the contractor prepares his own internal schedule of quantities, in operational order, from the specification and drawings, and prices labour and materials separately. This document, however, is private to the tenderer and does not form any part of the contract unless he is successful when he will normally hand over his estimate and sometimes the prices (but not the quantities) will be made part of the contract for the purpose of valuing the variations. The high cost of tendering, tradition, short tender periods and Contractors' own uncertainties regarding the accuracy of their tenders has encouraged pre-tender meetings, where all bids are disclosed and after determining a reasonable price for the works, rogue tenders are eliminated. Tenderers are then instructed by their association to add a percentage to cover all tendering costs, unsuccessful tenderers recovering a proportion of this sum from the successful Contractor.

THE NETHERLANDS

Often a considerable period of post-tender negotiation takes place between the successful Contractor and the Client before they enter into a formal contract, as it is very common for tenders received to exceed considerably clients initial estimates. Tender documents are endorsed by both parties and later become contract documents.

As in the United Kingdom, it is usual to employ a general contractor who is responsible for all work except for that carried out as a direct contract with the client. Examples of other direct contracts are for piling works and mechanical and electrical installations; nominated sub-contracting as we understand it in England is non-existent.

Interim payments on account may be made to the contractor and the frequency of these is set out in the specification. A common formula is to pay 5% of the contract sum once 15% of the work is complete, rising by increments of 5% on both scales thereafter; on completion of the contract a further $7\frac{1}{2}\%$ is paid followed by the final payment of $2\frac{1}{2}\%$ at the end of the three months maintenance period.

Clauses may be included in the specification to cover the reimbursement of variations in the costs of labour and materials. Where fluctuations are allowed they are usually paid according to a Government index. Authorized variations to the contract works are permitted and the prices for these are negotiated.

Variations are normally agreed before the works are carried out as required by the U.A.V. 68. The final account therefore takes the form of a summary of the previously agreed variations.

THE NETHERLANDS

PRICE DATA

Unless otherwise stated all prices and rates exclude Value Added Tax, for which an allowance of 18% is currently added as a separate item to tenders.

The prices for materials are approximate and include delivery to site in reasonable loads.

For the purposes of comparing Dutch prices with those in the United Kingdom it has been assumed that 1 Guilder = 22·1p approximately.

BASIC WAGE RATES (Second Quarter 1980)

Group	Guilders per hour	Sterling equivalent £
3	10·43	2·31
4	10·70	2·36
5	11·10	2·45
6	11·36	2·51
7a	11·75	2·60
7b	12·28	2·71
7c	12·46	2·75
7d	13·05	2·88

The above rates which are grouped according to trade and skill are basic rates exclusive of a differential addition of 5% for certain home-based areas and exclusive of percentage additions used to encourage mobility of labour and paid to operatives working away from the home base.

ADDITIONS TO BASIC WAGE RATES

Allowance is usually made to the basic rates for the following:

Additional allowance for length of service.

Social Insurance.

Holiday Stamps.

Administration of Holiday Scheme.

Accident Insurance.

Pension Scheme.

Contribution to O and O funds.

Clothing Allowance.

Bicycle Allowance.

The effect of the above is to add approximately 145% to the basic rate, although this percentage will vary slightly depending on the group and differential percentage applicable. In addition, a contractor tendering adds his own on-costs and profit.

THE NETHERLANDS

MATERIALS PRICES (Second Quarter 1980)

	Unit	Price Guilders	Sterling equivalent £
Hardcore (broken stone/brick)	m³	39·00	8·62
Ready mixed concrete			
With Portland cement type A (325 kg cement/m³), size of load			
up to 10 m³	,,	105·00	23·21
11–50 ,,	,,	100·00	22·10
51–150 ,,	,,	98·00	21·66
151–300 ,,	,,	95·00	21·00
With blast furnace cement type A (325 kg cement/m³), size of load			
up to 10 m³	,,	102·00	22·54
11– 50 ,,	,,	97·00	21·44
51–150 ,,	,,	95·00	21·00
151–300 ,,	,,	60·00	13·26
Cement			
Portland cement type A			
In bulk to silos, size of load 7–10 tonnes	tonne	111·60	24·66
In bags 7–10 tonnes	,,	145·00	32·05
Blast furnace cement type A			
In bulk to silos, size of load 7–10 tonnes	,,	103·00	22·76
In bags 7–10 tonnes	,,	136·00	30·06
Coarse aggregate (river) ⎫ for concrete	m³	37·00 (average)	8·18
Fine aggregate (river) ⎭	,,	23·75 (average)	5·25
Mild steel bar reinforcement	100 kg	101·25 (average)	22·38
Mild steel fabric reinforcement weighing 3 kg/m²	,,	132·00	29·17
Machine made clay bricks			
Plintklinkers (for use below DPC)	1000	278·00	61·44
Kleurigeklinkers	,,	272·00	60·11
Miskleurige Klinkers	,,	250·00	55·25
Kleurig hardgrauw	,,	270·00	59·67
Miskleurig hardgrauw	,,	264·00	58·34
Internal wall bricks	,,	239·25	52·87
Poriso Bricks (porous clay blocks for internal use)			
Size 215 × 105 × 70 mm	,,	302·00	66·74
Size 215 × 105 × 90 ,,	,,	390·00	86·19
6 mm blue/black asbestos slates	m²	25·75	5·69
Asphalt	100 kg	50·00	11·05
Softwood – carcassing quality	per m for scantlings		
50 × 150 mm	5 m long	4·05	0·90
63 × 175 ,,	,,	6·00	1·33
75 × 200 ,,	,,	8·37	1·85
– joinery quality (unplaned)			
63 × 100 mm	,,	4·35	0·96
75 × 100 ,,	,,	5·26	1·16
75 × 150 ,,	,,	7·80	1·72
Hardwood			
Afrormosia	m³	450·00	99·45
West African Mahogany	,,	374·00	82·65
Teak	,,	690·00	152·49
20 mm blockboard	m²	34·00	7·51
50 mm lightweight strawboards	,,	12·00	2·65

THE NETHERLANDS

MATERIALS PRICES (Second Quarter 1980) – *continued*

	Unit	Price Guilders	Sterling equivalent £
Aluminium (sheets or sections)	kg	12·25	2·71
Structural Steel	100 kg	124·00 (average)	27·40
100 mm PVC pipe	m	4·50	0·99
15 mm copper pipe	,,	2·35	0·52
Sanitary Fittings (Prices vary according to quality)			
WC Suite complete	No.	154·00	34·03
Lavatory basin 63 cm wide complete with all fittings and wall mirror	,,	178·00	39·34
Belgian Gypsum Plaster	1000 kg	205·00	45·31
9·5 mm plasterboard	m²	4·75	1·05
3 mm vinyl floor tiles	,,	17·00	3·76
White glazed wall tiles (15 × 15 cm) first quality .	,,	16·90	3·73
Unglazed red ceramic floor tiles size 21·5 × 10·5 cm . . .	,,	22·00	4·86
Ordinary glazing quality glass (up to 40 cm wide) 4 mm thick . .	,,	23·50	5·19
Paint			
Undercoat for interior work	kg	16·60	3·67
,, ,, exterior work	,,	20·70	4·57
Standard white interior/exterior	,,	20·50	4·53
Polyurethane lacquer	,,	22·05	4·87
Emulsion paint	,,	7·75	1·71
Cast iron pipes			
150 mm diameter	per 3 m length	109·45	24·19
100 mm clay drain pipes	m	6·80	1·50
60 mm pre-cast concrete paving slabs (500 × 500 mm) . . .	m²	13·60	3·01

THE NETHERLANDS

ESTIMATE OF COST FOR OFFICE TYPE BUILDING

The following estimate relates to a reinforced concrete framed four storey office block with a total floor area of 2400 m² and a car park of 500 m² forming part of the ground floor.

It has been assumed that the work would take about 13 months to complete and that access to the site in the Utrecht area is good.

	Quantity	Unit	Rate Guilders	Total Guilders	Sterling equivalent £
SUBSTRUCTURE					
Excavate					
to reduce levels	585	m³	7·70	4,504·50	995·49
to form trenches	413	,,	8·50	3,510·50	775·82
Extra for breaking out:					
brickwork or concrete	90	,,	487·40	43,866·00	9,694·39
Remove surplus material from site . .	846	,,	5·50	4,653·00	1,028·31
Hardcore (stone/brick)					
infilling	84	,,	16·00	1,344·00	297·02
in bed 150 mm thick	220	m²	4·10	902·00	199·34
Plain concrete 8·00 N/mm²					
in foundations	220	m³	181·00	39,820·00	8,800·22
in bed 50 mm thick	603	m²	12·00	7,236·00	1,599·16
Reinforced concrete 26·00 N/mm²					
in foundations	190	m³	160·00	30,400·00	6,718·40
in bed 200 mm thick	265	m²	31·20	8,268·00	1,827·23
Mild steel bar reinforcement					
in foundations	16,730	kg	2·35	39,315·50	8,688·73
Mild steel fabric reinforcement					
3·00 kg/m² in bed	258	m²	5·40	1,393·20	307·90
Formwork					
to sides of foundations	270	,,	60·00	16,200·00	3,580·20
SUPERSTRUCTURE					
Reinforced concrete 31·00 N/mm²					
in beams	196	m³	153·00	29,988·00	6,627·35
in columns	50	,,	182·00	9,100·00	2,011·10
Mild steel bar reinforcement					
in beams and columns	81,850	kg	2·30	188,255·00	41,604·36
Formwork					
in beams	900	m²	69·00	62,100·00	13,724·10
in columns	550	,,	80·00	44,000·00	9,724·00
Wrot formwork					
in beams	400	,,	73·00	29,200·00	6,453·20
in columns	54	,,	88·00	4,752·00	1,050·19
Structural steelwork in roof beams etc. including all fabrication, placing in position, fixing and protective coat of paint . . .	20,200	kg	2·85	57,570·00	12,722·97
			C/F	626,377·70	138,429·48

THE NETHERLANDS

	Quantity	Unit	Rate Guilders	Total Guilders	Sterling equivalent £
			B/F	626,377·70	138,429·48
225 mm thick concrete 31 N/mm² trough suspended floor consisting of 165 mm (av.) thick × 175 mm deep ribs at 600 mm centres one way, between 600 mm wide × 175 mm deep special trough forms supported as necessary, with a 50 mm concrete topping, the whole reinforced with 8·15 kg of steel bar reinforcement per m² and 2·20 kg of fabric reinforcement per m², including all necessary formwork to soffit OR: Suitable equivalent concrete floor spanning 4·80 m with a dead load of 4·57 N/mm² and a superimposed load of 4·60 N/mm² including all necessary reinforcement and formwork	745	m²	99·25	73,941·25	16,341·02
200 mm thick concrete 31 N/mm² hollow pot suspended floor consisting of 295 × 295 × 150 mm deep clay blocks at 400 mm centres with clay filler tiles between and a 50 mm concrete topping OR: Suitable equivalent concrete floor spanning 4·80 m with a dead load of 4·57 N/mm² and a superimposed load of 4·60 N/mm²	1,617	,,	53·90	87,156·30	19,261·54
Reinforced concrete 31 N/mm² in 200 mm suspended floor slab	85	,,	31·10	2,643·50	584·21
Mild steel bar reinforcement in suspended floors	13,900	kg	2·35	32,665·00	7,218·97
Mild steel fabric reinforcement 2·2 kg/m² in suspended slab	1,702	m²	4·00	6,808·00	1,504·57
Formwork to soffit of slab	1,450	,,	47·50	68,875·00	15,221·38
External rendering to underside of exposed slab and beams	1,060	,,	10·25	10,865·00	2,401·17
20 mm mastic asphalt roofing on felt on 50 mm thick lightweight channel reinforced composite decking spanning 4·80 m	675	,,	106·00	71,550·00	15,812·55
20 mm asphalt skirtings average 175 mm girth	215	m	20·00	4,300·00	950·30
4 mm blue/black asbestos cement slates size 500 × 250 mm laid to slopes on and including battens and 25 mm insulation quilt	460	m²	69·00	31,740·00	7,014·54
Ditto but fixed vertically on and including 20 mm blockboard	100	,,	116·00	11,600·00	2,563·60
15 mm blue/black asbestos cement sheet, 300 mm wide fixed to softwood soffit	130	m	45·00	5,850·00	1,292·85
Aluminium flashings	1,350	,,	58·00	78,300·00	17,304·30
Treated sawn softwood in roof members, plates etc.	25	m³	1,780·00	44,500·00	9,834·50
100 mm diameter PVC flat roof outlet with PVC pipe 12 m long including all fixings	7	No.	192·00	1,344·00	297·02
Dog leg stairs and half landing rising three levels (concrete 6 m³, reinforcement 440 kg, formwork 50 m²)	2	,,	9,000·00	18,000·00	3,978·00
			C/F	1,176,515·75	260,010·00

THE NETHERLANDS

	Quantity	Unit	Rate Guilders	Total Guilders	Sterling equivalent £
			B/F	1,176,515·75	260,010·00
Mild steel balustrade with plastic handrail to six flights of stairs (handrail 17 m, standards 36 No.)	1	No.	5,470·00	5,470·00	1,208·87
Mild steel balustrading with hardwood handrail to six flights of stairs, string to profile of treads and risers and wall string (handrail 19 m, standards 26 No.)	1	,,	7,740·00	7,740·00	1,710·54
Granolithic finish 25 mm thick to treads and risers	30	m²	129·75	3,892·50	860·24
Vinyl tiles 3 mm thick ditto . . .	22	,,	54·50	1,199·00	264·98
Exposed aggregate pre-cast concrete panels 100 mm thick on wall cladding including all fixings and joint sealing	345	,,	122·30	42,193·50	9,324·76
Ditto but coping panels 520 mm overall width	97	,,	185·17	17,961·49	3,969·49
Ditto but parapet stones 620 mm wide × 470 mm deep including damp proof course	15	m	203·00	3,045·00	672·95
75 × 500 mm pre-cast concrete (26·00 N/mm²) cill or head finished fair on all exposed faces .	52	,,	99·36	5,166·72	1,141·85
Reinforced concrete 26·00 N/mm² in walls 150 mm thick (155 m²) and 240 mm thick (25 m²)	30	m³	158·00	4,740·00	1,047·54
Mild steel bar reinforcement in walls	2,000	kg	2·30	4,600·00	1,016·60
Formwork to walls	230	m²	59·40	13,662·00	3,019·30
Wrot formwork to walls	110	,,	66·25	7,287·50	1,610·54
Brickwork, low quality bricks 115 mm wall	175	,,	53·00	9,275·00	2,049·78
240 mm wall	115	,,	93·60	10,764·00	2,378·84
Brickwork, good quality facings 115 mm wall	560	,,	72·75	40,740·00	9,003·54
Lightweight building blocks 100 mm wall	650	,,	42·00	27,300·00	6,033·30
Hardwood two panel door, 762 × 1981 × 50 mm, with glass in top panel and louvres in bottom panel with and including hardwood frame 76 × 114 mm, including all fixings and ironmongery	5	No.	930·00	4,650·00	1,027·65
Hardwood window board 400 mm wide fixed to concrete cill	26	m	86·00	2,236·00	494·16
Plastic faced blockboard window lining 25 × 150 mm fixed to concrete . . .	500	,,	47·50	23,750·00	5,248·75
Black anodized aluminium windows, 50% horizontal sliding, 1700 × 1600 mm including all fixing glass and ironmongery . . .	440	m²	560·00	246,400·00	54,454·40
Ditto but 12,250 × 675 mm . .	70	,,	595·00	41,650·00	9,204·65
Black anodized aluminium doors and screens including all fixings, 6 mm obscured glass and ironmongery	50	,,	550·00	27,500·00	6,077·50
			C/F	1,727,738·46	381,830·23

THE NETHERLANDS

	Quantity	Unit	Rate Guilders	Total Guilders	Sterling equivalent £
			B/F	1,727,738·46	381,830·23
Range of 6 No. prefabricated WC compartments 4·80 m long × 1·96 m high × 1·5 m deep consisting of 5 No. partitions and 6 No. doors	4	No.	4,568·00	18,272·00	4,038·11
100 × 150 mm pre-cast concrete (26·00 N/mm²) lintel reinforced with one 12 mm mild steel rod .	44	m	35·60	1,566·40	346·17
Hardwood faced solid core flush door, 762 × 1981 × 40 mm, with and including softwood frame 100 × 35 mm including all fixing and ironmongery .	28	No.	496·75	13,909·00	3,073·89
Pair of hardwood two panel doors, 512 × 1981 mm, one hour fire resisting, panels glazed fire resisting glass, with and including hardwood frame, 135 × 90 mm, including all fixing and ironmongery	11	,,	1,320·00	14,520·00	3,208·92
Treated wrot softwood fillet fixed to concrete	2,300	m	16·00	36,800·00	8,132·80
Two coats plaster, 12 mm thick to walls, columns and soffits .	1,150	m²	17·17	19,745·50	4,363·76
Foil backed plaster board fixed to softwood, with and including one coat plaster . .	560	,,	33·50	18,760·00	4,145·96
Two coats emulsion paint on walls and soffits	1,500	,,	25·20	37,800·00	8,353·80
Two coats oil paint ditto .	400	,,	19·20	7,680·00	1,697·28
Three coats oil paint on wood or metal .	70	,,	30·00	2,100·00	464·10
Two coats polyurethane lacquer on aluminium or hardwood .	270	,,	21·60	5,832·00	1,288·87
Insulation quilt 25 mm thick covered with cement and sand screed 48 mm thick reinforced with chicken wire . .	750	,,	26·50	19,875·00	4,392·38
Cement and sand screed 48 mm thick .	1,700	,,	10·60	18,020·00	3,982·42
Granolithic finish 50 mm thick to floors .	70	,,	35·10	2,457·00	543·00
Vinyl tiles 3 mm thick to floors . .	2,300	,,	31·10	71,530·00	15,808·13
Brown quarry tiles 150 × 150 × 16 mm to floors	75	,,	79·40	5,955·00	1,316·06
OR: Terrazzo tiles 300 × 300 × 25 mm .	75	,,	—	—	—
Hardwood skirting 100 × 25 mm on and including softwood grounds fixed to wall	820	m	12·60	10,332·00	2,283·37
Mineral board ceiling tiles 9·5 mm thick laid in and including aluminium tee section grid supported by metal hangers 400 mm long fixed to soffit, including all decorations .	2,350	m²	79·30	186,355·00	41,184·46
Lavatory basin, coloured, 560 × 400 mm, including fittings but excluding associated pipework	24	No.	403·00	9,672·00	2,137·51
Sink, white, 455 × 380 × 205 mm ditto .	4	,,	247·00	988·00	218·35
WC suite, low level, coloured, excluding associated pipework	24	,,	230·00	5,520·00	1,219·92
Towel dispensers	8	,,			
Soap dispensers	24	,,		3,389·00	748·97
Toilet roll holders	24	,,			
Mirrors 1750 × 450 × 6 mm . . .	8	,,			
			C/F	2,238,816·36	494,778·46

THE NETHERLANDS

	Quantity	Unit	Rate Guilders	Total Guilders	Sterling equivalent £
			B/F	2,238,816·36	494,778·46
Disposal installation to ground floor level for sanitary fittings listed above including all builders work in connection . . .				24,460·00	5,405·66
Hot and cold water installation for sanitary fittings listed above including all builders work in connection				18,650·00	4,121·65
Hot water and wall radiator heating system for offices (2400 m² of floor area) including all builders work in connection . .				287,000·00	63,427·00
OR:					
Low cost full air conditioning (6000 m³) .				—	—
Electric power and lighting installation including mains, transformers and all builders work in connection to the following:					
90 lighting points without fittings					
10 „ „ with fittings					
50 power points					
5 control points				154,750·00	34,199·75
Passenger lift, electric, eight persons, four stops, travelling through 9·50 m. Including all builders work in connection but excluding cost of shaft				82,000·00	18,122·00
Allow for 7 No. 30 m hose reels and all builders work in connection				4,360·00	963·56
Car park slab consisting of 175 mm bed of hardcore and 75 mm tarmacadam paving in two coats	880	m²	60·00	52,800·00	11,668·80
Total building cost				2,862,836·36	632,686·88
Add for site establishment, supervision and head office overheads where not included in above rates			18·00%	515,310·54	113,883·64
				3,378,146·90	746,570·52
Add for any other estimated expenditure .			0·60%	20,268·88	4,479·42
				3,398,415·78	751,049·94
Add for V.A.T. or equivalent tax . .			18·00%	611,714·84	135,188·99
TOTAL ESTIMATED TENDER FIGURE				4,010,130·62	886,238·93

THE NETHERLANDS

AVERAGE BUILDING PRICES PER SQUARE METRE

(Second Quarter 1980)

	Price Guilders	Sterling equivalent £
Offices for letting, including lifts and simple air conditioning.	1,860·00	411·06
Offices for owner occupation, including lifts and air-conditioning	2,450·00	541·45
Shops (excluding lighting and heating)	1,000·00	221·00
Flats, four storey:		
State assisted schemes	875·00	193·38
Private	910·00	201·11
Private houses	1,015·00	224·32
Hotels	2,960·00	654·16

N.B. The above prices include Value Added Tax.

BUILDING COST INDEX
1965 = 100

Year	Annual average
1965	100
1966	105
1967	109
1968	115
1969	133
1970	153
1971	172
1972	187
1973	206
1974	233
1975	251
1976	271
1977	295
1978	315
1979	340

N.B. The above index reflects the effect of Value Added Tax, introduced in 1969.

NORWAY

TENDERING PROCEDURES

The three main methods of obtaining tenders in Norway are by public advertisement, selective tendering and negotiation. Of these three, public advertisement is by far the most common, even for private work. Cost reimbursement contracts are occasionally used but in the main are limited to industrial building.

Contract documents usually comprise the following:

1. Drawings.
2. Specification, description of the works and Norwegian 'Standard' documents.
3. Contract conditions (N.S. 3401) – a standard form of contract.
4. The deed of agreement between the employer and the contractor (also N.S. 3401).
5. Form of tender.

Unit quantities are often prepared for jobs in Norway, usually by one of the Architects' employees to assist the contractor in submitting a lump sum tender. However, these 'bills of quantities' are not usually regarded as one of the 'contract' documents but once the successful contractor has been chosen and often after further negotiation, he is asked to 'check' the 'bill of quantities' before signing the contract. The quantities are later used for valuing variations.

It is usual for a general contractor to be employed, as in the United Kingdom, though in some regions trade contractors still flourish. When a general contractor is employed he is responsible for site co-ordination. He is usually allowed to sub-let work of a specialist nature which he has neither the staff nor experience to do himself.

Less frequently, and in certain regions only, the employer enters into contract with individual trade contractors; on small jobs site co-ordination is usually undertaken by the architect, on large projects a special 'project supervisor' is often engaged. Mechanical and Electrical contracts are nearly always separate contracts entered into by the employer.

Most contracts are subject to 'fluctuations' but variations to labour and material rates have to be 'officially' recognized; the only accepted labour increases being those brought about by a new agreement between the National Employers Federation and the Building Workers' Union. The prices of materials are subject to strict Government control.

Variations are permitted, providing the scope of the work is not changed, existing 'unit' rates being used as far as possible. If the work is so different that it bears no relationship to previous work, new rates are agreed, but in the event of no agreement the work is charged as daywork. The contractor (or contractors) have to prepare detailed statements to support claims for payment throughout the progress of the works. These interim payments, usually subject to 10% retention, are either stage or minimum valuation payments and are checked either by the architect or 'project supervisor'. Once 95% of the contract is completed the retention percentage falls to 2% and this is held over until the end of the maintenance period, which is usually one year. Alternatively the contractor may be allowed to take out a guarantee to cover his contractual liability and this entitles him to be paid for all work done without deduction.

The final account is prepared by the contractor and checked by the architect.

NORWAY

PRICE DATA

The prices for materials are those applicable in the Oslo region and include delivery to site. Allowance should be made to the rates shown for delivery outside this region.

Unless otherwise stated all prices and rates exclude Value Added Tax for which an allowance of 20% is currently added as a separate item to tenders.

For the purpose of comparing Norwegian prices with those in the United Kingdom it has been assumed that 1 Kroner = 8·83p approximately.

AVERAGE EARNINGS

Approximate average earnings for the first quarter of 1980, excluding overtime, shift pay and holiday supplements amounted to:

	Kroner per hour	*Sterling equivalent* £
All building workers	51·70	4·57
Bricklayers	50·00	4·42
Bricklayers' labourers . . .	48·00	4·24
Carpenters.	48·50	4·28

ADDITIONS TO AVERAGE EARNINGS

Allowance is usually made to the basic rates for the following:

Sick pay.
Social Security.
Industrial accidents and workers' protection insurance.
Holidays.
Supplementary Pensions.

The effect of the above is to add approximately 42% to average earnings. In addition, a Contractor tendering would add his own on-costs and profit.

NORWAY

MATERIALS PRICES (First Quarter 1980)

	Unit	Price Kroner	Sterling equivalent £
Norwegian Portland Cement			
In 50 kg bags	bag	16·50	1·46
In bulk	tonne	320·00	28·26
Rapid hardening Cement			
in 50 kg bags	bag	18·00	1·59
Danish White Portland Cement			
in 50 kg bags	,,	45·00	3·97
Aggregates, etc.			
Pit Sand	hl	4·00	0·35
River Sand	,,	4·10	0·36
Pit Gravel 32 mm	,,	4·50	0·40
,,　,,　22 mm	,,	4·50	0·40
Beach Gravel	,,	4·65	0·41
Lightweight Concrete aggregate	m³	165·00	14·57
Ready mixed Concrete			
Ordinary Quality type B250	,,	220·00	19·43
Rapid hardening Quality type B250	,,	215·00	18·98
Steel reinforcing rods	100 kg	283·00 (average)	24·99
Steel mesh reinforcement	,,	435·00 (average)	38·41
Hydrated Lime in 25 kg bags	bag	25·50	2·25
Ready mixed Mortar in 40 kg bags	,,	20·00	1·77
Clay Bricks			
Grade 1 facings	1000	1,360·00 (varies)	120·09
Grade 2　,,	,,	1,090·00 (varies)	96·25
Common Bricks	,,	945·00 (varies)	83·44
Danish facing Bricks			
size 23 × 11 × 5·5 cm	,,	1,950·00 (varies)	172·19
Clay roofing Tiles (good quality)	1000	3,900·00	344·37
Lightweight concrete blocks	m³	380·00	33·55
Siporex slabs size 60 × 200 × 10 cm	No.	65·00	5·74
General carcassing timbers (unconverted)	m³	245·00	21·63
Structural Steel			
Plates	100 kg	236·00 (average)	20·84
Universal sections	,,	285·00	25·17
Plaster in 33⅓ kg bags	bag	25·00	2·21
Linoleum flooring	m²	72·00	6·36
Paint	litre	25·00	2·21

NORWAY

BUILDING COSTS INDEX
1968 = 100

	First quarter	Second quarter	Third quarter	Fourth quarter	Annual average
1968 . . .	100	101	103	103	102
1969 . . .	104	105	106	108	106
1970 . . .	111	113	114	115	113
1971 . . .	116	118	120	120	119
1972 . . .	123	124	125	126	125
1973 . . .	130	136	139	144	137
1974 . . .	148	154	160	164	157
1975 . . .	167	172	174	176	172
1976 . . .	181	191	195	197	191
1977 . . .	203	208	208	208	207
1978 . . .	217	225	227	235	225
1979 . . .	232	245	252	261	248

N.B. The above index excludes Value Added Tax.

SWEDEN

TENDERING PROCEDURES

Excluding investment in civil engineering work, more than half the total annual expenditure of the building industry is accounted for by the housing sector. The number of new houses currently being built per 1000 inhabitants is twice as high in Sweden as in the United Kingdom. Over 50% of the work in this field stems from the private sector and housing societies, though some of this housing is Government subsidized.

The 'Statute of Tendering Procedure' lays down official procedures to be followed in obtaining and accepting tenders. It is the basis for all Government construction work but is not compulsory for work carried out by Local Authorities or private developers, though commonly used by them as a guide. Even for publicly advertised Government tenders, there is no requirement that the lowest tender should automatically be accepted.

General conditions of contract, supplemented by further conditions, instructions relating to the particular project, a comprehensive specification, drawings and a bill of approximate 'unit' quantities are the principle tender documents. A handbook of typical standard specification clauses exists (AMA 72) as well as a national building products catalogue and there are also recommended rules for contract drawings and project specification. It is also quite common nowadays for architects to engage the services of a '*beskrivninskonsult*' to be responsible for the preparation of the specification document.

If quantities form part of the contract they are usually measured by a '*massberäknar*' (quantity reckoner). It is widely accepted in Sweden that subject to a stipulated number of days during which the successful contractor can check these quantities thereafter only authorized variations will be accepted. A standard form of building contract (issued by the Swedish Association of Engineers and Architects) is used for about 80% of contracts. The tender documents together with this contract, when signed, become the contract documents with copies being held by both the client and the contractor.

The majority of Government contracts are placed as a result of public advertisement, but for Local Authority and private work selective tendering is more common. Cost reimbursement and other negotiated contracts are not popular although package deals are being increasingly used, especially in the housing field where 50% of buildings is undertaken using either industrialized methods or prefabricated components. This development is the outcome of the rationalization of timber components, Sweden's traditional building material, but lightweight concrete prefabricated systems are now frequently to be found in urban development.

A fairly recent innovation in the building industry has been the introduction of a '*byggnadskontrollant*' (building controller). On large schemes he has replaced the Architect as project co-ordinator and is responsible for financial control, site co-ordination, progress payments and variations. Often, he is a building engineer with special training in cost accounting as well as in structures and architecture. The building controller seems to have evolved because the Swedish practice of single trade contracting left all site co-ordination and responsibility in the hands of the architect and an increase in consultants and the greater degree of planning required by more sophisticated buildings, made the architects' job even more complicated and far-reaching. The fact that a considerable proportion of building work is carried out by labour-only gangs further aggravated the problem.

At one time most contracts in Sweden were let to separate trade contractors but general contracting with the general contractor being given responsibility for site management and liability for all sub-contractors' work is now very common. However, because of the specialization of contractors in the past it is quite common for the main contractor to be

SWEDEN

responsible for completing only the structure, sub-letting all other trades. Interim payments made to the contractor during the contract period are calculated either on the basis of a valuation of the completed work or by agreed stage payments.

Adjustments are made to the contract sum for fluctuations in wage rates and materials prices, normally by means of an agreed index, providing clauses to this effect are included in the contract documents.

Variations are usually permitted and these are either valued on the basis of the rates submitted with the tender or for an agreed lump sum.

On completion of the contract and until final settlement, 10% of the contract sum is retained, this being released at final settlement or when any defects have been made good. It is the duty of the contractor to prepare the final account.

SWEDEN

PRICE DATA

Unless otherwise stated all prices and rates exclude Value Added Tax, for which an allowance of 20·63% is currently added as a separate item to tenders.

The prices for materials which are approximate include delivery to site in reasonable loads, unless otherwise indicated. These prices are based on rates applicable to the Stockholm and Gothenburg areas only.

For the purpose of comparing Swedish prices with those in the United Kingdom it has been assumed that 1 Kroner = 10·29p approximately.

AVERAGE EARNINGS

In Sweden nearly 70% of 'construction workers' earnings are based on a results system.

A limited number of hours are paid for at basic wage rates every week, but usually only in respect of standing time, accident time and overpayments. Employers are expected to pay by results whenever this can be arranged and to provide a continuous supply of materials and equipment to enable workers to maintain their piece-rate earnings. Rates appear to be assessed according to a nationally agreed price list which runs to some 700 items or more.

Average hourly earnings throughout Sweden for the fourth quarter of 1979 excluding overtime, shift pay and holiday supplements were approximately:

	Kroner per hour	*Sterling equivalent £*
All building workers . . .	34·70	3·57
Bricklayers	39·25	4·04
Carpenters	39·25	4·04
Concrete workers ⎫ General labourers ⎬ . . .	31·00	3·19

There is a good deal of interchangeability among Swedish building operatives, many operatives having knowledge of two or more trades and this has led to a 'levelling out' of earnings between construction trades (e.g. between Concrete Workers and General Labourers).

SWEDEN

ADDITIONS TO AVERAGE EARNINGS

Allowance is usually made to average earnings for the following:

Supplementary Pension (general and special).
Sickness Insurance.
Industrial Injury Insurance.
Group Life Insurance.
Severance Pay.
Holiday Pay.
Employment Tax.
Unemployment Insurance.
Social Insurance.
General Illness Insurance.
Industrial safety, health control charges, etc.
Building Research Tax.

The effect of the above is to add approximately 37% to average earnings. In addition a contractor tendering adds his own on-costs and profit.

SWEDEN

MATERIALS PRICES (First Quarter 1980)

	Unit	Price Kroner	Sterling equivalent £
Cement			
Bags	50 kg	27·00	2·78
Loose	1000 kg	360·00	37·04
Ready mixed concrete			
Quality K 150	m³	180·00	18·52
Quality K 250	,,	192·00	19·76
Quality K 400	,,	230·00	23·67
Concrete aggregates			
Grade 0–8 mm	,,	53·00	5·45
Grade 16–32 mm	,,	55·00	5·66
Steel reinforcing rods			
Plain rods (10 mm diameter)	100 kg	255·00	26·24
Ribbed rods Ks 40 (10 mm diameter) (most commonly used)	,,	285·00	29·33
Ribbed rods Ks 60 (10 mm diameter)	,,	294·00	30·25
Common bricks (solid clay type)	1000	1,270·00	130·68
Common bricks (hollow clay type)	,,	1,020·00	104·96
Common bricks (sand-lime)	,,	1,070·00	110·10
Facing bricks	,,	1,070·00	110·10
Lightweight concrete blocks (irrespective of thickness)	m³	310·00	31·90
'Leca' blocks (irrespective of thickness)	,,	315·00	32·41
Clay roofing tiles	1000	1,780·00	183·16
General carcassing timbers	ft³	50·00	5·15
Softwood battens	,,	56·00	5·76
Softwood standard window with two opening lights, glazed and including ironmongery, 1800 × 1300 mm	No.	890·00	91·58
Structural Steel			
INP Sections 200–300 mm high	100 kg	230·00	23·67
HE/A Sections 200–300 mm high	,,	277·00	28·50
Higher quality IPE Sections 270 mm high	,,	280·00	28·81
Plaster (ready mixed-ordinary quality)	m³	380·00	39·10
Sand for plastering	,,	62·00	6·38
Hydrated lime	25 kg	17·00	1·75
Plasterboard 9 mm thick	m²	11·00	1·13
Paving slabs 150 × 150 × 7 mm thick	,,	75·50	7·77
Hollow glass blocks 196 × 196 × 96 mm thick	No.	26·00	2·68
Paint			
Internal latex paint	litre	13·50	1·39
External steel primer	,,	19·50	2·01
External finishing paint for steelwork	,,	26·00	2·68
External paint for concrete surfaces	,,	19·25	1·98
Drainage pipes			
Cast iron pipes (100 mm diameter) for internal use in 3 m lengths	per length	115·00	11·83
Cast iron pipes (150 mm diameter) for external use	m	100·00	10·29
Clay pipes (100 mm diameter) in 300 mm lengths	1000	2,225·00	228·95
Plastic pipes (90 mm diameter)	m	14·00	1·44

SWEDEN

BUILDING COST INDEX (for multi-storey housing)

1965 = 100

Year	Annual average
1962	86
1963	90
1964	95
1965	100
1966	105
1967	108
1968	111
1969	111
1970	119
1971	124
1972	132
1973	147
1974	154
1975	173
1976	192
1977	213
1978	239
1979	268

N.B. The above index excludes Value Added Tax.

COMPARATIVE COST OF BUILDING

The foregoing approximate estimates for an office block may be used to provide a rough guide to the comparative cost of building in the countries concerned.

The cost of the scheme in England in April 1980 is estimated to be £739,750·00. Taking this value as a datum costs for the other countries can be shown as follows:

	Approximate cost £	Index
Italy	462,060·00	62
Republic of Ireland	650,229·00	88
England	739,750·00	100
France	784,448·00	106
Belgium	884,060·00	120
The Netherlands	886,239·00	120
West Germany	935,764·00	126
Denmark	969,832·00	131

It will be appreciated that the above index is no more than a crude indication of the comparative costs of building in the countries listed; building practice and user requirement differ from country to country but these factors do not invalidate the comparison.

PART V

Property Insurance

The problem of adequately covering by insurance the loss and damage caused to buildings by fire and other perils has been highlighted in recent years by the increasing rate of inflation.

There are a number of schemes available to the building owner wishing to insure his property against the usual risks. Traditionally the insured value must be sufficient to cover the actual cost of reinstating the building. This means that in addition to assessing the current value an estimate has also to be made of the increases likely to occur during the period of the policy and of rebuilding which, for a moderate size building, could amount to a total of three years. Obviously such an estimate is difficult to make with any degree of accuracy, if it is too low the insured may be penalized under the terms of the policy and if too high will result in the payment of unnecessary premiums. There are variations on the traditional method of insuring which aim to reduce the effects of over estimating and details of these are available from the appropriate offices. For the convenience of readers who may wish to make use of the information contained in this publication in calculating insurance cover required the following may be of interest.

1. Present cost

The current rebuilding cost may be ascertained in a number of ways:

(a) Where the actual building cost is known this may be updated by reference to tender indices (page 156);

(b) By reference to average published prices per square metre of floor area (page 392). In this case it is important to clearly understand the method of measurement used to calculate the total floor area on which the rates have been based which were current at November 1980;

(c) By a professional valuation;

(d) By comparison with the known cost of another similar building.

Whichever of these methods is adopted regard must be paid to any special conditions that may apply i.e. a confined site, complexity of design or any demolition and site clearance that may be required.

2. Allowance for inflation

The 'Present Cost' when established will usually, under the conditions of the policy, be the rebuilding cost on the first day of the policy period. To this must be added a sum to cover future increases. For this purpose, using the historical indices on pages 155 and 156, as a base and taking account of the likely change in building costs and tender climate the following annual average indices are predicted for the future.

	Cost Index	Tender Index
1979	343	371
1980	409	475
1981	474	543
1982	521	597
1983	573	657
1984	630	723

3. Fees

To the total of 1 and 2 above must be added an allowance for fees.

4. Example

An assessment for insurance cover is required in mid 1980 for a property which cost £200,000 when completed in mid 1974.

	£
Present Cost	
Known cost at mid 1974	200,000·00
Predicted tender index mid 1980 = 475	
Tender index mid 1974 = 237	
Increase in tender index = 100·42%	
applied to known cost =	200,840·00
Present cost at day one of policy =	400,840·00

	£
Allowance for inflation	
Present cost at day one of policy	400,840·00
Allow for changes in tender levels during 12 months currency of policy	
Predicted tender index at mid 1981 = 543	
Predicted tender index at mid 1980 = 475	
Increase in tender index = 14·31%	
applied to present cost =	57,360·00
	458,200·00

Assuming that total damage is suffered on the last day of the currency of the policy and that planning and documentation would require a period of twelve months before re-building could commence then a further similar allowance must be made

Predicted tender index at mid 1982 = 597	
Predicted tender index at mid 1981 = 543	
Increase in tender index = 9·94%	
applied to adjusted present cost =	45,545·00
	503,745·00

Assuming that total reinstatement would take two years allowance must be made for the increases in costs which would directly or indirectly be met under a building contract.

Predicted cost index at mid 1984 = 630	
Predicted cost index at mid 1982 = 521	
Increase in cost index = 20·92%. This is the total increase at the end of two years and the amount applicable to the contract cost incurred over this period might be about half, say 10%	50,374·50
Estimated cost of reinstatement	554,119·50

	£
Fees	
Estimated cost of reinstatement	554,119·00
Add for professional fees at say 16%	88,659·12
(Note: No allowance made for VAT on fees)	
	642,778·62
Total insurance cover required, say	£ 643,000·00

INDEX TO ADVERTISERS

Advertising agent:
T. G. Scott & Son Ltd
30–32 Southampton Street
London WC2E 7HR